Lecture Notes in Computer Science 15572

The series Lecture Notes in Computer Science (LNCS), including its subseries Lecture Notes in Artificial Intelligence (LNAI) and Lecture Notes in Bioinformatics (LNBI), has established itself as a medium for the publication of new developments in computer science and information technology research, teaching, and education.

LNCS enjoys close cooperation with the computer science R & D community, the series counts many renowned academics among its volume editors and paper authors, and collaborates with prestigious societies. Its mission is to serve this international community by providing an invaluable service, mainly focused on the publication of conference and workshop proceedings and postproceedings. LNCS commenced publication in 1973.

Claudia Hauff · Craig Macdonald ·
Dietmar Jannach · Gabriella Kazai ·
Franco Maria Nardini · Fabio Pinelli ·
Fabrizio Silvestri · Nicola Tonellotto
Editors

Advances in Information Retrieval

47th European Conference on Information Retrieval, ECIR 2025
Lucca, Italy, April 6–10, 2025
Proceedings, Part I

Springer

Editors
Claudia Hauff
Spotify
Amsterdam, The Netherlands

Craig Macdonald
University of Glasgow
Glasgow, UK

Dietmar Jannach
University of Klagenfurt
Klagenfurt, Austria

Gabriella Kazai
Amazon
London, UK

Franco Maria Nardini
ISTI-CNR
Pisa, Italy

Fabio Pinelli
IMT Lucca
Lucca, Italy

Fabrizio Silvestri
Sapienza University of Rome
Rome, Italy

Nicola Tonellotto
University of Pisa
Pisa, Italy

ISSN 0302-9743 ISSN 1611-3349 (electronic)
Lecture Notes in Computer Science
ISBN 978-3-031-88707-9 ISBN 978-3-031-88708-6 (eBook)
https://doi.org/10.1007/978-3-031-88708-6

This Springer imprint is published by the registered company Springer Nature Switzerland AG
The registered company address is: Gewerbestrasse 11, 6330 Cham, Switzerland

Preface

The 47th European Conference on Information Retrieval (ECIR 2025) was held in Lucca, Italy, from 6 to 10 April 2025, and brought together hundreds of researchers from Europe and abroad. The conference was organised by the Information Science and Technologies Institute "A. Faedo" of the National Research Council of Italy, the Sapienza University of Rome, and the University of Pisa, in cooperation with the British Computer Society's Information Retrieval Specialist Group (BCS IRSG).

These proceedings contain the papers related to the presentations, workshops, tutorials, doctoral consortium, and other satellite tracks that took place during the conference. This year's ECIR program boasted a variety of novel work from contributors from all around the world. ECIR 2025 included a "Findings" track, which was offered to some full papers that were deemed to be solid, but which could not make the main conference track. Moreover, ECIR 2025 ran a special IR4Good track that presented high-quality, high-impact, original IR-related research on societal issues (such as algorithmic bias and fairness, privacy, and transparency) at the interdisciplinary level (e.g., philosophy, law, sociology, civil society), which go beyond the purely technical perspective. ECIR 2025 also featured the "Collab-a-thon", intended to provide an opportunity for participants to foster new collaborations that could lead to exciting new research, and forge lasting relationships with like-minded researchers.

The ECIR 2025 program featured a total of 530 papers from authors in 55 countries in its various tracks. The final program included 52 full papers (23% acceptance rate), an additional 11 findings papers, 42 short papers (26% acceptance rate), 16 IR4Good papers (29% acceptance rate), 25 demonstration papers (58% acceptance rate), 9 reproducibility papers (50% acceptance rate), 12 doctoral consortium papers (75% acceptance rate), and 14 invited CLEF papers. All submissions were peer-reviewed by at least three international Program Committee members to ensure that only submissions of the highest relevance and quality were included in the final ECIR 2025 program. The acceptance decisions were further informed by discussions among the reviewers for each submitted paper, led by a Senior Program Committee member. Each track had a final PC meeting where final recommendations were discussed and made, trying to reach a fair and equal outcome for all submissions.

The accepted papers cover the state of the art in information retrieval and recommender systems: user aspects, system and foundational aspects, artificial intelligence & machine learning, applications, evaluation, new social and technical challenges, and other topics of direct or indirect relevance to search and recommendation. As in previous years, the ECIR 2025 program contained a high proportion of papers with students as first authors, as well as papers from various universities, research institutes, and commercial organisations.

In addition to the papers, the program also included 3 keynotes, 7 tutorials, 9 workshops, a doctoral consortium, and an industry day. Keynote talks were given by Alon Halevy (Google), Hannah Bast (University of Freiburg), and this year's BCS IRSG

Karen Spärck Jones Award winner, Isabelle Augenstein. The tutorials covered a range of topics, including explainability, fairness, replicability and reproducibility, visual IR, conversational search, LLM, and combination techniques in IR, RecSys, NLP, and AI. The workshops brought together participants to discuss query performance prediction (QPP++); search futures; online misinformation (ROMCIR); narrative extraction from texts (Text2Story); geographic information extraction from texts (GeoExT); knowledge-enhanced IR (KEIR); open web search (WOWS); evaluation of retrieval-augmented generation systems (Eval4RAG); and scholarly information access (SCOLIA).

The success of ECIR 2025 would not have been possible without all the help from the strong team of volunteers and reviewers. We wish to thank all the reviewers and metareviewers who helped to ensure the high quality of the program. We also wish to thank: the reproducibility track chairs Vito Walter Anelli and Cristina Ioana Muntean, the IR4Good track chairs Ludovico Boratto, Mirko Marras, and Suzan Verberne, the demo track chairs Dawn Lawrie and Andrew Yates, the industry day chairs Amin Mantrach and Aleksandra Piktus, the doctoral consortium chairs Marinella Petrocchi and Gabriele Tolomei, the CLEF Labs chair Guglielmo Faggioli, the workshop chairs Sean MacAvaney and Ida Mele, the tutorial chairs Christine Bauer and Giovanni Trappolini, the Collab-a-thon chairs Cosimo Rulli and Federico Siciliano, the best paper awards committee chair Charles Clarke, the sponsorship chairs Dyaa Albakour and Edgar Meji, the proceeding chair Salvatore Trani, the local organisation chair Fabio Pinelli, the integration chair Maria Maistro, and the Web and social media chair Beatrice Rapisarda. We would also like to thank all the student volunteers who worked hard to ensure an excellent and memorable experience for participants and attendees.

ECIR 2025 was sponsored by a range of learned societies, research institutes, and companies. We thank them all for their support.

Finally, we wish to thank all of the authors and contributors to the conference.

April 2025

Claudia Hauff
Craig Macdonald
Dietmar Jannach
Gabriella Kazai
Franco Maria Nardini
Fabio Pinelli
Fabrizio Silvestri
Nicola Tonellotto

Organization

General Chairs

Franco Maria Nardini	ISTI-CNR, Italy
Fabrizio Silvestri	Sapienza University of Rome, Italy
Nicola Tonellotto	University of Pisa, Italy

Program Chairs – Full Papers

Claudia Hauff	Spotify, Netherlands
Craig Macdonald	University of Glasgow, UK

Program Chairs – Short Papers

Dietmar Jannach	University of Klagenfurt, Austria
Gabriella Kazai	Amazon, UK

Reproducibility Track Chairs

Vito Walter Anelli	Polytechnic University of Bari, Italy
Cristina Ioana Muntean	ISTI-CNR, Italy

IR4Good Chairs

Ludovico Boratto	University of Cagliari, Italy
Mirko Marras	University of Cagliari, Italy
Suzan Verberne	University of Leiden, Netherlands

Demo Chairs

Dawn Lawrie	Johns Hopkins University, USA
Andrew Yates	University of Amsterdam, Netherlands

Industry Day Chairs

Amin Mantrach	Amazon, Luxembourg
Aleksandra Piktus	Cohere, France

Doctoral Consortium Chairs

Marinella Petrocchi	ISTI-CNR, Italy
Gabriele Tolomei	Sapienza University of Rome, Italy

CLEF Labs Chair

Guglielmo Faggioli	University of Padua, Italy

Workshop Chairs

Sean MacAvaney	University of Glasgow, UK
Ida Mele	IASI-CNR, Italy

Tutorial Chairs

Christine Bauer	Paris Lodron University of Salzburg, Austria
Giovanni Trappolini	Sapienza University of Rome, Italy

Collab-a-thon Chairs

Cosimo Rulli	ISTI-CNR, Italy
Federico Siciliano	Sapienza University of Rome, Italy

Best Paper Awards Committee Chair

Charles Clarke	University of Waterloo, Canada

Sponsorship Chairs

Dyaa Albakour	Signal AI, UK
Edgar Meji	Bloomberg, UK

Proceedings Chair

Salvatore Trani	ISTI-CNR, Italy

Local Organisation Chair

Fabio Pinelli	IMT School for Advanced Studies Lucca, Italy

Integration Chair

Maria Maistro	University of Copenhagen, Denmark

Web and Social Media Chair

Beatrice Rapisarda	ISTI-CNR, Italy

Senior Program Committee

Omar Alonso	Amazon, USA
Ioannis Arapakis	Telefónica Research, Spain
Jaime Arguello	University of North Carolina at Chapel Hill, USA
Krisztian Balog	University of Stavanger & Google Research, Norway
Michael Bendersky	Google, USA
Mohand Boughanem	IRIT University Paul Sabatier Toulouse, France
Jamie Callan	Carnegie Mellon University, USA
Charles Clarke	University of Waterloo, Canada
Fabio Crestani	Università della Svizzera Italiana, Switzerland
Maarten de Rijke	University of Amsterdam, Netherlands
Arjen de Vries	Radboud University, Netherlands
Tommaso Di Noia	Politecnico di Bari, Italy

Laura Dietz	University of New Hampshire, USA
Shiri Dori-Hacohen	University of Connecticut, USA
Carsten Eickhoff	University of Tübingen, Germany
Tamer Elsayed	Qatar University, Qatar
Hui Fang	University of Delaware, USA
Nicola Ferro	University of Padua, Italy
Lorraine Goeuriot	Université Grenoble Alpes, CNRS, France
Nazli Goharian	Georgetown University, USA
Julio Gonzalo	UNED, Spain
Jiafeng Guo	Institute of Computing Technology, CAS, China
Matthias Hagen	Friedrich-Schiller-Universität Jena, Germany
Martin Halvey	University of Strathclyde, UK
Allan Hanbury	TU Wien, Austria
Faegheh Hasibi	Radboud University, Netherlands
Ben He	University of Chinese Academy of Sciences, China
Jiyin He	Signal AI, UK
Jimmy Huang	York University, Canada
Joemon Jose	University of Glasgow, UK
Jaap Kamps	University of Amsterdam, Netherlands
Evangelos Kanoulas	University of Amsterdam, Netherlands
Jussi Karlgren	Silo AI, Sweden
Bart Knijnenburg	Clemson University, USA
Udo Kruschwitz	University of Regensburg, Germany
Oren Kurland	Technion, Israel
Jochen L. Leidner	Coburg University of Applied Sciences, Germany/University of Sheffield, UK/KnowledgeSpaces UG, Germany
Francesco Lettich	ISTI-CNR, Italy
Elisabeth Lex	Graz University of Technology, Austria
Yiqun Liu	Tsinghua University, China
Sean MacAvaney	University of Glasgow, UK
Joao Magalhaes	Universidade NOVA de Lisboa, Portugal
Philipp Mayr	GESIS, Germany
Graham McDonald	University of Glasgow, UK
Edgar Meij	Bloomberg L.P., USA
Josiane Mothe	Univ. Toulouse, France
Henning Müller	HES-SO, Switzerland
Cataldo Musto	University of Bari, Italy
Julia Neidhardt	TU Wien, Austria
Jian-Yun Nie	University of Montreal, Canada
Iadh Ounis	University of Glasgow, UK

Javier Parapar	University of A Coruña, Spain
Gabriella Pasi	University of Milano-Bicocca, Italy
Raffaele Perego	ISTI-CNR, Italy
Benjamin Piwowarski	CNRS/ISIR/Sorbonne Université, France
Alan Said	University of Gothenburg, Sweden
Mark Sanderson	RMIT University, Australia
Rodrygo Santos	Federal University of Minas Gerais, Brazil
Philipp Schaer	TH Köln, Germany
Ralf Schenkel	Trier University, Germany
Gianmaria Silvello	University of Padua, Italy
Fabrizio Silvestri	Sapienza University of Rome, Italy
Mark Smucker	University of Waterloo, Canada
Torsten Suel	New York University, USA
Hussein Suleman	University of Cape Town, South Africa
Paul Thomas	Microsoft, USA
Nicola Tonellotto	University of Pisa, Italy
Julián Urbano	Delft University of Technology, Netherlands
Suzan Verberne	Leiden University, Netherlands
Ryen White	Microsoft, USA
Andrew Yates	University of Amsterdam, Netherlands
Hamed Zamani	University of Massachusetts Amherst, USA
Min Zhang	Tsinghua University, China
Shuo Zhang	Bloomberg, USA
Yi Zhang	University of California Santa Cruz, USA
Justin Zobel	University of Melbourne, Australia
Guido Zuccon	University of Queensland, Australia

Program Committee

Mustafa Abualsaud	Thomson Reuters Labs, USA
Ayush Agarwal	Walmart Global Tech, USA
Potito Aghilar	Polytechnic University of Bari, Italy
Dirk Ahlers	Norwegian University of Science and Technology, Norway
Qingyao Ai	Tsinghua University, China
Dyaa Albakour	Signal AI, UK
Mohammad Aliannejadi	University of Amsterdam, Netherlands
Ismail Sengor Altingovde	Middle East Technical University, Turkey
Negar Arabzadeh	University of Waterloo, Canada
Arian Askari	Leiden University, Netherlands
Matteo Attimonelli	Politecnico di Bari, Italy

Sandeep Avula	Amazon, USA
Sushil Awale	Leibniz Information Centre for Science and Technology, Germany
Hosein Azarbonyad	Elsevier, Netherlands
Ebrahim Bagheri	Ryerson University, Canada
Seyed-Ali Bahrainian	Brown University, USA
Alberto Barrón-Cedeño	Università di Bologna, Italy
Christine Bauer	Paris Lodron University of Salzburg, Austria
Andrea Bellandi	Institute for Computational Linguistics, Italy
Alejandro Bellogin	Universidad Autónoma de Madrid, Spain
Patrice Bellot	Aix-Marseille Université - CNRS (LIS), France
Alessandro Benedetti	Sease, Italy
Klaus Berberich	Saarbrücken University of Applied Sciences, Germany
Nolwenn Bernard	University of Stavanger, Norway
Filippo Betello	Sapienza University of Rome, Italy
Sumit Bhatia	Adobe Inc., USA
Giovanni Maria Biancofiore	Polytechnic University of Bari, Italy
Veronika Bogina	Tel Aviv University, Israel
Ludovico Boratto	University of Cagliari, Italy
Luís Borges	Instituto Superior Técnico, Portugal and Carnegie Mellon University, USA
Martin Braschler	Zurich University of Applied Sciences, Switzerland
Timo Breuer	TH Köln, Germany
George Buchanan	University of Melbourne, Australia
Salvatore Bufi	Polytechnic University of Bari, Italy
Francesco Busolin	Università Ca' Foscari, Venezia, Italy
Arthur Câmara	Zeta Alpha Vector, Netherlands
Ricardo Campos	University of Beira Interior and Porto Business School, Portugal
Iván Cantador	Universidad Autónoma de Madrid, Spain
Maria Cassese	ISTI-CNR, Italy
Emanuele Cavenaghi	Free University of Bozen-Bolzano, Italy
Bhavik Chandna	Indian Institute of Technology Guwahati, India
Shubham Chatterjee	Missouri University of Science and Technology, USA
Gullal Cheema	L3S Research Center, Leibniz University Hannover, Germany
Catherine Chen	Brown University, USA
Xiaoyang Chen	University of Chinese Academy of Sciences, China

Xuanang Chen	Institute of Software, Chinese Academy of Sciences, China
Adrian-Gabriel Chifu	Aix-Marseille University, France
Philipp Christmann	Max Planck Institute for Informatics, Germany
Evgenia Christoforou	CYENS Centre of Excellence, Cyprus
Charles Clarke	University of Waterloo, Canada
Paul Clough	University of Sheffield, UK
Fabio Crestani	Università della Svizzera Italiana, Switzerland
Florin Cuconasu	Sapienza University of Rome, Italy
Washington Cunha	Federal University of Minas Gerais, Brazil
Alfredo Cuzzocrea	University of Calabria, Italy
Danilo Danese	Politecnico di Bari, Italy
Suchana Datta	University College Dublin, Ireland
Alessandro De Bellis	Politecnico di Bari, Italy
Maarten de Rijke	University of Amsterdam, Netherlands
Amey Porobo Dharwadker	Meta, USA
Rishit Dholakia	Cohere, USA
Emanuele Di Buccio	University of Padua, Italy
Angela Di Fazio	Politecnico di Bari, Italy
Giorgio Maria Di Nunzio	University of Padua, Italy
Dario Di Palma	Politecnico di Bari, Italy
Linus W. Dietz	King's College London, UK
Peter Dolog	Aalborg University, Denmark
Gregor Donabauer	University of Regensburg, Germany
Ana Maria Dragulinescu	University Politehnica of Bucharest, Romania
Sajad Ebrahimi	University of Guelph, Canada
Michael Ekstrand	Drexel University, USA
Sohaila Eltanbouly	Qatar University, Qatar
Elias Entrup	TIB – Leibniz Information Centre for Science and Technology, Germany
Liana Ermakova	HCTI, Université de Bretagne Occidentale, France
Marina Ernst	University Koblenz, Germany
Marwa Essam	Qatar University, Qatar
Andrea Esuli	ISTI-CNR, Italy
Ralph Ewerth	Leibniz Universität Hannover, Germany
Guglielmo Faggioli	University of Padua, Italy
Fabrizio Falchi	ISTI-CNR, Italy
Kim Falk	Binary Vikings, Denmark
Jinyuan Fang	University of Glasgow, UK
Hossein Fani	University of Windsor, Canada
Michael Färber	ScaDS.AI & TU Dresden, Germany

Antonio Ferrara	Politecnico di Bari, Italy
Maurizio Ferrari Dacrema	Politecnico di Milano, Italy
Andres Ferraro	Pandora/SiriusXM, USA
Rafael Ferreira	Universidade Nova de Lisboa, Portugal
Nicola Ferro	University of Padua, Italy
Thibault Formal	Naver Labs Europe, France
Maik Fröbe	Friedrich-Schiller-Universität Jena, Germany
Ingo Frommholz	University of Wolverhampton, UK
Petra Galuščáková	University of Stavanger, Norway
Debasis Ganguly	University of Glasgow, UK
Alba García Seco de Herrera	UNED, Spain
Dario Garigliotti	University of Bergen, Norway
Eric Gaussier	Université Grenoble Alpes, France
Angelo Geninatti Cossatin	Università di Torino, Italy
Kripabandhu Ghosh	Indian Institute of Science Education and Research, Kolkata, India
Anastasia Giachanou	Utrecht University, Netherlands
Lukas Gienapp	Leipzig University, Germany
Daniela Godoy	ISISTAN Research Institute (UNCPBA-CONICET), Argentina
Lorraine Goeuriot	Université Grenoble Alpes, CNRS, France
Tim Gollub	Bauhaus-Universität Weimar, Germany
Gabriela González Sáez	Université Grenoble Alpes, France
Wolfgang Gritz	TIB – Leibniz Information Centre for Science and Technology, Germany
Vinesh Gudla	Instacart, USA
Adrien Guille	Université de Lyon, France
Shuyu Guo	Shandong University, China
Fabian Haak	TH Köln, Germany
Tim Hagen	University of Kassel, Germany
Matthias Hagen	Friedrich-Schiller-Universität Jena, Germany
Philipp Hager	University of Amsterdam, Netherlands
Radin Hamidi Rad	Toronto Metropolitan University, Canada
Morgan Harvey	University of Sheffield, UK
Claudia Hauff	Spotify, USA
Mariya Hendriksen	University of Amsterdam, Netherlands
Maria Heuss	University of Amsterdam, Netherlands
Phuong Hoang	Kaiser Permanente, USA
Tracy Holloway King	Adobe, USA
Frank Hopfgartner	Universität Koblenz, Germany
Anett Hoppe	Leibniz Universität Hannover, Germany
Oana Inel	University of Zurich, Switzerland

Kevin Innerebner	Graz University of Technology, Austria
Bogdan Ionescu	Politehnica University of Bucharest, Romania
Ornella Irrera	University of Padua, Italy
Adam Jatowt	University of Innsbruck, Austria
Faizan Javed	Kaiser Permanente, USA
Tianbo Ji	Nantong University, China
José L. Jorro-Aragoneses	Universidad Autónoma de Madrid, Spain
Saurabh Vishwas Joshi	Pinterest Inc., USA
Chris Kamphuis	Spinque, Netherlands
Jaap Kamps	University of Amsterdam, Netherlands
Jasmin Kareem	University of Amsterdam, Netherlands
Mesut Kaya	IT University of Copenhagen, Denmark
Jüri Keller	TH Köln (University of Applied Sciences), Germany
Johannes Kiesel	Bauhaus-Universität Weimar, Germany
Benjamin Kille	Norwegian University of Science and Technology, Norway
Tracy Holloway King	Adobe, USA
Styliani Kleanthous	Open University of Cyprus, Cyprus
Anastasiia Klimashevskaia	University of Bergen, Norway
Noam Koenigstein	TelAviv University, Israel
Michal Kompan	Kempelen Institute of Intelligent Technologies, Slovakia
Deguang Kong	Google, USA
Simone Kopeinik	TU Graz, Austria
Anton Korikov	University of Toronto, Canada
Dominik Kowald	Know-Center, Austria
Hrishikesh Kulkarni	Georgetown University, USA
Saar Kuzi	Amazon, USA
Emanuel Lacić	Infobip, Croatia
Weronika Łajewska	University of Stavanger, Norway
Monica Landoni	Università della Svizzera italiana, Switzerland
Martha Larson	Radboud University and Delft University of Technology, Netherlands
Carlos Lassance	Naver Labs Europe, France
Dawn Lawrie	Johns Hopkins University, USA
Tai Le Quy	University of Koblenz, Germany
Nicola Leschke	Paris Lodron University of Salzburg, Austria
Francesco Lettich	ISTI-CNR, Italy
Jake Lever	University of Glasgow, UK
Elisabeth Lex	Graz University of Technology, Austria
Jing Li	Google, USA

Wei Li	University of Roehampton, UK
Chen Liang	Apple, USA
Siwei Liu	University of Aberdeen, UK
Xuebing Liu	University of Science and Technology Beijing, China
Benedikt Loepp	Fraunhofer IMS, Germany
Domenico Lofù	Politecnico di Bari, Italy
Andreas Lommatzsch	TU Berlin, Germany
Teng Long	University of Amsterdam, Netherlands
Andrea Lops	Politecnico di Bari, Italy
David Losada	University of Santiago de Compostela, Spain
Jiyun Luo	Pinterest, USA
Yougang Lyu	University of Amsterdam, Netherlands
Chenyang Lyu	MBZUAI, United Arab Emirates
Sean MacAvaney	University of Glasgow, UK
Craig Macdonald	University of Glasgow, UK
Joel Mackenzie	University of Queensland, Australia
Maria Maistro	University of Copenhagen, Denmark
Daniele Malitesta	Université Paris-Saclay, France
Antonio Mallia	Pinecone, USA
Watheq Mansour	University of Queensland, Australia
Behrooz Mansouri	University of Southern Maine, USA
Masoud Mansoury	Delft University of Technology, Netherlands
Marcelo G. Manzato	University of São Paulo, Brazil
Jiaxin Mao	Renmin University of China, China
Stefano Marchesin	University of Padua, Italy
Sara Vera Marjanovic	University of Copenhagen, Denmark
Mirko Marras	University of Cagliari, Italy
Monica Marrero	Europeana Foundation, Netherlands
Silvio Martinico	Università di Pisa & ISTI-CNR, Italy
Noemi Mauro	University of Torino, Italy
David Maxwell	Delft University of Technology, Netherlands
James Mayfield	Johns Hopkins University, USA
Richard McCreadie	University of Glasgow, UK
Graham McDonald	University of Glasgow, UK
Jack McKechnie	University of Glasgow, UK
Paul McNamee	Johns Hopkins University, USA
Giacomo Medda	University of Cagliari, Italy
Parth Mehta	Parmonic, USA
Ida Mele	IASI-CNR, Italy
Chuan Meng	University of Amsterdam, Netherlands
Zaiqiao Meng	University of Glasgow, UK

Laura Menotti	University of Padua, Italy
Felice Antonio Merra	Amazon, USA
Nicola Messina	ISTI-CNR, Italy
Paul Missault	Amazon, USA
Jose Moreno	IRIT/UPS, France
Maryam Mousavian	Università della Svizzera italiana, Switzerland
Henning Müller	HES-SO, Switzerland
Peter Müllner	Know-Center, Austria
Cristina Ioana Muntean	ISTI-CNR, Italy
Cataldo Musto	University of Bari, Italy
Seung-Hoon Na	Chonbuk National University, South Korea
Franco Maria Nardini	ISTI-CNR, Italy
Fedelucio Narducci	Politecnico di Bari, Italy
Evelyn Navarrete	L3S Research Center, Leibniz University Hannover, Germany
Julia Neidhardt	TU Wien, Austria
Anil Nelakanti	Amazon, USA
Anastasios Nentidis	NSCR Demokritos, Greece
Sérgio Nunes	University of Porto, Portugal
Harrie Oosterhuis	Radboud University, Netherlands
Salvatore Orlando	Università Ca' Foscari Venezia, Italy
Rebekah Overdorf	University of Lausanne, Switzerland
Panagiotis Papadakos	ERCIM, France
Vincenzo Paparella	Politecnico di Bari, Italy
Andrea Papenmeier	University of Twente, Netherlands
Monica Lestari Paramita	University of Sheffield, UK
Javier Parapar	University of A Coruña, Spain
Andrew Parry	University of Glasgow, UK
Andrea Pasin	University of Padua, Italy
Gustavo Penha	Spotify, USA
Maria Soledad Pera	TU Delft, Netherlands
Raffaele Perego	ISTI-CNR, Italy
João Pereira	NOVA School of Science and Technology, Portugal
Marinella Petrocchi	IIT-CNR, Italy
Aleksandr Petrov	University of Glasgow, UK
Fabio Pinelli	IMT Lucca, Italy
Florina Piroi	TU Wien, & ZFDM, Austria
Gabrielle Poerwawinata	University of Amsterdam, Netherlands
Claudio Pomo	Politecnico di Bari, Italy
Lorenzo Porcaro	Sapienza University of Rome, Italy
Amey Porobo Dharwadker	Meta, USA

Martin Potthast	University of Kassel, hessian.AI, and ScaDS.AI, Germany
Shishir Kumar Prasad	Instacart, USA
Chiara Pugliese	ISTI - CNR, Italy
Antonio Purificato	Sapienza University of Rome, Italy
Erasmo Purificato	Joint Research Centre European Commission, Italy
Shanbao Qioa	Jeonbuk National University, South Korea
Fiana Raiber	Technion, Israel
Amifa Raj	Microsoft, USA
Zhaochun Ren	Leiden University, Netherlands
Yongli Ren	RMIT University, Australia
Chiara Renso	ISTI-CNR, Italy
Guido Rocchietti	Università di Pisa & ISTI-CNR, Italy
Adam Roegiest	Zuva, Canada
Julien Rossi	Amsterdam Business School, Netherlands
Davide Rucci	ISTI-CNR, Italy
Cosimo Rulli	ISTI-CNR, Italy
Tony Russell-Rose	Goldsmiths, University of London, UK
Sara Salamat	Toronto Metropolitan University, Canada
Abbas Saleminezhad	Toronto Metropolitan University, Canada
Eric Sanjuan	Université d'Avignon, France
Javier Sanz-Cruzado	University of Glasgow, UK
Maya Sappelli	HAN University of Applied Sciences, Netherlands
Mahmoud Sayed	University of Maryland, USA
Harrisen Scells	Leipzig University, Germany
Markus Schedl	Johannes Kepler University Linz, Austria
Ferdinand Schlatt	Friedrich-Schiller-Universität Jena, Germany
Falk Scholer	RMIT University, Australia
Ratan Sebastian	TIB Leibniz Information Centre for Science and Technology, Germany
Procheta Sen	University of Liverpool, UK
Giovanni Servedio	Politecnico di Bari, Italy
Shirin Seyedsalehi	Toronto Metropolitan University, Canada
Faisal Shehzad	University of Klagenfurt, Austria
Federico Siciliano	Sapienza University of Rome, Italy
Madhav Sigdel	Kaiser Permanente, USA
Gianmaria Silvello	University of Padua, Italy
Clemencia Siro	University of Amsterdam, Netherlands
Paolo Sorino	Polytechnic University of Bari, Italy
Laure Soulier	Sorbonne Université-ISIR, France
Marc Spaniol	Université de Caen Normandie, France

Francesca Spezzano	Boise State University, USA
Damiano Spina	RMIT University, Australia
Moritz Staudinger	TU Wien, Austria
Fabio Stella	University of Milano-Bicocca, Italy
Ting Su	University of Glasgow, UK
Lynda Tamine	IRIT, France
Junichi Tatemura	Google, USA
Leila Tavakoli	RMIT University, Australia
Carla Teixeira Lopes	University of Porto, Portugal
Anna Tigunova	Amazon, USA
Nava Tintarev	University of Maastricht, Netherlands
Antonela Tommasel	ISISTAN Research Institute, CONICET-UNCPBA, Argentina
Salvatore Trani	ISTI-CNR, Italy
Giovanni Trappolini	Sapienza University of Rome, Italy
Johanne R. Trippas	RMIT University, Australia
Andrew Trotman	University of Otago, New Zealand
Chun-Hua Tsai	University of Omaha, USA
Theodora Tsikrika	Information Technologies Institute, CERTH, Greece
Md Zia Ullah	Edinburgh Napier University, UK
Moshe Unger	Tel Aviv University, Israel
Svitlana Vakulenko	Amazon Alexa AI, Spain
Marco Valentini	Politecnico di Bari, Italy
Christophe Van Gysel	Apple, USA
Alberto Veneri	Ca' Foscari University of Venice, Italy
Rossano Venturini	Università di Pisa, Italy
João Vinagre	Joint Research Centre - European Commission, Spain
Marco Viviani	Università degli Studi di Milano-Bicocca, Italy
Shuai Wang	University of Queensland, Australia
Xi Wang	University of Sheffield, UK
Xiao Wang	University of Glasgow, UK
Yingxu Wang	Mohamed bin Zayed University of Artificial Intelligence, UAE
Matti Wiegmann	Bauhaus-Universität Weimar, Germany
Wolfgang Wörndl	Technical University of Munich, Germany
Diji Yang	University of California Santa Cruz, USA
Eugene Yang	Johns Hopkins University, USA
Andrew Yates	University of Amsterdam, Netherlands
Zixuan Yi	University of Glasgow, UK
Ran Yu	University of Bonn, Germany

Meng Yuan	University of Melbourne, Australia
Shuzhou Yuan	Karlsruhe Institute of Technology, Germany
Eva Zangerle	University of Innsbruck, Austria
Richard Zanibbi	Rochester Institute of Technology, USA
Fattane Zarrinkalam	University of Guelph, Canada
Oleg Zendel	RMIT University, Australia
Hansi Zeng	University of Massachusetts Amherst, USA
Xi Zhang	Fuzhou University, China
Zhilu Zhang	Xi'an Jiaotong-Liverpool University, China

Keynotes

Crossing the Structure Chasm – Querying Data Without Limits

Alon Halevy

Google
alonhalevy@gmail.com

Abstract. Over the past few decades, we've seen significant progress in querying structured data through database systems and unstructured data via information retrieval. Yet, the vision of a unified system capable of querying all types of data has remained elusive. Recent advances in AI have not only amplified the need for such a system but also provided the foundational tools to make it a reality. Indeed, the future success of AI may hinge on our ability to finally cross this "structure chasm." In this talk, I will describe several recent works that attempt to combine the strengths of database systems, namely, set-based processing of structured data and those of large language models and IR systems that can retrieve and answer questions from data in other modalities. Building a system that combines these capabilities opens up fundamental questions about the semantics of querying all data to many practical questions about how to build scalable and usable such systems.

Understanding the Interplay between LLMs' Utilisation of Parametric and Contextual Knowledge

Isabelle Augenstein

University of Copenhagen
augenstein@di.ku.dk

Abstract. Language Models (LMs) acquire parametric knowledge from their training process, embedding it within their weights. The increasing scalability of LMs, however, poses significant challenges for understanding a model's inner workings and further for updating or correcting this embedded knowledge without the significant cost of retraining. Moreover, when using these language models for knowledge-intensive language understanding tasks, LMs have to integrate relevant context, mitigating their inherent weaknesses, such as incomplete or outdated knowledge. Nevertheless, studies indicate that LMs often ignore the provided context as it can be in conflict with the pre-existing LM's memory learned during pre-training. Conflicting knowledge can also already be present in the LM's parameters, termed intra-memory conflict. This underscores the importance of understanding the interplay between how a language model uses its parametric knowledge and the retrieved contextual knowledge. In this talk, I will aim to shed light on this important issue by presenting our research on evaluating the knowledge present in LMs, diagnostic tests that can reveal knowledge conflicts, as well as on understanding the characteristics of successfully used contextual knowledge.

Knowledge Graphs Are Dead, Long Live Knowledge Graphs

Hannah Bast

University of Freiburg
bast@informatik.uni-freiburg.de

Abstract. Knowledge graphs have been around for over 50 years, under various names and definitions, falling in and out of fashion multiple times. I will explain the basics, talk about the history and the current state of the art and open problems, and present my stance on whether AI will eventually render knowledge graphs obsolete once and for all or whether the two will enter a happy marriage. The talk will be full of examples and demos, accessible to anyone with a basic grasp of computer science, and it should be entertaining, instructive, and discussion-inspiring. If time permits, I might also drop a few words on the challenges of doing systems work in academia.

Contents – Part I

LIBRA: Measuring Bias of Large Language Model from a Local Context

Bo Pang[1(✉)], Tingrui Qiao[1], Caroline Walker[2], Chris Cunningham[3], and Yun Sing Koh[1]

[1] School of Computer Science, University of Auckland, Auckland, New Zealand
{bpan882,tqia361}@aucklanduni.ac.nz, y.koh@auckland.ac.nz
[2] The Liggins Institute, University of Auckland, Auckland, New Zealand
caroline.walker@auckland.ac.nz
[3] Research Centre for Māori Health and Development, Massey University, Wellington, New Zealand
C.W.Cunningham@massey.ac.nz

Abstract. Large Language Models (LLMs) have significantly advanced natural language processing applications, yet their widespread use raises concerns regarding inherent biases that may reduce utility or harm for particular social groups. Despite the advancement in addressing LLM bias, existing research has two major limitations. First, existing LLM bias evaluation focuses on the U.S. cultural context, making it challenging to reveal stereotypical biases of LLMs toward other cultures, leading to unfair development and use of LLMs. Second, current bias evaluation often assumes models are familiar with the target social groups. When LLMs encounter words beyond their knowledge boundaries that are unfamiliar in their training data, they produce irrelevant results in the local context due to hallucinations and overconfidence, which are not necessarily indicative of inherent bias. This research addresses these limitations with a Local Integrated Bias Recognition and Assessment Framework (LIBRA) for measuring bias using datasets sourced from local corpora without crowdsourcing. Implementing this framework, we develop a dataset comprising over 360,000 test cases in the New Zealand context. Furthermore, we propose the Enhanced Idealized CAT Score ($EiCAT$), integrating the $iCAT$ score with a beyond knowledge boundary score (bbs) and a distribution divergence-based bias measurement to tackle the challenge of LLMs encountering words beyond knowledge boundaries. Our results show that the BERT family, GPT-2, and Llama-3 models seldom understand local words in different contexts. While Llama-3 exhibits larger bias, it responds better to different cultural contexts. The code and dataset are available at: https://github.com/ipangbo/LIBRA.

Keywords: Bias · Large Language Model · Dataset

1 Introduction

Large Language models (LLMs) have become a cornerstone in natural language processing (NLP) applications, providing substantial advancements in

C. Hauff et al. (Eds.): ECIR 2025, LNCS 15572, pp. 1–16, 2025.
https://doi.org/10.1007/978-3-031-88708-6_1

tasks ranging from chat to text generation [7]. The use of LLMs in real-world applications has raised significant concerns about their potential biases and the impact these biases may have on different social groups [10]. Bias in LLMs is typically influenced by the data they are trained on, which consists of internet-sourced corpora reflecting the dominant cultural stereotypes from all over the world [22,29]. When used in diverse cultural settings, this leads to unfair and potentially harmful outcomes. However, current research on stereotypical bias in LLMs focuses on bias based on United States' culture [20,21,24,26], while there is a lack of research on bias based on cultures from other parts of the world. We refer to this stereotypical bias contained in corpora in a region as the local bias.

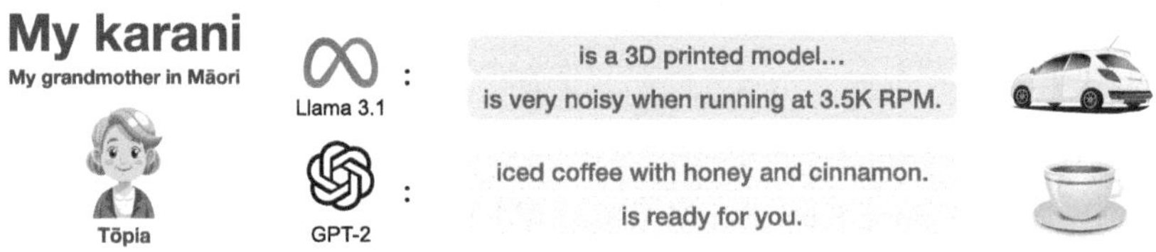

Fig. 1. The comparative responses of different LLMs to prompts such as "My karani" or "My karani Tōpia", which is a transliteration between English and Māori "My granny Tōpia", illustrate the challenges faced in local contexts. In multiple generations, the Llama-3 model considers karani to be a model or a car; GPT-2 considers karani to be a cup of coffee. Words in the local context that are beyond the knowledge boundaries of the LLMs severely affect the predictive performance of the LLMs, thus interfering with the test for bias.

Developing methodologies to detect local biases in region-specific contexts accurately is essential, ensuring that LLMs are evaluated and improved with a sensitivity to cultural diversity. In localized contexts, a significant challenge in evaluating LLMs is the presence of words beyond the models' knowledge boundaries [28]. LLMs are often trained on extensive internet-sourced corpora, but these datasets may lack sufficient representation of region-specific terms, including those from local languages and dialects [9,11,15,31]. This deficiency can lead to unrepresentative and culturally inappropriate outputs [14]. For example, as illustrated in Fig. 1, let LLMs generate text after "My karani", which contains the Māori word "karani" meaning grandmother. However, Llama-3 [31] and GPT 2 [27] misinterpreted "karani" as a model, a car, or a cup of coffee, generating unrelated content. Given these challenges, we ask the following research question: How can we effectively measure local biases in LLMs by leveraging region-specific corpora while addressing the challenges posed by local words that exceed the knowledge boundaries of LLMs?

To tackle the challenge of measuring the local bias of LLMs, we propose a novel framework called **L**ocal **I**ntegrated **B**ias **R**ecognition and **A**ssessment (LIBRA) for efficiently constructing datasets and measuring bias using local corpora without relying on crowdsourcing. We address the issue of words beyond LLMs' knowledge boundaries within the bias testing dataset by detecting beyond knowledge boundaries words and minimizing the effects of hallucinatory results from them during bias testing, thereby ensuring the authenticity and reliability of the test results. Additionally, we create a New Zealand-based dataset to reveal the bias of LLMs in the New Zealand context. When designing the dataset structure, we used the triplet structure of StereoSet [20], where each test case contains three similar sentences with different words describing the target social group. The description regarding target social groups requires manual annotation from crowdsourced workers. Instead of changing the descriptive words, as shown in Fig. 2, we identify and replace the words that denoted the target social group in our collected corpora, thus allowing the entire dataset to be generated automatically. When dealing with sentences with words beyond the LLM's knowledge boundary, we identify local words by constructing local vocabulary lists that differ from those of general English. We add extra evaluations for local words to ensure that LLMs' interpretation of these local words matches their formal definition. A key innovation of our approach is the introduction of the Enhanced Idealised CAT Score ($EiCAT$), which integrates the traditional $iCAT$ score [20] with a beyond LLM's knowledge boundary score (bbs). This new metric uses the Jensen-Shannon divergence to analyze bias by examining the distribution of logits associated with stereotypical and anti-stereotypical choices in the dataset. By incorporating the bbs, the $EiCAT$ score penalizes models that fail to understand local words, thus ensuring an accurate assessment of bias in localized contexts.

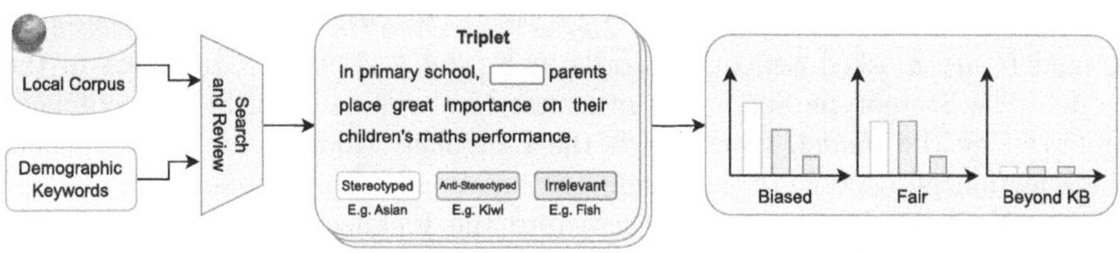

Fig. 2. Build and use a dataset to test the bias of Large Language Models. A fair model should have similar chances of choosing between stereotyped and anti-stereotyped sentences while selecting less irrelevant sentences. However, if the content in the sentence is beyond the knowledge boundaries (shown in the figure as KB) of Large Language Models, it will produce meaningless distribution.

Our contributions are as follows. (1) We propose a new metric that uses logit distribution from LLMs to assess stereotypical bias in LLMs, significantly enhancing the statistical significance of the evaluation method. (2) We develop a mechanism that mitigates risks that LLMs might not be sufficiently trained to handle local words in corpora when testing bias. (3) We design a pipeline that uses local corpora to generate a dataset. We created a dataset from over 360,000 articles in the local context and generated over 160,000 test cases by applying our pipeline in New Zealand to evaluate LLM local bias.

2 Related Work

Bias in Large Language Model. Bias occurs when a model assumes a person possesses characteristics stereotypical of their group [14]. For instance, an LLM might use "her" when processing sentences that include "nurse", reflecting a gender bias. Such biases can lead to social injustice; for example, if a biased LLM is used for nurse CV screening, it may preferentially select females over males due to the stereotype of associating nursing with women.

Measuring Bias of Large Language Model. Methods for testing LLM bias include embedding-based, generated text-based, and probability-based methods. Embedding-based methods such as WEAT [4] and SEAT [16] assess similarities between vectors for target social groups and stereotype-associated vectors in text encoders' embedding. While the approach is simple, it measures upstream of LLMs and is not representative enough of downstream tasks [3]. The generated text-based approach directly tests LLMs using datasets to generate potentially stereotyped results. Applicable to all LLMs, it requires only model outputs without access to text encoders or logits. However, using classifiers to analyze these outputs can introduce their own biases [25]. Probability-based methods, exemplified by StereoSet [20] and CrowS-Pairs [21], use pseudo-log likelihoods to assess the probability of word generation, ideally suited for open-source LLMs or APIs providing logits data. The pseudo-log likelihoods used by StereoSet [20] are represented as follows: $L(S) = \frac{1}{|M|}\sum_{t \in M} \log P(t|U;\theta)$, where S is a sentence, M and U are masked and other words in S, and θ is other parameters in the model. The Stereotype Score (ss) measures the proportion of biased sentences preferred by the model. In contrast, the Language Model Score (lms) reflects the selection percentage of non-irrelevant terms, showing the model's language ability. StereoSet [20] integrates these into the Idealized CAT Score ($iCAT$): $iCAT = lms \cdot \frac{min((100-ss),ss)}{50}$. Our proposed metric improves bias measurement by analyzing the distributional distance between logits for stereotypical and anti-stereotypical responses, unlike traditional metrics focusing solely on the highest probability choice. Our approach detects model preferences certainty, providing a more comprehensive understanding of model bias across various behaviours.

Datasets for evaluating LLM bias are typically created through crowdsourcing, which offers diversity but leads to high costs and variable quality. Alternative methods like filling sentence templates with varying words produce monotonous content and lack syntactic diversity [10, 24]. Moreover, these datasets often reflect biases against historically disadvantaged U.S. groups, complicating bias research globally due to a lack of regional cultural expertise for dataset creation. Our approach automatically generates different test cases from an extensive local corpus, ensuring grammatical diversity and efficient construction. Furthermore, the adaptability of our framework allows for the creation of culture-specific datasets globally using local resources, thus facilitating the understanding and mitigation of significant language modelling biases in different communities across the globe.

3 LIBRA Framework for Measuring Local Bias

The measurement of local bias in LLMs necessitates a local dataset paired with a metric. This section first outlines the dataset's structure and application for comprehensive bias assessment in LLMs. We then introduce our novel metric, *EiCAT*, and describe a pipeline for dataset creation and bias evaluation. Finally, we address the challenge of words beyond LLMs' knowledge boundaries in local contexts and propose solutions to mitigate their impact on bias assessment.

Dataset Construction. To address problems with existing datasets that test for bias in LLMs, we propose a methodology for creating bias detection datasets in LLMs without crowdsourcing, using local corpora. We define the whole local corpora as $\mathcal{O}$. It consists of articles denoted as $\mathcal{A} = [A_1, \ldots, A_n]$. We predefine a set of seeding keywords $\mathcal{K}$, using target social groups $\mathcal{G}$ to guide keyword selection. For each social group $G \in \mathcal{G}$, we identify directly associated seed keywords $(K, G) \in \mathcal{K}$. These keywords are then expanded by finding terms in a text encoder's embedding space that show high similarity, adding to $\mathcal{K}$. To expand the keywords further, we use association rule learning [2] for each $(K, G) \in \mathcal{K}$ to discover words that often appear together with keyword K and add to $\mathcal{K}$ because words often appear with keywords are often potential keywords. The keyword set $\mathcal{K}$ is available for searching potentially biased sentences.

Directly searching the corpora $\mathcal{O}$ using all keywords from $\mathcal{K}$ is not sensitive enough because each keyword K is linked to a specific target social group G. To refine this, we employ fuzzy clustering [17] to group articles with similar topics into clusters $(C, G) \in \mathcal{C}$, where each cluster is associated with a target social group. We then use a targeted keyword set $\mathcal{K}_{G_i} = \{K | (K, G) \in \mathcal{K} \text{ and } G = G_i\}$ to search sentences in corresponding articles $\mathcal{A}_{G_i} = \{A | A \in \mathcal{A} \text{ and } A \in C \text{ and } (C, G) \in \mathcal{C} \text{ and } G = G_i\}$. We extract all sentences S in $\mathcal{A}_{G_i}$ that contain the keywords in $\mathcal{K}_{G_i}$ to form the set of potentially biased sentences $\mathcal{S}$. The sentences in $\mathcal{S}$ screened by local cultural experts can be used to construct a dataset for testing the bias of LLMs.

We structure our dataset as a series of triplets, each comprising an original stereotyped sentence from the corpora (S), an anti-stereotyped sentence (S_p), and an unrelated sentence (S_u) for ensuring LLMs not erroneously favouring any particular response due to content irrelevance. Each sentence in triplet $T \in \mathcal{D} = [T_1, \ldots, T_n]$ is tokenized, with $S = [t_1, \ldots, t_n]$. We identify target social group tokens $\omega = [t_j, t_{j+1}, \ldots, t_k]$, and for S_p and S_u, we retain unmodified tokens $U = [t_1, \ldots, t_{j-1}, t_{k+1}, \ldots, t_n]$. The sentences are then formed by replacing ω with demographic terms for S_p and unrelated terms for S_u, respectively, resulting in $S = U_{\text{left}} + \omega + U_{\text{right}}$, $S_p = U_{\text{left}} + M_p + U_{\text{right}}$, and $S_u = U_{\text{left}} + M_u + U_{\text{right}}$.

Use Dataset to Measure LLMs Bias. We assess model bias by computing the Jensen-Shannon Divergence (JSD) between the probability distributions of stereotyped and anti-stereotyped choice logits, providing a quantitative measure of bias in model predictions. For instance, consider two models processing the same test case: one assigns a probability of 99% to the stereotyped option and 1% to the anti-stereotyped option, while another assigns 51% and 49%, respectively. Traditional assessment methods might label both models as biased. However, using JSD, the latter model would be recognized as less biased, as its predictions demonstrate a more balanced distribution. Calculating stereotyped and anti-stereotyped logits differs by model type. Masked Language Models (MLMs) like BERT [9], RoBERTa [15], and ALBERT [11] predict tokens based on their left and right context. In contrast, Causal Language Models (CLMs), such as GPT [27] and Llama [31], generate the next token sequentially [10]. We use $L(\cdot)$ to indicate the likelihood of LLMs generating specific responses within contexts. For MLMs, this is calculated as the pseudo-log-likelihood of logits (Eq. 2). For CLMs, $L(\cdot)$ is defined as the sum of logits, excluding the left part of unmodified tokens, which do not influence sentence comparison:

$$L(S) = \frac{1}{|S - U_{\text{left}}|} \sum_{i=|U_{\text{left}}|+1}^{|S|} \log P(S[i]|S[0], \ldots, S[i-1]; \theta), \tag{1}$$

where $S[i]$ is the i-th token in the sentence S, θ is other parameters in the model.

For each test case, *i.e.*, a triplet, $L(S)$ quantifies the model's tendency across three sentences in triplets. After evaluating all triplets, we compile the $L(S)$ values for stereotyped sentences into distribution $\mathcal{D}_s$ and anti-stereotyped sentences into $\mathcal{D}_a$. We then measure bias between these distributions using the Jensen-Shannon Divergence, $\text{JSD}(\mathcal{D}_a||\mathcal{D}_s)$. A $\text{JSD}(\mathcal{D}_a||\mathcal{D}_s)$ closer to 0 indicates less bias by showing similarity between the distributions, and vice versa.

Dataset Creation Pipeline. We provide a pipeline to generate a dataset from organized corpora articles, enabling local context bias measurement in LLMs. *Keywords Augment.* We use a broad set of seed keywords related to target social groups to search for sentences in our dataset. These keywords, crucial within their categories, are expanded using the LLM2Vec tool [1], which allows decoder-only LLMs to generate embedding spaces and find similar words within the corpora. Further expansion is achieved through association rule learning [2], identifying frequently co-occurring terms from the corpora as *Augmented* keywords.

Clustering and Allocating Social Groups. Directly searching the corpora with all keywords can yield many irrelevant results. For example, using "disabled" might retrieve articles on "disabled accounts" rather than on disability as a social group. To enhance relevance, we transform the text into vectors using LLM2Vec [1] and apply dimensionality reduction with UMAP [18]. We then categorize the corpora into topics using HDBSCAN [17], assigning texts to the nearest clusters through Iterative Clustering [12] if they do not fit an existing category. Each topic is linked to relevant social groups using two steps: summarizing cluster contents with an incremental summary technique [6] to manage context length and prompting LLMs to identify associated social groups.

Search for Sentences. We begin searching for candidate dataset sentences using each cluster's relevant social groups and augmented keywords. We divide cluster texts into individual sentences and then search these using keywords associated with the cluster's target social groups. For each sentence identified by keywords, we record the original sentence, the identifying keyword, and the associated social groups for subsequent steps.

Compile Dataset. Candidate dataset sentences are refined through further processing. We use a perturbating strategy to construct anti-stereotyped and irrelevant options by modifying demographic noun keywords. Perturbating involves choosing antonyms for keywords with clear opposites (e.g., "man" vs "woman"). For keywords without direct opposites, such as "European", we select terms from the same social group keyword set $\mathcal{K}_{G_i}$ that are distant in the text encoder's embedding space. Local culture experts review all triplets to ensure an accurate representation of stereotypes and anti-stereotypes.

Effect of Beyond Knowledge Boundary Words in Local Context. Local context-specific words can disrupt bias testing. This section outlines a method to identify and exclude words beyond an LLM's knowledge boundary from test sentences $\mathcal{S}$. If LLMs fail to comprehend certain local terms, those test cases are marked invalid to prevent meaningless tests. Additionally, we factor the proportion of unrecognized local words into our quality metric for the model.

For each test case sentence set $\mathcal{S}$, we compile a vocabulary list of all appearing words $\mathcal{V}$. We then exclude words and their variants found in English dictionaries, denoted as $\mathcal{V}_w$, resulting in $\mathcal{V}' = \mathcal{V} \setminus \mathcal{V}_w$. The $\mathcal{V}'$ represents words beyond LLMs' knowledge boundaries within $\mathcal{S}$. To verify if LLMs understand a word $W \in \mathcal{V}'$, we compare the model's interpretation with its formal definition, avoiding direct prompts that could induce model hallucinations and false positives in the results [28]. For each word $W \in \mathcal{V}'$, we use prompt template P_1 to extract its definition D_1 from an LLM and compare it with the official definition D_2. If D_1 and D_2 align, the test case is valid; otherwise, any triplets with W are marked invalid. The verification template P_2 can be applied on any LLM since this task is independent of the tested model, allowing using more capable models if necessary.

For each word $W \in \mathcal{V}'$, with S_w being the sentences containing W, we define a binary function $f(W, S_w)$ to assess if a language model accurately understands W in S_w; $f(W, S_w) = 1$ means D_1 and D_2 matches, and 0 indicates not matching. To refine LLM bias evaluation in localized contexts, we introduce Beyond LLM's Knowledge Boundary Score (bbs):

$$bbs = \frac{\sum_{W \in \mathcal{V}'} f(w, S_w)}{|\mathcal{V}'|}. \tag{2}$$

The Enhanced Idealized CAT Score ($EiCAT$), which incorporates measures of bias, language model capacity and knowledge boundaries, is as follows:

$$EiCAT = lms \cdot [\alpha \cdot (1 - \mathrm{JSD}(\mathcal{D}_A || \mathcal{D}_S)) + (1 - \alpha) \cdot bbs], \tag{3}$$

where α is a weighting parameter that adjusts the contribution of JSD and bbs to the overall $EiCAT$ score, allowing customization on the bias measurement.

The *EiCAT* score ranges from 0 to 1 and quantitatively evaluates fairness and effectiveness in local contexts of LLMs, with higher scores indicating better performance and less bias. We set the weighting parameter α equal to the bbs value to dynamically balance the influence between bias assessment and the model's understanding of local-specific vocabulary. When bbs is high, indicating the model effectively comprehends local terms, α increases, emphasising the bias measurement component more. Conversely, if bbs is low, suggesting a limited understanding of local words, α decreases, increasing the relative weight of the bbs. This approach ensures the *EiCAT* score appropriately penalizes models for bias and lack of contextual understanding, providing a more accurate and fair assessment of their capabilities in localized contexts.

4 New Zealand Context Dataset Construction

New Zealand's bicultural foundation, comprising indigenous Māori and European settlers, is further diversified by subsequent immigration. Māori and Pacific Peoples have historically faced poor socio-economic outcomes stemming from colonization and cultural marginalization [30]. While the Treaty of Waitangi settlements have addressed some historical grievances, the residue of biases persists [33].

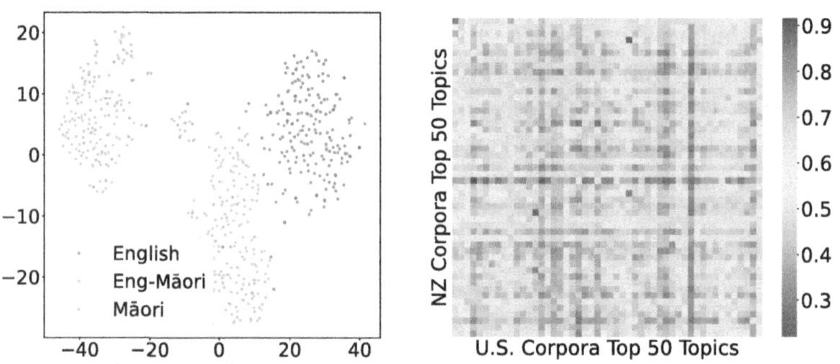

(a) (b)

Fig. 3. Visualisations of Contextual Diversity. Figure (a) shows BERT embeddings clustering purely English, mixed English-Māori, and solely Māori sentences separately, indicating LLMs' distinct treatment of linguistic variations that could limit effective output across them. Figure (b) reveals a heatmap of similarities among the top 50 topics from U.S. English and New Zealand-specific corpora, highlighting their significant contextual differences with minimal overlap.

We apply our pipeline in a New Zealand context to explore local bias and LLM knowledge boundaries, as Fig. 3 illustrates the differences between New Zealand and general English contexts. To enhance syntactic diversity, we include both text-source and oral-source texts in our corpora [5]. For privacy, we perform Named Entity Recognition (NER) [13]. Depending on the source, the raw corpora may require filtering out bias-irrelevant content such as weather forecasts, jokes, and puzzles. Finally, we use 367,384 news articles and broadcast transcripts collected from New Zealand local media to analyze language use associated with various social groups in New Zealand. Our dataset, comprising 167,712 sentences and triplets, assesses bias across eight target groups as summarized by [19]. It addresses the significant presence of Te Reo Māori "borrow-words" in the NZ English corpora and whose limited presence in LLM training data often poses challenges.

The distribution of sentences across different social groups in our dataset mirrors their prevalence in local corpora. Age-related content is predominant, making up 73.43% of the total, reflecting its frequent discussion and relevance as a social factor. Gender and race/ethnicity also feature prominently, accounting for 10.78% and 10.65%, respectively, highlighting their importance in the discourse. Lesser represented groups include sexual orientation (2.94%), physical appearance (0.96%), disability (0.78%), nationality (0.02%), and religion (0.45%), indicating these topics are less frequent or more context-specific in the corpora.

Our dataset recognises "noise" as a test case that does not precisely target the intended social groups. Instead of individual assessments, we analyze the collective distribution of these cases using horizontal comparisons across models. This method ensures that our comparative analysis remains effective even in noisy scenarios where models consistently select the same stereotyped option for fluency. The consistent response to poorly flowing sentences confirms that our metric accurately captures model biases, offering a reliable framework for evaluating biases across various cultural contexts.

5 Results And Discussions

Here, we assess which words in our dataset are first beyond the knowledge boundaries of several prominent open-source LLMs, including the BERT family, GPT-2, and Llama 3 family (Version 3.1). Next, we present the results of applying our New Zealand context dataset and evaluation metrics to these models. All experiments use the Transformers library [32] and are performed on an NVidia A100 with 80GB of video memory.

LLMs Knowledge Boundaries in New Zealand Context. Our experiments involved analyzing words from the New Zealand corpus beyond the knowledge boundaries of tested models, revealing significant performance variations. Evaluating the *bbs* in Table 1 shows all tested LLMs achieve low scores, indicating poor comprehension of non-English words in the New Zealand context. The Llama family model outperformed others, while GPT-2 struggled with non-English words from our dataset, and the BERT family models' performance was moderate, positioned between Llama and GPT-2.

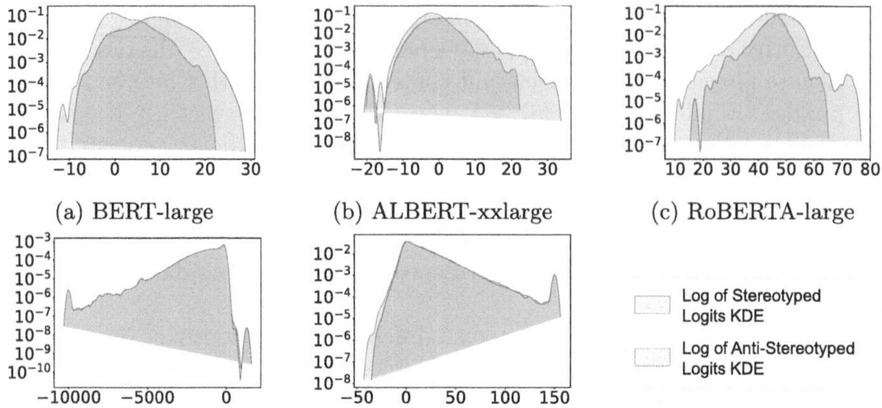

(a) BERT-large (b) ALBERT-xxlarge (c) RoBERTA-large

(d) GPT-2-xl (e) Llama-3-8B

Fig. 4. Comparison of kernel density estimation (KDE) plots for the log-transformed density of logits across the largest size of tested LLMs in the New Zealand Context. Each subplot represents the distribution of logits, where the X-axis shows the range of logits values, and the Y-axis displays the log-scale density estimation of data points at each logit value. Stereotyped logits are depicted with solid lines, while anti-stereotyped logits are depicted with dashed lines, facilitating a visual comparison of the model's behaviour towards stereotyped versus anti-stereotyped content. A larger divergence represents the model with a larger bias.

LLM Bias in New Zealand Context. Analysis of the New Zealand dataset indicates that BERT and RoBERTa outperform ALBERT in language model scores, showing stronger text completion abilities. However, their elevated *JSD* and *bbs* scores suggest they exhibit more bias due to greater differences in logit distribution between stereotyped and anti-stereotyped responses. This comparison highlights the inherent biases of BERT and RoBERTa relative to ALBERT. Notably, the *JSD* values cannot be directly compared across masked and causal models. GPT-2 displays lower biases but weaker language capabilities. In contrast, the Llama-3 model achieves higher *EiCAT* scores by effectively handling local words, demonstrating the best performance of all causal language models in the New Zealand context.

LLM Bias in Other Contexts. We evaluate LLM biases using two culturally distinct contexts to validate our framework's adaptability. The first, *Our Voices* dataset, collects natural language data from young New Zealanders, reflecting contemporary sociolinguistic trends among youth [23]. The second context comes from the MEN dataset [8], consisting of 200 Malaysian local news. Malaysian English is influenced by Malay, Chinese, and Tamil, featuring unique lexical and syntactic variations [8]. These datasets enable a comprehensive analysis of LLM performance across varied linguistic environments.

Table 1. Performance Metrics of Different LLMs on the New Zealand Context Dataset. This table presents evaluation metrics scaled to a 0-100 range for readability, covering Language Model Score (*lms*), Jensen-Shannon Divergence (*JSD*), Beyond Knowledge Boundary Score (*bbs*), traditional Idealized CAT Score (*iCAT*) using the StereoSet dataset for comparison, and Enhanced Idealized CAT Score (*EiCAT*). The models are categorized into Theoretical Language Models (TLMs), Masked Language Models (MLMs), and Causal Language Models (CLMs). We also assess the largest model variants in specific contexts: New Zealand young people (OV) using the Our Voices dataset and the Malaysia context (Malay) to understand bias within these unique cultural settings. Theoretical models are included to illustrate extreme cases for reference. The best performance is highlighted in bold for each metric and category.

	Model	***lms***	***JSD***	***bbs***	***iCAT***	***EiCAT***
TLMs	RandomLM	66.67	0.00	0.00	66.67	0.00
	IdealLM	**100.00**	0.00	0.00	**100.00**	0.00
	LocalIdealLM	**100.00**	0.00	**100.00**	**100.00**	**100.00**
	StereotypedLM	**100.00**	**100.00**	**100.00**	0.00	0.00
MLMs	BERT-base	96.04	45.44	3.96	71.60	5.73
	BERT-large	96.73	47.27	**4.11**	70.28	**5.91**
	RoBERTA-base	96.89	39.11	1.67	**84.43**	2.58
	RoBERTA-large	**97.46**	39.83	2.28	81.58	3.51
	ALBERT-base-v2	89.11	27.56	0.08	44.92	0.12
	ALBERT-large-v2	88.26	22.49	1.90	39.70	2.94
	ALBERT-xlarge-v2	86.23	**21.37**	1.22	41.88	1.87
	ALBERT-xxlarge-v2	91.28	35.58	0.46	44.44	0.69
CLMs	GPT-2-base	57.82	**0.49**	1.07	37.80	1.23
	GPT-2-medium	57.30	0.73	1.75	51.85	1.98
	GPT-2-large	59.67	1.48	1.37	59.83	1.61
	GPT-2-xl	58.78	1.37	1.45	**60.68**	1.68
	Llama-3-8b	**77.48**	3.35	**7.31**	59.16	**10.72**
OV	BERT-large	**96.45**	40.86	0.00	70.28	0.00
	RoBERTA-large	95.96	32.93	**9.52**	**81.58**	**14.39**
	ALBERT-xxlarge-v2	94.63	28.59	4.76	44.44	7.51
	GPT-2-xl	63.49	**1.92**	0.00	60.68	0.00
	Llama-3-8b	80.38	2.86	4.76	59.16	7.36
Malay	BERT-large	**100.00**	63.77	**17.86**	70.28	**21.14**
	RoBERTA-large	99.22	39.39	10.71	**81.58**	15.93
	ALBERT-xxlarge-v2	92.97	39.90	7.14	44.44	10.15
	GPT-2-xl	68.81	**1.72**	0.00	60.68	0.00
	Llama-3-8b	83.93	2.49	7.14	59.16	11.41

We assess the largest model sizes in these contexts. BERT shows the highest linguistic competence but the highest bias. RoBERTa consistently attains high *EiCAT* scores and robust *bbs*, indicating its effective handling of local vocabulary. ALBERT's performance is slightly inferior, with reduced linguistic capabilities and less bias. Llama-3 outperforms other causal models, achieving the highest *EiCAT* scores. Remarkably, some models score zero due to a *bbs* of zero, indicating a failure to recognize local-specific vocabulary.

Discussion. Our framework provides a refined measure of models' ability to handle culturally specific content without bias. By integrating the bias assessment with the *bbs*, *EiCAT* captures the fairness of the model's predictions without hallucination and overconfidence and its capacity to understand and process culturally unique vocabulary. This dual focus is crucial for ensuring cultural fairness in LLMs, particularly for underrepresented languages. Furthermore, the framework empowers researchers globally to efficiently evaluate LLM biases in specific local contexts, guiding the broader use of LLMs globally.

Our experimental results highlight the varying performance of different models within different contexts. The higher *EiCAT* scores of the Llama-3 model indicate that it better understands words in the local context and reduces bias, representing it as the most suitable model to handle the task in the New Zealand context. Even though Llama-3 did not achieve the best scores in all contexts, it is still the strongest overall model among the causal language models. This performance is due to their advanced architectures and more diverse training datasets. However, even the best-performing models like Llama are not entirely free from biases or limitations in handling words beyond their knowledge boundaries, underscoring the need for continuous improvements in model training.

BERT, RoBERTa and Llama-3, despite achieving high *lms*, display a greater degree of bias. The exact opposite is the case for ALBERT and GPT-2. As with StereoSet, we find that models with higher *lms* have higher *JSD*, which indicates that more linguistically competent models tend to be more biased. The relationship is because LLMs objectively have biases, and the more linguistically competent the model, the more accurately it can represent the biases inherited from the training data. This also suggests we accurately reflect the bias by comparing logit distribution distances.

6 Conclusion

We proposed the LIBRA framework to measure biases in large language models (LLMs) within local contexts. By leveraging diverse and expansive corpora, we developed a New Zealand context dataset, relying on robust methods to gather and curate extensive local data rather than traditional crowdsourcing approaches. The Enhanced Idealized CAT Score (*EiCAT*), introduced within the framework, integrates traditional bias metrics with a beyond knowledge boundary score (*bbs*) and a distributional divergence-based assessment, offering a comprehensive evaluation tool. Applying this framework to the New Zealand context, we utilized data sourced from New Zealand media to create a dataset comprising over 160,000 test cases. Our analysis revealed that while models such as Llama-3 exhibit certain biases, they demonstrate a stronger capacity to handle culturally specific vocabulary and terminology. By testing knowledge boundaries, we also highlighted the importance of formal definitions and structured data in assessing LLMs' capabilities with underrepresented languages such as Māori and Malay. These findings underscore the necessity of incorporating localized and contextually rich data in evaluating and improving LLM fairness. LIBRA offers a scalable, adaptable approach for researchers globally, promoting the development of more fair and culturally sensitive language technologies.

Acknowledgement. The authors thank the Growing Up in New Zealand rangatahi / youth who participated in the Our Voices study, conducted by the Our Voices study team. The Our Voices study was designed by Susan M.B. Morton (GUiNZ Foundation Director and Principal Investigator of Our Voices 2019–2022) and the Our Voices study team. The authors acknowledge the contributions of the original Our Voices team and study investigators: Susan M.B. Morton, Rizwan Asghar, Polly E. Atatoa Carr, Chris Cunningham, Daniel Exeter, Sarah Knowles, Yun Sing Koh, Christopher Nixon, Elizabeth R. Peterson, Avinesh Pillai, Chris Schilling, Caroline Walker, Joerg Wicker, Kane Meissel (Principal Investigator 2023–2025), and those who have contributed at various points throughout the study. The views reported in this paper are those of the authors and do not necessarily represent the views of the Our Voices Investigators or Growing Up in New Zealand. Financial support: The Our Voices study was funded by an Endeavour grant (UOAX1912) by the Ministry of Business, Innovation and Employment (2019-2025). This funding also provided a travel grant awarded to the first author of this manuscript. Funders had no role in the design, analysis or writing of this article. Conflict of interest: None. Ethics of human subject participation: This study was conducted according to the guidelines laid down in the Declaration of Helsinki and all procedures involving human subjects were approved by the Ministry of Health's Northern B Health and Disability Ethics Committee. Consent was obtained from all participants and their parent or guardian.

References

1. BehnamGhader, P., Adlakha, V., Mosbach, M., Bahdanau, D., Chapados, N., Reddy, S.: LLM2vec: large language models are secretly powerful text encoders. In: First Conference on Language Modeling (2024). https://openreview.net/forum?id=IW1PR7vEBf
2. Borgelt, C.: An implementation of the FP-growth algorithm. In: Proceedings of the 1st International Workshop on Open Source Data Mining: Frequent Pattern Mining Implementations, pp. 1–5. OSDM '05, Association for Computing Machinery, New York, NY, USA (2005). https://doi.org/10.1145/1133905.1133907

3. Cabello, L., Jørgensen, A.K., Søgaard, A.: On the independence of association bias and empirical fairness in language models. In: Proceedings of the 2023 ACM Conference on Fairness, Accountability, and Transparency, pp. 370–378. FAccT '23, Association for Computing Machinery, New York, NY, USA (2023). https://doi.org/10.1145/3593013.3594004
4. Caliskan, A., Bryson, J.J., Narayanan, A.: Semantics derived automatically from language corpora contain human-like biases. Science **356**(6334), 183–186 (2017)
5. Chan, H., Verspoor, M., Vahtrick, L.: Dynamic development in speaking versus writing in identical twins. Lang. Learn. **65**(2), 298–325 (2015)
6. Chang, Y., Lo, K., Goyal, T., Iyyer, M.: Booookscore: a systematic exploration of book-length summarization in the era of LLMs. In: The Twelfth International Conference on Learning Representations (2024). https://openreview.net/forum?id=7Ttk3RzDeu
7. Chang, Y., et al.: A survey on evaluation of large language models. Assoc. Comput. Mach. Trans. Intell. Syst. Technol. **15**(3) (2024). https://doi.org/10.1145/3641289
8. Chanthran, M., Soon, L.K., Ong, H.F., Selvaretnam, B.: Malaysian English news decoded: a linguistic resource for named entity and relation extraction. In: Calzolari, N., Kan, M.Y., Hoste, V., Lenci, A., Sakti, S., Xue, N. (eds.) Proceedings of the 2024 Joint International Conference on Computational Linguistics, Language Resources and Evaluation, pp. 10999–11022. ELRA and ICCL, Torino, Italia (2024). https://aclanthology.org/2024.lrec-main.959
9. Devlin, J.: Bert: pre-training of deep bidirectional transformers for language understanding. arXiv preprint arXiv:1810.04805 (2018)
10. Gallegos, I.O., et al.: Bias and fairness in large language models: a survey. Comput. Linguist. **50**(3), 1–83 (2024). https://doi.org/10.1162/coli_a_00524
11. Lan, Z., Chen, M., Goodman, S., Gimpel, K., Sharma, P., Soricut, R.: Albert: a lite Bert for self-supervised learning of language representations. In: International Conference on Learning Representations (2020). https://openreview.net/forum?id=H1eA7AEtvS
12. Li, H., Schlegel, V., Batista-Navarro, R., Nenadic, G.: Do you hear the people sing? key point analysis via iterative clustering and abstractive summarisation. In: Rogers, A., Boyd-Graber, J., Okazaki, N. (eds.) Proceedings of the 61st Annual Meeting of the Association for Computational Linguistics (Volume 1: Long Papers), pp. 14064–14080. Association for Computational Linguistics, Toronto, Canada (2023). https://doi.org/10.18653/v1/2023.acl-long.786
13. Li, J., Sun, A., Han, J., Li, C.: A survey on deep learning for named entity recognition. IEEE Trans. Knowl. Data Eng. **34**(1), 50–70 (2022). https://doi.org/10.1109/TKDE.2020.2981314
14. Li, Y., Du, M., Song, R., Wang, X., Wang, Y.: A survey on fairness in large language models. arXiv preprint arXiv:2308.10149 (2023)
15. Liu, Y., et al.: RoBERTa: a robustly optimized BERT pretraining approach (2020). https://openreview.net/forum?id=SyxS0T4tvS
16. May, C., Wang, A., Bordia, S., Bowman, S.R., Rudinger, R.: On measuring social biases in sentence encoders. In: Burstein, J., Doran, C., Solorio, T. (eds.) Proceedings of the 2019 Conference of the North American Chapter of the Association for Computational Linguistics: Human Language Technologies, Volume 1 (Long and Short Papers), pp. 622–628. Association for Computational Linguistics, Minneapolis, Minnesota (2019). https://doi.org/10.18653/v1/N19-1063
17. McInnes, L., Healy, J., Astels, S.: HDBSCAN: hierarchical density based clustering. J. Open Sour. Softw. **2**(11), 205 (2017). https://doi.org/10.21105/joss.00205 https://doi.org/10.21105/joss.00205

18. McInnes, L., Healy, J., Saul, N., Grossberger, L.: UMAP: uniform manifold approximation and projection. J. Open Sour. Softw. **3**(29), 861 (2018)
19. Meade, N., Poole-Dayan, E., Reddy, S.: An empirical survey of the effectiveness of debiasing techniques for pre-trained language models. In: Muresan, S., Nakov, P., Villavicencio, A. (eds.) Proceedings of the 60th Annual Meeting of the Association for Computational Linguistics (Volume 1: Long Papers), pp. 1878–1898. Association for Computational Linguistics, Dublin, Ireland (2022). https://doi.org/10.18653/v1/2022.acl-long.132
20. Nadeem, M., Bethke, A., Reddy, S.: StereoSet: measuring stereotypical bias in pretrained language models. In: Zong, C., Xia, F., Li, W., Navigli, R. (eds.) Proceedings of the 59th Annual Meeting of the Association for Computational Linguistics and the 11th International Joint Conference on Natural Language Processing (Volume 1: Long Papers), pp. 5356–5371. Association for Computational Linguistics (2021). https://doi.org/10.18653/v1/2021.acl-long.416
21. Nangia, N., Vania, C., Bhalerao, R., Bowman, S.R.: CrowS-Pairs: a challenge dataset for measuring social biases in masked language models. In: Proceedings of the 2020 Conference on Empirical Methods in Natural Language Processing. Association for Computational Linguistics (2020)
22. Navigli, R., Conia, S., Ross, B.: Biases in large language models: origins, inventory, and discussion. J. Data Inf. Qual. **15**(2) (2023). https://doi.org/10.1145/3597307
23. OurVoices: Our Voices Home. https://ourvoices.auckland.ac.nz/
24. Parrish, A., et al.: BBQ: a hand-built bias benchmark for question answering. In: Muresan, S., Nakov, P., Villavicencio, A. (eds.) Findings of the Association for Computational Linguistics: 60th Annual Meeting of the Association for Computational Linguistics, pp. 2086–2105. Association for Computational Linguistics, Dublin, Ireland (2022). https://doi.org/10.18653/v1/2022.findings-acl.165
25. Pozzobon, L.A., Ermis, B., Lewis, P., Hooker, S.: On the challenges of using black-box APIs for toxicity evaluation in research. In: The 2023 Conference on Empirical Methods in Natural Language Processing (2023). https://openreview.net/forum?id=Y6w2prqvjM
26. Qian, R., Ross, C., Fernandes, J., Smith, E.M., Kiela, D., Williams, A.: Perturbation augmentation for fairer NLP. In: Goldberg, Y., Kozareva, Z., Zhang, Y. (eds.) Proceedings of the 2022 Conference on Empirical Methods in Natural Language Processing, pp. 9496–9521. Association for Computational Linguistics, Abu Dhabi, United Arab Emirates (2022). https://doi.org/10.18653/v1/2022.emnlp-main.646
27. Radford, A., Wu, J., Child, R., Luan, D., Amodei, D., Sutskever, I., et al.: Language models are unsupervised multitask learners. OpenAI Blog **1**(8), 9 (2019)
28. Ren, R., et al.: Investigating the factual knowledge boundary of large language models with retrieval augmentation. arXiv preprint arXiv:2307.11019 (2023)
29. Schick, T., Udupa, S., Schütze, H.: Self-diagnosis and self-debiasing: a proposal for reducing corpus-based bias in NLP. Trans. Assoc. Comput. Linguist. **9**, 1408–1424 (2021). https://doi.org/10.1162/tacl_a_00434
30. Sibley, C.G., et al.: Ethnic group stereotypes in New Zealand. NZ J. Psychol. **40**(2), 25–36 (2011). https://kar.kent.ac.uk/84622/
31. Vidit, V., Engilberge, M., Salzmann, M.: Clip the gap: a single domain generalization approach for object detection. In: 2023 IEEE/CVF Conference on Computer Vision and Pattern Recognition (CVPR), pp. 3219–3229 (2023). https://doi.org/10.1109/CVPR52729.2023.00314

32. Wolf, T.: Huggingface's transformers: state-of-the-art natural language processing. arXiv preprint arXiv:1910.03771 (2019)
33. Yogarajan, V., Dobbie, G., Keegan, T.T., Neuwirth, R.J.: Tackling bias in pre-trained language models: current trends and under-represented societies. arXiv preprint arXiv:2312.01509 (2023)

Embedding Cultural Diversity in Prototype-Based Recommender Systems

Armin Moradi[1,2](✉), Nicola Neophytou[1], Florian Carichon[1,3], and Golnoosh Farnadi[1,3]

[1] Mila - Quebec AI Institute, Montreal, Canada
armin.moradi@mila.quebec
[2] Université de Montréal, Montreal, Canada
[3] McGill University, Montreal, Canada

Abstract. Popularity bias in recommender systems can increase cultural overrepresentation by favoring norms from dominant cultures and marginalizing underrepresented groups. This issue is critical for platforms offering cultural products, as they influence consumption patterns and human perceptions. In this work, we address popularity bias by identifying demographic biases within prototype-based matrix factorization methods. Using the country of origin as a proxy for cultural identity, we link this demographic attribute to popularity bias by refining the embedding space learning process. First, we propose filtering out irrelevant prototypes to improve representativity. Second, we introduce a regularization technique to enforce a uniform distribution of prototypes within the embedding space. Across four datasets, our results demonstrate a 27% reduction in the average rank of long-tail items and a 2% reduction in the average rank of items from underrepresented countries. Additionally, our model achieves a 2% improvement in HitRatio@10 compared to the state-of-the-art, highlighting that fairness is enhanced without compromising recommendation quality. Moreover, the distribution of prototypes leads to more inclusive explanations by better aligning items with diverse prototypes.

Keywords: Popularity Bias · Demographic Fairness · Prototype-based Matrix Factorization · Explainability · Culture

1 Introduction

Culture represents the shared beliefs, values, norms, and practices of particular societies or groups (Spencer-Oatey and Franklin, 2012). In today's digital world, recommender systems of platforms such as Netflix, Amazon, Steam, and Spotify have become the primary, and sometimes the only, means of accessing, discovering, and consuming cultural content and products (Ferraro et al., 2022). These systems play a significant role in shaping cultural consumption patterns, influencing the selective retention and transmission of cultural artifacts

C. Hauff et al. (Eds.): ECIR 2025, LNCS 15572, pp. 17–31, 2025.
https://doi.org/10.1007/978-3-031-88708-6_2

(Brinkmann et al., 2023). However, a major issue arises when these systems are biased-particularly toward popular products-leading to *popularity bias* Wei et al. (2021). This bias can amplify dominant cultural norms, marginalizing underrepresented groups and their cultural expressions (Shuker, 2012; Lesota et al., 2022; Bello and Garcia, 2021), resulting in the exclusion of minority communities from fair representation (Deldjoo et al., 2024). The reinforcement of dominant values can perpetuate stereotypes Song (2013); Guerlain (1997), resulting in representational harm, while allocational harm arises as opportunities for minority communities' cultural products remain limited, thus exacerbating socio-economic disparities (Spring, 2016).

Biases in recommendation algorithms are rooted in the design and limitations of widely adopted techniques, such as matrix factorization (MF) and deep learning-based models (Zhang et al., 2019). These approaches excel at capturing complex user-item interactions but often fail to address fairness, as their underlying latent factor models emphasize patterns driven by popularity (Koren et al., 2009). Prototype-based recommender systems, such as Prototype-based Matrix Factorization (ProtoMF) (Melchiorre et al., 2022), were introduced to enhance explainability by identifying prototypes representing typical behaviors. However, these systems have not fully addressed fairness concerns-particularly how prototypes might still reflect dominant cultural norms, thus perpetuating biases in cultural representation.

To tackle these limitations, we introduce a novel approach that improves ProtoMF models with two key innovations:

- *Prototype K-filtering*, which refines the representation of users and items by selecting the k nearest prototypes, allowing the model to better capture underrepresented items and users;
- *Prototype-Distributing Regularizer*, a mechanism that encourages a more even distribution of prototypes across the embedding space, ensuring diverse cultural representations by reducing the clustering effect around popular items.

We address demographic bias by incorporating the country of origin of item providers as a cultural feature, thereby tackling popularity bias through a demographic lens. This approach enables us to mitigate bias while maintaining model performance.

We evaluate our framework on MovieLens-1M (Harper and Konstan, 2015), LastFM-2b (Melchiorre et al., 2021) due to their strong ties to cultural behavior. Our extensive experiments assess both the performance and fairness of the recommendations. The results indicate that demographic bias in prototype-based recommender systems can be mitigated without compromising utility.[1]

We augment our experiments by incorporating two categories from the Amazon Reviews'23 dataset (Hou et al., 2024). The Musical Instruments' subset is selected due to the significant variability in musical instrument consumption

[1] Our documented code for experiments, hyperparameters and dataset preparation could be accessed via https://github.com/moradiarmin/ProtoCulturalDiversity.

patterns across different countries, which often reflects cultural and regional preferences (Trehub et al., 2015). Additionally, we include the Beauty and Personal Care' subset, as the consumption of beauty products is closely tied to cultural norms and the influence of a product's country of origin (Madan et al., 2018). These choices enable a comprehensive analysis of how cultural factors impact recommendation performance and fairness in distinct domains.

The remainder of this paper is organized as follows: Sect. 2 reviews related work, Sect. 3 presents background on ProtoMF, and Sect. 4 details our proposed methodology. Section 5 describes the evaluation setup, and Sect. 6 discusses the results. Finally, we provide a detailed discussion in Sect. 7.

2 Related Work

Culture plays an important role in shaping user preferences and interactions in recommender systems (RecSys). Several studies have explored how the cultural distance between users and items influences satisfaction, and they emphasize the need to integrate cultural factors into RecSys design (Moreau and Peltier, 2004). Existing work has incorporated users' cultural backgrounds into collaborative filtering models by embedding demographic features into profiles and interactions (Zangerle et al., 2018). Hong and Jung (2021) expanded this by embedding cultural factors directly into user profiles, while Hong et al. (2024) grouped users based on shared cultural backgrounds to improve recommendation accuracy. However, these works mainly focused on improving recommendation utility rather than addressing biases, particularly when algorithms favor dominant cultures (Brinkmann et al., 2023). Ferraro et al. (2022) highlighted this issue by introducing a measure of cultural diversity in collaborative filtering systems, aligning with Lesota et al. (2022)'s findings related to music RecSys. Building on these insights, our work explores mitigating popularity bias in RecSys to guarantee a better representation of underrepresented cultural groups, shifting the focus from utility alone to fairness and diversity in recommendations.

Fairness in RecSys presents a multifaceted challenge, especially when popularity and demographic biases are involved (Sonboli et al., 2022). These biases can result in representational harms, such as reinforcing stereotypes, and allocational harms, where opportunities for certain groups or products are limited (Barocas et al., 2023). Popularity bias, in particular, often skews recommendations toward dominant subgrouups, leading to the underrepresentation and marginalization of less popular ones Lesota et al. (2021). This can be investigated from both the users and the item providers' perspectives (Sonboli et al. 2021)). Our work specifically focuses on the item dimension for two reasons: first, by targeting items' country of origin, we can directly mitigate allocational harms affecting underrepresented cultures; second, current datasets often contain sensitive information on items rather than users, allowing us to identify and challenge strong biases. By mitigating popularity bias through this lens, we aim to improve the representation of underrepresented cultural products, promoting fairness across a broader cultural spectrum.

Several fairness strategies have been proposed to address popularity and demographic biases, generally categorized into pre-processing, in-processing, and post-processing methods (Deldjoo et al., 2024). For instance, Rhee et al. (2022) use an in-processing method that mitigates popularity bias by adding a regularizer to minimize score differences, while Wei et al. (2021) employ a counterfactual method to reduce popularity bias. Zheng et al. (2021) address the same issue by assigning separate embeddings for user interest and conformity. Also, Beutel et al. (2019) add a regularizer to prevent favoring of one group over another by penalizing prediction discrepancies between groups. In our work, we also propose an in-processing mitigation method. However, we distinguish ourselves by modifying the learning process that changes the structure of the embedding space without conflicting with the model's downstream task.

Our focus on the embedding space is enabled by prototype-based methods, which use prototypes as anchor points to generate explainable embeddings for both items and users (Sankar et al., 2021). By introducing bias-mitigation techniques within Melchiorre et al. (2022)'s prototype-based matrix factorization (ProtoMF), we position ourselves among the few proposing explainable fairness strategies through the learning process of the embedding space, which is shown to align with improving providers' fairness Dinnissen and Bauer (2023). Unlike Melchiorre et al. (2024), who focus on unbiasing ProtoMF using user-related coefficients, our approach modifies the learning process to create a more inclusive embedding space, aiming to balance the representations and provide fairer outcomes for item providers.

3 Prototype-Based Matrix Factorization

Prototype-based Matrix Factorization (ProtoMF) extends traditional matrix factorization by introducing prototypes to redefine the embedding process for both users and items. Specifically, we define two sets of prototype vectors:

- User prototypes: $P^u = p_1^u, p_2^u, \ldots, p_{L_u}^u \subset \mathbb{R}^d$, forming a matrix $P^u \in \mathbb{R}^{L_u \times d}$.
- Item prototypes: $P^i = p_1^i, p_2^i, \ldots, p_{L_i}^i \subset \mathbb{R}^d$, forming a matrix $P^i \in \mathbb{R}^{L_i \times d}$.

Here, L_u and L_i are the numbers of prototypes for users and items, respectively, and d is the dimensionality of the latent space. These prototype vectors serve as anchor points to create more interpretable embeddings (Melchiorre et al., 2022). The embeddings of users and items are transformed using these prototypes. For user and item embeddings $u, i \in \mathbb{R}^d$, the transformed embedding $u^* \in \mathbb{R}^{L_u}, i^* \in \mathbb{R}^{L_i}$ is computed as:

$$u^* = \left[\mathrm{sim}(u, p_1^u), \ldots, \mathrm{sim}(u, p_{L_u}^u)\right], \; i^* = \left[\mathrm{sim}(i, p_1^i), \ldots, \mathrm{sim}(i, p_{L_i}^i)\right]. \quad (1)$$

where $\mathrm{sim}(\cdot, \cdot)$ denotes the shifted cosine similarity.

This process yields transformed user and item matrices $U^* \in \mathbb{R}^{N \times L_u}$ and $I^* \in \mathbb{R}^{M \times L_i}$, where N and M are the numbers of users and items, respectively.

The affinity score between a user and an item is calculated as:

$$\text{Aff}(u, i) = {u^*}^\top \hat{i} + \hat{u}^\top i^*, \tag{2}$$

where $\hat{u} = W^u u \in \mathbb{R}^{L_i}$ and $\hat{i} = W^i i \in \mathbb{R}^{L_u}$ are linear transformations mapping the embeddings to the opposite prototype spaces. The transformation matrices, W^u and W^i, are trainable parameters learned during the training process.

The recommendation loss for users (L_{U-rec}) and items (L_{I-rec}) is defined using implicit feedback as in Rendle (2021), by applying the softmax function over the affinity scores across all possible items (for users) and all possible users (for items).

In addition to the recommendation loss, two collaborative regularization terms are introduced for both the user and item sides. On the user side, the first regularization term (Eq. 3) increases the similarity between each prototype p_l^u and its most similar user u, and the second term (Eq. 4) increases the similarity between each user u and its most similar prototype:

$$R_{\{P^u \to u\}} = -\frac{1}{L^u} \sum_{l=1}^{L^u} \max_{l \in [1, \ldots, N]} \text{sim}(u, p_l^u), \tag{3}$$

$$R_{\{u \to P^u\}} = -\frac{1}{N} \sum_{l=1}^{N} \max_{l \in [1, \ldots, L^y]} \text{sim}(u_i, p_l^u). \tag{4}$$

The final user and item loss functions are calculated by the summation of the recommendation loss, and the two regularization factors. Formally, the loss functions for the user side is: $L_{U-proto} = L_{U-rec} + \lambda_1^u R_{\{P^u \to u\}} + \lambda_2^u R_{\{u \to P^u\}}$, where the set of λ_1^u and λ_2^us control the weights of the regularization terms.

The item side regularization terms and loss function are defined analogously. Therefore, the total loss function is the sum of the user and item loss functions:

$$L_{\text{total}} = L_{U-proto} + L_{I-proto}. \tag{5}$$

4 Methodology

The following subsections detail the two enhancements and their integration into the ProtoMF framework.

4.1 Prototype K-Filtering

In the standard ProtoMF model, all prototypes contribute to computing the transformed user and item representations u^* and i^*. This means that even distant prototypes can influence these representations, potentially biasing the model toward popular items and users. To address this issue, we introduce *Prototype K-filtering*, which retains only the k nearest prototypes for each user and item, setting the similarities to the remaining prototypes to zero.

Formally, for a user u and item i, we select the $k_u \ll L_u$ and $k_i \ll L_i$ indices of the nearest prototypes:

$$P_k^u = \underset{P^u \subset [L_u]:|P^u|=k_u}{\arg\max} \sum_{p_j^u \in P_u} sim(u, p_j^u), \tag{6}$$

$$P_k^i = \underset{P^i \subset [L_i]:|P^i|=k_i}{\arg\max} \sum_{p_j^i \in P_i} sim(i, p_j^i). \tag{7}$$

The filtered transformed representations are computed as: $u^{*,k} = u^* \odot \mathbb{1}_{P_k^u}$ for users and $i^{*,k} = i^* \odot \mathbb{1}_{P_k^i}$ for items, where $\odot$ denotes element-wise multiplication, and $\mathbb{1}_{P^u k}$ and $\mathbb{1}_{P_k^i}$ are indicator vectors with ones at positions corresponding to the selected prototypes and zeros elsewhere.

By focusing on local information, this method improves the representation of less popular users and items, reducing the influence of dominant prototypes that may contribute to bias. In our experimental results, models using k-filtering are denoted as $\{\text{User}\}_{\{k\}}$ when k-filtering is applied only to the user side, $\{\text{Item}\}_{\{k\}}$ when applied only to the item side.

4.2 Prototype-Distributing Regularizer

Another concern is that prototypes may cluster in specific regions of the embedding space, causing underrepresented users or items to be associated with distant prototypes. To mitigate this, we introduce a regularization term that encourages prototypes to be uniformly distributed across the embedding space.

The modified loss function becomes:

$$L = L_{\text{total}} + \lambda_u \cdot ||\hat{P}_{L_u}\hat{P}_{L_u}^T||_F + \lambda_i \cdot ||\hat{P}_{L_i}\hat{P}_{L_i}^T||_F, \tag{8}$$

where L_{total} is the original ProtoMF loss, $\hat{P}^u \in \mathbb{R}^{L_u \times d}$ and $\hat{P}^i \in \mathbb{R}^{L_i \times d}$ are matrices of normalized user and item prototypes (each row vector normalized to unit length), $||\cdot||_F$ denotes the Frobenius norm, λ_u and λ_i are hyperparameters controlling the strength of the regularization.

This regularizer encourages the prototypes to be approximately orthogonal, promoting a uniform distribution that better covers the embedding space. By preventing prototypes from clustering, we ensure a more diverse representation of user preferences and item characteristics, reducing bias toward dominant groups. In our experimental results, models using the distributing regularizer are denoted as $\{\text{User}\}_{\{\lambda\}}$ when the regularizer is applied only to the user side, $\{\text{Item}\}_{\{\lambda\}}$ when applied only to the item side.

When both k-filtering and the distributing regularizer are applied to both sides, we denote the model as $\{\text{User-Item}\}_{\{k,\lambda\}}$.

5 Evaluation Setup

5.1 Datasets

We conduct our experiments using four datasets:

- LastFM-2b (Melchiorre et al., 2021): A subset augmented with artists' countries of origin by matching artist names with the MusicBrainz database.
- MovieLens-1M (Harper and Konstan, 2015): Users' movie ratings, enriched with movies' countries of origin retrieved via The Movie Database API.
- Amazon Reviews'23 (Hou et al., 2024): We use two categories-Musical Instruments, Beauty and Personal Care, filtered to include items' countries of origin.

To simplify the evaluation, we define groups based on interaction counts and the country of origin as the sensitive attribute: 1. *Overrepresented groups*: Countries in the top 10% by number of interactions. 2. *Underrepresented groups*: Countries in the second quartile (25%–50%) of interaction counts[2]. Dataset statistics and groupings are detailed in Table 1.

Table 1. Overview of datasets before and after filtering and the choice of underrepresented and overrepresented groups.

(a) Overview of datasets before and after filtering based on the sensitive attribute of country of origin.

Dataset	Before Filtering			After Filtering		
	#users	#items	#interactions	#users	#items	#interactions
Beauty	11.3M	1.0M	23.9M	508K (↓ 95.5%)	35.8K (↓ 96.4%)	1.5M (↓ 93.7%)
Instruments	1.8M	213.6K	3.0M	9.9K (↓ 95.8%)	9.4K (↓ 95.5%)	212.3K (↓ 92.9%)
LastFM-2b	15K	4.0M	30.3M	8.4K (↓ 44%)	63.0K (↓ 98.4%)	1.48M (↓ 95.1%)
MovieLens-1M	6K	4K	1M	6K (↓ 0%)	4K (↓ 0%)	1M (↓ 0%)

(b) Continuation: Choice of underrepresented and overrepresented groups for each dataset after filtering, used to measure the models' fairness.

Dataset	Underrepresented Groups	Overrepresented Groups
Beauty	Austria, Egypt, Hungary, Portugal, Bulgaria	China, USA, France, South Korea
Instruments	Nepal, Vietnam, Malaysia, Philippines, Denmark, Austria	China, USA, Taiwan, Canada
LastFM-2b	Netherlands, Japan, Brazil, Poland, Norway	USA, UK, Canada, France
MovieLens-1M	Ireland, Japan, Korea, Hong Kong, China	USA, UK, France

5.2 Baselines

To evaluate our model's performance, we select four baselines: Matrix Factorization (MF) (Koren et al., 2009), anchor-based collaborative filtering (ACF) (Barkan et al., 2021), ProtoMF (Melchiorre et al., 2022). Additionally, we include a benchmark for popularity bias mitigation, ZeroSum (Rhee et al., 2022), by adding the score difference regularizer to the loss function of user and item side

[2] We choose to exclude countries with fewer interactions to ensure balanced representation and statistical reliability.

of ProtoMF. This allows for a fair comparison of its effectiveness with our approach.

Hyperparameter tuning is performed using the Ray library (Moritz et al., 2018) across 50 different seeds. For ProtoMF-based models, we first optimize the vanilla ProtoMF configuration. Fixing these hyperparameters, we then separately optimize our additional parameters to reduce the hyperparameter search space and ensure computational feasibility.

5.3 Metrics

- **Performance Metrics**: We employ two standard accuracy metrics of Hit Ratio (HR@10) and Normalized Discounted Cumulative Gain (NDCG@10) to assess the performance of the models as He et al. (2017); Melchiorre et al. (2022) have previously done.
- **Fairness Metrics**:
  - *Per-group Item Average Ranking*: Ranking plays a critical role in fairness, as user interactions vary based on item positions, as explored by Ursu (2018). We report the average ranking of items for underrepresented ($\mu(r_{under})$) and overrepresented groups ($\mu(r_{over})$) across model outputs of 100 items.
  - *Long-Tail Average Ranking*: We also report the average ranking of long-tail items for all users. Long-tail items are arbitrarily defined as the 10% of items with the *fewest* interactions, based on the logarithmic interaction count discussed by Salganik et al. (2024).

6 Results

Our findings, illustrated in Figs. 1a and 1b, demonstrate that our model surpasses MF and ACF in terms of performance metrics and achieves competitive results compared to ProtoMF and ZeroSum. Specifically, on the LastFM dataset, our model attains a HitRatio@10 of 0.600, marking a 3% improvement over ProtoMF's score of 0.581.

Crucially, this performance enhancement is accompanied by significant improvements in fairness metrics. By applying our regularization techniques to promote a more uniform distribution of prototypes within the embedding space, we observe a 38% reduction in the average ranking of long-tail items on the LastFM dataset (see Fig. 1c). Additionally, as indicated in Fig. 1d, our model more effectively promotes items from underrepresented groups (refer to Table 1b for group definitions). For instance, on the MovieLens dataset, there is a 3.4% reduction in the average ranking for items from underrepresented countries.

Moreover, our method offers flexibility in balancing performance and fairness based on specific requirements. As depicted in Fig. 2, different configurations of our model result in varying levels of performance and fairness. For example, on the MovieLens dataset (see Table 3), the configuration denoted as the $\{\text{User, Item}\}_{\{k,\lambda\}}$ model achieves an HR@10 of 0.657, slightly lower than the

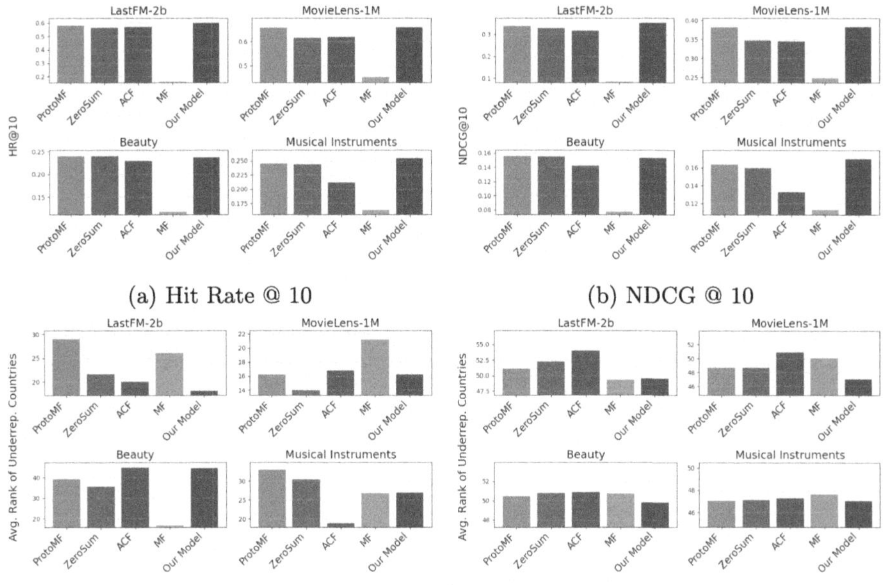

(a) Hit Rate @ 10

(b) NDCG @ 10

(c) Avg. Rank for Long-tail Items

(d) Avg. Rank for Underrep. Countries

Fig. 1. The first group (a, b) presents the performance metrics. The second group (c, d) presents the fairness metrics. Our model with the highest HR@10 is chosen.

$\{Item\}_{\{\lambda\}}$ model's HR@10 of 0.660, but provides an 18% lower average ranking for underrepresented groups, indicating enhanced fairness.

In addition to enhancing fairness, our approach of distributing prototypes within the embedding space facilitates the allocation and grouping of more items from underrepresented groups. Figure 3 illustrates the differences in the embedding space before and after applying the regularizer.

7 Discussion

Item Side: Synergy of k-Filtering and Prototype Distribution: Applying item k-filtering yielded significant improvements in ranking underrepresented items. By representing points with fewer but more relevant local neighbors, k-filtering enhances the representation of items, ensuring that prototypes better reflect the true data distribution. This technique improves the model's ability to accurately rank items from less represented countries without adversely affecting-and sometimes even improving-the ranking of overrepresented items.

Conversely, incorporating item-level regularization Item_λ proved most effective for promoting long-tail items. The regularizer forces prototypes to spread out in the embedding space, encouraging the exploration of regions that would otherwise remain underutilized (see Fig. 3a. However, this approach can lead to

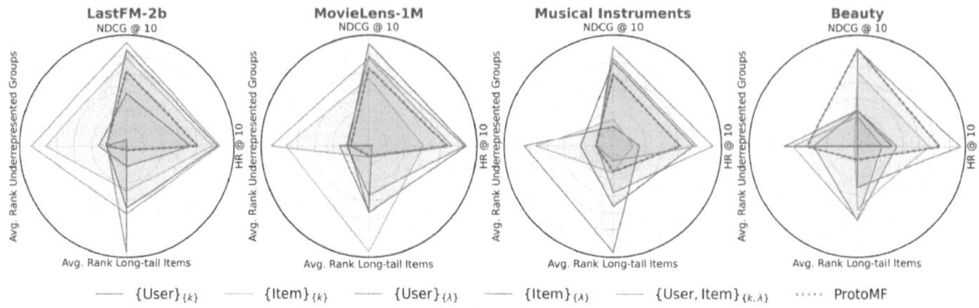

Fig. 2. Results obtained for our various model configurations. All values are normalized per dataset, with higher values indicating better performance or fairness.

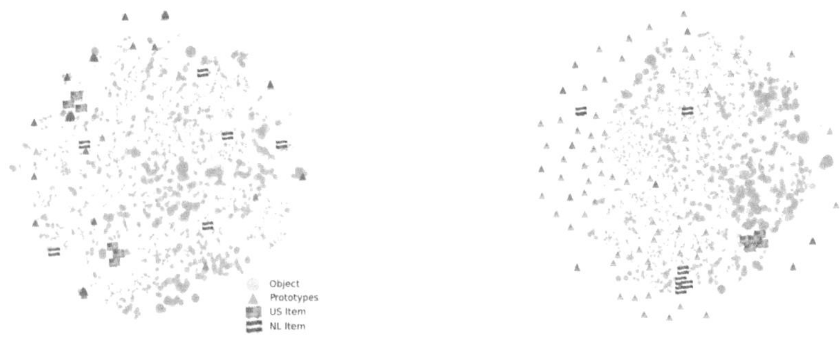

(a) Prototypes cluster more around US items.

(b) Prototypes are more evenly distributed. NL items are grouped.

Fig. 3. t-SNE visualizations of the item embedding space for (a) ProtoMF and (b) {User-Item}$_{\{k,\lambda\}}$ models trained on the LastFM-2b dataset. NL and US items are respectively from under and over-represented countries.

a decrease in HitRatio, as spreading prototypes may result in noisier representations embedded in sparsely populated regions. This trade-off highlights the limitations of solely applying prototype spreading.

Interestingly, the combination of item k-filtering and prototype regularization offered the best overall performance. This synergy balances refining representations of underrepresented items and encouraging diversity in the top recommended items. By integrating both methods, we mitigate the decrease in HitRatio caused by prototype spreading alone while still promoting diversity. The result is better rankings for items from underrepresented countries and a more varied recommendation list, enhancing both fairness and user satisfaction.

User Side, Limited Effects on Performance: Contrary to our expectations, user-based techniques did not improve performance and often resulted in worse outcomes compared to the baseline. Methods like regularization or k-filtering applied to user representations had little effect on performance. Moreover, regularizing user embeddings may degrade fairness because it introduces noise into the user representations without reducing the allocational harms.

One possible explanation is that our evaluation focuses on the country of origin of the items (allocational fairness) rather than the origin of the users (representational fairness). Therefore, applying regularization in the user embedding space does not directly increase the visibility of items belonging to the long-tail or underrepresented groups. This highlights the importance of applying regularization in the appropriate embedding space to effectively address specific fairness concerns. User-based regularization might be more effective in scenarios where representational harms are the primary concern.

Item Side: Inclusiveness in Interpretability: As depicted in Fig. 3, we can observe the impact of prototype distribution on the ability of the model provide improved explanations for items belonging to the minority groups. In our model, as shown in Fig. 3b, the prototypes become more evenly distributed across the item embedding space. This redistribution ensures that items from underrepresented cultures are represented by closer prototypes. As a result, the model can offer more *culturally-aware* explanations for its recommendations. Table 2 shows nearest prototype examples of a Japanese animation and an Aztec musical instrument: The prototypes associated with our model take up not only the category of the product but also its cultural proximity, making the recommendations more intuitive, representative, and inclusive.

Table 2. Closest items to the five nearest prototypes for each product, listed in decreasing order of similarity (values in parentheses). Bolded items indicate those from the same country as the original product. Products are sourced from the Amazon Movies and Instruments datasets.

Product	Model	Proto. 1	Proto. 2	Proto. 3	Proto. 4	Proto. 5
Princess Mo- nonoke (JP)	Baseline	The Lion King (1.58) (US)	**Pokémon** (1.51) (JP)	Aladdin (1.50) (US)	Small Soldiers (1.48) (US)	Toy Story 2 (1.47) (US)
	Ours	**Pokémon (1.64)** (JP)	**Akira (1.59)** (JP)	The Lion King (1.55) (US)	**Perfect Blue (1.52)** (JP)	The Iron Giant (1.31) (US)
Aztec Death Whistle (CO)	Baseline	Flute (1.56) (CN)	Ocarina (1.53) (CN)	Drumset (1.47) (US)	Saxophone (1.40) (CN)	Trumpet (1.32) (US)
	Ours	Ocarina (1.61) (CN)	Flute (1.58) (CN)	**Screaming Whistle** (1.53) (CO)	Trumpet (1.43) (US)	**Tambora** (1.42) (TW)

8 Conclusion and Future Directions

We introduced enhancements to Prototype-based Matrix Factorization to mitigate demographic biases in RecSys, focusing on cultural overrepresentation. By implementing Prototype K-filtering and a Prototype-Distributing Regularizer, our model improves the representation of underrepresented items without compromising recommendation quality. Experimental results show that our approach achieves higher performance than baseline models and significantly reduces the average ranking of long-tail and underrepresented items. This promotes fairness and inclusiveness in recommendations and explanations, contributing to a more equitable user experience.

However, our study has several limitations. First, the theoretical justification for the effectiveness of k-filtering and the regularization techniques is not thoroughly established; more rigorous analysis is needed to fully understand their impact on performance and fairness. Second, while we observed enhancements in explainability, we did not conduct a systematic analysis of this aspect. Additionally, the significant reduction of data due to filtering for country information may affect the generalizability of our findings. Finally, by simplifying culture to country of origin, we overlook its complex and multifaceted nature that extends beyond national boundaries; culture is influenced by factors such as ethnicity, language, and religion. This oversimplification may limit the effectiveness of our bias mitigation strategies and fail to capture the diversity within and across countries.

Acknowledgements. Funding support for project activities has been partially provided by Canada CIFAR AI Chair, and Facebook award. We also express our gratitude to Compute Canada and Mila for their support in providing compute resources, software and technical help for our evaluations.

A Detailed Results

Table 3. Results of Models Across All Datasets. Bold numbers show the best value with respect to each dataset. μ_{over} to investigate how each model treats the overrepresented groups and in order to compare it against the μ_{under}. We also include all the results on the 'Amazon Grocery and Gourmet Food' dataset to provide more depth in our experiments.

Model	Dataset	↑HR@10	↑NDCG@10	↓ μ_{under}	μ_{over}	↓ μ_{LT}
ZeroSum Rhee et al. (2022)	LastFM-2b	0.564	0.327	52.233	44.733	21.543
	MovieLens-1M	0.615	0.347	48.660	49.381	13.964
	Amazon Beauty	0.238	0.154	50.771	49.634	35.372
	Amazon Instruments	0.243	0.159	47.099	49.997	30.416
	Amazon Grocery	0.261	0.170	50.582	49.088	18.413
ProtoMF Melchiorre et al. (2022)	LastFM-2b	0.581	0.337	51.106	44.300	29.008
	MovieLens-1M	0.656	0.381	48.674	49.582	16.215
	Amazon Beauty	0.239	0.155	50.425	49.646	39.129
	Amazon Instruments	0.245	0.164	47.012	50.002	32.965
	Amazon Grocery	0.259	0.168	50.350	49.132	27.991
ACF Barkan et al. (2021)	LastFM-2b	0.569	0.317	53.970	43.458	20.009
	MovieLens-1M	0.618	0.345	50.891	53.923	16.783
	Amazon Beauty	0.229	0.142	50.885	49.615	44.690
	Amazon Instruments	0.211	0.132	47.250	49.924	18.849
	Amazon Grocery	0.250	0.159	50.530	49.137	24.343

(*continued*)

Table 3. (*continued*)

Model	Dataset	↑HR@10	↑NDCG@10	↓ μ_{under}	μ_{over}	↓ μ_{LT}
MF Koren et al. (2009)	LastFM-2b	0.159	0.083	49.299	49.755	25.991
	MovieLens-1M	0.452	0.248	50.098	49.340	21.201
	Amazon Beauty	0.117	0.077	50.711	49.595	16.660
	Amazon Instruments	0.163	0.113	47.620	49.926	26.825
	Amazon Grocery	0.132	0.086	50.338	49.208	21.541
$\{\text{user}\}_{\{k\}}$	LastFM-2b	0.574	0.326	50.952	44.587	26.136
	MovieLens-1M	0.640	0.365	48.452	50.956	13.330
	Amazon Beauty	**0.240**	**0.155**	51.846	49.573	28.715
	Amazon Instruments	0.234	0.152	**45.949**	50.077	**9.865**
	Amazon Grocery	**0.262**	**0.170**	50.765	49.118	19.098
$\{\text{item}\}_{\{k\}}$	LastFM-2b	0.594	0.344	49.849	43.852	21.995
	MovieLens-1M	0.651	0.379	47.357	49.898	**11.263**
	Amazon Beauty	0.238	0.154	51.762	49.588	20.693
	Amazon Instruments	0.233	0.154	46.123	50.044	32.674
	Amazon Grocery	0.259	0.167	50.560	49.116	**16.457**
$\{\text{user}\}_{\{\lambda\}}$	LastFM-2b	0.520	0.302	51.115	45.007	21.391
	MovieLens-1M	0.660	0.383	48.596	51.154	14.179
	Amazon Beauty	0.237	0.153	50.613	49.630	16.327
	Amazon Instruments	**0.254**	**0.170**	47.024	50.012	26.963
	Amazon Grocery	0.262	0.169	50.564	48.855	28.072
$\{\text{item}\}_{\{\lambda\}}$	LastFM-2b	0.598	0.348	51.111	44.085	18.995
	MovieLens-1M	**0.660**	**0.386**	48.622	51.089	13.293
	Amazon Beauty	0.237	0.153	**49.776**	49.641	44.378
	Amazon Instruments	0.249	0.166	46.781	49.957	23.017
	Amazon Grocery	0.261	0.169	**50.480**	48.999	23.561
$\{\text{user-item}\}_{\{k,\lambda\}}$	LastFM-2b	**0.600**	**0.352**	**49.552**	44.296	**18.013**
	MovieLens-1M	0.657	0.383	**47.048**	49.571	16.158
	Amazon Beauty	0.236	0.153	50.018	49.643	**16.684**
	Amazon Instruments	0.250	0.167	47.017	49.993	36.062
	Amazon Grocery	0.261	0.168	50.568	49.047	26.788

References

Barkan, O., Hirsch, R., Katz, O., Caciularu, A., Koenigstein, N.: Anchor-based collaborative filtering. In: Proceedings of the 30th ACM International Conference on Information & Knowledge Management, pp. 2877–2881 (2021)

Barocas, S., Hardt, M., Narayanan, A.: Fairness and Machine Learning: Limitations and Opportunities. MIT Press (2023)

Bello, P., Garcia, D.: Cultural Divergence in popular music: the increasing diversity of music consumption on Spotify across countries. Hum. Soc. Sci. Commun. **8**(1), 1–8 (2021)

Beutel, A., et al.: Fairness in recommendation ranking through pairwise comparisons. In: Proceedings of the 25th ACM SIGKDD International Conference on Knowledge Discovery & Data Mining, pp. 2212–2220 (2019)

Brinkmann, L., et al.: Machine culture. Nat. Hum. Behav. **7**(11), 1855–1868 (2023)
Deldjoo, Y., Jannach, D., Bellogin, A., Difonzo, A., Zanzonelli, D.: Fairness in recommender systems: research landscape and future directions. User Model. User-Adapted Interact. **34**(1), 59–108 (2024)
Dinnissen, K., Bauer, C.: How control and transparency for users could improve artist fairness in music recommender systems. In: Ismir 2023 Hybrid Conference (2023)
Ferraro, A., Ferreira, G., Diaz, F., Born, G.: Measuring commonality in recommendation of cultural content: recommender systems to enhance cultural citizenship. In: Proceedings of the 16th ACM Conference on Recommender Systems, pp. 567–572 (2022)
Guerlain, P.: The ironies and dilemmas of America's cultural dominance: a transcultural approach. Am. Stud. Int. **35**(2), 30–51 (1997)
Harper, F.M., Konstan, J.A.: The movielens datasets: history and context. ACM Trans. Interact. Intell. Syst. (TIIS) **5**(4), 1–19 (2015)
He, X., Liao, L., Zhang, H., Nie, L., Hu, X., Chua, T.-S.: Neural collaborative filtering. In: Proceedings of the 26th International Conference on World Wide Web, pp. 173–182 (2017)
Hong, M., An, S., Akerkar, R., Camacho, D., Jung, J.J.: Cross-cultural contextualisation for recommender systems. J. Ambient. Intell. Humaniz. Comput. **2024**, 1–12 (2024)
Hong, M., Jung, J.J.: Multi-criteria tensor model for tourism recommender systems. Expert Syst. Appl. **170**(2021), 114537 (2021)
Hou, Y., Li, J., He, Z., Yan, A., Chen, X., McAuley, J.: Bridging Language and Items for Retrieval and Recommendation. arXiv preprint arXiv:2403.03952 (2024)
Koren, Y., Bell, R., Volinsky, C.: Matrix factorization techniques for recommender systems. Computer **42**(8), 30–37 (2009)
Lesota, O., et al.: Analyzing item popularity bias of music recommender systems: are different genders equally affected?. In: Proceedings of the 15th ACM Conference on Recommender Systems, pp. 601-606 (2021)
Lesota, O., Parada-Cabaleiro, E., Brandl, S., Lex, E., Rekabsaz, N., Schedl, M.: Traces of globalization in online music consumption patterns and results of recommendation algorithms. In: ISMIR, pp. 291–297 (2022)
Madan, S., Basu, S., Ng, S., Lim, E.A.C.: Impact of culture on the pursuit of beauty: Evidence from five countries. J. Int. Mark. **26**(4), 54–68 (2018)
Melchiorre, A.B., Masoudian, S., Kumar, D., Schedl, M.: Modular debiasing of latent user representations in prototype- based recommender systems. In: Joint European Conference on Machine Learning and Knowledge Discovery in Databases, pp. 56–72. Springer (2024)
Melchiorre, A.B., Rekabsaz, N., Ganhör, C., Schedl, M.: ProtoMF: prototype-based matrix factorization for effective and explainable recommendations. In: Proceedings of the 16th ACM Conference on Recommender Systems, pp. 246–256 (2022)
Melchiorre, A.B., Rekabsaz, N., Parada-Cabaleiro, E., Brandl, S., Lesota, O., Schedl, M.: Investigating gender fairness of recommendation algorithms in the music domain. Inf. Process. Manag. **58**(5), 102666 (2021). https://doi.org/10.1016/j.ipm.2021.102666
Moreau, F., Peltier, S.: Cultural diversity in the movie industry: a cross-national study. J. Media Econ. **17**(2), 123–143 (2004)
Moritz, P., et al.: Ray: a distributed framework for emerging AI applications. In: 13th USENIX Symposium on Operating Systems Design and Implementation (OSDI 18), pp. 561–577 (2018)

Rendle, S.: Item recommendation from implicit feedback. In: Recommender Systems Handbook, pp. 143–171. Springer (2021)

Rhee, W., Cho, S.M., Suh, B.: Countering popularity bias by regularizing score differences. In: Proceedings of the 16th ACM Conference on Recommender Systems, pp. 145–155 (2022)

Salganik, R., Diaz, F., Farnadi, G.: Fairness through domain awareness: mitigating popularity bias for music discovery. In: European Conference on Information Retrieval, pp. 351–368. Springer (2024)

Sankar, A., Wang, J., Krishnan, A., Sundaram, H.: Protocf: prototypical collaborative filtering for few-shot recommendation. In: Proceedings of the 15th ACM Conference on Recommender Systems, pp. 166–175 (2021)

Shuker, R.: Popular Music Culture: The Key Concepts. Routledge (2012)

Sonboli, N., Burke, R., Ekstrand, M., Mehrotra, R.: The multisided complexity of fairness in recommender systems. AI Magazine **43**(2), 164–176 (2022)

Sonboli, N., Smith, J.J., Berenfus, F.C., Burke, R., Fiesler, C.: Fairness and transparency in recommendation: the users' perspective. In: Proceedings of the 29th ACM Conference on User Modeling, Adaptation and Personalization, pp. 274–279 (2021)

Song, H.: Deconstruction of cultural dominance in Korean EFL textbooks. Intercultural Educ. **24**(4), 382–390 (2013)

Spencer-Oatey, H., Franklin, P.: What is culture. A compilation of quotations. GlobalPAD Core Concepts **1**(22), 1–21 (2012)

Spring, J.: Deculturalization and the Struggle for Equality: A Brief History of the Education of Dominated Cultures in the United States. Routledge (2016)

Trehub, S.E., Becker, J., Morley, I.: Cross-cultural perspectives on music and musicality. Philos. Trans. Roy. Soc. B: Biol. Sci. **370**(1664), 20140096 (2015)

Ursu, R.M.: The power of rankings: quantifying the effect of rankings on online consumer search and purchase decisions. Mark. Sci. **37**(4), 530–552 (2018)

Wei, T., Feng, F., Chen, J., Wu, Z., Yi, J., He, X.: Model-agnostic counterfactual reasoning for eliminating popularity bias in recommender system. In: Proceedings of the 27th ACM SIGKDD Conference on Knowledge Discovery & Data Mining. 1791–1800 (2021)

Zangerle, E., Pichl, M., Schedl, M.: Culture-aware music recommendation. In: Proceedings of the 26th Conference on User Modeling, Adaptation and Personalization, pp. 357–358 (2018)

Zhang, S., Yao, L., Sun, A., Tay, Y.: Deep learning based recommender system: a survey and new perspectives. ACM Comput. Surv. (CSUR) **52**(1), 1–38 (2019)

Zheng, Y., Gao, C., Li, X., He, X., Li, Y., Jin, D.: Disentangling user interest and conformity for recommendation with causal embedding. In: Proceedings of the Web Conference 2021, pp. 2980–2991 (2021)

Is Relevance Propagated from Retriever to Generator in RAG?

Fangzheng Tian(✉), Debasis Ganguly, and Craig Macdonald

University of Glasgow, Glasgow, UK
f.tian.1@research.gla.ac.uk,
{debasis.ganguly,craig.macdonald}@glasgow.ac.uk

Abstract. Retrieval Augmented Generation (RAG) is a framework for incorporating external knowledge, usually in the form of a set of documents retrieved from a collection, as a part of a prompt to a large language model (LLM) to potentially improve the performance of a downstream task, such as question answering. Different from a standard retrieval task's objective of maximising the relevance of a set of top-ranked documents, a RAG system's objective is rather to maximise their total *utility*, where the utility of a document indicates whether including it as a part of the additional contextual information in an LLM prompt improves a downstream task. Existing studies investigate the role of the relevance of a RAG context for knowledge-intensive language tasks (KILT), where relevance essentially takes the form of answer containment. In contrast, in our work, relevance corresponds to that of topical overlap between a query and a document for an information seeking task. Specifically, we make use of an IR test collection to empirically investigate whether a RAG context comprised of topically relevant documents leads to improved downstream performance. Our experiments lead to the following findings: (a) there is a small positive correlation between relevance and utility; (b) this correlation decreases with increasing context sizes (higher values of k in k-shot); and (c) a more effective retrieval model generally leads to better downstream RAG performance.

Keywords: Retrieval Augmented Generation (RAG) · Context Utility · Topical Relevance · RAG Evaluation

1 Introduction

Retrieval Augmented Generation (RAG) is a framework to help Large Language Models (LLMs) generate more accurate answers for tasks that require domain-specific knowledge [24]. This knowledge is passed to an LLM as context - usually in the form of a ranked list of retrieved documents, which are then inserted in a prompt template designed for a specific downstream task [11,14]. Usually, these retrieved documents are topically similar to the input instance, e.g., for a particular input question in a question-answering (QA) task, they can be constituted

C. Hauff et al. (Eds.): ECIR 2025, LNCS 15572, pp. 32–48, 2025.
https://doi.org/10.1007/978-3-031-88708-6_3

of semantically similar questions along with their answers as retrieved from a collection [30].

A challenging problem in RAG is to find an effective way to construct this context for each input (query) [8,28]. In contrast to the standard ranking task, where the aim is to maximise the likelihood of relevance of the retrieved documents, the task in RAG is to maximise the likelihood of the usefulness of this retrieved context for a downstream task [35].

While existing research has investigated the role of relevance in generating quality output for QA systems [8,28,39], the notion of relevance in their work is largely different from the notion of topical relevance required in an information seeking task (e.g., query-focused summarisation [1]), where relevance is obtained with manual assessments via depth-pooling [5,7,15,40]. More precisely speaking, the notion of 'relevance' in existing studies applies specifically to knowledge-intensive language tasks (KILT benchmark [37]), e.g., the CommonQA [44] or the NaturalQA benchmarks [21], where a Boolean indicator determines whether the correct answer is present in a retrieved context (a passage of text from which the correct answer needs to be extracted out - commonly called 'provenance' in these benchmarks).

Different from the existing studies for **KILT tasks**, where the relevance is evaluated in the form of **answer containments** [8,28,39], in our paper, **relevance of retrieved context** corresponds to the **topical match** between a query and the included documents for an **information seeking task**, i.e., the standard IR sense of the term. Inspired by the work of [2], we propose an experiment setup for RAG using an IR test collection, where a set of queries is used as input questions to an LLM - the task is to generate an answer that satisfies the information need underlying the question [1]. The quality of the generated answer is then measured by its similarity with the set of assessed relevant documents for the query (which is a part of the IR test collection).

A major difference of our work from [2] is that while they investigated only the 0-shot answer generation (no context is involved), we investigate the role of retrieved contexts for effective answer generation in a k-shot setup. We provide a precise definition of the usefulness of a retrieved context as the relative change in answer effectiveness induced by including the corresponding context. This **utility measure** follows the general counterfactual attribution methodology, i.e. quantifying the varying inputs' influence upon a model's outputs (as used in e.g. model explanations [45,46]). More specifically, we investigate the relation between the relevance of the ranked list of k documents in k-shot RAG and the quality of the generated answer as measured by its semantic similarity with the known relevant documents for a query [2]. In other words, we are interested in finding out if a context comprised of more relevant documents can lead to generating more relevant answers, or is the **relevance lost in transmission** across the generation phase as the data flows from the 'retriever' to the 'generator' components of a RAG system?

Figure 1 presents a schematic overview of our proposed evaluation framework to investigate the effect of RAG contexts' topical relevance on the quality

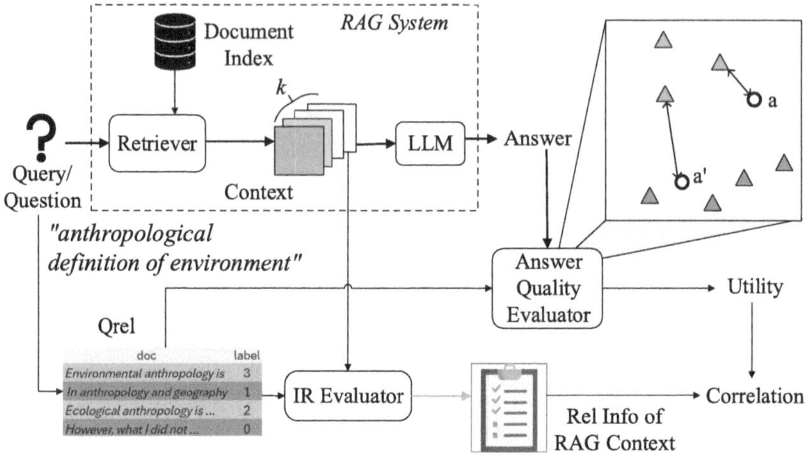

Fig. 1. The illustration of the generator (LLM) with k top-retrieved documents comprising a RAG context. Conducting experiments with an IR test collection enables the 'retriever' component to be evaluated with standard IR metrics using available ground truth of known relevance assessments. The 'generator' component is evaluated by computing the similarity of a generated answer's embedding w.r.t. the embeddings of the judged relevant documents (shown as the green triangles). For instance, generated answer $\mathbf{a}$ has a shorter distance to its closest relevant document than $\mathbf{a}'$ does, so $\mathbf{a}$ gets a higher F_{BERT} than $\mathbf{a}'$. (Color figure online)

(measured in terms of relevance) of the generated answers. More details about the evaluation framework are presented in Sect. 3.

Contributions. The overall contribution of our paper is the proposal of a novel experiment framework for RAG systems, in which:

- We introduce a measure for context utility that focuses on the relative gain brought by the retrieved context wrt. the 0-shot answer quality;
- We correlate the relevance of the retrieved context (e.g. nDCG) and their utility in downstream RAG tasks and use this correlation to indicate whether relevant information is effectively transmitted from the retrieval results to the generated answers.

In our experiments examining the relevance-utility correlation, we obtain interesting findings from the experiments based on our framework, including:

- The correlation between nDCG and the corresponding utility is positive but small, which means that *a part of relevance is indeed lost in transmission* from the retrieval result to the generated output;
- This correlation decreases as the context size increases and is relatively insensitive to the order of the documents in the context for the tested LLM.

To help further studies in this domain and the reproduction of the results, we make our experiment framework implementation publicly available at: https://github.com/DanielTian97/rag-utility.git

2 Related Work

RAG Background. The concept of retrieval-augmented generation (RAG) was first introduced in [24]. Subsequent studies have adapted it for a number of different applications, e.g., for improving the performance of knowledge-intensive language tasks (KILT) [37] which include question answering [21,44], hallucination and factual incorrectness mitigation [19], summarisation [13], software code generation [25]. For a more comprehensive review on RAG literature, see [14,16].

Optimal context selection plays an important role in determining RAG's performance. While the most common approach to select this context is based on lexical or semantic similarities with the input text, e.g., question, instruction, query etc. [27,30], other selection methodologies with different objectives have also been explored, e.g., for example selecting diverse documents as retrieved context [23], or learning a task-specific ranking model to maximise the likelihood of the retrieved context improving the downstream performance [38]. It has also been empirically verified that RAG performance depends not only on the set of documents selected as the context, but also on the order in which these are appended to the input text of an LLM [20,29].

Existing research, however, has not explicitly investigated the *effect of topical relevance of the RAG context* on the *quality of the generated output* (in our case, an answer), which is the core objective of this paper.

Analysing the Influence of Documents on Downstream RAG Performance. Sauchuk et al. [41] reported that topically non-relevant documents lead to substantial degradation of downstream performance of cascaded IR-NLP systems (precursor to an end-to-end RAG system but characteristically similar to it). More recently, [8,28] also showed that not only does the presence of relevant documents lead to improved performance in the downstream tasks compared with 0-shot generation, but the order of those relevant documents also affects the performance. In particular, the authors of [28] reported that "*model performance is highest when relevant information occurs at the beginning or end of its input context*" and coined the term "lost in the middle". Their observation was further corroborated by the work of [8], with some further analysis on the effects of non-relevant vs. randomly sampled off-topic documents.

In our work, we explore if the observations reported in [8,28] for a notion of relevance restricted to *answer containment* are consistent across topical relevance in the form of manually assessed ground truths. Similar to [39], we also report the correlation between retrieval effectiveness and downstream RAG performance.

Evaluation Methods of RAG Performance. A RAG model's performance measure depends on the specific downstream tasks [4], e.g., in QA or summarization tasks, the common evaluation metrics measure the overlap between the ground-truth and a generated output either at a lexical level or at a semantic level. Lexical measures include METEOR [22], ROUGE [26], BLEU [34], whereas semantic ones include BERTScore [17,47]. Other overlap measures have also been proposed, e.g., FActScore [32] employs aspect-based matching to measure the factual correctness of generated answers, whereas the studies reported in [18,48] used LLMs to evaluate the quality of generated output.

In contrast to evaluating factual correctness, the authors of [2] proposed a *relevance-based* evaluation strategy for measuring the quality of generated answers by conducting their experiments on an IR dataset where relevance ground-truths are available. Different from evaluation methodologies that compute the overlap of a generated output with a ground-truth one, the authors of [2] used the centroid of the relevant documents for a query to construct a pseudo-ground truth output - an approach that we also adapt in our work.

Research Gap. Overall, the closest prior work concerned with the relation between RAG context and RAG performance are [8,28], but neither directly investigated the problem with a conventional IR definition of topical relevance, as is addressed in our work. Moreover, those studies focused on overall RAG performance; in contrast, our work distinguishes the LLM's inherent performance in the downstream RAG task and the relative change brought by the retrieval context, which we define as utility. To investigate this problem systematically, we extended the framework that leverages IR benchmarks to evaluate RAG [2]. We introduce our RAG Utility evaluation framework next.

3 Evaluating Relevance vs. Utility in RAG

In this section, we formally define a RAG evaluation framework that uses an IR test collection to measure the relation between the relevance of retrieved documents and that of the output generated in a downstream task.

3.1 Performance Measure of a Downstream Generation Task

As the first step in our evaluation framework, we define the performance measure of a RAG downstream task. Similar to the experimental settings of [2], we use the passage-level relevance information of queries in an IR test collection. Formally, we can define an IR test collection as $\langle Q, \mathcal{C}, R_Q \rangle$, where Q is a set of queries, $\mathcal{C}$ is a collection of documents, and R_Q is the set of available relevance assessments (qrels) of the queries.

The RAG task in this setup corresponds to *generating* an answer $a_q = A(q; \phi_{\text{LLM}}, \theta(q)_k)$ that seeks to provide relevant information to the query $q \in Q$, where the answer generation function A depends on: (a) an LLM model ϕ_{LLM} with frozen parameters, and (b) contextual information, $\theta(q)_k$, comprised of a list of top-k documents retrieved by an IR model θ. In contrast to a standard IR task, where the objective is to retrieve the relevant documents R_q for query q, the objective here is to generate a single answer that aligns with the relevant information contained in the set of documents R_q.

Similar to [2], we define the performance measure of an answer a_q as

$$\mathcal{P}(a_q) = \max_{r \in R_q} \sigma(\mathbf{a}_q, \mathbf{r}), \tag{1}$$

where σ is an abstract similarity function, and $\mathbf{a}_q$ and $\mathbf{r}$, respectively, denote the embedded representations of the answer text a_q and a relevant document text $r \in R_q$, with R_q being the set of assessed relevant documents for q.

In particular, we employ BERTScore [47] as the similarity function σ, as prescribed in [2]. The BERTScore measure between a pair of text is a semantic extension of the BLEU score [34] metric that, instead of aggregating their term overlaps, rather aggregates the cosine similarities between their constituent token embeddings. Following the standard practice, we work with the F-score version from the BERTScore paper [47], which we refer to as F_{BERT} from here on.

Although previous work [36] has shown that BERTScore may not be entirely effective for out-of-domain question answering, our experiments on an IR test collection only involve in-domain queries, with each query having at least one relevant document.

3.2 Utility of Contextual Information

We now define the utility of contextual information as the relative change in the performance measure (Eq. (1)) between a non-contextual (i.e., 0-shot) generation and a contextual (k-shot RAG) one. Formally,

$$U(\theta(q)_k) = \frac{\mathcal{P}(A(q; \phi_{\text{LLM}}, \theta(q)_k)) - \mathcal{P}(A(q; \phi_{\text{LLM}}, \varnothing))}{\mathcal{P}(A(q; \phi_{\text{LLM}}, \varnothing))}, \tag{2}$$

where $\theta : q \mapsto \{d_1, \ldots, d_k\}$ is a retrieval model that, given a query q, retrieves a ranked list of top-k documents, which we denote by $\theta(q)_k$.

Intuitively speaking, the notion of utility as defined in Eq. (2) captures the usefulness of a set of top-retrieved documents relative to the 0-shot performance (an LLM's inherent capability of generating relevant output). A positive value of utility indicates that the contextual information retrieved from a collection improves the answer quality over 0-shot generation, and, conversely, a negative utility value indicates that the retrieved context information is detrimental to 0-shot answer generation quality. A reason to work with relative gains instead of absolute ones is that it can correctly capture the improvements over ineffective 0-shot predictions.

3.3 Measuring Relevance Transmission from Retriever to Generator

To see how much of relevant information actually gets transmitted via the generation process, we propose to measure the correlation between the utility values (Eq. (2)) and the relevance of the retrieved context comprised of the top-k documents retrieved for a query q. Formally speaking, we measure

$$\rho(U(\theta(Q)_k), \mu(\theta(Q)_k)), \text{ where } \theta(Q)_k = \cup_{q \in Q} \theta(q)_k, \tag{3}$$

with the notations being explained as follows.

- $U(\theta(Q)_k)$ denotes a set of utility values computed for each top-retrieved set of documents $\theta(q)_k$;
- $\mu(\theta(Q)_k))$ denotes a set of IR performance measure values (e.g., nDCG, which we use for our experiments) computed for the contextual information $\theta(q)_k$ retrieved for each query q in Q;
- ρ denotes a correlation metric (e.g. Pearson's r in our experiments).

The correlation between the relevance measured at the retriever and at the generator, as shown in Eq. (3), indicates how effectively an LLM can translate the relevant context retrieved from a collection of documents into a generated answer text that is also relevant to the question.

This correlation measure of Eq. (3) depends on several explicit and latent factors. While the explicit factors include the LLM being used, the retrieval model θ, and the RAG context size k, the more latent factors may include the specificity of the information needs of the questions etc. Our research questions, as described in the next section, investigate how the more explicit factors, namely the retrieval model θ and the RAG context size k, affect the correlation between the relevance in the retrieved context and the relevance in the generated answer.

4 Experiment Setup

Research Questions. Our first research question is about analysing the effects of the retrieval model θ on the downstream utility in Eq. (2). The second research question is about the correlation as described in Eq. (3). Usually, retrieval models have different levels of abilities in ranking documents based on their estimated relevance to a topic. In particular, we study one lexical (unsupervised) and another neural (supervised) model - specifically, BM25 and MonoT5 [33]. These models are compared with two extreme cases: an oracle and an *adversarial* oracle. Specifically, by using the available relevance information R_q for each query q, the oracle constructs a completely relevant RAG context by sampling k *judged relevant* documents from R_q, while the adversarial oracle constructs a completely non-relevant RAG context by sampling k *judged non-relevant* documents. This setup leads to the following research questions.

- **RQ-1**: Is the contextual information of a RAG system obtained with a more effective ranking model of higher utility (i.e., better improvements over 0-shot generation) than that of less effective ones?
- **RQ-2**: How does the correlation between the relevance of the RAG context and the downstream utility change with the number of documents constituting the context?

Datasets. To study the effects of relevance on downstream RAG performance using an IR-based experiment setup, we employ the TREC DL'19 [7] and DL'20 [6] test collections, constituting 43 and 54 queries, respectively, with depth-pooled manually assessed relevance judgements. The underlying document collection for the TREC DL queries is the MS MARCO passage collection, which contains 8.8 million passages [3].

In addition to evaluating our results on TREC DL (deep relevance assessments over a small number of queries), we also conduct experiments on the MS MARCO dev_small queryset, which has a much larger number of queries (6980 in total), but with much shallower relevance assessments.

LLM and Prompt Details. Our experiments were conducted on an 8-bit quantised instruction-tuned model of Llama-3-8b[1] [12]. To account for the stochastic nature of the generation process of an LLM, we conduct our analysis on an average across 5 runs.

The prompt template used in our experiments instructs an LLM to generate a single answer to a question (TREC DL query) as per both its internal knowledge and the contextual information provided to aid the generation. Figure 2 shows the prompt used in our experiments along with an example query and BM25-retrieved documents about it. The average length of each document in the MS MARCO passage collection is about 56 words, and we found that 16 such documents fit within a 2048-token input buffer. We thus set the maximum value of the RAG context size to 15 in our experiments.

You are an expert at answering questions based on your own knowledge and related context. Please answer this question based on the given context. End your answer with STOP.
Context 1: {Definition of SIGMET. plural. ... before takeoff — compare airmet.}
Context 2: {sigmet SIGMET, or Significant ... convective and non-convective.}
Question: {definition of a sigmet}
Now start your answer.
Answer:

Fig. 2. The prompt template (shown bold-faced) used in our experiments with example values substituted for the placeholders (shown enclosed within {}). The query shown is a sample from the TREC DL'19 queryset, and the RAG context is comprised of the top 2 documents retrieved with BM25.

IR Models Investigated. To obtain the RAG contexts, we employ a lexical ranker (namely BM25) and a supervised neural ranker (namely MonoT5 [33]), as provided by PyTerrier [31]. The documents used as context constitute the top-k results of a retrieval model θ at different cut-offs to yield different context sizes (denoted as $\theta(q)_k$ in Eqs. (1) and (2)). More details about the IR models and oracles follow.

- **BM25**: A lexical retriever based on a variant of TF-IDF-based term weighting coupled with document length normalisation. The k top-ranked documents

[1] https://huggingface.co/QuantFactory/Meta-Llama-3-8B-GGUF.

are appended to the RAG context (see Fig. 2) in decreasing order of retrieval score, i.e., the top-ranked document is 'Context 1', and so on.
- **MonoT5**: A cross-encoder re-ranker based on the contextual embedding of query-document pairs [33]. In our setup, the MonoT5 top-k were obtained by re-ranking the top-100 BM25 results. Similar to BM25, the context documents are appended to the LLM prompt in descending order of retrieval score.
- **BM25-r** and **MonoT5-r**: For both the aforementioned IR models, to investigate the potential adverse effects of not conforming to the order induced by the retrieval scores, we construct the RAG context by appending each document from the top-k list $\theta(q)_k$ in ascending order of retrieval score. In this way, the most likely relevant documents are put at the end of the context, i.e., the top-ranked document is mapped to 'Context k' (see Fig. 2). We denote these contexts with an 'r' as the suffix.
- **Rel (Oracle)**: To study the upper bound on the RAG performance, we experiment with an oracle approach that samples a set of k relevant documents from the known set of assessed documents, thus appending them to the LLM prompt in an arbitrary order. If the number of relevant documents for a query is less than k then we take all of them.
- **NRel (Adversarial Oracle)**: In contrast to the relevant-only approach, we also study the worst-case settings by including a sample of k non-relevant documents (assessed non-relevant documents from the qrels file of TREC DL'19 and TREC DL'20) into the RAG context.

We do not explore oracles for the dev_small queryset due to the shallowness of the relevance assessments (only 1.1 relevant documents per query, on average).

Evaluation Metrics. As described in Sect. 3.3, we measure the quality of the retrieved list of k documents constituting a RAG context with nDCG@k. The downstream performance is measured using the F1-based BERTScore [47] between the generated answer and the known relevant documents (denoted F_{BERT}, see also Fig. 3). We employ two different settings for constructing the ground-truth of relevant documents for the TREC DL query sets. In the first setting, following the standard practice of IR evaluation, we consider the documents with relevance labels (r_{DL}) of 2 and 3 to compute F_{BERT}. In the second setting, we apply a more restrictive notion of relevance where we consider only the highly relevant documents ($r_{\text{DL}} = 3$) for F_{BERT} computation. Only one configuration is available for the MS MARCO dev_small queryset as the relevance labels are binary (results shown as $r_{\text{dev\_small}} = 1$ in Table 1).

5 Results

In this section, we present our findings corresponding to the research questions as defined in Sect. 4.

RQ-1: Relation Between Retrieval Effectiveness and Downstream Performance. Table 1 shows the average utility achieved by different RAG context

Table 1. Average utilities of different RAG contexts retrieved by BM25 and MonoT5 (denoted MT5), including their reversed-order (-r) variants, and 'relevant-only' (Rel) and 'non-relevant only' (NRel) for context sizes 2, 5, 10 and 15. To compute the relative gains in the F_{BERT} scores, we use two configurations for TREC DL: (1) the documents with relevance labels (r_{DL}) of 2 and 3 are included in the computation; and (2) only the documents with label 3 are considered relevant. A † alongside a result (relative change of F_{BERT} w.r.t. 0-shot) indicates statistical significance compared to the relevant-only (oracle) method as per the paired t-test. Similarly, the significance of an IR model's ordered context (e.g., BM25) relative to its reversal (e.g., BM25-r) is indicated by '‡'.

Query Set	0-shot F_{BERT}	IR Model	Utility of RAG Context: $r_{\text{DL}} \in \{2,3\}$, $k=2$	$k=5$	$k=10$	$k=15$	$r_{\text{DL}}=3$, $r_{\text{dev\_small}}=1$, $k=2$	$k=5$	$k=10$	$k=15$
		BM25	.0582†	.1004	.1065	.0997	.0314†	.0798	.0987	.0901†
		BM25-r	.0736	.1026	.0902†	.0864†	.0123†‡	.0844	.0713†	.0698†
DL'19	.6427	MT5	**.1161**	.0988	.1201	.1201	.0809	.0878	.1010†	.0902†
		MT5-r	.1023	.1149	.0928	.0996	.0819	.0966	.0988†	.0886†
		Rel	.1034	**.1245**	**.1424**	**.1520**	**.1236**	**.1386**	**.1613**	**.1642**
		NRel	−.0359†	−.0059†	−.0028†	.0013†	−.0082†	−.0256†	.0025†	−.0256†
		BM25	.0963	.1204	.1235	.1333	.1100†	.1298	.1342	.1424
		BM25-r	.0877	.1139	.0948†‡	.0922†‡	.0997†	.1047†	.0849†‡	.0867†‡
DL'20	.6159	MT5	.1366	**.1550**	**.1443**	.1323	.1385	.1537	**.1565**	.1320†
		MT5-r	**.1526**	.1118†‡	.1056	.1066†	.1454	.0953†‡	.0857†‡	.0866†‡
		Rel	.1401	.1503	.1359	**.1602**	**.1675**	**.1627**	.1490	**.1707**
		NRel	−.0224†	−.0147†	.0138†	−.0033†	−.0020†	−.0256†	.0043†	−.0031†
		BM25					.0414	.0617	.0700	.0721
		BM25-r					.0394	.0567‡	.0623‡	.0580‡
dev small	.5574	MT5					**.0915**	**.0944**	**.0965**	**.0916**
		MT5-r					.0852‡	.0798‡	.0677‡	.0614‡

sizes with two IR models (along with their reversal variants). We enlist the following important observations from Table 1.

First, we observe that the context comprised of relevant documents only (Rel-Only) almost always achieves the highest utility, outperformed only by MonoT5 for small context sizes. These observations are consistent for two different ways of constructing the ground-truth to measure downstream effectiveness, i.e., both for $r_{\text{DL}} \geq 2$ and the more restrictive $r_{\text{DL}} = 3$. Similar trends are also observed for dev_small - a much larger set of queries with shallow relevance ground truth.

Second, we observe that the more effective ranker, MonoT5, which is reported in the literature to yield much higher values of nDCG@k than BM25 for the TREC DL benchmarks [33], also yields higher downstream utility than the weaker ranker (BM25). This shows the importance of employing a more effective ranking model as the 'retriever' component of a RAG system. Nevertheless, despite being a relatively weak ranker, using BM25 as a traditional unsupervised

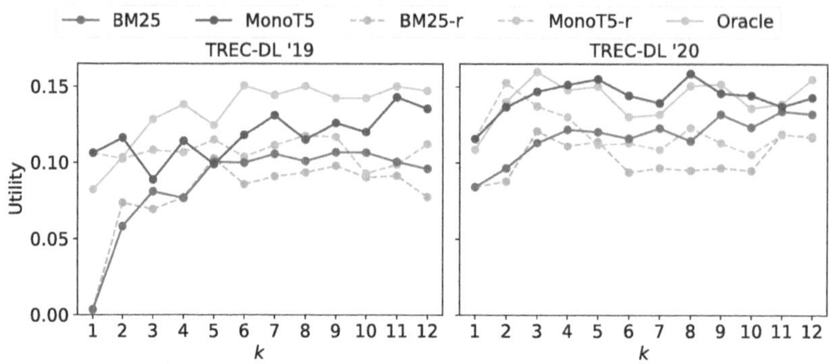

Fig. 3. Average utility of the different RAG contexts with different sizes (k) for TREC DL-19 (left) and DL-20 (right). The relevant-only oracle method almost always yields the best utilities (only outperformed by MonoT5 for small context sizes). This indicates that it is not always the case that topical relevance of the context leads to better downstream performance. The utility values of the NRel-Only method (RAG context with non-relevant documents) are not shown in the plots as these values are significantly lower than all the other reported values.

model is still appealing because it leads to positive gains over 0-shot results, especially with larger context sizes.

Thirdly, the results obtained with an effective IR model, such as MonoT5, are statistically indistinguishable from those achieved with a perfect ranker (paired t-test between them). This means that improving the topical relevance of retrieved contexts beyond a saturation point will likely have marginal incremental effects on RAG downstream effectiveness.

Finally, the results with the reverse ordering of the top-k documents indicate the importance of including relevant documents towards the top ranks. When context sizes are small (e.g., $k \leq 5$), following or reversing the order induced by the retrieval scores does not significantly affect the utility, as shown by the fewer significant differences between them for smaller ks (viz. the number of ‡s present in Table 1). The most likely reason being that for an effective ranker the very top-retrieved documents are likely to be relevant. However, as expected, with an increase in context size, there is a higher likelihood of including non-relevant documents in the context, which means that ranking documents by their retrieval scores potentially plays a key role in determining the downstream task effectiveness.

Figure 3 presents a wider range of the spectrum of context size variations, which mostly corroborates the observations in Table 1. Comparing the two plots, it can be observed that the utility, generally speaking, also depends on the queries themselves. We observe that the gains achieved by BM25-retrieved RAG contexts with smaller sizes are almost negligible for DL'19, which also suggests that the choice of the optimal IR model for a given query can potentially depend on latent factors such as query specificity [35].

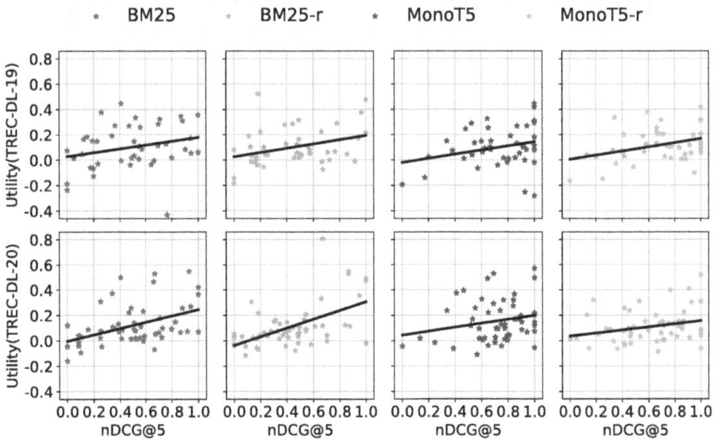

Fig. 4. Examples of the relationship between nDCG@k and utility when $k = 5$. The upper plots are for TREC DL-19, while the lower ones are for DL-20. There is generally a positive correlation between the effectiveness and utility, with different numbers of outliers in each subplot. Linear fits are shown as the black lines over the scatter plots.

To conclude RQ-1, the contexts retrieved by more effective models yield higher utility. Reversing the order given by effective retrievers decreases utility.

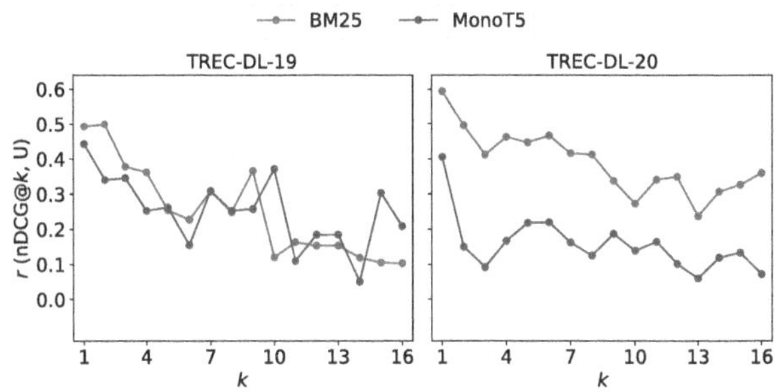

Fig. 5. The correlation between nDCG@k and context utility in terms of Pearson's r on TREC DL'19 and DL'20.

RQ-2: Correlation Values Between Contextual Relevance and Utility. Figure 4 reports the correlation values as a scatter plot (Eq. (3) with Pearson's r) for a fixed context size of $k = 5$. We report values for this particular value of k as it belongs to neither of the two extremes - of too small or too large a context. We report the correlations for the two different retrieval approaches - BM25 and MonoT5 and their reversed permutations (for analysing the adversarial effects).

From Fig. 4, we observe that there is a small positive correlation between relevance and utility: Pearson's r can be visualised as the slope of the linear fits in the plots. The scatter plots show that a number of outliers exist for both IR models, i.e., where high relevance leads to low downstream performance.

However, we find no evidence of low relevance leading to large gains in RAG performance.

Different from Fig. 4, which shows the per-query relevance-utility values for a fixed context size of $k = 5$, Fig. 5 plots the relevance-utility correlation values over a range of context sizes. An interesting finding from Fig. 5 is that although the average utility keeps increasing with larger context sizes, the role of the relevance in ensuring a quality output keeps decreasing, reflected by the decrease in the correlation values. With higher context sizes, the likely presence of non-relevant documents potentially leads to this decrease in correlation between relevance and utility.

To conclude RQ-2 - the correlation between the relevance of context and the corresponding utility is small but positive. It decreases as the number of documents in the context increases.

6 Concluding Remarks

In this paper, we investigate the correlation between the topical relevance of a RAG context and its utility, thus capturing how effectively can the *generator component* of a RAG workflow translate the relevance, as obtained by its *retriever component*, into a relevant answer at its output. The observed small positive correlations, which decrease as the context gets larger, indicate that relevance is not often and adequately transmitted from the retriever to the generated answers. This finding leaves the scope of devising novel retrieval models in the future, which will be aware of the downstream task objective in addition to topical relevance - an idea described in [35].

An important difference of our evaluation compared to those of [8,28,39] is in the nature of the ground truth. While the notion of relevance in our setup is more robust (manually judged), the downside is that there is no manually formulated ground-truth answer as in the QA datasets [21,44]. With approximated ground truths for the generator, our observations on an IR test collection agree with the ones reported on KILT benchmarks [8,28,39] in the sense that more relevant RAG context leads to better downstream performance. In contrast to the findings of [8], who found that the downstream performance improves either when relevant documents are towards the top or the bottom of a ranked list, we observe that a reverse permutation of the context leads to a substantially detrimental effect, especially with large context sizes (see Table 1). Moreover, as the relevance of retrieved contexts in our setup is not simply 'answer containment', it is difficult to construct a context that is 'middle-heavy' in relevance, analogous to the "Lost in the Middle" paper [28]. Our study, therefore, cannot comment on the effect of including relevant documents in the middle of a RAG context.

To address the limitation that our work considers approximated ground-truth of answers (as prescribed in [2]), we plan to work with manually generated reference answers for TREC DL queries. We would also like to apply the findings of this work to a more practical RAG workflow setup without the knowledge of any relevance assessments, which means that it is not possible to (exactly) compute the relevance of a RAG context. Query performance prediction (QPP) [9,42,43] methods may then be applied to *estimate the relevance of a RAG context* towards achieving per-query control of the RAG workflow [10].

Acknowledgments. We would like to express our gratitude to the anonymous reviewers for their valuable feedback, which were helpful for us to improve this paper.

Disclosure of Interests. There is no competing interest.

References

1. Alaofi, M., Arabzadeh, N., Clarke, C.L.A., Sanderson, M.: Generative Information Retrieval Evaluation (2024). https://arxiv.org/abs/2404.08137
2. Arabzadeh, N., Bigdeli, A., Clarke, C.L.A.: Adapting standard retrieval benchmarks to evaluate generated answers. In: Advances in Information Retrieval: 46th European Conference on Information Retrieval, ECIR 2024, Glasgow, UK, March 24–28, 2024, Proceedings, Part II. p. 399-414. Springer, Heidelberg (2024). https://doi.org/10.1007/978-3-031-56060-6_26
3. Bajaj, P., et al.: MS MARCO: a human generated machine reading comprehension dataset (2018). https://arxiv.org/abs/1611.09268
4. Chang, Y., et al.: A survey on evaluation of large language models. ACM Trans. Intell. Syst. Technol. **15**(3) (2024). https://doi.org/10.1145/3641289
5. Cormack, G.V., Palmer, C.R., Clarke, C.L.A.: Efficient construction of large test collections. In: Proceedings of the 21st Annual International ACM SIGIR Conference on Research and Development in Information Retrieval. SIGIR '98, New York, NY, USA, pp. 282–289. Association for Computing Machinery (1998). https://doi.org/10.1145/290941.291009
6. Craswell, N., Mitra, B., Yilmaz, E., Campos, D.: Overview of the TREC 2020 deep learning track (2021). https://arxiv.org/abs/2102.07662
7. Craswell, N., Mitra, B., Yilmaz, E., Campos, D., Voorhees, E.: Overview of the TREC 2019 deep learning track (2020). https://arxiv.org/abs/2003.07820
8. Cuconasu, F., et al.: The power of noise: redefining retrieval for RAG systems. In: Proceedings of the 47th International ACM SIGIR Conference on Research and Development in Information Retrieval. SIGIR '24, New York, NY, USA, pp. 719–729. Association for Computing Machinery (2024). https://doi.org/10.1145/3626772.3657834
9. Datta, S., Ganguly, D., Greene, D., Mitra, M.: Deep-QPP: a pairwise interaction-based deep learning model for supervised query performance prediction. In: Proceedings of the Fifteenth ACM International Conference on Web Search and Data Mining. WSDM '22, New York, NY, USA, pp. 201–209. Association for Computing Machinery (2022). https://doi.org/10.1145/3488560.3498491
10. Datta, S., Ganguly, D., MacAvaney, S., Greene, D.: A deep learning approach for selective relevance feedback. In: Advances in Information Retrieval: 46th European Conference on Information Retrieval, ECIR 2024, Glasgow, UK, March 24–28, 2024, Proceedings, Part II, pp. 189-204. Springer, Heidelberg (2024). https://doi.org/10.1007/978-3-031-56060-6_13
11. Dong, Q., et al.: A Survey on In-context Learning (2024). https://arxiv.org/abs/2301.00234
12. Dubey, A., et al.: The Llama 3 herd of models (2024). https://arxiv.org/abs/2407.21783

13. Edge, D., et al.: From local to global: a graph RAG approach to query-focused summarization (2024). https://arxiv.org/abs/2404.16130
14. Fan, W., et al.: A survey on RAG meeting LLMs: towards retrieval-augmented large language models. In: Proceedings of the 30th ACM SIGKDD Conference on Knowledge Discovery and Data Mining. KDD '24, New York, NY, USA, pp. 6491–6501. Association for Computing Machinery (2024). https://doi.org/10.1145/3637528.3671470
15. Ganguly, D., Yilmaz, E.: Query-specific variable depth pooling via query performance prediction. In: Proceedings of the 46th International ACM SIGIR Conference on Research and Development in Information Retrieval. SIGIR '23, New York, NY, USA, pp. 2303–2307. Association for Computing Machinery (2023). https://doi.org/10.1145/3539618.3592046
16. Gao, Y., et al.: Retrieval-augmented generation for large language models: a survey (2024). https://arxiv.org/abs/2312.10997
17. Hanna, M., Bojar, O.: A fine-grained analysis of BERTScore. In: Proceedings of the Sixth Conference on Machine Translation, pp. 507–517. Association for Computational Linguistics (2021). https://aclanthology.org/2021.wmt-1.59
18. Huang, H., et al.: An empirical study of LLM-as-a-judge for LLM evaluation: fine-tuned judge model is not a general substitute for GPT-4 (2024). https://arxiv.org/abs/2403.02839
19. Jiang, Z., et al.: Active retrieval augmented generation. In: Proceedings of the 2023 Conference on Empirical Methods in Natural Language Processing, Singapore, pp. 7969–7992. Association for Computational Linguistics (2023). https://doi.org/10.18653/v1/2023.emnlp-main.495
20. Kumar, S., Talukdar, P.: Reordering examples helps during priming-based few-shot learning. In: Findings of the Association for Computational Linguistics: ACL-IJCNLP 2021, pp. 4507–4518. Association for Computational Linguistics (2021). https://aclanthology.org/2021.findings-acl.395/
21. Kwiatkowski, T., et al.: Natural questions: a benchmark for question answering research. Trans. Assoc. Comput. Linguistics **7**, 452–466 (2019). https://aclanthology.org/Q19-1026
22. Lavie, A., Agarwal, A.: METEOR: an automatic metric for MT evaluation with high levels of correlation with human judgments. In: Proceedings of the Second Workshop on Statistical Machine Translation, Prague, Czech Republic, pp. 228–231. Association for Computational Linguistics (2007). https://aclanthology.org/W07-0734/
23. Levy, I., Bogin, B., Berant, J.: Diverse demonstrations improve in-context compositional generalization. In: Proceedings of the 61st Annual Meeting of the Association for Computational Linguistics (Volume 1: Long Papers), Toronto, Canada, pp. 1401–1422. Association for Computational Linguistics (2023). https://aclanthology.org/2023.acl-long.78
24. Lewis, P., et al.: Retrieval-augmented generation for knowledge-intensive NLP tasks. In: Proceedings of the 34th International Conference on Neural Information Processing Systems. NIPS '20, Red Hook, NY, USA. Curran Associates Inc. (2020). https://arxiv.org/abs/2005.11401
25. Li, J., Zhao, Y., Li, Y., Li, G., Jin, Z.: Acecoder: utilizing existing code to enhance code generation (2023). https://arxiv.org/abs/2303.17780
26. Lin, C.Y.: ROUGE: A package for automatic evaluation of summaries. In: Text Summarization Branches Out, Barcelona, Spain, pp. 74–81. Association for Computational Linguistics (2004). https://aclanthology.org/W04-1013

27. Liu, J., Shen, D., Zhang, Y., Dolan, B., Carin, L., Chen, W.: What makes good in-context examples for GPT-3? In: Proceedings of Deep Learning Inside Out (DeeLIO 2022): The 3rd Workshop on Knowledge Extraction and Integration for Deep Learning Architectures, pp. 100–114, Dublin, Ireland. Association for Computational Linguistics (2022). https://aclanthology.org/2022.deelio-1.10
28. Liu, N.F., et al.: Lost in the middle: how language models use long contexts. Trans. Assoc. Comput. Linguistics **12**, 157–173 (2024). https://doi.org/10.1162/tacl_a_00638
29. Lu, Y., Bartolo, M., Moore, A., Riedel, S., Stenetorp, P.: Fantastically ordered prompts and where to find them: overcoming few-shot prompt order sensitivity. In: Proceedings of the 60th Annual Meeting of the Association for Computational Linguistics (Volume 1: Long Papers), Dublin, Ireland, pp. 8086–8098. Association for Computational Linguistics (2022). https://aclanthology.org/2022.acl-long.556
30. Luo, M., et al.: Dr. ICL: Demonstration-retrieved in-context learning (2023). https://arxiv.org/abs/2305.14128
31. Macdonald, C., Tonellotto, N., MacAvaney, S., Ounis, I.: PyTerrier: declarative experimentation in python from bm25 to dense retrieval. In: Proceedings of the 30th ACM International Conference on Information & Knowledge Management. CIKM '21, New York, NY, USA, pp. 4526–4533. Association for Computing Machinery (2021). https://doi.org/10.1145/3459637.3482013
32. Min, S., et al.: FActScore: fine-grained atomic evaluation of factual precision in long form text generation. In: Proceedings of the 2023 Conference on Empirical Methods in Natural Language Processing, Singapore, pp. 12076–12100. Association for Computational Linguistics (2023). https://doi.org/10.18653/v1/2023.emnlp-main.741
33. Nogueira, R.F., Jiang, Z., Pradeep, R., Lin, J.: Document ranking with a pretrained sequence-to-sequence model. In: Findings of the Association for Computational Linguistics: EMNLP 2020, Online Event, 16-20 November 2020. Findings of ACL, vol. EMNLP 2020, pp. 708–718. Association for Computational Linguistics (2020). https://doi.org/10.18653/V1/2020.FINDINGS-EMNLP.63
34. Papineni, K., Roukos, S., Ward, T., Zhu, W.J.: BLEU: a method for automatic evaluation of machine translation. In: Proceedings of the 40th Annual Meeting on Association for Computational Linguistics. ACL '02, USA, pp. 311–318. Association for Computational Linguistics (2002). https://doi.org/10.3115/1073083.1073135
35. Parry, A., Ganguly, D., Chandra, M.: "In-context learning" or: how I learned to stop worrying and love "Applied Information Retrieval". In: Proceedings of the 47th International ACM SIGIR Conference on Research and Development in Information Retrieval. SIGIR '24, New York, NY, USA, pp. 14–25. Association for Computing Machinery (2024). https://doi.org/10.1145/3626772.3657842
36. Pathiyan Cherumanal, S., et al.: Walert: putting conversational information seeking knowledge into action by building and evaluating a large language model-powered chatbot. In: Proceedings of the 2024 Conference on Human Information Interaction and Retrieval. CHIIR '24, New York, NY, USA, pp. 401-405. Association for Computing Machinery (2024). https://doi.org/10.1145/3627508.3638309
37. Petroni, F., et al.: KILT: a benchmark for knowledge intensive language tasks. In: Proceedings of the 2021 Conference of the North American Chapter of the Association for Computational Linguistics: Human Language Technologies, pp. 2523–2544. Association for Computational Linguistics (2021). https://aclanthology.org/2021.naacl-main.200/

38. Rubin, O., Herzig, J., Berant, J.: Learning to retrieve prompts for in-context learning. In: Proceedings of the 2022 Conference of the North American Chapter of the Association for Computational Linguistics: Human Language Technologies,, Seattle, United States, pp. 2655–2671. Association for Computational Linguistics (2022). https://aclanthology.org/2022.naacl-main.191
39. Salemi, A., Zamani, H.: Evaluating retrieval quality in retrieval-augmented generation. In: Proceedings of the 47th International ACM SIGIR Conference on Research and Development in Information Retrieval. SIGIR '24, New York, NY, USA, pp. 2395–2400. Association for Computing Machinery (2024). https://doi.org/10.1145/3626772.3657957
40. Sanderson, M.: Test collection based evaluation of information retrieval systems. Foundations and Trends® in Information Retrieval **4**(4), 247–375 (2010). https://doi.org/10.1561/1500000009, http://dx.doi.org/10.1561/1500000009
41. Sauchuk, A., Thorne, J., Halevy, A., Tonellotto, N., Silvestri, F.: On the role of relevance in natural language processing tasks. In: Proceedings of the 45th International ACM SIGIR Conference on Research and Development in Information Retrieval. SIGIR '22, New York, NY, USA, pp. 1785–1789. Association for Computing Machinery (2022). https://doi.org/10.1145/3477495.3532034
42. Shtok, A., Kurland, O., Carmel, D.: Using statistical decision theory and relevance models for query-performance prediction. In: Proceedings of the 33rd International ACM SIGIR Conference on Research and Development in Information Retrieval. SIGIR '10, New York, NY, USA, pp. 259–266. Association for Computing Machinery (2010). https://doi.org/10.1145/1835449.1835494
43. Shtok, A., Kurland, O., Carmel, D., Raiber, F., Markovits, G.: Predicting query performance by query-drift estimation. ACM Trans. Inf. Syst. **30**(2) (2012). https://doi.org/10.1145/2180868.2180873
44. Talmor, A., Herzig, J., Lourie, N., Berant, J.: CommonsenseQA: a question answering challenge targeting commonsense knowledge. In: Proceedings of the 2019 Conference of the North American Chapter of the Association for Computational Linguistics: Human Language Technologies, Volume 1 (Long and Short Papers), Minneapolis, Minnesota, pp. 4149–4158. Association for Computational Linguistics (2019). https://doi.org/10.18653/v1/N19-1421
45. Verma, M., Ganguly, D.: LIRME: locally interpretable ranking model explanation. In: Proceedings of the 42nd International ACM SIGIR Conference on Research and Development in Information Retrieval. SIGIR'19, New York, NY, USA, pp. 1281–1284. Association for Computing Machinery (2019). https://doi.org/10.1145/3331184.3331377
46. Verma, S., Boonsanong, V., Hoang, M., Hines, K.E., Dickerson, J.P., Shah, C.: Counterfactual explanations and algorithmic recourses for machine learning: a review (2022). https://arxiv.org/abs/2010.10596
47. Zhang, T., Kishore, V., Wu, F., Weinberger, K.Q., Artzi, Y.: BERTScore: evaluating text generation with Bert (2020). https://arxiv.org/abs/1904.09675
48. Zhuo, T.Y.: ICE-Score: instructing large language models to evaluate code (2024). https://arxiv.org/abs/2304.14317

Measuring Actual Privacy of Obfuscated Queries in Information Retrieval

Francesco Luigi De Faveri(✉), Guglielmo Faggioli, and Nicola Ferro

Department of Information Engineering, University of Padova, Padova, Italy
francescoluigi.defaveri@phd.unipd.it,
{guglielmo.faggioli,nicola.ferro}@unipd.it

Abstract. Privacy is a fundamental right that could be threatened by Information Retrieval (IR) models when applied and trained on sensitive data and personal user information. Although mechanisms have been proposed to protect user privacy, the effectiveness of the privacy protections is typically assessed by studying the relations between performance and parameters of the mechanisms, such as the privacy budget in Differential Privacy (DP). This often causes a disconnection between formal privacy and the privacy experienced by the user, the actual privacy. In this paper, we present the Query Inference for Privacy and Utility (QuIPU) framework, a novel evaluation paradigm to assess actual privacy based on the risk that an *"honest-but-curious"* IR system can infer the original query from the obfuscated queries received. QuIPU represents the first attempt at measuring actual privacy for IR tasks beyond the comparison of formal privacy parameters. Our analysis shows that formal privacy parameters do not imply actual privacy, causing scenarios where, for the same privacy parameter values, two systems provide different utility, but also different actual privacy. Therefore, there is a necessity for a proper way of assessing the risk, represented by QuIPU.

Keywords: Evaluation Measures · Differential Privacy · Information Retrieval · Information Security · Privacy Risks

1 Introduction

Privacy is an essential right guaranteed by Article 12 of the Fundamental Declaration of Human Rights, which states, "No one shall be subjected to arbitrary interference with his privacy [...]". Natural Language Processing (NLP) models and Information Retrieval (IR) systems are developed using large textual datasets, including queries, documents, reviews, and online posts, frequently containing sensitive and personal user information. Including personal information in these texts, such as user profiles, personal opinions and information needs, could raise significant privacy concerns for individuals interacting with such systems. The privacy threats may compromise user safety following text analysis if not adequately addressed. Specifically, from the examination of browser search

C. Hauff et al. (Eds.): ECIR 2025, LNCS 15572, pp. 49–66, 2025.
https://doi.org/10.1007/978-3-031-88708-6_4

history and retrieved documents, malicious actors can unveil sensitive details, such as an individual's salary or medical conditions [4,15]. Heuristics strategies [3,26] have been studied for providing privacy for IR tasks. At the same time, progress in the field of NLP has shown the potentialities of Differential Privacy (DP) [21] in releasing privacy-preserving text for different purposes, spanning from text classification [25], authorship anonymization [6], and query obfuscation [22]. In the IR field, DP has emerged as the leading framework for safely obfuscating user queries.

Obfuscating a query means that the real information need of a user is protected in such a way that the obfuscated queries produced can still retrieve relevant documents yet not (fully) disclose that information need. For example, the query "`how cancer grows`" may be transformed into the obfuscated alternatives "`how myeloma grows`", "`how disease spreads`", "`how melanoma evolves`". Focusing on the mechanisms' privacy parameters represents a naïve way to evaluate privacy. To this end, several attempts to assess the privacy provided have been proposed by adapting information security measures based on entropy [11,43], and syntactic and semantic similarities [31,55] between original and obfuscated texts. However, all these measures fall short in assessing the actual privacy achieved by the mechanism [18,23,35]. Indeed, in the previous example, an adversarial system could easily infer the actual user information need from the obfuscated queries "`how myeloma grows`", "`how disease spreads`", "`how melanoma evolves`", i.e. the user is looking for information about tumours, by using available query logs on its side and generate potential guesses of the original query. Yet, some value of the privacy budget would lead to generating such obfuscated queries, giving formal and proven guarantees of privacy, and most of the state-of-the-art measures would consider those queries as properly obfuscated. Moreover, since such queries are still semantically similar to the original, they would probably deliver good retrieval performance, giving a somewhat false impression of a very effective yet properly private approach. By only studying the impact of the formal mechanism parameters, there is no assurance of the risk [19] the user faces by submitting the obfuscated queries.

In this paper, we introduce the Query Inference for Privacy and Utility (QuIPU) framework, a novel evaluation paradigm developed to evaluate the trade-offs between the actual privacy against potential information leakage in an obfuscation protocol represented by the inference of the original query committed by a malicious IR system and the utility gained by a user. QuIPU is based on the family of attacks known as Membership Inference Attack (MIA) [45], adapted to a query obfuscation protocol, namely Query Inference Attack (QuIA), and used against the obfuscated queries submitted to the system. It considers the potential risk that the original query is successfully inferred by a IR system after analyzing the alternative queries received and obfuscated using different configurations, i.e., using different formal privacy parameters, of an obfuscation mechanism. The measure considers the privacy-utility trade-off beyond the configuration parameters of the obfuscated mechanisms by computing a modified version of Area Under the Curve (AUC) of the risk *vs.* utility curve obtained. Therefore, the

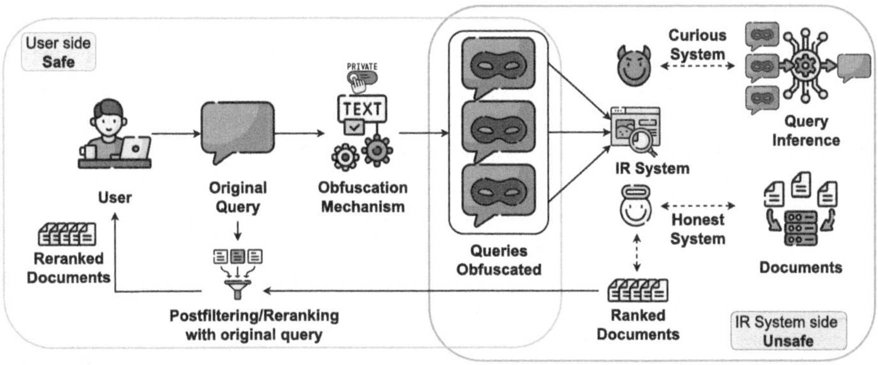

Fig. 1. Overview of the query obfuscation protocol in the presence of an *"honest-but-curious"* IR system. On the safe user side, the obfuscation mechanism takes in input the user query and produces N obfuscated queries. Such obfuscated queries are submitted to the *honest* IR System to retrieve documents. Then, if *curious*, the IR system might use the obfuscated queries to infer the original information need.

main contributions of this paper are: i) improving the privacy analysis paradigm by distinguishing between traditional and actual privacy guarantees within the query obfuscation process, thus studying a more comprehensive definition of privacy; ii) establishing a connection between MIAs in machine learning and its application to IR; and iii) a novel measure to assess privacy by evaluating the risk of information inference from the outputs of an obfuscation mechanism, providing insights into the formal privacy parameters.

The paper is organised as follows: Sect. 2 explains the context of the query obfuscation protocol; Sect. 3 introduces the formal definition of the QuIPU framework and the steps performed to evaluate the textual privacy provided by a mechanism. Moreover, Sect. 4 reports the results and discussion to investigate privacy from the *formal* privacy parameters to *actual* evaluation using the QuIPU framework. Finally, Sect. 5 presents the related works concerning privacy evaluation, and Sect. 6 draws the conclusion, outlining future directions.

2 Query Obfuscation Background

Privacy has been widely studied by the NLP and IR community [1,32,56,57]. The scenario discussed in this study assumes that the users are willingly paying part of the utility during the document retrieval phase to defend the privacy of their search activity. The system is considered non-cooperative, as it does not actively contribute to protecting user privacy, e.g., it does not provide any private API to mask the information need of the user. Figure 1 illustrates the general query obfuscation protocol, focus of this work and commonly used in IR [17,22]. The process considers two distinct environments: on the user (*safe*) side, the original query is generated by the user and privatized using an obfuscation mechanism, i.e., an algorithm that, given an original sensitive query q, generates N

non-sensitive obfuscated queries that (theoretically) prevent the unveiling of the original information need. These obfuscated queries are sent to the IR system without explicitly disclosing their information need, after the initial obfuscation process. During such operation, the user configures the mechanism parameters and privacy guarantees considering the amount of utility is lost on the downstream tasks [12,28,34]. On the (*unsafe*) IR system side, relevant documents are retrieved by the *"honest"* system considering the obfuscated queries received. The documents are then returned to the users for the post-processing described above. In order to prevent a *"curious"* IR system from discovering the real user query, the obfuscation methods employed are divided into two families of mechanisms, either based on heuristics or DP strategies, explained in the following paragraphs and in Table 1.

Table 1. ε-DP obfuscation mechanisms organised considering the obfuscation strategy.

Obfuscation Strategy	Mechansim	Description
Sampling	CusText [10]	Sampling of a new term is bounded to K possible terms picked using the scores computed using the distances among word embeddings.
	SanText [54]	Sampling of a new term is computed with a score based on the distances among embeddings, with terms closer to the obfuscation having a higher probability.
	TEM [7]	Noise sampled from an n - dimensional Gumbel distribution is added to the scores, and the final obfuscation term is sampled accordingly to the maximum noisy score.
	WBB [17]	Based on the word type, e.g., Nouns or Verbs, the mechanism finds the top-$(k+n)$ similar terms, excludes the first k, and samples the private word in the residuals n.
Embedding Perturbation	CMP [24]	The noise is sampled from an n - dimensional Laplace distribution of scale $\frac{1}{\varepsilon}$.
	Mhl [52]	The noise is sampled from an n - dimensional Normal distribution defined by the λ regularized Mahalanobis norm of the term embedding.
	Vickrey [53] (CMP/Mhl)	The noise is sampled as defined by the parent mechanism (CMP or Mhl) and the obfuscation term is set based on a free parameter t.

Heuristics Obfuscation. To protect privacy in IR tasks, non-formal obfuscation methods were proposed [3,26]. Arampatzis et al. [3] employed the WordNet [37] database to replace original terms within the query using synonyms, hypernyms, and holonyms. The obfuscation was performed based on a hierarchical degree, i.e., the level parameter, aligned with the desired obfuscation the user aims to achieve. Such an approach was further extended by Fröbe et al. [26]. More in detail, the obfuscation approach retrieves locally the top-k documents from a local corpus. Then, using a sliding window, the sequences of n terms within such documents are taken as candidate obfuscation queries, removing those queries that contain synonyms and holonyms. To decide which query to submit to the IR system, the top-k documents retrieved on the local corpus are considered as pseudo-relevant regarding the nDCG achieved.

Differential Privacy (DP) Obfuscation. Dwork et al. [21] introduced the ε-DP framework to formalize the privacy guarantees when releasing data. Given a privacy budget $\varepsilon \in \mathbb{R}^+$, and any pair of neighbouring datasets D, D', i.e., datasets that differ for only one entry, an obfuscation mechanism $\mathcal{M}$ is DP if it holds the inequality $\Pr[\mathcal{M}(D) \in S] \leq e^{\varepsilon} \cdot \Pr[\mathcal{M}(D') \in S] \ \forall S \subset \mathrm{Im}(\mathcal{M})$. DP introduces calibrated noise levels during output computation using the privacy budget ε, which controls the balance between data privacy and utility. The adoption of the DP framework for metric spaces, and therefore for NLP tasks, has been proposed in [8]. Metric-DP extends the traditional DP definition by ensuring that the probability of obfuscating two distinct points x, x' is proportional to the distance $d(x, x')$ between them. The DP formal framework has enabled the privacy research community to propose different strategies based on noisy sampling [7,10,17,54] and perturbed word embeddings [24,52,53], cf. Table 1.

3 The QuIPU framework

In this Section, we define the Query Inference for Privacy and Utility (QuIPU) framework: we report the threat model for an obfuscation protocol and the settings of QuIA. Finally, we report the risk evaluation of the attack.

3.1 Overview of the Threat Model

In this scenario, the adversary is represented by the IR system, which aims to understand the original user information need. In the query obfuscation protocol, the sweet spot for inferring the original queries is represented by the ones the system receives. The mechanism parameters, e.g., the ε privacy budget parameter of the DP obfuscation mechanisms, do not guarantee with absolute certainty that the original text is changed (or changed enough). Therefore, such queries may cause a leakage of the real information need. In addition, for the same parameters, different obfuscation strategies may produce texts with different obfuscation degrees. For instance, the effect of the parameter ε depends on the specific mechanism used [20,33]. As a result, two DP mechanisms (one embedding-perturbation based and the other sampling-based) both parametrized with $\varepsilon = 15$ and could lead to a situation where one method achieves an actual obfuscation while the other achieves only formal obfuscation. Therefore, the IR system aims to extract as much information as possible from the received queries, previously obfuscated on the user side, using this knowledge to infer the real text.

Consequently, the threat of a successful query inference stems not only from the obfuscation failure of the mechanism but also from the additional knowledge about the queries possessed by the adversary. The IR system might exploit its queries from the logs [9,30,46]: by producing a classification on the information needs carried by the obfuscated queries received and the information in its logs, it aims to improve the chances of a correct guess of the original user query. Note that if the original query is not an extremely *long tail* one [2], it is reasonable to assume that the original information need has been previously submitted to the IR system, and thus, the attack can succeed with high probability.

Finally, an important remark must be made regarding the use of cryptographic primitives in the protocol of the scenario we are analysing. Eavesdroppers or man-in-the-middle adversaries do not significantly threaten the user or the system. Cryptography can be employed while exchanging queries and documents between the client, i.e., the user, and server, i.e., the IR system, ensuring confidentiality among the internal parties of the protocol and security against external auditors. However, confidentiality does not imply privacy: if the IR system aims to disclose the user's original query, cryptography techniques alone are insufficient to safeguard privacy concerning an internal adversary.

3.2 Query Inference Attack (QuIA)

Algorithm 1: The Query Inference Attack (QuIA).

Data: $\mathcal{Q}_{\text{obf}}$ (obf. queries q_i'), $\mathcal{Q}_{\text{logs}}$ (query log q_i), $\mathcal{T}$ (transformer encoder).
Result: Ranked list of query logs $\mathcal{L}$.
Encoding $\mathcal{T}(\mathcal{Q}_{\text{obf}}) = \{\mathcal{T}(q_i') \in \mathbb{R}^n\}$ and $\mathcal{T}(\mathcal{Q}_{\text{logs}}) = \{\mathcal{T}(q_i) \in \mathbb{R}^n\}$;
Define $\hat{q}$ as the centroid of the vectors in $\mathcal{T}(\mathcal{Q}_{\text{obf}})$;
Compute $\mathcal{S} = [\cos(\hat{q}, \mathcal{T}(q_i)), \mathcal{T}(q_i) \in \mathcal{T}(\mathcal{Q}_{\text{logs}})]$;
Define $\mathcal{L} = [(s_i, q_i), s_i \in \mathcal{S}, q_i \in \mathcal{Q}_{\text{logs}}]$;
Sort $\mathcal{L}$ in descending order considering the similarity score s_i;
return $\mathcal{L}$;

The class of attacks known as Membership Inference Attack (MIA) was introduced by Shokri et al. [45] to investigate the information leakage stemming from the output of machine learning models. The attack is defined under the assumption that the attacker sees a data record but has no information about either the model parameters or the actual model architecture, i.e., a so-called *black-box* scenario. The attack is deemed successful when the attacker is able to correctly guess if the data record belongs to the model's training dataset or not.

In an obfuscation protocol, the Query Inference Attack (QuIA) uses the received obfuscated queries and the query logs to generate a ranked list of queries from the logs based on the similarity with the information need. Similarly to the *black-box* of the MIA scenario, the assumption is that the IR system does not know the obfuscation mechanism used on the user side and the privacy parameters of the obfuscation mechanism. Algorithm 1 reports the pseudo-code of the attack: the system receives the set of obfuscated queries $\mathcal{Q}_{\text{obf}}$ and knows its query logs $\mathcal{Q}_{\text{logs}}$. Firstly, it uses a Transformer [49] encoder $\mathcal{T}$ to obtain the embeddings of the queries in the sets[1]. Once the texts in $\mathcal{Q}_{\text{obf}}$ are encoded, it calculates the centroid $\hat{q}$ of the vectors in $\mathcal{T}(\mathcal{Q}_{\text{obf}})$, to capture the average contextual similarities among the obfuscated queries received. The system computes the cosine similarity between the embeddings of the queries from the logs $\mathcal{T}(q_i) \in \mathcal{T}(\mathcal{Q}_{\text{logs}})$ and the query $\hat{q}$ to understand which queries from the logs most closely represents the average information need carried by the obfuscated queries and saves it into the list $\mathcal{S}$. The algorithm finally generates a ranked list $\mathcal{L}$ of the queries in the logs $q_i \in \mathcal{Q}_{\text{logs}}$ by sorting the pairs of (s_i, q_i) in descending order based on

[1] Remark on the notation: with $\mathcal{T}(\mathcal{Q}_{\text{obf}}), \mathcal{T}(\mathcal{Q}_{\text{logs}})$ we indicate the sets of text embeddings, and with $\mathcal{T}(q_i'), \mathcal{T}(q_i)$ the singular vector embedding of the queries.

the similarities $s_i \in \mathcal{S}$. If the obfuscation was ineffective, then most likely, the higher a query from the logs is ranked in $\mathcal{L}$, the more likely it corresponds to the original user information need.

3.3 QuIPU Risk Modelling

Privacy is strictly linked with the definition of risk [40], i.e., the possibility that an action or event generates consequences that have an impact on what users value, in this scenario, disclosing sensitive information. The higher the risk, the lower the privacy. For example, DP obfuscation mechanisms offer the possibility that privacy and utility can be balanced by tuning the privacy budget ε. However, the framework does not provide any assurance against inference attacks [48]. To overcome this limitation, we need a formal definition of the risk against inference in the obfuscation protocol. After the QuIA algorithm has returned the ranked list $\mathcal{L}$, the IR system is tasked to guess the original query. This inference is based on the computed ranking, which considers the similarities between the obfuscated queries received (potentially leaking information) and the system's query logs (auxiliary knowledge for a correct guess of the original user query). At this point, the IR system strategy to guess the correct query is sequential: knowing that the first query is the most similar to the average information need carried by the obfuscated queries, it represents the best choice for the guess. If the first query in the logs $\mathcal{L}$ is the correct query, the attack is successful, and there is a 100% risk of correct inference. On the other hand, if the first one is not the correct guess, the adversary tries with the second query in the list, and so on, until the original query is guessed, decreasing the risk of success. Therefore, the risk r_t of successful QuIA in t guesses can be defined as the probability that the IR system correctly guesses $\bar{q}$ as the original query q, seeing the sets $\mathcal{Q}_{\text{obf}}$ and $\mathcal{Q}_{\text{logs}}$, i.e., $r_t = \mathbb{P}[\{\bar{q} = q\} \cap \{t \leq k\} | \mathcal{Q}_{\text{obf}}, \mathcal{Q}_{\text{logs}}]$, with k the maximum number of guessing attempts the IR system is willing to take. The upper bound for the value of k is determined by the size of the set $\mathcal{Q}_{\text{logs}}$. However, determining the precise threshold t and assessing the risk the user faces is impossible without access to the IR system's internal data and kind of attack. Therefore, we propose to model the malicious IR system with three kinds of attackers, representing relevant use cases: i) the "*lazy*" attacker, i.e., the one that looks only at the top position of the ranked list $\mathcal{L}$ and makes only one guess; ii) the "*active*" attacker, i.e., an adversary that selects the top-k queries and checks only with them if the guess is correct; and iii) the "*motivated*" attacker, i.e., the one that tries all the queries until the original one has been found. To model the probability of the risk a user faces against each of such attackers, we propose to use proxy indicators computed on where the original query appears in the ranked list $\mathcal{L}$: Precision at 1 ($P@1$) for the lazy attacker, Recall at k ($R@k$) for the active attacker, and Reciprocal Rank (RR) for the motivated attacker.

Drawing inspiration from the usual ROC AUC, Fig. 2 illustrates the evaluation plane that links the risk r of a successful QuIA and the utility u measure considering a set of queries $\mathcal{Q}_{\text{obf}}(p_i)$ – an effectiveness measure such as nDCG in the IR case – obfuscated by a certain formal parameter p_i – the ε parameter in case of DP. In the risk-utility plane, the *Risk-Utility Boundary line*, i.e., the diagonal, describes two regions where the risk-utility trend $f(r, u)$ can be: i) above the line indicates that the utility u exceeds the associated risk r, and ii) below the line, where u is less than r. Therefore, we can define the QuIPU score, Eq. 1, considering the different pairs (r, u) estimated by submitting to the IR system the set of obfuscated queries $\mathcal{Q}_{\text{obf}}(p_i)$.

$$\text{QuIPU} = 2(\mathcal{R}_+ + \mathcal{R}_-) = 2\int_{\mathcal{R}_+} f(r, u)\, d\nu + 2\int_{\mathcal{R}_-} f(r, u)\, d\nu \tag{1}$$

where $d\nu$ represents an infinitesimal variation on the *Risk-Utility Boundary line*, and the factor 2 is introduced to map the score from $\left[-\frac{1}{2}, \frac{1}{2}\right]$ to $[-1, 1]$ interval. The integrals are calculated with respect to the diagonal of the plane, such that regions where the curve lies below this diagonal, i.e., R_-, are assigned negative values, indicating that the risk r is greater than the utility u. Conversely, positive values are computed for regions where the utility u exceeds the risk r, i.e., R_+. In Fig. 2, four critical points are defined:

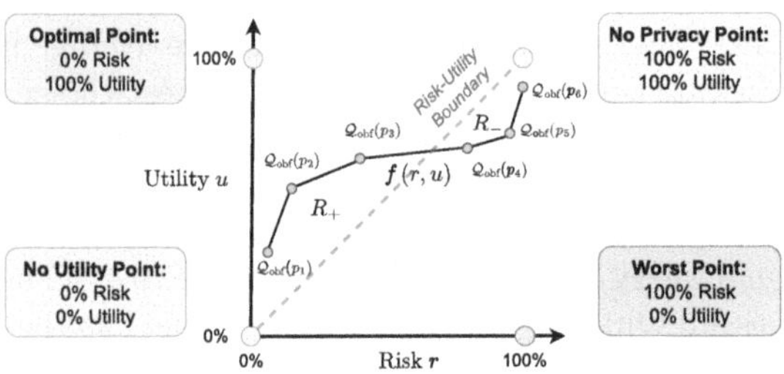

Fig. 2. Risk r *vs.* Utility u model to evaluate actual privacy. The four labels describe the plane's critical points, and the red dashed line shows the Boundary tracing two areas, R_+ and R_-, where the QuIPU score is positive or negative. The curve f shows the r *vs.* u trend when the system receives the queries $\mathcal{Q}_{\text{obf}}(p_i)$ masked with a parameter p_i. (Color figure online)

- **No Utility Point**: This point shows when the risk and utility are both reduced to 0. It corresponds to the situation where the obfuscation mechanism fully modifies the original query, completely stopping a QuIA. However, the user completely renounces the effectiveness of the task, i.e., the submitted queries failed to retrieve any relevant documents from the IR system.
- **No Privacy Point**: This point illustrates the effect of not using the obfuscation protocol. The queries are not obfuscated, meaning the original query is fully exposed to the IR system, resulting in 100% risk, i.e., the attacker has all the information to infer correctly. Yet, utility is fully achieved, as the system can use the original query to retrieve all the relevant documents.
- **Optimal Point**: The optimal point is the best we can *theoretically* obtain. The obfuscation mechanism provides complete protection against Query

Inference attacks, i.e., 0% of risk, while maintaining maximum utility. The user's information need are entirely met during the retrieval without exposing any information related to the original query.
- **Trash Point**: This is the opposite of the optimal point. The mechanism neither obfuscates the query nor these queries can retrieve any relevant documents. This case can happen if we change the *"honest-but-curious"* assumption to a *"fully-dishonest"* scenario, a.e., a phishing IR system [39,42].

4 Experimental Evaluation

We outline here the experimental setup and compare the results observed for the traditional and actual privacy analysis using the QuIPU framework. Further analyses, the data used, and the source code are publicly available[2].

4.1 Experimental Setup

User side. We test the QuIPU framework on three different TREC collections: Deep Learning (DL'19) [14] and Deep Learning (DL'20) [13], based on the MS MARCO [41] passages corpus, containing 43 and 54 queries respectively, and the Robust '04 (Robust '04) [50] which relies on disks 4 and 5 of the TIPSTER corpus and contains 249 queries. As obfuscation mechanisms, we consider those described in Sect. 2, using their implementation provided by the pyPANTERA framework [16]. As privacy budget ε, we followed the parametrization reported in the original papers, which is also the one used by the pyPANTERA framework and other recent experiments [22]. In detail, we select $\varepsilon \in \{1, 5, 10, 12.5, 15, 17.5, 20, 25, 30, 50\}$. The heuristics obfuscation mechanisms, i.e., AEA [3] and FEA [26], have been parametrized using different synonyms levels $\{3, 4, 5\}$, and sliding windows sizes $\{12, 14, 16\}$, respectively. We generated 50 obfuscation variants for each query and mechanism configuration. Finally, the IR system used as re-ranker in the post-retrieval phase of the protocol is the neural dense model Contriever [29], as proposed in [17,22].

IR system side. On the one hand, the *honest* aspect of the IR system, i.e., the part that performs the document retrieval task, is showcased by the Contriever model; on the other hand the *curious* part of the system uses as encoder model `distilbert-base-uncased` [44]. We use two different models for the two tasks to obtain unbiased results, in line with [17,22]. To simulate a realistic scenario for the curious IR to perform the QuIA, we use as query logs the AOL collection available in the `ir_datasets` catalog[3]. We sampled 750k queries, to which we added the original queries as explained in Sect. 3.

[2] https://github.com/Kekkodf/QuIPU_Framework.
[3] https://ir-datasets.com/aol-ia.html.

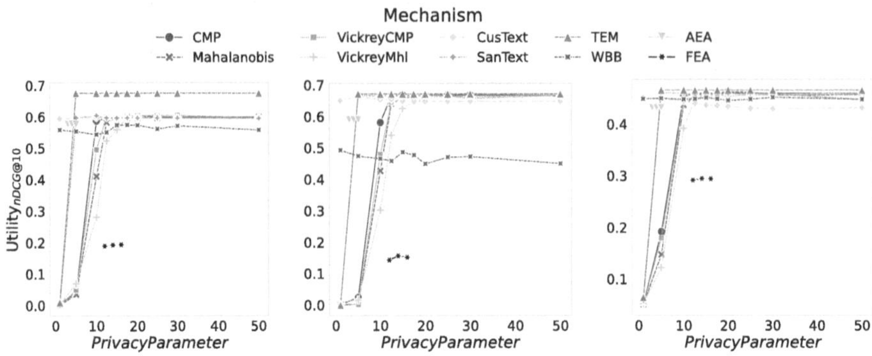

(a) DeepLearning '19 (b) DeepLearning '20 (c) Robust'04

Fig. 3. nDCG@10 (*Utility*) when varying privacy formal parameters (*PrivacyParameters*) of the obfuscation mechanisms: some mechanisms, such as TEM, achieve immediately high performance, with unclear consequences on actual privacy.

4.2 Privacy from the Formal Parameters Analysis

The traditional privacy analysis evaluates the utility as a function of the formal privacy parameters, e.g., ε. Figure 3 reports the results of the nDCG@10 *vs* the formal privacy parameters on the three different collections analysed. Note that the x-axis, representing the *PrivacyParameter*, considers both the values for the ε parameter of the DP mechanisms and the parameters of the heuristics [3,26]. From this traditional perspective, it emerges that with lower values of the privacy parameters, mechanisms based on a DP strategy, like TEM or SanText, achieve higher effectiveness for low values of the privacy parameter ε. On the other hand, obfuscation mechanisms based on the embedding obfuscation strategies perform with high effectiveness only if the formal parameter is high. Finally, the Heuristics show high nDCG@10 for AEA and the worst results for the FEA mechanism. These results show a misleading sense of privacy: high-performance results do not imply the actual privacy of the texts, i.e., the submitted queries are the original ones. To address this issue, we analyse the impact of obfuscation on denying a correct inference of the original query.

4.3 Privacy Analysis Using the QuIPU Framework

In this section, we report the evaluation of the actual privacy the obfuscation mechanisms provide to the original user query. Figure 4 show the different Risk r *vs.* Utility u evaluation planes on the MSMARCO Deep Learning'19 collection obtained for the different malicious IR system attacker nature that can perform the QuIA, i.e., "*lazy*" (Fig. 4(a)), "*active*" (Fig. 4(b)) and "*motivated*" (Fig. 4(c)). The plots show that the utility of the DP mechanisms, i.e., the nDCG@10, remains stable after an initial increase from the "No Utility Point",

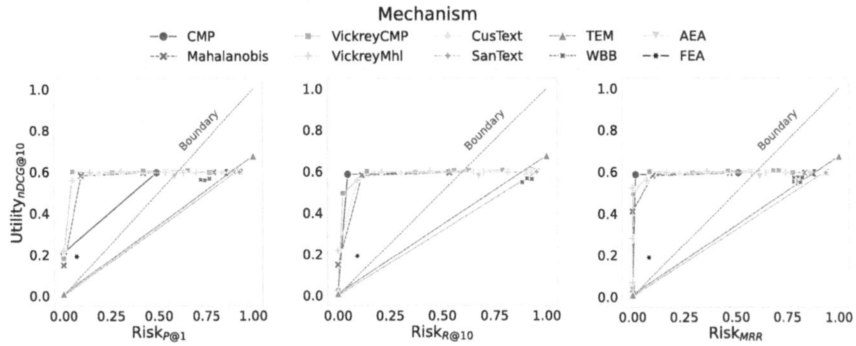

(a) *Lazy Attacker* (b) *Active Attacker* (c) *Motivated Attacker*

Fig. 4. Resulting Risk estimated *vs.* Utility (i.e., retrieval nDCG@10) achieved sending the $\mathcal{Q}_{\text{obf}}$ of the MSMARCO Deep Learning'19 collection. The risk is estimated as explained in Sect. 3, showing Risk$_{P@1}$, Risk$_{R@10}$, and Risk$_{MRR}$ for the different adversaries. The utility Utility$_{nDCG@10}$ shows the nDCG@10 achieved when retrieving the documents after the reranking phase.

see Fig. 2. On the other hand, the risk in all three scenarios increases, implying a high probability of discovering the real user information need from the obfuscated queries submitted. This confirms our hypothesis: although some privacy configurations give formal and theoretically proven guarantees of privacy, such obfuscated queries are vulnerable against the QuIA, even considering the "*Lazy*" attacker (Fig. 4(a)), under the false impression of effective obfuscation. Conversely, the heuristics AEA and FEA mechanisms achieve the same utility and risk on all the privacy configurations. The former is always allocated around the Risk *vs.* Utility Boundary line, implying a risk, thus a probability of correct inference of the original query, equal to the utility achieved. The latter is more prone to defending user privacy against the inference of the original information need, but renouncing more than 60% of nDCG@10.

Table 2 reports the QuIPU scores obtained analysing the Risk *vs.* Utility on each set of obfuscated queries of the collections. The results show three distinct patterns that can be traced back to the three different obfuscation strategies. The Sampling-based mechanisms show weaker defences against the three attackers, and especially against the "*active*" one, i.e., QuIPU score more negative. In contrast, the Embedding Perturbation mechanisms are designed to protect user information from attackers, yielding higher QuIPU scores even against a "*motivated*" attacker. This suggests that, when using as DP obfuscation mechanisms, if the user wants to achieve strong actual privacy guarantees against the QuIA, it should select an obfuscation relying on changing the word embeddings of the queries. Finally, the heuristics strategies obtain a null QuIPU score due to the stable risk and utility achieved. FEA reaches a slightly positive QuIPU score against the three attackers, implying that it is impractical for an attacker to guess the original query even if "*motivated*" to do so.

Table 2. QuIPU Score computed organizing the results by obfuscation strategy. computed using the Eq. 1, measured in terms of u = nDCG@10, and the risk r of a successful Query Inference considering different adversary models. Positive values correspond to a better Utility-Privacy trade-off, cf. Sect. 3.

Obfuscation Strategy	Mechanism	*Lazy Attacker*			*Active Attacker*			*Motivated Attacker*		
		DL'19	DL'20	Robust	DL'19	DL'20	Robust	DL'19	DL'20	Robust
Sampling	CusText	0.041	0.109	0.034	−0.014	0.010	−0.084	0.020	0.074	0.028
	SanText	−0.247	−0.222	0.046	−0.277	−0.252	−0.237	−0.255	−0.231	0.043
	TEM	−0.264	−0.274	0.028	−0.264	−0.274	−0.329	−0.264	−0.274	0.027
	WBB	−0.005	−0.002	0.001	−0.001	−0.022	−0.011	−0.006	−0.010	0.001
Embedding Perturbation	CMP	0.299	0.372	0.175	0.257	0.323	0.001	0.283	0.353	0.170
	Mahalanobis	0.280	0.381	0.202	0.258	0.363	0.090	0.272	0.371	0.200
	VickreyCMP	0.341	0.430	0.194	0.310	0.411	0.103	0.334	0.424	0.193
	VickreyMhl	0.342	0.426	0.199	0.318	0.410	0.119	0.335	0.421	0.199
Heuristics	AEA	0.000	0.000	0.000	0.000	0.000	0.000	0.000	0.000	0.000
	FEA	0.001	0.001	0.001	0.000	0.001	0.002	0.000	0.001	0.001

5 Related Works

Different works have been proposed to organize available privacy measures [47, 51]. Wagner and Eckhoff [51] systematically classified over eighty privacy metrics, offering a comprehensive framework for assessing privacy across different domains, e.g., communication, databases, and social networks. The survey underscores the significance of identifying the specific aspect of privacy that a metric aims to quantify, suggesting nine guiding questions for selecting the appropriate privacy measures. Specifically, the authors underlined the importance of considering the adversary's knowledge and capability when evaluating privacy. In addition, Sousa and Kern [47] described how different mechanisms developed for NLP tasks provide privacy for textual data and which can be the threats in such scenarios. Moreover, Habernal [27] stressed the importance of not relying strictly on formal analysis of DP and its application on NLP tasks, but to push research towards a concrete measurement of the privacy provided to texts.

Traditional methods for evaluating privacy primarily focus on calculating the failure rates of obfuscation mechanisms [11] or assessing the similarities between original and obfuscated texts [22,36]. On the one hand, uncertainty measures such as N_w and S_w [24,52] estimate the probability that a term w remains unchanged after obfuscation and the minimum cardinality of the set of words to which the mechanism maps w, respectively. However, such measures do not capture if the mechanism changes the original term with a closely related one. On the other hand, the similarity between the original and obfuscated texts is commonly estimated using metrics like the Jaccard index or cosine similarity between sentence embeddings computed by a Transformer model, drawing inspiration by the use of BERTScores used to evaluate the quality of generated texts [55]. Meisenbacher et al. [36] proposed the α-PUC score to compute an α weighted mean between uncertainty, similarity measures, and utility preserved. However, none of the above measures offer insights into the actual privacy afforded to the texts, nor do they assess the adversarial potential to infer the original meaning of the obfuscated text. Blanco-Justicia et al. [5] criticize the reliance on formal privacy analysis solely based on the privacy budget ε parameter. They argue that DP mechanisms employing configurations where $\varepsilon > 1$ lack a comprehensive analysis of actual privacy guarantees, raising concerns about the sufficiency of privacy protection in practice[4]. In addition, Damie et al. [15] introduced a novel indicator to assess the risk of successful query recovery attacks within search able encryption protocols. The study revealed that, even without additional background knowledge, an adversary could reconstruct the original queries with a success rate of 85%, encouraging analysis of privacy measures employed considering real adversarial scenarios of application.

To the best of our knowledge, this study represents an initial attempt to bridge the gap between the evaluation based on formal parameters and actual privacy analysis, proposing a novel effort to assess the risk of inferring the original query from the set of obfuscations produced to protect user privacy against a malicious IR system.

6 Conclusion and Future Work

Assessing the privacy guarantees provided to users during IR tasks remains an open challenge. Relying solely on a formal privacy analysis considering the mechanism parameters is insufficient for concretely evaluating the privacy of obfuscation mechanisms. In this study, we introduced the QuIPU framework, a new benchmark designed to assess actual privacy provided to queries in an obfuscation protocol. We empirically evaluated the risk that an "honest-but-curious" IR system can accurately infer the original query from the obfuscated ones received using its queries from the logs. Our findings demonstrate that strong formal privacy guarantees do not necessarily imply actual privacy protection. In future work, we plan to explore additional proxy measures, e.g., the Rank-biased precision [38], to investigate their correlation with the QuIPU score. In addition, we intend to explore the capabilities of Large Language Models in determining whether or not a query has been sufficiently obfuscated, adopting such models as defensive mechanisms against a successful QuIA.

[4] The DP configurations with $\varepsilon > 1$ deviate from the "theoretically safe" privacy setup, i.e., strong assurance about the formal privacy introduced, see DP definition [21].

References

1. Ahmad, W.U., Chang, K., Wang, H.: Intent-aware query obfuscation for privacy protection in personalized web search. In: Collins-Thompson, K., Mei, Q., Davison, B.D., Liu, Y., Yilmaz, E. (eds.) The 41st International ACM SIGIR Conference on Research & Development in Information Retrieval, SIGIR 2018, Ann Arbor, MI, USA, July 08-12, 2018, pp. 285–294. ACM (2018). https://doi.org/10.1145/3209978.3209983
2. Anderson, C.: The long tail. Mann, Ivanov & Ferber, Effective Business Model on the Internet-Moscow (2012)
3. Arampatzis, A., Drosatos, G., Efraimidis, P.: A versatile tool for privacy-enhanced web search. In: Serdyukov, P., et al. (eds.) Advances in Information Retrieval - 35th European Conference on IR Research, ECIR 2013, Moscow, Russia, March 24-27, 2013. Proceedings. Lecture Notes in Computer Science, vol. 7814, pp. 368–379. Springer (2013). https://doi.org/10.1007/978-3-642-36973-5_31
4. Bavadekar, S., et al.: Google COVID-19 search trends symptoms dataset: Anonymization process description (version 1.0). CoRR abs/2009.01265 (2020). https://arxiv.org/abs/2009.01265
5. Blanco-Justicia, A., Sánchez, D., Domingo-Ferrer, J., Muralidhar, K.: A critical review on the use (and misuse) of differential privacy in machine learning. ACM Comput. Surv. **55**(8), 160:1–160:16 (2023). https://doi.org/10.1145/3547139
6. Bo, H., Ding, S.H.H., Fung, B.C.M., Iqbal, F.: ER-AE: differentially private text generation for authorship anonymization. In: Toutanova, K., et al. (eds.) Proceedings of the 2021 Conference of the North American Chapter of the Association for Computational Linguistics: Human Language Technologies, NAACL-HLT 2021, Online, June 6-11, 2021, pp. 3997–4007. Association for Computational Linguistics (2021). https://doi.org/10.18653/V1/2021.NAACL-MAIN.314, https://doi.org/10.18653/v1/2021.naacl-main.314
7. Carvalho, R.S., Vasiloudis, T., Feyisetan, O., Wang, K.: TEM: high utility metric differential privacy on text. In: Shekhar, S., Zhou, Z., Chiang, Y., Stiglic, G. (eds.) Proceedings of the 2023 SIAM International Conference on Data Mining, SDM 2023, Minneapolis-St. Paul Twin Cities, MN, USA, April 27-29, 2023, pp. 883–890. SIAM (2023). https://doi.org/10.1137/1.9781611977653.CH99
8. Chatzikokolakis, K., Andrés, M.E., Bordenabe, N.E., Palamidessi, C.: Broadening the scope of differential privacy using metrics. In: Cristofaro, E.D., Wright, M.K. (eds.) Privacy Enhancing Technologies - 13th International Symposium, PETS 2013, Bloomington, IN, USA, July 10-12, 2013. Proceedings. Lecture Notes in Computer Science, vol. 7981, pp. 82–102. Springer (2013). https://doi.org/10.1007/978-3-642-39077-7_5
9. Chau, M., Fang, X., Sheng, O.R.L.: Analysis of the query logs of a web site search engine. J. Assoc. Inf. Sci. Technol. **56**(13), 1363–1376 (2005). https://doi.org/10.1002/ASI.20210
10. Chen, S., et al.: A customized text sanitization mechanism with differential privacy. In: Rogers, A., Boyd-Graber, J., Okazaki, N. (eds.) Findings of the Association for Computational Linguistics: ACL 2023, pp. 5747–5758. Association for Computational Linguistics, Toronto, Canada (2023). https://doi.org/10.18653/v1/2023.findings-acl.355, https://aclanthology.org/2023.findings-acl.355
11. Clauß, S., Schiffner, S.: Structuring anonymity metrics. In: Juels, A., Winslett, M., Goto, A. (eds.) Proceedings of the 2006 Workshop on Digital Identity Management, Alexandria, VA, USA, November 3, 2006, pp. 55–62. ACM (2006). https://doi.org/10.1145/1179529.1179539

12. Clifton, C., Tassa, T.: On syntactic anonymity and differential privacy. In: Chan, C.Y., Lu, J., Nørvåg, K., Tanin, E. (eds.) Workshops Proceedings of the 29th IEEE International Conference on Data Engineering, ICDE 2013, Brisbane, Australia, April 8-12, 2013, pp. 88–93. IEEE Computer Society (2013). https://doi.org/10.1109/ICDEW.2013.6547433
13. Craswell, N., Mitra, B., Yilmaz, E., Campos, D.: Overview of the TREC 2020 deep learning track. CoRR abs/2102.07662 (2021). https://arxiv.org/abs/2102.07662
14. Craswell, N., Mitra, B., Yilmaz, E., Campos, D., Voorhees, E.M.: Overview of the TREC 2019 deep learning track. CoRR abs/2003.07820 (2020). https://arxiv.org/abs/2003.07820
15. Damie, M., Hahn, F., Peter, A.: A highly accurate query-recovery attack against searchable encryption using non-indexed documents. In: Bailey, M.D., Greenstadt, R. (eds.) 30th USENIX Security Symposium, USENIX Security 2021, August 11-13, 2021, pp. 143–160. USENIX Association (2021). https://www.usenix.org/conference/usenixsecurity21/presentation/damie
16. De Faveri, F.L., Faggioli, G., Ferro, N.: py-PANTERA: a Python PAckage for Natural language obfuscaTion Enforcing pRivacy & Anonymization. In: Proceedings of the 33rd ACM International Conference on Information and Knowledge Management (CIKM '24), October 21-25, 2024, Boise, ID, USA. p. 6. Springer (2024). https://doi.org/10.1145/3627673.3679173, https://doi.org/10.1145/3627673.3679173
17. De Faveri, F.L., Faggioli, G., Ferro, N.: Words Blending Boxes. Obfuscating Queries in Information Retrieval using Differential Privacy. CoRR abs/2405.09306 (2024). https://doi.org/10.48550/ARXIV.2405.09306
18. Domingo-Ferrer, J., Sánchez, D., Blanco-Justicia, A.: The limits of differential privacy (and its misuse in data release and machine learning). Commun. ACM **64**(7), 33–35 (2021). https://doi.org/10.1145/3433638
19. Duncan, G., Keller-McNulty, S., Stokes, L.: Disclosure risk vs. data utility: the RU confidentiality map. A Los Alamos National Laboratory Technical Report **LA-UR-01-6428**, 1–30 (2001)
20. Dwork, C., Kohli, N., Mulligan, D.K.: Differential privacy in practice: expose your epsilons! J. Priv. Confidentiality **9**(2) (2019). https://doi.org/10.29012/JPC.689
21. Dwork, C., McSherry, F., Nissim, K., Smith, A.: Calibrating noise to sensitivity in private data analysis. In: Halevi, S., Rabin, T. (eds.) Theory of Cryptography, pp. 265–284. Springer, Berlin, Heidelberg (2006)
22. Faggioli, G., Ferro, N.: Query obfuscation for information retrieval through differential privacy. In: Goharian, N., Tonellotto, N., He, Y., Lipani, A., McDonald, G., Macdonald, C., Ounis, I. (eds.) Advances in Information Retrieval - 46th European Conference on Information Retrieval, ECIR 2024, Glasgow, UK, March 24-28, 2024, Proceedings, Part I. Lecture Notes in Computer Science, vol. 14608, pp. 278–294. Springer (2024). https://doi.org/10.1007/978-3-031-56027-9_17
23. Faveri, F.L.D., Faggioli, G., Ferro, N.: Beyond the parameters: Measuring actual privacy in obfuscated texts. In: Roitero, K., Viviani, M., Maddalena, E., Mizzaro, S. (eds.) Proceedings of the 14th Italian Information Retrieval Workshop, Udine, Italy, September 5-6, 2024. CEUR Workshop Proceedings, vol. 3802, pp. 53–57. CEUR-WS.org (2024). https://ceur-ws.org/Vol-3802/paper5.pdf
24. Feyisetan, O., Balle, B., Drake, T., Diethe, T.: Privacy- and utility-preserving textual analysis via calibrated multivariate perturbations. In: Caverlee, J., Hu, X.B., Lalmas, M., Wang, W. (eds.) Proceedings of the 13th International Conference on Web Search and Data Mining, pp. 178–186. ACM (2020). https://doi.org/10.1145/3336191.3371856

25. Feyisetan, O., Kasiviswanathan, S.: Private release of text embedding vectors. In: Pruksachatkun, Y., et al. (eds.) Proceedings of the First Workshop on Trustworthy Natural Language Processing, pp. 15–27. Association for Computational Linguistics, Online (2021). https://doi.org/10.18653/v1/2021.trustnlp-1.3, https://aclanthology.org/2021.trustnlp-1.3
26. Fröbe, M., Schmidt, E.O., Hagen, M.: Efficient query obfuscation with keyqueries. In: He, J., et al. (eds.) WI-IAT '21: IEEE/WIC/ACM International Conference on Web Intelligence, Melbourne VIC Australia, December 14–17, 2021, pp. 154–161. ACM (2021). https://doi.org/10.1145/3486622.3493950
27. Habernal, I.: When differential privacy meets NLP: the devil is in the detail. In: Moens, M., Huang, X., Specia, L., Yih, S.W. (eds.) Proceedings of the 2021 Conference on Empirical Methods in Natural Language Processing, EMNLP 2021, Virtual Event / Punta Cana, Dominican Republic, 7-11 November, 2021, pp. 1522–1528. Association for Computational Linguistics (2021). https://doi.org/10.18653/V1/2021.EMNLP-MAIN.114
28. Hsu, J., et al.: Differential privacy: an economic method for choosing epsilon. In: IEEE 27th Computer Security Foundations Symposium, CSF 2014, Vienna, Austria, 19-22 July, 2014, pp. 398–410. IEEE Computer Society (2014). https://doi.org/10.1109/CSF.2014.35, https://doi.org/10.1109/CSF.2014.35
29. Izacard, G., et al.: Unsupervised dense information retrieval with contrastive learning. Trans. Mach. Learn. Res. **2022** (2022). https://openreview.net/forum?id=jKN1pXi7b0
30. Jansen, B.J., Spink, A., Saracevic, T.: Real life, real users, and real needs: a study and analysis of user queries on the web. Inf. Process. Manag. **36**(2), 207–227 (2000). https://doi.org/10.1016/S0306-4573(99)00056-4
31. Kang, Y., Liu, Y., Niu, B., Tong, X., Zhang, L., Wang, W.: Input perturbation: a new paradigm between central and local differential privacy. CoRR abs/2002.08570 (2020). https://arxiv.org/abs/2002.08570
32. Klymenko, O., Meisenbacher, S., Matthes, F.: Differential privacy in natural language processing the story so far. In: Feyisetan, O., Ghanavati, S., Thaine, P., Habernal, I., Mireshghallah, F. (eds.) Proceedings of the Fourth Workshop on Privacy in Natural Language Processing, pp. 1–11. Association for Computational Linguistics, Seattle, United States (2022). https://doi.org/10.18653/v1/2022.privatenlp-1.1, https://aclanthology.org/2022.privatenlp-1.1
33. Kohli, N., Laskowski, P.: Epsilon voting: mechanism design for parameter selection in differential privacy. In: 2018 IEEE Symposium on Privacy-Aware Computing, PAC 2018, Washington, DC, USA, September 26-28, 2018, pp. 19–30. IEEE (2018). https://doi.org/10.1109/PAC.2018.00009
34. Lee, J., Clifton, C.: How much is enough? Choosing ϵ for differential privacy. In: Lai, X., Zhou, J., Li, H. (eds.) Information Security, 14th International Conference, ISC 2011, Xi'an, China, October 26-29, 2011. Proceedings. Lecture Notes in Computer Science, vol. 7001, pp. 325–340. Springer (2011). https://doi.org/10.1007/978-3-642-24861-0_22
35. Mattern, J., Weggenmann, B., Kerschbaum, F.: The limits of word level differential privacy. In: Carpuat, M., de Marneffe, M., Ruíz, I.V.M. (eds.) Findings of the Association for Computational Linguistics: NAACL 2022, Seattle, WA, United States, July 10-15, 2022, pp. 867–881. Association for Computational Linguistics (2022). https://doi.org/10.18653/V1/2022.FINDINGS-NAACL.65, https://doi.org/10.18653/v1/2022.findings-naacl.65

36. Meisenbacher, S.J., Nandakumar, N., Klymenko, A., Matthes, F.: A comparative analysis of word-level metric differential privacy: benchmarking the privacy-utility trade-off. In: Calzolari, N., Kan, M., Hoste, V., Lenci, A., Sakti, S., Xue, N. (eds.) Proceedings of the 2024 Joint International Conference on Computational Linguistics, Language Resources and Evaluation, LREC/COLING 2024, 20-25 May, 2024, Torino, Italy, pp. 174–185. ELRA and ICCL (2024). https://aclanthology.org/2024.lrec-main.16
37. Miller, G.A.: WordNet: a lexical database for English. Commun. ACM **38**(11), 39–41 (1995). https://doi.org/10.1145/219717.219748
38. Moffat, A., Zobel, J.: Rank-biased precision for measurement of retrieval effectiveness. ACM Trans. Inf. Syst. **27**(1), 2:1–2:27 (2008). https://doi.org/10.1145/1416950.1416952
39. Moore, T., Clayton, R.: Evil searching: compromise and recompromise of internet hosts for phishing. In: Dingledine, R., Golle, P. (eds.) Financial Cryptography and Data Security, 13th International Conference, FC 2009, Accra Beach, Barbados, February 23-26, 2009. Revised Selected Papers. Lecture Notes in Computer Science, vol. 5628, pp. 256–272. Springer (2009). https://doi.org/10.1007/978-3-642-03549-4_16
40. National Institute of Standards and Technology: Information security. Tech. Rep. National Institute of Standards and Technology Special Publication 800-60, Volume 1 Revision 1, August, 2008, U.S. Department of Commerce, Washington, D.C. (2008). https://doi.org/10.6028/NIST.SP.800-60v1r1
41. Nguyen, T., Rosenberg, M., Song, X., Gao, J., Tiwary, S., Majumder, R., Deng, L.: MS MARCO: a human generated machine reading comprehension dataset. In: Besold, T.R., Bordes, A., d'Avila Garcez, A.S., Wayne, G. (eds.) Proceedings of the Workshop on Cognitive Computation: Integrating neural and symbolic approaches 2016 co-located with the 30th Annual Conference on Neural Information Processing Systems (NIPS 2016), Barcelona, Spain, December 9, 2016. CEUR Workshop Proceedings, vol. 1773. CEUR-WS.org (2016). https://ceur-ws.org/Vol-1773/CoCoNIPS_2016_paper9.pdf
42. Rao, R.S., Pais, A.R.: Jail-Phish: an improved search engine based phishing detection system. Comput. Secur. **83**, 246–267 (2019). https://doi.org/10.1016/J.COSE.2019.02.011
43. Rényi, A.: On measures of entropy and information. In: Proceedings of the fourth Berkeley symposium on mathematical statistics and probability, volume 1: contributions to the theory of statistics, vol. 4, pp. 547–562. University of California Press (1961)
44. Sanh, V., Debut, L., Chaumond, J., Wolf, T.: Distilbert, a distilled version of BERT: smaller, faster, cheaper and lighter. CoRR abs/1910.01108 (2019). http://arxiv.org/abs/1910.01108
45. Shokri, R., Stronati, M., Song, C., Shmatikov, V.: Membership inference attacks against machine learning models. In: 2017 IEEE Symposium on Security and Privacy, SP 2017, San Jose, CA, USA, May 22-26, 2017, pp. 3–18. IEEE Computer Society (2017). https://doi.org/10.1109/SP.2017.41
46. Silvestri, F.: Mining query logs: turning search usage data into knowledge. Found. Trends Inf. Retr. **4**(1-2), 1–174 (2010). https://doi.org/10.1561/1500000013
47. Sousa, S., Kern, R.: How to keep text private? A systematic review of deep learning methods for privacy-preserving natural language processing. Artif. Intell. Rev. **56**(2), 1427–1492 (2023). https://doi.org/10.1007/S10462-022-10204-6

48. Truex, S., Liu, L., Gursoy, M.E., Wei, W., Yu, L.: Effects of differential privacy and data skewness on membership inference vulnerability. In: First IEEE International Conference on Trust, Privacy and Security in Intelligent Systems and Applications, TPS-ISA 2019, Los Angeles, CA, USA, December 12-14, 2019, pp. 82–91. IEEE (2019). https://doi.org/10.1109/TPS-ISA48467.2019.00019
49. Vaswani, A., et al.: Attention is all you need. In: Guyon, I., et al. (eds.) Advances in Neural Information Processing Systems 30: Annual Conference on Neural Information Processing Systems 2017, December 4-9, 2017, Long Beach, CA, USA, pp. 5998–6008 (2017)
50. Voorhees, E.M.: Overview of the TREC 2004 robust track. In: Voorhees, E.M., Buckland, L.P. (eds.) Proceedings of the Thirteenth Text REtrieval Conference, TREC 2004, Gaithersburg, Maryland, USA, November 16-19, 2004. NIST Special Publication, vol. 500-261. National Institute of Standards and Technology (NIST) (2004). http://trec.nist.gov/pubs/trec13/papers/ROBUST.OVERVIEW.pdf
51. Wagner, I., Eckhoff, D.: Technical privacy metrics: a systematic survey. ACM Comput. Surv. **51**(3), 57:1–57:38 (2018). https://doi.org/10.1145/3168389, https://doi.org/10.1145/3168389
52. Xu, Z., Aggarwal, A., Feyisetan, O., Teissier, N.: A differentially private text perturbation method using regularized mahalanobis metric. In: Proceedings of the Second Workshop on Privacy in NLP. Association for Computational Linguistics (2020). https://doi.org/10.18653/v1/2020.privatenlp-1.2
53. Xu, Z., Aggarwal, A., Feyisetan, O., Teissier, N.: On a utilitarian approach to privacy preserving text generation. CoRR abs/2104.11838 (2021). https://doi.org/10.48550/ARXIV.2104.11838
54. Yue, X., Du, M., Wang, T., Li, Y., Sun, H., Chow, S.S.M.: Differential privacy for text analytics via natural text sanitization. In: Zong, C., Xia, F., Li, W., Navigli, R. (eds.) Findings of the Association for Computational Linguistics: ACL-IJCNLP 2021, pp. 3853–3866. Association for Computational Linguistics, Online (2021). https://doi.org/10.18653/v1/2021.findings-acl.337, https://aclanthology.org/2021.findings-acl.337
55. Zhang, T., Kishore, V., Wu, F., Weinberger, K.Q., Artzi, Y.: Bertscore: evaluating text generation with BERT. In: 8th International Conference on Learning Representations, ICLR 2020, Addis Ababa, Ethiopia, April 26-30, 2020. OpenReview.net (2020). https://openreview.net/forum?id=SkeHuCVFDr
56. Zhao, Y., Chen, J.: A survey on differential privacy for unstructured data content. ACM Comput. Surv. **54**(10s), 207:1–207:28 (2022). https://doi.org/10.1145/3490237
57. Zimmerman, S., Thorpe, A., Fox, C., Kruschwitz, U.: Investigating the interplay between searchers' privacy concerns and their search behavior. In: Piwowarski, B., Chevalier, M., Gaussier, É., Maarek, Y., Nie, J., Scholer, F. (eds.) Proceedings of the 42nd International ACM SIGIR Conference on Research and Development in Information Retrieval, SIGIR 2019, Paris, France, July 21-25, 2019, pp. 953–956. ACM (2019). https://doi.org/10.1145/3331184.3331280

One Size Doesn't Fit All: Predicting the Number of Examples for In-Context Learning

Manish Chandra[(✉)], Debasis Ganguly, and Iadh Ounis

University of Glasgow, Glasgow, UK
m.chandra.1@research.gla.ac.uk,
{Debasis.Ganguly,iadh.ounis}@glasgow.ac.uk

Abstract. In-context learning (ICL) refers to the process of adding a small number of localized examples from a training set of labelled data to an LLM's prompt with an objective to effectively control the generative process seeking to improve the downstream task performance. Existing ICL approaches use an identical number of examples (a pre-configured hyper-parameter) for each data instance. Our work alleviates the limitations of this 'one fits all' approach by dynamically predicting the number of examples for each data instance to be used in few-shot inference with LLMs. In particular, we employ a multi-label classifier, the parameters of which are fitted using a training set, where the label for each instance in this training set indicates if using a specific value of k (number of most similar examples from 0 up to a maximum value) leads to correct k-shot downstream predictions. Our experiments on a number of text classification benchmarks show that AICL substantially outperforms standard ICL by up to 17%.

Keywords: Large Language Model · Adaptive In-context Learning · Multi-label Classification

1 Introduction

Large Language Models (LLMs) exhibit remarkable abilities to model text semantics in an abstract and general manner without any task specific training [35]. **In-context learning (ICL)**[1] makes use of this abstract representation and knowledge representation capabilities of LLMs [1,3,45] to address a range of different downstream tasks. For instance, such models can provide effective solutions for tasks including assessing reviews [28], answering questions [15], recommending relevant documents [33] etc., by using only a small number of (even zero) examples without any task-specific training.

[1] Also interchangeably known as few-shot learning or retrieval-augmented generation (RAG) with ground-truth labels.

C. Hauff et al. (Eds.): ECIR 2025, LNCS 15572, pp. 67–84, 2025.
https://doi.org/10.1007/978-3-031-88708-6_5

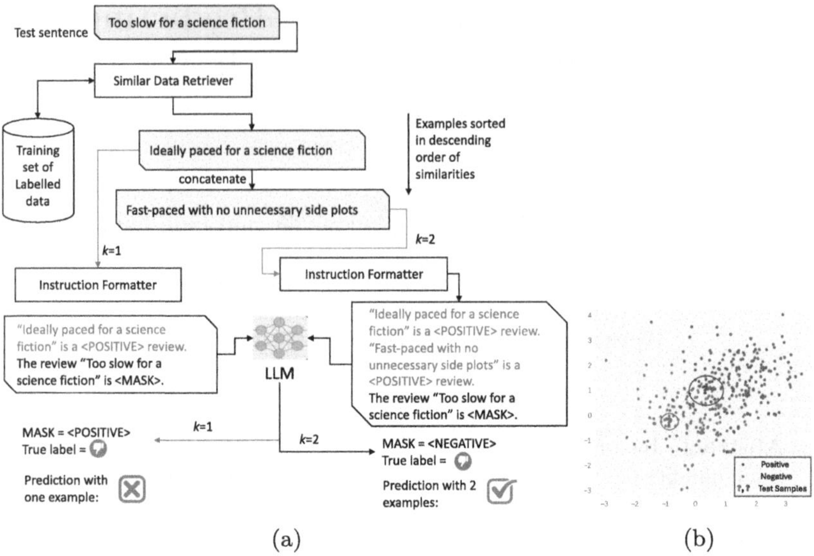

Fig. 1. a) *Example workflow of ICL for sentiment classification*: The example shows a test instance for which a single demonstration (as retrieved from the training set) does not result in a correct prediction (prediction workflow of the red arrows). It also shows that increasing the number of demonstrations from one to two results in a correct prediction (green arrows). We propose a method to estimate the number of examples that are likely to yield a correct prediction. **b)** *Motivation behind using a variable number of examples for ICL across the test instances*: The test sample '?' is located within a homogeneous neighborhood of negative data points indicating that an LLM may perform well with only a few nearest neighbour demonstrations. The test instance '?', on the other hand, is located within a heterogeneous neighborhood, as a result of which, an LLM may require a higher number of such demonstrations to correctly predict its class. (Color figure online)

More formally, ICL refers to the process of conditioning an LLM's decoder (frozen parameters) towards generating potentially relevant outputs that could then be interpreted in the context of a specific task, e.g., words such as 'great' and 'wow' can be mapped to a positive sentiment [38]. The output of an LLM's decoder is controlled by varying the input to the LLM, which is usually structured in the form of an instruction (the task description), and a small number of representative examples [29]. Figure 1a shows an example of how ICL works for the downstream task of movie review sentiment analysis. It can be seen that both demonstrations (shown in blue) and the current instance form a part of the input to an LLM. Researchers have worked towards adapting an ICL workflow in several ways, which range from using localised examples [19] instead of random ones, to diversifying these examples [13], and also ideally ordering them [12,37]. What prior work has not addressed so far is the effect of **dynamically selecting the number of examples** in an ICL workflow. Our idea takes motivation from information retrieval (IR), where different queries exhibit different levels of retrieval performance mainly due to the inherent characteristic of the information need itself [11], or how well-formulated the query is [5]. Drawing the parallel that a query in IR is analogous to a test instance in ICL [32] and that the localised examples are potentially relevant documents [37], we hypothesise that some test instances are associated with better candidates for training examples (i.e., examples which are useful in the sense that including them as a part of the prompt leads to correct predictions), and hence including a small number of these examples should be adequate. On the other hand, the retrieval quality for some test instances used as queries do not yield good candidates, as a result of which, one needs to further look down the ranked list to collect the useful ones [2,8]. The parallel with IR means that the notion of relevance of a document to a query needs to be replaced by the downstream usefulness of an example to a test instance.

The idea of choosing a variable number of localised examples is also somewhat similar to selecting a variable-sized neighborhood for k-NN classification [48], where the key idea is that a homogeneous neighborhood is likely to require a relatively small-sized neighborhood for a correct prediction, and a heterogeneous one would likely require a larger one (see Fig. 1b). The same idea can be applied to ICL, where a test instance that is similar to a number of training instances with conflicting labels may require a larger number of examples.

In our work, we apply a supervised learning based workflow to learn a mapping between the features of a test instance (embedding plus the label distribution of its neighborhood) and the ideal number of examples that should be used in ICL for the downstream prediction. More specifically, for each instance of the training set, we vary k - the number of ICL examples - within a range of 0 to a pre-configured maximum value (say M) and store each indicator of whether a particular value of k leads to a correct prediction as an $(M+1)$-length Boolean vector (e.g., the indicator for $k = 1$ in Fig. 1a is negative, whereas the one for $k = 2$ is positive). We then train a multi-label classifier on these pairs of instances and the Boolean indicators, the assumption being that similar instances with similar label distributions should also exhibit a similar distribution over the values of k leading to correct k-shot learning. During inference time, for each test instance, we apply the multi-label classifier to obtain a candidate list of predicted k values, and we select the one for which the prediction confidence is the highest. We then use these many examples for the ICL-based downstream prediction. This means that as per the schematics of Fig. 1a, we can potentially find out a 'green path' leading to a correct prediction for each test instance among several other 'red paths' that lead to incorrect ones.

2 Related Work

Prompt Tuning and Searching. LLMs, when scaled from millions to billions of parameters, have been demonstrated to be adaptable to a broad set of tasks due to instruction tuning [3,31], in the sense that they are not only able to produce semantically correct and coherent text, but are also able to adapt themselves surprisingly well with small changes in contexts supplied as inputs, commonly called prompts [1]. Previous research studies the problem of constructing an appropriate prompt for LLMs from two broad perspectives: a) *prompt tuning* in the embedding space [16,20,21,34], and b) *prompt searching* in the text space [6,19,22,40,47]. Prompt tuning is a lightweight alternative to fine-tuning, which keeps language model parameters frozen and instead optimizes a sequence of continuous task-specific vectors. The key idea of prompt tuning is to inject task-specific embedding into hidden layers and then tune these embeddings using gradient-based optimization. However, these methods require the modification of the original inference process of the model, which is impractical for the case of black-box LLM services, such as GPT3 [3]. Furthermore, prompt tuning introduces additional computational and storage costs, which is typically expensive for LLMs. A more efficient way is to optimize prompting via searching appropriate demonstration samples and ordering them in the original text space.

In-context Learning (ICL). ICL has advanced the use of LLMs for task-specific applications with minimal examples, forming a solid foundation for subsequent investigations into optimizing this learning paradigm [7]. In ICL, a small number of labeled examples from a training set are appended to a prompt instruction to control the text generation of LLM so that it is beneficial to a downstream task [14,28,30,33]. In addition to leveraging ICL for a purely generative task, e.g., question answering or abstractive summarisation [3,15,41], a more common use is in a predictive task, such as text classification [22,26,44], where each class is specified by a set of words, commonly called a verbaliser [39]. Once each class for a predictive task is well-defined, the generated text can be mapped to the most likely class by using the probabilities of the tokens generated by the decoder.

A crucial aspect of ICL is the adept selection and utilization of the demonstrations for task comprehension. Recent works have explored the dynamics of in-context examples, elucidating how ICL enhances the language model's performance across various NLP tasks by providing a minimal set of examples at inference time [10]. Furthermore, the effectiveness of ICL is significantly shaped by the strategies for selecting and ordering demonstrations. For instance, the work in [23] expanded the applicability of retrieval-based ICL approaches by demonstrating that even simple word-overlap similarity measures such as BM25 outperform randomly selected demonstrations. Methods such as KATE [19] also showed that localised examples work better than random ones. Different from KATE that uses a fixed-size neighbourhood for each data instance, we propose to modify the standard ICL workflow by employing a variable number of examples.

3 Proposed Methodology

3.1 Standard In-Context Learning (ICL)

In-context learning (ICL), unlike supervised learning, does not involve training a set of parameters on labeled examples. Rather, the posteriors are now a function

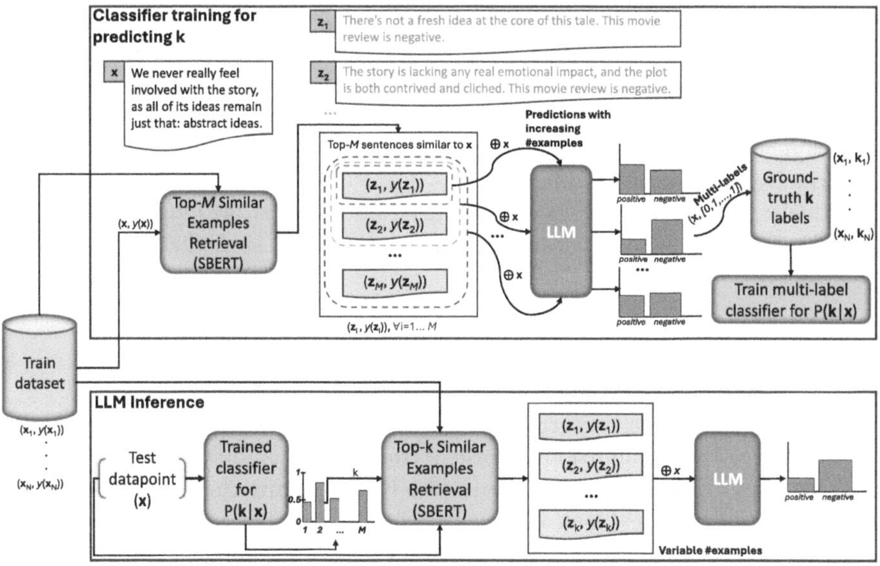

Fig. 2. Schematic diagram of Adaptive In-Context Learning (AICL) workflow.

of the following: a) text of the input test instance, b) the decoder parameters of a pre-trained LLM, c) a prompt instruction, and d) optionally, a set of k input examples, commonly called k-shot prompting [19]. Formally,

$$P(y|\mathbf{x}, k) = f(\mathbf{x}, \mathcal{N}_k(\mathbf{x}); \phi_{\text{LLM}}), \tag{1}$$

where, different from a supervised setup, the function f does not have a parameterized representation that can be learned using a training set with gradient descent. The function itself depends on the pre-trained frozen parameters ϕ_{LLM} of an LLM, the current inputs for which a label is to be predicted, and a prompt comprising a set of k text units denoted by $\mathcal{N}_k(\mathbf{x})$. This set $\mathcal{N}_k(\mathbf{x}) = \{\mathbf{z}_1, \ldots, \mathbf{z}_k\}$ of Eq. 1 (where each $\mathbf{z}_j \in \mathcal{T}$, a training set of labelled data instances) in ICL is usually comprised of localised examples, i.e., examples that are topically similar to the current instance [19,24]. Particularly for our experiments, we employ SBERT [36] as the neighborhood similarity computation function as per the findings of [19].

3.2 Adaptive ICL

We now describe our methodology that uses a variable number of examples by extending the standard ICL workflow of Eq. 1. We call our method 'Adaptive In-Context Learning', or **AICL** in short. The idea of AICL centres around choosing the context $\mathcal{N}_k(\mathbf{x})$ in a data-driven manner, i.e., making k a function of the data, i.e., the current instance $\mathbf{x}$ itself. This is analogous to choosing a different value of k for a k-NN based non-parametric model or choosing a different rank cut-off for top-retrieved set of documents for different queries [2,8]. The motivation is that classifying some instances would be more difficult than others, in which cases they are potentially to benefit from a larger value of k (more context). On the other hand, for relatively easier data instances, using too much context may be detrimental for an effective prediction.

Formally speaking, the difference of AICL with that of ICL (Eq. (1)) is that the value k, indicating the size of the neighborhood, is no longer a constant. Instead, we denote it by a parameterised function $\kappa(\mathbf{x})$ such that

$$P(y|\mathbf{x}, \kappa) = f(\mathbf{x}, \mathcal{N}_{\kappa(\mathbf{x})}(\mathbf{x}); \phi_{\text{LLM}}), \tag{2}$$

where $\kappa : \mathbf{x} \mapsto \{0, \ldots, M\}$, and M is an upper bound on the number of example instances.

Figure 2 presents an overarching view of the AICL workflow. In contrast to ICL, AICL involves an additional phase of training a classifier, $\kappa(\mathbf{x})$ of Eq. (2), to predict an appropriate number of examples by leveraging the training instances (the 'classifier training' block in Fig. 2). The 'LLM inference' block shows the inference phase, where, given a test instance, we first predict the number of examples to be used and then follow the standard ICL procedure with those many examples fed as part of the additional context to the LLM.

Obtaining the Ground-Truth Values of the Number of ICL Examples for Each Training Set Instance. For each training set instance $\mathbf{x} \in \mathcal{T}$, we first execute k-shot inference with progressively increasing values of k within $\{0, \ldots, M\}$, where M is a pre-configured threshold (specifically, $M = 10$ in our experiments). This is depicted by the progressively increasing colored dotted regions within the 'classifier training' block of Fig. 2. After obtaining $M + 1$ posteriors $P(\hat{y}|\mathbf{x})$ for an instance $\mathbf{x}$ of the training set, we construct an $(M + 1)$ dimensional Boolean vector, the ith component of which indicates if using an i number of ICL examples lead to a correct downstream prediction for the current instance $\mathbf{x}$ (note that since $\mathbf{x}$ is a training set instance, the knowledge of its true label $y(\mathbf{x})$ is available). Formally speaking,

$$\mathcal{K}_i(\mathbf{x}) = \mathbb{I}[y(\mathbf{x}) = \arg\max P(y|\mathbf{x}, i, \phi_{\text{LLM}})], \tag{3}$$

where $\mathcal{K}(\mathbf{x}) = \{\mathcal{K}_0(\mathbf{x}), \ldots, \mathcal{K}_M(\mathbf{x})\}$ is an $(M + 1)$ dimensional Boolean vector (note that we allow provision for the number of examples to be 0 as well, thus allowing provision for zero-shot prediction).

Training a Multi-label Classifier. As a next step, we learn a parameterized map from the inputs to the $M + 1$ dimensional indicators on the number of ICL examples, i.e.,

$$\theta : \mathbf{x} \mapsto \mathcal{K}(\mathbf{x}), \tag{4}$$

where $\mathcal{K}(\mathbf{x}) \in \{0, 1\}^{M+1}$ as defined in Eq. 3 and θ is a set of parameters learned via cross-entropy loss. Our decision to use a multi-label classification approach is motivated from a set of initial experiments, where we found that it outperformed a multi-class classifier approach (which we thus do not report in the paper). The most likely reason why a multi-label classifier works more effectively for this predictive task is that it allows provision to select the most likely candidate number of ICL examples from a number of alternatives rather than relying on a single predicted value of $\theta(\mathbf{x})$.

Of course, this also means that there needs to be a precise selection criteria to choose the number of examples to be used in the inference time from a list of choices, i.e., the result of the multi-label prediction where the predicted label is 1 (sigmoid posterior is higher than 0.5). We tried two different heuristics for this selection - i) choosing the smallest index from the $M+1$ dimensional vector with a predicted value of 1 (sigmoid $>$ 0.5), and ii) selecting the component with the highest confidence (sigmoid posterior). Again after conducting a set of initial experiments, we found that the second heuristics almost always outperforms the first one, and hence our workflow of Fig. 2 employs the max-confidence heuristic for predicting the number of ICL examples to be used during inference (see Eq. 2). More formally,

$$\kappa(\mathbf{x}) = \arg\max \theta(\mathbf{x}), \tag{5}$$

where $\theta(\mathbf{x}) \in [0, 1]^{M+1}$ denotes the posteriors (sigmoid values) obtained from a multi-label classifier trained with the labels as defined in Eq. 3.

Distribution of the Downstream-task Class Labels as Additional Features. The multi-label classifier $\theta : \mathbf{x} \mapsto \mathcal{K}(\mathbf{x})$ means that the number of ICL examples depend only on the document content, or in other words, topically similar content potentially requires a similar number of ICL examples for accurate predictions. However, this does not take into account the class label distribution, i.e., the $y(\mathbf{x})$ values of the examples. In fact, it is reasonable to assume that an instance with a more homogeneous neighborhood (in terms of the class label distribution) likely indicates an easier instance and hence may require a smaller number of examples, whereas a more heterogeneous neighborhood may indicate a larger number of examples (e.g., see Fig. 1b).

This hypothesis means that the use of neighborhood class distribution as additional features to encode each training data instance ($\mathbf{x} \in \mathcal{T}$) is likely to lead to a better predictor θ, which in turn, is likely to lead to better estimations of $\kappa(\mathbf{x})$ for AICL. Specifically, to incorporate the class prior information to aid predicting the number of ICL examples, we append an M dimensional vector $\boldsymbol{l} \in \mathbb{Z}_p^M$ (p is the number of class labels) to each SBERT embedded input vector $\mathbf{x} \in \mathcal{T}$. The ith component of this vector contains the class label of the ith neighbor in $\mathcal{N}_M(\mathbf{x})$, i.e., $\boldsymbol{l} = \{y(\mathbf{z}_1), \ldots, y(\mathbf{z}_M)\}$. With this modification, the multi-label classifier now learns:

$$\theta : \mathbf{x} \oplus \boldsymbol{l} \mapsto \mathcal{K}(\mathbf{x}), \tag{6}$$

where $\boldsymbol{l}$ denotes the class label distribution of M nearest neighbors, and $\oplus$ denotes the concatenation operator.

Table 1. The characteristics of the datasets used in our experiments.

Dataset	Task	#Classes	Document Type	Avg Len (#words)
SST2	Sentiment Analysis	2	Movie Reviews	19.3
TREC	Question classification	6	Open Domain Questions	10.2
CoLA	Grammatical Error Detection	2	News and Wikipedia	6.7
RTE	Language Inference	2	Miscellaneous	52.4

4 Evaluation

4.1 Research Questions and Datasets

In Sect. 3, we proposed a classifier-based approach to learn the optimal number of examples. In our experiments, we compare this approach for adaptive k-shot with fixed k-shot on standard datasets for text classification. In particular, we investigate the following research questions.

- **RQ-1**: Does adaptively selecting the number of examples in ICL (AICL) lead to improved downstream effectiveness?
- **RQ-2**: Do the class labels of the neighbouring training set examples act as useful features to improve AICL performance?
- **RQ-3**: Does AICL generalise well across different LLMs and datasets?
- **RQ-4**: Can the multi-label classification model (Eq. 6) be trained effectively on a small subsample of the training set, thus reducing the number of LLM calls during the training?

As per the setup of [25,46], we conducted our experiments on four standard text classification datasets, namely SST2, TREC [17], CoLA and RTE. Except TREC, the other three datasets are part of the GLUE benchmark [43] exhibiting diversity in: i) domain, e.g., from movie reviews to news, ii) number of classes, ranging from 2 to 6, iii) average length of the documents ranging from sentences (e.g., 6.7 words on an average in CoLA) to relatively large text (e.g., 52.4 words on an average in RTE). This diversity in the characteristics of the respective datasets (see Table 1) thus allows provision to assess the generalizability of our proposed approach. More details on each dataset is as follows.

- **SST2**: The Stanford Sentiment Treebank (SST) corpus consists of 11855 sentences extracted from movie reviews. The SST2 (also called SST-binary) dataset is a subset of SST, specifically prepared for the task of binary sentiment classification by discarding neutral sentences.
- **TREC**: The task is that of open-domain question classification of fact-based questions categorized into a total of 6 types [25], constituting 5500 labeled questions in the training set and 500 in the test set.
- **CoLA**: The Corpus of Linguistic Acceptability (CoLA) dataset consists of 10657 sentences from 23 linguistics publications annotated for grammatical correctness. It has a wide coverage of syntactic and semantic phenomena.

- **RTE**: The Recognizing Textual Entailment (RTE) dataset comes from a series of annual textual entailment challenges. The authors of the GLUE benchmark [43] combined the data from RTE1, RTE2, RTE3 and RTE5.

4.2 Methods Investigated

Baselines. We compare our proposed approach[2] with the following baselines:

- **0-shot**: This approach tests a model's innate capabilities, i.e., the capability to operate without any training data (information on class labels). Specifically, this requires setting $k = 0$ in Eq. 1.
- **FICL** [19] (Fixed ICL): This refers to the standard method of applying a static (fixed) number of localized (semantically similar) examples as input. The number of examples to use in FICL is determined by optimizing $k \in \{1, \ldots, 10\}$ by means of grid search on a validation set, which comprises 10% of the training set (the same training set is also used in AICL).

Variants of AICL. We employ the following variants for our proposed methodology of employing a variable number of ICL examples.

- **AICL with Embedding-only - AICL(E)**: This is an ablation study, where the number of examples for each data instance is only a function of its textual content - more specifically its SBERT [36] embedding (see Eq. 4).
- **AICL with Embedding + Neighborhood class labels - AICL(E+N)**: Similar to AICL(E) except that the input to the multi-label classifier is now the embedding of a training data instance along with the class labels of its neighbors (see Eq. 6).

We also analyze the performance of AICL under an ideal setting, i.e., when one can choose the best prediction from among a set of predictions obtained with different values of k in k-shot LLM-based inference by making use of the test set labels. Despite being non-pragmatic, this method provides an upper bound on the effectiveness that could, in principle, be obtained with the AICL framework. To avoid unfair comparisons of AICL* with AICL, we gray out the AICL* results indicating that these results are only to be used as an apex-line.

Corresponding to **RQ-4**, since constructing the ground-truth via Eq. 3 requires a total of $(M + 1)$ LLM calls for each training instance, for efficiency reasons, we first train the multi-label classifier of the AICL workflow only on a subset of training data obtained by randomly sampling 50% of data from each class. Later on, as post-hoc analysis towards answering **RQ-4**, we also investigate if AICL results further improve (or degrade) with larger (or smaller) subsets sampled from the training data.

[2] Code available at https://github.com/ManishChandra12/adaptiveICL.

4.3 Model and Hyper-Parameter Settings

To learn the number of examples to be used during inference, the AICL workflow relies on training the $\kappa(\mathbf{x})$ predictor (multi-label classifier of Eqs. 4 and 6), which in turn, depends on the ground-truth labels ($\mathcal{K}$ of Eq. 3) of the number of examples that lead to correct predictions via LLM-based few-shot inference. But this means that these ground-truth labels depend on a particular choice of the LLM that is used during the training phase to check the accuracy of the downstream task labels (see Eq. 3).

To analyze the variations that may be caused due to this choice of LLMs, we conduct our experiments on three different LLMs - two each from the same family with different sizes, and one from a different family. The objective is to analyze the effects of variations in AICL's performance corresponding to the size of a model from the same family, or variations in characteristics in models across different families. In particular, as instances of LLMs from the same family we use Meta's Llama-2-7b and Llama-2-13b [42], whereas as another LLM from a different family, we use Microsoft's Phi-2 [18]. Details on these models are as follows: Llama-2 is a collection of open-source pretrained and fine-tuned text models with a maximum context length of 4096 tokens (we use the 7B and the 13B parameter models). Phi-2 [18] is a transformer with 2.7 billion parameters. It is trained on documents comprising Python code, StackOverflow, websites and natural language text generated by GPT-3.5. Phi-2 with less than 3B parameters has been shown to yield comparable performance with those achieved with much larger number of parameter, e.g., the 13B variant of Llama-2. The maximum input length of Phi-2 is 2048 tokens.

In our experiments, we set M (the maximum number of ICL examples) to 10 (see Eq. 3). Furthermore, for training the multi-label classifier parameters $\kappa(\mathbf{x})$ of Eqs. 4 and 6, we choose a single hidden layer network with 64 neurons. These decision choices are based on initial experiments of hyper-parameter tuning conducted on the SST2 dataset with the Llama-2-7b model.

4.4 Results

Main Observations. Table 2 shows a comparison between different ICL approaches on the four datasets. The reported precision, recall and F-scores are averaged over 5 runs of the experiments. The number of examples in FICL, as shown in the corresponding column, were obtained by a grid search over k from 1 to 10 on a validation set. From Table 2, it can be seen that AICL(E+N) turns out to be the best among the competing approaches - this method being statistically distinguishable than the FICL results (McNemar's test with 95% confidence interval). The likely reason it outperforms the baselines is that AICL is able to effectively adapt the number of examples to use, thereby preventing itself from the degradation effects of non-relevant (not useful) examples. In effect, it learns a latent relationship between the topical content and the quantity of context required to guide the decoder's output towards improving a downstream task. Our observations reveal that **RQ-1** is answered in the affirmative,

Table 2. A comparison between the two variants of our proposed Adaptive ICL approach and the baselines on the 4 datasets used in our experiments. The column 'k' denotes the number of few-shot examples. In particular, for AICL, this denotes the average of the number of examples used for each test instance. In our experiments, the maximum value of k (i.e., M of Eq. 3) is set to 10 for both AICL and FICL. We found that the AICL(E+N) results are statistically distinguishable with respect to the FICL results (McNemar's test with $p = 95\%$ confidence interval).

		Llama-2-7b				Phi-2				Llama-2-13b			
	Method	k	Prec	Rec	F1	k	Prec	Rec	F1	k	Prec	Rec	F1
SST2	0-shot	0	.7227	.6123	.5589	0	.8988	.8919	.8914	0	.7341	.6443	.5891
	FICL	9	.8744	.8687	.8682	10	.9279	.9253	.9252	9	.8912	.8733	.8809
	AICL(E)	8.63	.9089	.8946	.8925	6.39	.9307	.9300	.9300	5.73	.9175	.9065	.9028
	AICL(E+N)	8.23	**.9099**	**.8964**	**.8954**	5.64	**.9350**	**.9345**	**.9345**	4.89	**.9189**	**.9071**	**.9034**
	AICL*	1.27	.9765	.9753	.9753	.17	.9865	.9863	.9863	.84	.9866	.9859	.9848
TREC	0-shot	0	.1697	.1687	.0100	0	.6289	.4273	.3526	0	.3212	.3685	.3016
	FICL	9	.7388	.7324	.6608	8	.6842	.6582	.6192	9	.7612	.7922	.7071
	AICL(E)	9.60	.7654	.8049	.7325	4.72	.7146	.7549	.7196	5.51	.7659	.8218	.7383
	AICL(E+N)	9.58	**.7682**	**.8075**	**.7357**	5.57	**.7254**	**.7673**	**.7291**	4.93	**.7737**	**.8288**	**.7532**
	AICL*	3.35	.8496	.9334	.8616	1.27	.9337	.9294	.9313	2.25	.9513	.9367	.9413
CoLA	0-shot	0	.5585	.5323	.3699	0	.4315	.4899	.2558	0	.6474	.4321	.3455
	FICL	10	.6240	.6415	.6008	9	.7071	.6306	.6433	8	.7167	.6338	.6472
	AICL(E)	8.32	.6556	.6741	.6582	7.73	.7289	.6471	.6601	2.33	.7412	.6447	.6667
	AICL(E+N)	7.43	**.6580**	**.6765**	**.6616**	7.93	**.7392**	**.6486**	**.6613**	1.99	**.7432**	**.6558**	**.6714**
	AICL*	2.42	.9417	.9281	.9299	.69	.9663	.9232	.9413	.78	.9408	.9466	.9483
RTE	0-shot	0	.4913	.4992	.3985	0	.6741	.6741	.6741	0	.5345	.5444	.4403
	FICL	10	.6095	.6051	.5967	1	.7347	.7239	.7240	4	.6287	.6233	.6214
	AICL(E)	8.57	.6304	.6277	.6227	2.44	.7536	.7412	.7415	5.16	.7592	.6441	.6700
	AICL(E+N)	8.32	**.6342**	**.6311**	**.6252**	1.15	**.7551**	**.7465**	**.7471**	3.68	**.7631**	**.6485**	**.6738**
	AICL*	.95	.9783	.9783	.9783	.81	.9296	.9214	.9234	.65	.9734	.9733	.9733

i.e., an adaptive selection of the number of examples in ICL does improve the downstream effectiveness.

For the AICL approaches, we report the average number of examples, as predicted by the output of the multi-label classifier (Eqs. 4 and 6), in the column named k. In contrast to the grid-searched integer k values in FICL, for AICL the number of examples (k) is an average over the test instances, and hence is a non-integer.

One interesting observation is that the oracle version of AICL (AICL*) performs substantially better than AICL with much smaller values of k, which testifies the fact that selecting the correct value of k independently for each test instance can lead to large improvements in few-shot prediction effectiveness. This also means that there is a scope to further improve the AICL workflow.

Sensitivity Analysis of FICL (Fixed-k ICL). Figure 3 shows the F1-scores of FICL on the test splits of the different datasets with a varying number of demonstrations. Firstly, it can be observed from Fig. 3 that there is no consistent improvement in the performance of FICL with an increase in k. Since the FICL results are sensitive to k, it is difficult to choose a particular value of k for a given downstream task because such a grid search is not a practical solution as the ground-truth is not supposed to be known priority. We already showed in Table 2 that optimizing k by grid search on a validation set produces worse results than AICL. Additionally, Fig. 3 shows that the AICL results also outperform FICL with k optimized on the test set. This indicates that AICL can, in principle, be deployed on any downstream task without requiring any hyper-parameter optimization of k.

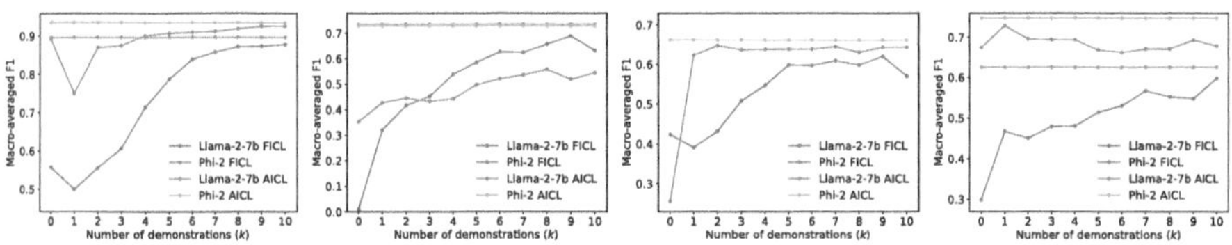

(a) SST2 (b) TREC (c) CoLA (d) RTE

Fig. 3. FICL (macro-averaged) F-scores for different context sizes on the test splits of the different datasets (AICL results also included for comparison). These plots demonstrate that AICL can be applied on any dataset without requiring to optimize any hyper-parameter (e.g., k in FICL).

Effect of Neighbourhood. Among the two variants of AICL, Table 2 shows that the use of neighbor class labels (the 'E+N' variant) yields better results than the ablated ('E' variant), which means that **RQ-2** is answered in affirmative. In relation to **RQ-3**, Table 2 shows that the improvements in AICL with respect to FICL is consistent across different datasets and also across different LLMs.

Multi-label Classifier in AICL with Reduced Training Data. It is evident from Table 2 that for each dataset, the best value of k in FICL differs from one LLM to another. This means to construct the correct ground-truth for training the multi-label classifier in AICL (Eq. 6), $M+1$ LLM predictions need to be made for each instance, which means that the total number of LLM invocations is $|\mathcal{T}|(M+1)$, where $\mathcal{T}$ denotes the training set.

A solution to reduce training time is to use a small subsample of the training data to learn the parameters of Eq. 6. Figure 4 shows the variations in F1 scores for different proportions of the training set used to learn $\kappa(\mathbf{x})$. While instance selection based sampling methods have been proposed in the literature [4], for

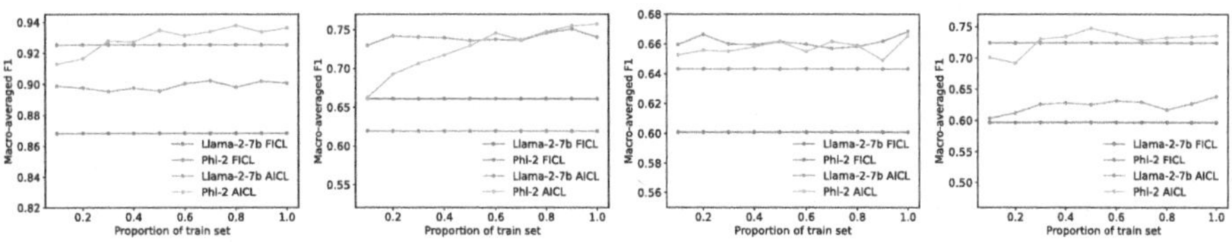

(a) SST2 (b) TREC (c) CoLA (d) RTE

Fig. 4. Macro-averaged F1 scores with different proportions of the training set used to train $\kappa(\mathbf{x})$ - the multi-label classifier of AICL (Eq. 6).

this work we simply use uniform sampling and report the results averaged over 5 runs of the experiments.

An interesting observation is that AICL is able to outperform the baseline with as small a proportion as 30% (or even less in some cases) of the training set being used to train $\kappa(\mathbf{x})$, which means that **RQ-4** is answered in the affirmative.

Sensitivity to Labels. We now analyze how AICL behaves in the absence of any labelled training data [27] - a setup commonly known as retrieval-augmented generation (RAG) [9]. A way to simulate this configuration in our experiment setup is to keep only the content and exclude the labels from the demonstrations added to the prompt. Equation 3, in its current form, cannot be used to get all the numbers of examples that work, because of the unavailability of labels. We therefore modify Eq. 3 to set $\mathcal{K}_i(\mathbf{x}) = 1$, where $0 \leq i \leq M$ denotes the number of examples that lead to the maximum mutual information of the Softmax class posteriors, as obtained with i-shot LLM output. The rest of the components of $\mathcal{K}_i(\mathbf{x})$ are set to 0, which means that instead of a multi-label classification setting, for the unlabeled setup the k predictor is now a multi-class classifier. Also note that due to the unavailability of class labels, the AICL(E+N) version (Eq. 6) cannot be executed in this RAG setup.

Table 3. A comparison between the F1 scores obtained by AICL (Phi-2) on two different settings - a) w/ labels (same as Table 2), and b) w/o labels, as in a RAG pipeline. Best performing RAG results are bold-faced.

		RAG setup (w/o Labels)			ICL setup (w/ Labels)		
Dataset	0-shot	FICL	AICL(E)	AICL*	FICL	AICL(E)	AICL*
SST2	.8914	.7339	**.9119**	.9610	.9252	.9300	.9863
TREC	.3526	.4287	**.4752**	.4922	.6192	.7196	.9313
CoLA	.2558	.2469	**.2679**	.7937	.6433	.6601	.9413
RTE	**.6741**	.6144	.6688	.8655	.7240	.7415	.9234

The results of these additional experiments, carried out on the Phi-2 model, are shown in Table 3. From Table 3, it can be seen that absence of labels hurt performance - however, not by large margins, which is consistent with the findings of [27]. AICL still continues to mostly be the best performing method even without the presence of any class labels. The only dataset where AICL does not yield the best performance is RTE, where the best performance is achieved with 0-shot indicating that labels are crucial for this dataset, and hence all RAG approaches fail on this particular dataset. It is encouraging to see, however, that AICL* (the oracle setting of AICL) without labels still continues to be substantially better than FICL (ICL with fixed context sizes).

Concluding Remarks. To improve the effectiveness of LLM-based few-shot inference, we proposed a supervised multi-label classification approach that learns how many examples to use for which data instance. The classifier itself is trained on a set of labeled examples, where the ground-truth for each training data instance constitutes the set of k (number of examples) that lead to correct downstream task prediction via k-shot LLM inference.

Disclosure of Interests. The authors have no competing interests to declare that are relevant to the content of this article.

References

1. Arora, S., et al.: Ask me anything: a simple strategy for prompting language models (2022)
2. Bahri, D., Tay, Y., Zheng, C., Metzler, D., Tomkins, A.: Choppy: cut transformer for ranked list truncation. In: Proceedings of the 43rd International ACM SIGIR Conference on Research and Development in Information Retrieval, pp. 1513-1516. SIGIR '20, Association for Computing Machinery, New York, NY, USA (2020). https://doi.org/10.1145/3397271.3401188
3. Brown, T., et al.: Language models are few-shot learners. In: Larochelle, H., Ranzato, M., Hadsell, R., Balcan, M., Lin, H. (eds.) Advances in Neural Information Processing Systems, vol. 33, pp. 1877–1901. Curran Associates, Inc. (2020). https://proceedings.neurips.cc/paper_files/paper/2020/file/1457c0d6bfcb4967418bfb8ac142f64a-Paper.pdf
4. Cunha, W., Viegas, F., França, C., Rosa, T., Rocha, L., Gonçalves, M.A.: A comparative survey of instance selection methods applied to non-neural and transformer-based text classification. ACM Comput. Surv. **55**(13s) (2023). https://doi.org/10.1145/3582000
5. Datta, S., Ganguly, D., Greene, D., Mitra, M.: Deep-QPP: a pairwise interaction-based deep learning model for supervised query performance prediction. In: Candan, K.S., Liu, H., Akoglu, L., Dong, X.L., Tang, J. (eds.) WSDM '22: The Fifteenth ACM International Conference on Web Search and Data Mining, Virtual Event / Tempe, AZ, USA, February 21 - 25, 2022, pp. 201–209. ACM (2022). https://doi.org/10.1145/3488560.3498491
6. Diao, S., Wang, P., Lin, Y., Zhang, T.: Active prompting with chain-of-thought for large language models (2023)
7. Dong, Q., et al.: A survey for in-context learning. arXiv preprint arXiv:2301.00234 (2022)

8. Ganguly, D., Yilmaz, E.: Query-specific variable depth pooling via query performance prediction. In: SIGIR, pp. 2303–2307. ACM (2023)
9. Gao, Y., et al.: Retrieval-augmented generation for large language models: a survey (2024). https://arxiv.org/abs/2312.10997
10. Han, X., Simig, D., Mihaylov, T., Tsvetkov, Y., Celikyilmaz, A., Wang, T.: Understanding in-context learning via supportive pretraining data. arXiv preprint arXiv:2306.15091 (2023)
11. Kanoulas, E., Carterette, B., Clough, P.D., Sanderson, M.: Evaluating multi-query sessions. In: Proceedings of the 34th International ACM SIGIR Conference on Research and Development in Information Retrieval, pp. 1053–1062. SIGIR '11, Association for Computing Machinery, New York, NY, USA (2011). https://doi.org/10.1145/2009916.2010056
12. Kumar, S., Talukdar, P.P.: Reordering examples helps during priming-based few-shot learning. In: ACL/IJCNLP (Findings). Findings of ACL, vol. ACL/IJCNLP 2021, pp. 4507–4518. Association for Computational Linguistics (2021)
13. Levy, I., Bogin, B., Berant, J.: Diverse demonstrations improve in-context compositional generalization. In: Proceedings of the 61st Annual Meeting of the Association for Computational Linguistics (Volume 1: Long Papers), pp. 1401–1422. Association for Computational Linguistics, Toronto, Canada (2023). https://doi.org/10.18653/v1/2023.acl-long.78, https://aclanthology.org/2023.acl-long.78
14. Li, M., Ma, X., Lin, J.: An encoder attribution analysis for dense passage retriever in open-domain question answering. In: Proceedings of the 2nd Workshop on Trustworthy Natural Language Processing (TrustNLP 2022), pp. 1–11. Association for Computational Linguistics, Seattle, U.S.A. (2022). https://doi.org/10.18653/v1/2022.trustnlp-1.1, https://aclanthology.org/2022.trustnlp-1.1
15. Li, T., Ma, X., Zhuang, A., Gu, Y., Su, Y., Chen, W.: Few-shot in-context learning on knowledge base question answering. In: Proceedings of the 61st Annual Meeting of the Association for Computational Linguistics (Volume 1: Long Papers), pp. 6966–6980. Association for Computational Linguistics, Toronto, Canada (2023). https://doi.org/10.18653/v1/2023.acl-long.385, https://aclanthology.org/2023.acl-long.385
16. Li, X.L., Liang, P.: Prefix-Tuning: optimizing continuous prompts for generation. In: Proceedings of the 59th Annual Meeting of the Association for Computational Linguistics and the 11th International Joint Conference on Natural Language Processing (Volume 1: Long Papers), pp. 4582–4597. Association for Computational Linguistics, Online (2021). https://doi.org/10.18653/v1/2021.acl-long.353, https://aclanthology.org/2021.acl-long.353
17. Li, X., Roth, D.: Learning question classifiers. In: Proceedings of the 19th International Conference on Computational Linguistics, vol. 1, pp. 1–7. COLING '02, Association for Computational Linguistics, USA (2002). https://doi.org/10.3115/1072228.1072378
18. Li, Y., Bubeck, S., Eldan, R., Del Giorno, A., Gunasekar, S., Lee, Y.T.: Textbooks are all you need ii: phi-1.5 technical report. arXiv preprint arXiv:2309.05463 (2023)
19. Liu, J., Shen, D., Zhang, Y., Dolan, B., Carin, L., Chen, W.: What makes good in-context examples for GPT-3? In: Proceedings of Deep Learning Inside Out (DeeLIO 2022): The 3rd Workshop on Knowledge Extraction and Integration for Deep Learning Architectures, pp. 100–114. Association for Computational Linguistics, Dublin, Ireland and Online (2022). https://doi.org/10.18653/v1/2022.deelio-1.10, https://aclanthology.org/2022.deelio-1.10

20. Liu, X., et al.: P-Tuning: prompt tuning can be comparable to fine-tuning across scales and tasks. In: Proceedings of the 60th Annual Meeting of the Association for Computational Linguistics (Volume 2: Short Papers), pp. 61–68. Association for Computational Linguistics, Dublin, Ireland (2022). https://doi.org/10.18653/v1/2022.acl-short.8, https://aclanthology.org/2022.acl-short.8
21. Liu, X., et al.: GPT understands, too. AI Open (2023). https://doi.org/10.1016/j.aiopen.2023.08.012, https://www.sciencedirect.com/science/article/pii/S2666651023000141
22. Lu, Y., Bartolo, M., Moore, A., Riedel, S., Stenetorp, P.: Fantastically ordered prompts and where to find them: overcoming few-shot prompt order sensitivity. In: Proceedings of the 60th Annual Meeting of the Association for Computational Linguistics (Volume 1: Long Papers), pp. 8086–8098. Association for Computational Linguistics, Dublin, Ireland (2022). https://doi.org/10.18653/v1/2022.acl-long.556, https://aclanthology.org/2022.acl-long.556
23. Luo, M., et al.: DR. ICL: Demonstration-retrieved in-context learning. arXiv preprint arXiv:2305.14128 (2023)
24. Luo, M., Xu, X., Liu, Y., Pasupat, P., Kazemi, M.: In-context learning with retrieved demonstrations for language models: a survey (2024)
25. Ma, H., et al.: Fairness-guided few-shot prompting for large language models. In: Thirty-seventh Conference on Neural Information Processing Systems (NeurIPS) (2023)
26. Milios, A., Reddy, S., Bahdanau, D.: In-context learning for text classification with many labels. In: Hupkes, D., et al. (eds.) Proceedings of the 1st GenBench Workshop on (Benchmarking) Generalisation in NLP, pp. 173–184. Association for Computational Linguistics, Singapore (Dec 2023). https://doi.org/10.18653/v1/2023.genbench-1.14, https://aclanthology.org/2023.genbench-1.14
27. Min, S., et al.: Rethinking the role of demonstrations: what makes in-context learning work? In: EMNLP (2022)
28. Mysore, S., Mccallum, A., Zamani, H.: Large language model augmented narrative driven recommendations. In: Association for Computing Machinery, pp. 777–783. RecSys '23, New York, NY, USA (2023). https://doi.org/10.1145/3604915.3608829
29. Ni, J., et al.: Large dual encoders are generalizable retrievers. In: Proceedings of the 2022 Conference on Empirical Methods in Natural Language Processing, pp. 9844–9855. Association for Computational Linguistics, Abu Dhabi, United Arab Emirates (2022). https://doi.org/10.18653/v1/2022.emnlp-main.669, https://aclanthology.org/2022.emnlp-main.669
30. Ni, J., et al.: Large dual encoders are generalizable retrievers (2021)
31. Ouyang, L., et al.: Training language models to follow instructions with human feedback. In: Koyejo, S., Mohamed, S., Agarwal, A., Belgrave, D., Cho, K., Oh, A. (eds.) Advances in Neural Information Processing Systems, vol. 35, pp. 27730–27744. Curran Associates, Inc. (2022). https://proceedings.neurips.cc/paper_files/paper/2022/file/b1efde53be364a73914f58805a001731-Paper-Conference.pdf
32. Parry, A., Ganguly, D., Chandra, M.: In-context learning or: how i learned to stop worrying and love "applied information retrieval". In: Proceedings of the 47th International ACM SIGIR Conference on Research and Development in Information Retrieval, pp. 14–25. SIGIR '24, Association for Computing Machinery, New York, NY, USA (2024). https://doi.org/10.1145/3626772.3657842, https://doi.org/10.1145/3626772.3657842
33. Pradeep, R., et al.: How does generative retrieval scale to millions of passages? (2023)

34. Qin, G., Eisner, J.: Learning how to ask: querying LMs with mixtures of soft prompts. In: Proceedings of the 2021 Conference of the North American Chapter of the Association for Computational Linguistics: Human Language Technologies, pp. 5203–5212. Association for Computational Linguistics, Online (2021). https://doi.org/10.18653/v1/2021.naacl-main.410, https://aclanthology.org/2021.naacl-main.410
35. Radford, A., Narasimhan, K., Salimans, T., Sutskever, I.: Improving language understanding with unsupervised learning. OpenAI (2018). https://openai.com/research/language-unsupervised
36. Reimers, N., Gurevych, I.: Sentence-BERT: sentence embeddings using Siamese BERT-networks. In: Inui, K., Jiang, J., Ng, V., Wan, X. (eds.) Proceedings of the 2019 Conference on Empirical Methods in Natural Language Processing and the 9th International Joint Conference on Natural Language Processing (EMNLP-IJCNLP), pp. 3982–3992. Association for Computational Linguistics, Hong Kong, China (2019). https://doi.org/10.18653/v1/D19-1410, https://aclanthology.org/D19-1410
37. Rubin, O., Herzig, J., Berant, J.: Learning to retrieve prompts for in-context learning. In: Proceedings of the 2022 Conference of the North American Chapter of the Association for Computational Linguistics: Human Language Technologies, pp. 2655–2671. Association for Computational Linguistics, Seattle, United States (2022). https://doi.org/10.18653/v1/2022.naacl-main.191, https://aclanthology.org/2022.naacl-main.191
38. Schick, T., Schmid, H., Schütze, H.: Automatically identifying words that can serve as labels for few-shot text classification. In: Proceedings of the 28th International Conference on Computational Linguistics, pp. 5569–5578. International Committee on Computational Linguistics, Barcelona, Spain (Online) (2020). https://doi.org/10.18653/v1/2020.coling-main.488, https://aclanthology.org/2020.coling-main.488
39. Schick, T., Schütze, H.: Exploiting cloze-questions for few-shot text classification and natural language inference. In: Proceedings of the 16th Conference of the European Chapter of the Association for Computational Linguistics: Main Volume, pp. 255–269. Association for Computational Linguistics, Online (2021). https://doi.org/10.18653/v1/2021.eacl-main.20, https://aclanthology.org/2021.eacl-main.20
40. Shi, F., et al.: Large language models can be easily distracted by irrelevant context. In: Proceedings of the 40th International Conference on Machine Learning. ICML'23, JMLR.org (2023)
41. Tang, Y., Puduppully, R., Liu, Z., Chen, N.: In-context learning of large language models for controlled dialogue summarization: a holistic benchmark and empirical analysis. In: Dong, Y., Xiao, W., Wang, L., Liu, F., Carenini, G. (eds.) Proceedings of the 4th New Frontiers in Summarization Workshop, pp. 56–67. Association for Computational Linguistics, Singapore (2023). https://doi.org/10.18653/v1/2023.newsum-1.6, https://aclanthology.org/2023.newsum-1.6
42. Touvron, H., et al.: Llama 2: open foundation and fine-tuned chat models (2023)
43. Wang, A., Singh, A., Michael, J., Hill, F., Levy, O., Bowman, S.: GLUE: a multi-task benchmark and analysis platform for natural language understanding. In: Linzen, T., Chrupała, G., Alishahi, A. (eds.) Proceedings of the 2018 EMNLP Workshop BlackboxNLP: Analyzing and Interpreting Neural Networks for NLP, pp. 353–355. Association for Computational Linguistics, Brussels, Belgium (2018). https://doi.org/10.18653/v1/W18-5446, https://aclanthology.org/W18-5446

44. Wei, J., Huang, C., Vosoughi, S., Cheng, Y., Xu, S.: Few-shot text classification with triplet networks, data augmentation, and curriculum learning. arXiv preprint arXiv:2103.07552 (2021)
45. Weidinger, L., et al.: Taxonomy of risks posed by language models. In: 2022 ACM Conference on Fairness, Accountability, and Transparency, pp. 214–229 (2022)
46. Zhang, Y., Feng, S., Tan, C.: Active example selection for in-context learning. In: Proceedings of the 2022 Conference on Empirical Methods in Natural Language Processing, pp. 9134–9148. Association for Computational Linguistics, Abu Dhabi, United Arab Emirates (2022). https://doi.org/10.18653/v1/2022.emnlp-main.622, https://aclanthology.org/2022.emnlp-main.622
47. Zhang, Z., Zhang, A., Li, M., Smola, A.: Automatic chain of thought prompting in large language models. In: The Eleventh International Conference on Learning Representations (ICLR 2023) (2023)
48. Zhong, X.F., Guo, S.Z., Gao, L., Shan, H., Zheng, J.H.: An improved K-NN classification with dynamic k. In: Proceedings of the 9th International Conference on Machine Learning and Computing, pp. 211–216. ICMLC '17, Association for Computing Machinery, New York, NY, USA (2017). https://doi.org/10.1145/3055635.3056604

MURR: Model Updating with Regularized Replay for Searching a Document Stream

Eugene Yang[1(✉)], Nicola Tonellotto[2], Dawn Lawrie[1], Sean MacAvaney[3], James Mayfield[1], Douglas W. Oard[1,4], and Scott Miller[5]

[1] HLTCOE, Johns Hopkins University, Baltimore, MD, USA
{eugene.yang,lawrie,mayfield}@jhu.edu
[2] University of Pisa, Pisa, Italy
nicola.tonellotto@unipi.it
[3] University of Glasgow, Glasgow, UK
sean.macavaney@glasgow.ac.uk
[4] University of Maryland, College Park, MD, USA
oard@umd.edu
[5] Information Sciences Institute, University of Southern California, Marina del Rey, CA, USA
smiller@isi.edu

Abstract. The Internet produces a continuous stream of new documents and user-generated queries. These naturally change over time based on events in the world and the evolution of language. Neural retrieval models that were trained once on a fixed set of query-document pairs will quickly start misrepresenting newly-created content and queries, leading to less effective retrieval. Traditional statistical sparse retrieval can update collection statistics to reflect these changes in the use of language in documents and queries. In contrast, continued fine-tuning of the language model underlying neural retrieval approaches such as DPR and ColBERT creates incompatibility with previously-encoded documents. Re-encoding and re-indexing all previously-processed documents can be costly. In this work, we explore updating a neural dual encoder retrieval model without reprocessing past documents in the stream. We propose MURR, a model updating strategy with regularized replay, to ensure the model can still faithfully search existing documents without reprocessing, while continuing to update the model for the latest topics. In our simulated streaming environments, we show that fine-tuning models using MURR leads to more effective and more consistent retrieval results than other strategies as the stream of documents and queries progresses.

Keywords: Streaming · Model Updating · Memory Replay

C. Hauff et al. (Eds.): ECIR 2025, LNCS 15572, pp. 85–100, 2025.
https://doi.org/10.1007/978-3-031-88708-6_6

1 Introduction

As a large amount of content is created every day, search engines that aim to serve queries targeting old, new, or even emerging topics need to incorporate new documents continuously. Such a *stream of documents* can address a fluid set of topics, and in news, for example, we would expect that documents created at different times would discuss different issues. Public-facing search engines serving ad hoc queries, such as Baidu, Bing or Google, need to account not just for topic changes in their document streams, but also topical changes in their query streams over time [15,39]. For example, a query of "Harris" in 2024 likely targets a different "Harris" than was being searched for in 2014. Fine-tuning a query-document relevance model—such as a BERT cross-encoder [30] or a ColBERT dual encoder [19]—only once would not work well in such cases; some approach to periodic, incremental, or continuous updating is needed to reflect these changes.

These kinds of distributional shifts are less of a problem for earlier lexical retrieval systems, such as BM25, which can easily update their statistics as term distributions change; they rely on directly-observable term statistics such as document frequency or (for unigram language models) collection frequency. As a result, lexical retrieval systems such as Lucene can easily incorporate emerging topics in a stream of documents. However, providing similar capabilities for dense retrieval systems, such as those that use FAISS [17], PLAID's cluster-based dot product approximation [33], or other nearest-neighbor indexes, is much more challenging. Popular approximate nearest neighbor (ANN) search algorithms (e.g., HNSW [27] and Product Quantization [16])) support fast vector searches and allow incremental ingestion of vectors to an index. However, the parameters of the models that provide the vectors require training to reflect distributional changes. Given that these models produce vectors in an arbitrary latent space, an updated model is not guaranteed to be compatible with vectors from another model (though they can be moderately compatible by chance [22]). Therefore, existing approaches either need to use the same model over the lifespan of a dense index (limiting effectiveness) or re-index the entire collection when a new model is trained (a costly endeavor). We address these limitations in this work.

There has been some work on this problem of nonstationary query and collection statistics under dense retrieval in which the document and query encoders are fixed, and adaptation to a new domain or genre uses few-shot training [36]. In this paper, we broaden that perspective to explore model updating strategies that adapt to emerging topics through larger-scale fine-tuning on newly-collected training triples. Since models trained on different training examples provide no guarantees that existing documents will be represented using similar vectors, we also study the effect of distributional shifts on our ability to retrieve older documents without reindexing.

In practical applications, we expect that search engines might use implicit feedback to obtain the training signals that are needed to support model updating in the presence of changing distributional statistics [32]. In this work, we simulate document and query streams from an existing multi-domain retrieval collection to experiment with various streaming scenarios.

We introduce MURR, a model updating strategy with regularized training data replay on their document representations. By assuming distributional shifts between known domains, we can compare the effect of training data replay against that of representation regularization. We show that dense retrieval models trained with MURR are more effective than models without replay or representation regularization, and that they provide more consistent retrieval effectiveness to the same set of queries as distributional statistics change.

Our contributions are threefold: (1) an effective model updating strategy that does not require reprocessing existing documents, (2) a set of simulated query and document streams, and (3) experimental results demonstrating MURR's effectiveness and stability.

2 Background and Related Work

Neural retrieval models are generally based on a pretrained language model, such as BERT [4] or RoBERTa [24], and further fine-tuned with a retrieval objective, such as contrastive [2,31] or distillation loss [6,23,34], on a set of queries and documents, such as MS MARCO [28]. For dual encoders, because of their architecture, queries and documents can be processed separately and only require a lightweight interaction, such as vector dot product [7,18] or token level MaxSim [19]; documents can thus be preprocessed and indexed for supporting fast searches [17,33]. How well the language model is tuned for the specific query and document language distributions is crucial for designing an effective dense retrieval model [20].

Since proper training data is difficult to collect and the fine-tuning process is also computationally expensive, prior work in ad hoc retrieval typically aims for a one-size-fits-all general-purpose model and evaluates its robustness on various retrieval domains and tasks [6,29,38]. This perspective drove the creation of the BEIR benchmark [36]. Work on domain adaptation either aims for better zero-shot ability [3,35,41], which is the same as developing a general-purpose model, or it exploits additional unsupervised or semi-supervised fine-tuning [9,14,37]. However, with a stream of queries and documents arriving, even if one can afford to continuously fine-tune the retrieval model, the risk of catastrophic forgetting increases [12,25], which destroys the purpose of adaptation.

As mentioned in the previous section, statistical sparse retrieval models such as BM25 are capable of incrementally updating the index and related statistics with some careful lexical feature selection to account for the language drift over time [5,26]. Early work around TREC filtering task, which dates back to TREC 4 [13], also requires monitoring the lexical differences over time for serving standing queries [1]. In contrast, there is little work on neural dual encoders in such a streaming scenario. Recent work by Lawrie et al. [21] investigated incrementally indexing a large corpus using PLAID-X [40] with a fixed model, but only evaluated on a fixed set of queries at the end of the stream. How the model performs during and across the stream remains unknown. Furthermore, updating the underlying model to adapt to emerging topics and language distributions is definitely beneficial for retrieval effectiveness, but it can be challenging. The recent LongEval CLEF Lab [8] studies a similar problem but with queries on a set of persistent (standing) topics rather than a query stream that shifts topically over time. While closer to our setting, that setup limits us to studying cases in which the query topic distribution is stable.

3 MURR: Model Updating with Regularized Replay

In this section, we describe the training losses used for model updating with regularized training data replay – MURR. The objective is to avoid reprocessing previously-indexed documents while adapting the retrieval model to newly-arrived documents and queries with backward compatibility that allows queries to target both new and old documents.

Without loss of generality, we break the continuous stream of documents and queries into discrete periods, named *sessions*. Modeling and operational decisions are made between sessions but not within them.

In each session, we continue to fine-tune the trained retrieval model from the previous session (MURR-CF). The trained model is then used to encode and index the documents that arrive during the session, resulting in multiple indexes encoded by *different* models throughout the document stream. As an alternative, one can fine-tune from a language model for each session (MURR-LM), which inherits no knowledge from the previous session (discussed more in Sect. 4.2). We use the model trained at the current session to encode incoming queries and search each existing vector index. Regularized replay, which is discussed in this section, ensures that the models trained later in time are still compatible with existing indexes created by previous models.

3.1 Contrastive Loss

The typical training paradigm for dense retrieval models is contrastive learning [31], which is often expressed using the SimCLR loss [2]. For each training query $q_s \in \mathcal{Q}_s$ in the current session $s \in \mathbb{N}^0$, we randomly sample one relevant and one non-relevant document, denoted d_s^+ and d_s^- respectively, from the training collection to form the training triple $(q_s, d_s^+, d_s^-) \in \mathcal{D}_s$. With the triples, we formulate the contrastive loss for training the model $\mathcal{M}_s$ as

$$\mathcal{L}_C(\mathcal{D}_s) = \frac{1}{|\mathcal{D}_s|} \sum_{(q,d^+,d^-)\in\mathcal{D}_s} -\log \frac{\exp\left(f(\mathbf{q}_{(s)}, \mathbf{d}^+_{(s)})\right)}{\exp\left(f(\mathbf{q}_{(s)}, \mathbf{d}^+_{(s)})\right) + \exp\left(f(\mathbf{q}_{(s)}, \mathbf{d}^-_{(s)})\right)} \quad (1)$$

where $\mathbf{q}_{(s)}$ and $\mathbf{d}_{(s)}$ denote the vector representations encoded by the model $\mathcal{M}_s$ of the query and document, respectively, and the function f denotes the similarity function between the query and document vectors. In this work, we use the transformer model's embedding of the `[CLS]` token as the vector representation of the text sequence; we discuss this in detail in the next section.

3.2 Representation Regularized Replay

At the end of training in each session i, we sample a subset of training triples $\tilde{\mathcal{D}}_i \subset \mathcal{D}_i$, along with their vector representations encoded by the trained model $\mathcal{M}_i$, for memory replay in the future sessions through simple random sampling. At the current session s, we construct an aggregated replay training set $\mathcal{D}_s^R = \cup_{i \in [0,s-1]} \tilde{\mathcal{D}}_i$.

For replay triples, we construct an additional representation regularization loss to ensure that the current model $\mathcal{M}_s$ is still capable of matching queries with existing document vectors produced by past models by encouraging the current model to recreate the same document representation as was produced by the model that originally indexed the document. Specifically, the regularization loss is defined as

$$\mathcal{L}_R(\mathcal{D}_s^R) = \frac{1}{2|\mathcal{D}_s^R|} \sum_{(q,d^+,d^-) \in \mathcal{D}_s^R} \left\| \mathbf{d}_{(s)}^+ - \mathbf{d}_{(i)}^+ \right\|_2 + \left\| \mathbf{d}_{(s)}^- - \mathbf{d}_{(i)}^- \right\|_2 \tag{2}$$

where $\|\cdot\|_2$ denotes the L^2 norm of a vector.

Therefore, the overall training loss of for model $\mathcal{M}_s$ is the combination of the contrastive loss on both the training triples from the current session and the replay history along with the regularization; this can be expressed as

$$\mathcal{L}_s = \mathcal{L}_C(\mathcal{D}_s \cup \mathcal{D}_s^R) + \alpha \mathcal{L}_R(\mathcal{D}_s^R) \tag{3}$$

where α is a hyperparameter for regularization strength.

3.3 Continue Indexing and Distributed Search

At each session, we use model $\mathcal{M}_s$ to encode and index the documents coming in at session s. Each document is encoded into a vector and indexed into the approximate nearest neighbor search index $\mathcal{I}_s$.

At search time, queries are encoded by the current model $\mathcal{M}_s$. We search all existing indexes, i.e., $\mathcal{I}_0, \mathcal{I}_1, ...\mathcal{I}_{s-1}$, and merge the resulting ranked lists by document scores. With this approach, we do not need to retain or serve any previously trained model, triage the queries to different models, or reprocess any documents. In the remainder of the paper, we discuss the compatability of models trained using regularized replay with document vectors encoded by different models.

4 Experiments

In this section, we describe our experiment setup for evaluating the proposed MURR strategy.

4.1 Dataset and Stream Simulation

We simulate the query and document streams using the forum subset of the LoTTe dataset [34]. The LoTTE forum subset contains documents and queries from five domains, with each domain split into training (named *dev* in the original paper) and test sets. The number of documents in each domain varies, while the number of queries per domain is roughly equal. The statistics of the dataset are summarized in Table 1.

Sampling Distribution. For our experiments, we simulate four streams of data. Each simulated data stream is split into five sessions, representing five model or index updating decision points; this is similar to prior work in document streaming [21]. Each domain follows a beta-binomial sampling distribution over the sessions for both queries and documents. The sampling distributions are summarized in Fig. 1. We start by sampling the queries. Each session contains an emerging domain to simulate the topical shift of search queries over time. The ordering of the emerging domain is the same across all four simulated streams: science, followed by recreation, technology, lifestyle, and writing.

We assume that search topics follow aspects of the world that are reflected in the document stream, so in this work we simulate the same topical shift. This is implemented by having documents follow the same sampling distribution as queries across sessions. However, for evaluation, documents relevant to a specific query should appear in the current or earlier session in which the query was

Table 1. Dataset Statistics of LoTTe Forum Datasets.

Split		Science	Recreation	Technology	Lifestyle	Writing
Train	Documents	343,642	263,025	1,276,222	268,893	277,072
	Queries	2,013	2,002	2,003	2,076	2,003
Test	Documents	1,694,164	166,975	638,509	119,461	199,994
	Queries	2,017	2,002	2,004	2,002	2,000

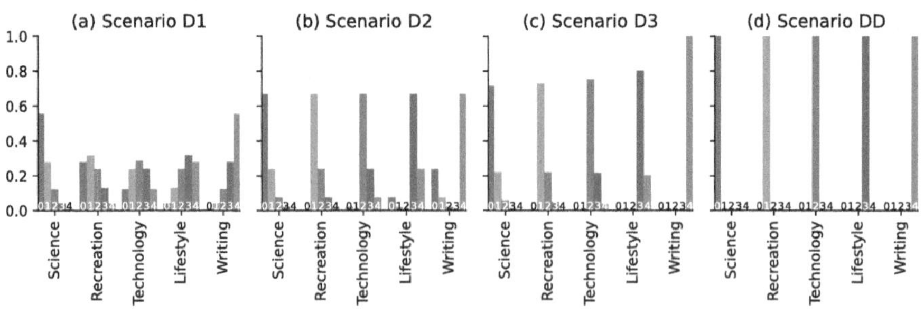

Fig. 1. Sampling distribution of each domain in each session and simulated stream. The color of the bars represents the session, along with the session ID marked at the bottom of each bar.

sampled. This creates a sampling constraint on that document, so document sampling is not purely random. Therefore, to ensure this quality, a document relevant to a query follows a truncated sampling distribution at the time of the session in which its query was sampled. Recall that while the number of queries in each domain is similar, the number of documents in each domain is not. Therefore, when sampling documents, the number of documents will not mirror the distribution, while the number of queries does. Thus, the resulting numbers of queries and documents in each session will naturally differ.

Scenario Simulated. Each simulated stream represents a different streaming scenario.

- **D1** contains a constant flow of technology queries and documents, with a symmetric and smooth distribution peaking at session 2, while other domains shift in different sessions with a skewed distribution. This simulates a mixture of different topical shifts in a combined stream.
- **D2**. Each domain in D2 follows a five-session decay with a different starting point and rotates to the front of the sessions, illustrated in Fig. 1(b). While the documents follow the same sampling distribution, since the number of documents varies across domains, the sampled number of documents in each session is not proportional to the number of sampled queries. This scenario simulates a seasonal shift in topics, with one topic dominating a session but continuing to appear in the rest of the stream. Session 0 is, therefore, clipped from an infinite seasonal stream.
- **D3** simulates a finite stream with each domain following a decaying distribution. Since the stream ends at Session 4, the length of the decay also decreases as the sessions progress, leaving the last domain (writing) to appear only in Session 4.
- **DD** (where the second D indicates "Default"), simulates discrete topic shifts, with each session containing only one domain. Despite being artificial, this stream is diagnostic for investigating the effectiveness and compatibility of model updating strategies.

The script for creating the simulated scenarios and the specific simulated query and document streams used in this work are released on Huggingface.[1]

4.2 Dense Retrieval Model

We use Tevatron [11] to train DPR [18] models with a feed-forward layer attached to the [CLS] token to project the final hidden layer to a vector of 768 dimensions for the experiments. At each session, we continue to fine-tune the model trained in the previous session using our proposed regularized replay (MURR-CF). At Session 0, which is the very first session, we initialize the DPR model with a retrieval-prepared language model trained with CoCondenser [10].[2] As

[1] https://huggingface.co/datasets/hltcoe/lotte-streams-for-murr.
[2] https://huggingface.co/Luyu/co-condenser-marco.

a variant, we also evaluate a strategy that always initializes the model from the CoCondenser model instead of the one from the previous session (MURR-LM). We use all training triples from the emerging domain (training set) at each session to train the model, which is a different set from the set of queries and documents that the model would be indexing and searching on (the test set). The statistics of the dataset are summarized in Table 1. At the end of each session, we retain 200 replay triples. We fine-tune the model for 20,000 gradient steps with an effective batch size of 256 using 8 NVIDIA V100 GPUs. We optimize the model with an Adam optimizer, a learning rate of 3×10^{-6}, and a regularization strength of 0.01.

Each document arriving during the current session is encoded with the latest DPR model, which is produced differently when using MURR-CF and MURR-LM, into a 768-dimension vector. The vectors are indexed using FAISS [17] with product quantization [16] using a 4-bit fast scan that relies on SIMD instructions for distance computations. At search time, the queries are also encoded by the latest model but distributed to all existing indexes to support searching documents that arrived in previous sessions. Ranked lists from indexes are combined based on the similarity scores of each document. Since no document is indexed more than once, we do not need fusion approaches (e.g., reciprocal rank fusion or min-max normalization).

4.3 Baselines

Under the streaming setup, there are three baseline strategies that we compare with the proposed MURR:

- **Same Model**. In most prior studies and applications, the underlying DPR model is not updated, and the previously-trained model continues to encode newly-arrived queries and documents.
- **LM without Replay**. At the other extreme, the DPR model is always up-to-date by training only on the latest emerging domain. This strategy biases the model toward the latest topics but will be less effective in serving past ones.
- **CF without Replay**. Similar to the previous one but continuously fine-tune the DPR model, which retains the knowledge learned from the previous sessions but may quickly be washed out by fine-tuning focused on the latest emerging domain.

4.4 Evaluation

We evaluate the retrieval effectiveness by Success@5, which is 1 if at least one relevant document is at or above rank five and 0 otherwise, as used in the work that introduced LoTTe [34]. To assess the significance of the effectiveness differences between strategies, we report the macro-average of Success@5 and use a paired t-test on Success@5 values on each query set across all sessions with a 95% confidence level. Such significance tests essentially ask the question, "Is the effectiveness of the resulting retrieval models when using strategy A different from using strategy B over the entire stream?"

However, since the focus of this work is not only on the absolute effectiveness of the model but also the quality of retrieval from session to session, quantitatively, we also report the relative difference between two consecutive sessions of the average Success@5 on a set of queries. This relative metric provides indicators of how well the retrieval model serves *each set of search queries* (realized as query sets in this work) throughout the stream with additional documents and updated retrieval models. In our experiments, the query set evaluated in each session is a union of the query sets from prior sessions and the new query set. This allows us to observe the effectiveness of queries targeting previously indexed documents and documents indexed by the latest model. Specifically, the relative Success@5 gain is calculated by

$$\frac{2}{S(S-1)} \sum_{s=0}^{S} \sum_{i=0}^{s-1} \frac{\text{Success@5}(\mathcal{Q}_i \mid \cup_{j=0}^{s} \mathcal{I}_j)}{\text{Success@5}(\mathcal{Q}_i \mid \cup_{j=0}^{s-1} \mathcal{I}_j)} \tag{4}$$

where S is the number of sessions (five in our experiments), and Success@5($\mathcal{Q}_i \mid \cup_{j=0}^{s} \mathcal{I}_j$) indicates the Success@5 value evaluated on the test queries in Session i with the accumulated indexes up to Session s. To assess the difference between the relative Success@5 gain, we also report the standard deviation of the $\frac{S(S-1)}{2}$ values that we average over, which we argue is more informative than merely reporting significance levels.

This metric does *not* consider how effective the model is on a specific query set; instead, we measure how much the model *retains* effectiveness throughout the stream. Theoretically, a model can perform extremely poorly on a set of queries and continue to do so, so under this gain measure, the value would be higher than a model that performs very effectively on the set but only slightly degrades as the stream progresses. Therefore, this gain measure is orthogonal to the effectiveness measure and evaluates the consistency of the strategy in serving a set of queries over time.

5 Results and Analysis

Compared to other strategies, MURR-CF, which continues to fine-tune the dense retrieval model from the previous session with regularized replay, provides consistent improvements across sessions in all four simulated streams. Summarized in Fig. 2, MURR-CF (purple with pluses) is statistically significantly more effective than other strategies across all sessions except the very first one (where all strategies result in the same model for that session). We show that our proposed simple regularization on memory replay provides a significant boost compared to only continued fine-tuning of models from the previous session (CF w/o Replay, in green with squares).

Among the five strategies, training only on the latest training data (LM w/o Replay, in orange with triangles) is consistently and statistically significantly the

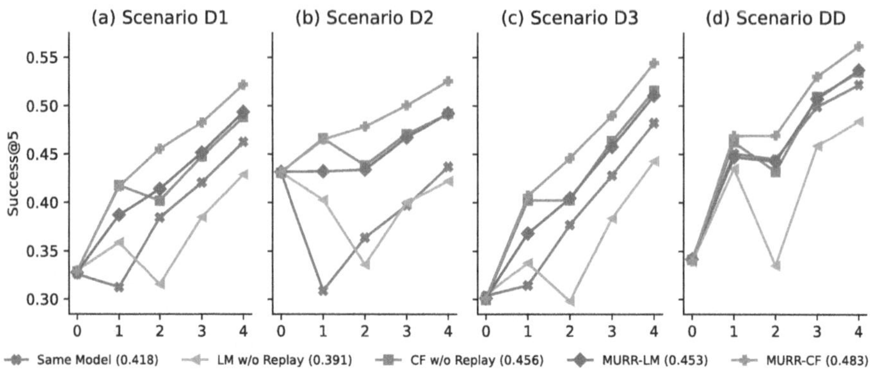

Fig. 2. Success@5 of each model updating strategy where the x-axis is the running session ID. Values in parenthesis in the legend are macro-average of Success@5 across all query sets and streams. Differences between each strategy are all statistically significant using the paired t-test described in Sect. 4.4 except for the pair of *CF w/o Replay* and *MURR-LM* after multiple test corrections. (Color figure online)

least effective; it is even worse than not updating the model (Same Model, in blue with crosses), which is the obvious baseline for a streaming scenario. Only training on the latest data loses compatibility with previous sessions, making old indexes incompatible with the current query representation. However, to our surprise, the representations are not totally incompatible. Even in Scenario DD, where domain shifts are dramatic, LM w/o Replay still achieves reasonable Success@5 values (within 0.2 of the best in most cases) despite being the worst among others.

Continued fine-tuning without training data replay (CF w/o Replay) performs similarly to MURR-LM (red with diamonds), which always initializes the DPR model with the CoCondenser model during fine-tuning with replay data.[3] The two share similar trajectories throughout the sessions in all streams except for the second session. This indicates a small amount of training data replay with regularization on the document representations is sufficient to replicate training from the previous session. Combining both, which is MURR-CF, leads to more effective models than either alone, indicating that, although similar in net effect, they improve the model in different ways.

5.1 Effectiveness on Individual Query Sets

For a specific set of queries, MURR-CF is also more effective than other strategies in terms of both absolute effectiveness and stability from session to session. As shown in Fig. 3, MURR-CF is consistently more effective in all five query sets

[3] We conduct additional equivalence test between *CF w/o Replay* and *MURR-LM* using two one-sided t-tests (TOST) with a 5% relative difference as the threshold and conclude that the two strategies are statistically equivalent with 95% confidence.

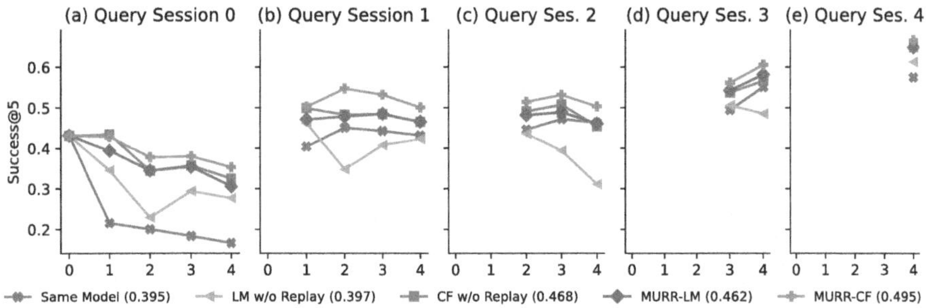

Fig. 3. Query set breakdown of five model updating strategies in Scenario D2. Each subgraph is the effectiveness over sessions of the set of queries introduced in the session indicated in the title. Values in parenthesis are macro-average Success@5 on query sets in only D2. Values at the x-axis is the running session ID.

in Scenario D2. Due to space constraints, we only present graphs on Scenario D2, but we observe similar relationships between strategies in other simulated scenarios despite the differences in stream characteristics. In fact, all four simulated scenarios demonstrate similar strategy orders, while some exhibit larger variations in absolute effectiveness than others. Interestingly, in almost all cases, the Success@5 of the second session after the query set is introduced is higher than the session when it first appeared, but that measure gradually decreases as the stream progresses. This is likely an artifact of the stream simulation since the sampling probability of a domain is similar between the peak and the very next session.

Table 2. Relative Success@5 Gain. Reported values are macro-averaged over the query sets sampled in each session (for example, the queries sampled for Session 0 have four gains over the subsequent four sessions, but the queries sampled for session 1 have three), as discussed in Sect. 4.4. The standard deviation is in parentheses.

Scenario	Same Model	LM w/o Replay	CF w/o Replay	MURR-LM	MURR-CF
D1	0.046 (0.17)	0.049 (0.32)	0.041 (0.20)	0.043 (0.17)	0.055 (0.18)
D2	−0.052 (0.17)	−0.070 (0.19)	−0.034 (0.08)	−0.030 (0.07)	−0.013 (0.07)
D3	−0.032 (0.15)	−0.020 (0.42)	−0.016 (0.12)	−0.023 (0.09)	0.005 (0.07)
DD	−0.008 (0.01)	−0.065 (0.27)	−0.046 (0.07)	−0.040 (0.06)	−0.020 (0.02)
Overall	−0.012 (0.14)	−0.027 (0.30)	−0.014 (0.13)	−0.012 (0.11)	0.007 (0.10)

To characterize the consistency of retrieval quality from session to session, we measure the relative Success@5 gain, as described in Sect. 4.4. Summarized in Table 2, on average, all strategies maintain roughly the same effectiveness after their introduction to the stream, with MURR-CF being the only strategy providing positive gains. This indicates that, on average, effectiveness is improving

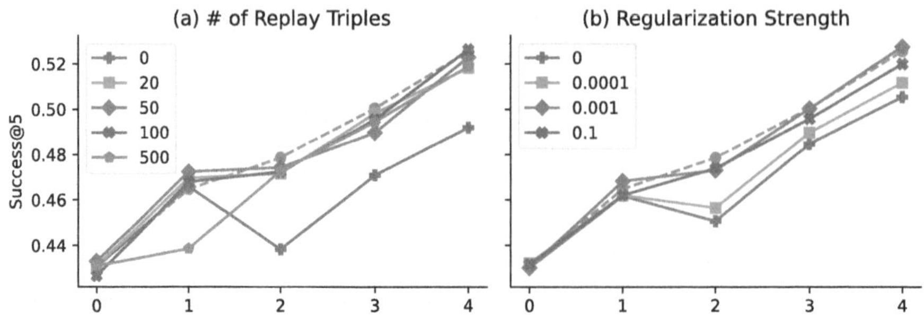

Fig. 4. Ablation study on the number of replay triples and the regularization strength under MURR-CF in Scenario D2. The x-axis is the session ID. The dashed gray line is the setup (200 triples, $\alpha = 0.01$) used in the main experiments. Using 0 replay triples is essentially *CF w/o Replay* in the main results. (Color figure online)

with additional off-domain training. While the other strategies offer negative gains, they all maintain roughly the same level of effectiveness. However, in terms of stability, LM w/o Replay, which has no modeling connections from session to session, is the least stable strategy with the highest standard deviation (overall 0.30). With more explicit connections, such as continued fine-tuning and regularized replay, MURR-CF becomes more stable (overall standard deviation 0.10).

Again, we observe similar effects between CF w/o Replay and MURR-LM in both means and standard deviations over the four simulated scenarios, with MURR-LM being slightly more stable than CF w/o Replay. This indicates that the model memory likely has a limit and will eventually move on from knowledge acquired during the early sessions that are no longer relevant to later sessions. However, since our experimental setup only simulates a short stream, it cannot clearly demonstrate this effect.

5.2 Ablation Studies

We study the robustness of the two hyperparameters we introduced in MURR-CF: number of replay triples and regularization strength. Summarized in Fig. 4, we observe that MURR-CF is robust to both the number of replay triples and the regularization strength, although the existence of both is still critical. Using smaller or stronger regularization strength than 0.01 results in slight but noticeable degradation. Using 0 or 0.0001 as the regularization strength, essentially only replaying triples but no representation regularization, leads to less effective results. Particularly for Session 2, both using no replay and no regularization led to a degradation in effectiveness from the previous session; using some replay triples and a moderate regularization strength provided a stable trajectory. Surprisingly, while more replay triples provide stability to the process, a small number of replay triples are sufficient to calibrate the models trained in the following sessions to be compatible with document representations from the previous sessions. However, the 0 replay triples case is significantly worse than any runs that use some replay. We observe a very similar gap in Scenario D1 and much smaller gaps in Scenarios D3 and DD, suggesting that such representation regularization depends on the stream, which we leave for future investigation.

6 Conclusion and Future Work

This work proposes a strategy for updating a neural retrieval model that adapts to changes in document topicality over time without requiring documents already seen to be reindexed. We simulated four scenarios that contain streams of queries and documents to evaluate our proposed MURR model updating strategy. We showed that our proposed method provides more effective and more stable retrieval results throughout the streams without reprocessing the existing documents. These results demonstrate the feasibility of deploying a dense retrieval model and continuing to modify it on-the-fly for emerging topics.

While simulation experiments provide critical evidence for developing such model-updating strategies, experimenting on real document streams with real language distribution shifts can further empirically verify our findings. We plan to evaluate on LongEval to verify our findings in this work. Additionally, collecting documents from a stream, such as Common Crawl, and developing queries at different snapshots (or assigning existing queries to appropriate time points) is also in our plan to evaluate evolving information needs over time.

Acknowledgement. This work was partially supported by the Spoke "FutureHPC & BigData" of the ICSC – Centro Nazionale di Ricerca in High-Performance Computing, Big Data and Quantum Computing funded by the Italian Government, the FoReLab and CrossLab projects (Departments of Excellence), the NEREO PRIN project (Research Grant no. 2022AEFHAZ) funded by the Italian Ministry of Education and Research (MUR).

References

1. Allan, J.: Incremental relevance feedback for information filtering. In: Proceedings of the 19th Annual International ACM SIGIR Conference on Research and Development in Information Retrieval, pp. 270–278 (1996)
2. Chen, T., Kornblith, S., Norouzi, M., Hinton, G.: A simple framework for contrastive learning of visual representations. In: International Conference on Machine Learning, pp. 1597–1607, PMLR (2020)
3. Chen, T., Zhang, M., Lu, J., Bendersky, M., Najork, M.: Out-of-domain semantics to the rescue! Zero-shot hybrid retrieval models. In: European Conference on Information Retrieval, pp. 95–110. Springer (2022)
4. Devlin, J., Chang, M.W., Lee, K., Toutanova, K.: BERT: pre-training of deep bidirectional transformers for language understanding. In: Proceedings of the 2019 Conference of the North American Chapter of the Association for Computational Linguistics: Human Language Technologies, Volume 1 (Long and Short Papers), pp. 4171–4186. Association for Computational Linguistics, Minneapolis, Minnesota (2019)

5. Florio, K., Basile, V., Polignano, M., Basile, P., Patti, V.: Time of your hate: the challenge of time in hate speech detection on social media. Appl. Sci. **10**(12), 4180 (2020)
6. Formal, T., Lassance, C., Piwowarski, B., Clinchant, S.: From distillation to hard negative sampling: making sparse neural IR models more effective. In: Proceedings of the 45th International ACM SIGIR Conference on Research and Development in Information Retrieval, pp. 2353–2359 (2022)
7. Formal, T., Piwowarski, B., Clinchant, S.: Splade: sparse lexical and expansion model for first stage ranking. In: Proceedings of the 44th International ACM SIGIR Conference on Research and Development in Information Retrieval, pp. 2288–2292 (2021)
8. Galuscáková, P., et al.: Longeval-retrieval: French-English dynamic test collection for continuous web search evaluation. In: Proceedings of the 46th International ACM SIGIR Conference on Research and Development in Information Retrieval, pp. 3086–3094 (2023)
9. Gangi Reddy, R., et al.: Synthetic target domain supervision for open retrieval QA. In: Proceedings of the 44th International ACM SIGIR Conference on Research and Development in Information Retrieval, pp. 1793–1797 (2021)
10. Gao, L., Callan, J.: Unsupervised corpus aware language model pre-training for dense passage retrieval. In: Muresan, S., Nakov, P., Villavicencio, A. (eds.) Proceedings of the 60th Annual Meeting of the Association for Computational Linguistics (Volume 1: Long Papers), pp. 2843–2853, Association for Computational Linguistics, Dublin, Ireland (2022). https://doi.org/10.18653/v1/2022.acl-long.203, https://aclanthology.org/2022.acl-long.203
11. Gao, L., Ma, X., Lin, J.J., Callan, J.: Tevatron: an efficient and flexible toolkit for dense retrieval. ArXiv abs/2203.05765 (2022)
12. Goodfellow, I.J., Mirza, M., Xiao, D., Courville, A., Bengio, Y.: An empirical investigation of catastrophic forgetting in gradient-based neural networks. arXiv preprint arXiv:1312.6211 (2013)
13. Harman, D.: Overview of the fourth text retrieval conference (TREC-4). In: Proceedings of the Fourth Text Retrieval Conference, pp. 1–24 (1996)
14. Hashemi, H., Zhuang, Y., Kothur, S.S.R., Prasad, S., Meij, E., Croft, W.B.: Dense retrieval adaptation using target domain description. In: Proceedings of the 2023 ACM SIGIR International Conference on Theory of Information Retrieval, pp. 95–104 (2023)
15. Jatowt, A., Sato, M., Draxl, S., Duan, Y., Campos, R., Yoshikawa, M.: Is this news article still relevant? Ranking by contemporary relevance in archival search. Int. J. Digit. Libr. **25**(2), 197–216 (2024)
16. Jegou, H., Douze, M., Schmid, C.: Product quantization for nearest neighbor search. IEEE Trans. Pattern Anal. Mach. Intell. **33**(1), 117–128 (2010)
17. Johnson, J., Douze, M., Jégou, H.: Billion-scale similarity search with GPUs. IEEE Trans. Big Data **7**(3), 535–547 (2019)
18. Karpukhin, V., et al.: Dense passage retrieval for open-domain question answering. arXiv preprint arXiv:2004.04906 (2020)
19. Khattab, O., Zaharia, M.: Colbert: efficient and effective passage search via contextualized late interaction over BERT. In: Proceedings of the 43rd International ACM SIGIR Conference on Research and Development in Information Retrieval, pp. 39–48 (2020)
20. Lassance, C., Déjean, H., Clinchant, S.: An experimental study on pretraining transformers from scratch for IR. In: European Conference on Information Retrieval, pp. 504–520, Springer (2023)

21. Lawrie, D., Kayi, E., Yang, E., Mayfield, J., Oard, D.W.: PLAID SHIRTTT for large-scale streaming dense retrieval. In: Proceedings of the 47th International ACM SIGIR Conference on Research and Development in Information Retrieval, pp. 2574–2578 (2024)
22. Li, M., Lin, J.: Encoder adaptation of dense passage retrieval for open-domain question answering. arXiv preprint arXiv:2110.01599 (2021)
23. Lin, Z., et al.: PROD: progressive distillation for dense retrieval. In: Proceedings of the ACM Web Conference 2023, pp. 3299–3308 (2023)
24. Liu, Y., et al.: RoBERTa: a robustly optimized BERT pretraining approach (2019)
25. Lovón-Melgarejo, J., Soulier, L., Pinel-Sauvagnat, K., Tamine, L.: Studying catastrophic forgetting in neural ranking models. In: Hiemstra, D., Moens, M.-F., Mothe, J., Perego, R., Potthast, M., Sebastiani, F. (eds.) ECIR 2021. LNCS, vol. 12656, pp. 375–390. Springer, Cham (2021). https://doi.org/10.1007/978-3-030-72113-8_25
26. Lukes, J., Søgaard, A.: Sentiment analysis under temporal shift. In: Proceedings of the 9th Workshop on Computational Approaches to Subjectivity, Sentiment and Social Media Analysis, pp. 65–71 (2018)
27. Malkov, Y.A., Yashunin, D.A.: Efficient and robust approximate nearest neighbor search using hierarchical navigable small world graphs. IEEE Trans. Pattern Anal. Mach. Intell. **42**(4), 824–836 (2018)
28. Nguyen, T., et al.: MS MARCO: A human generated machine reading comprehension dataset. arXiv preprint arXiv:1611.09268 (2016). http://arxiv.org/abs/1611.09268
29. Ni, J., et al.: Large dual encoders are generalizable retrievers. arXiv preprint arXiv:2112.07899 (2021)
30. Nogueira, R., Yang, W., Cho, K., Lin, J.: Multi-stage document ranking with BERT. arXiv preprint arXiv:1910.14424 (2019)
31. van den Oord, A., Li, Y., Vinyals, O.: Representation learning with contrastive predictive coding. arXiv preprint arXiv:1807.03748 (2018)
32. Radlinski, F., Joachims, T.: Evaluating the robustness of learning from implicit feedback. In: Learning in Web Search (LWS 2005), p. 42 (2005)
33. Santhanam, K., Khattab, O., Potts, C., Zaharia, M.: PLAID: an efficient engine for late interaction retrieval. arXiv preprint arXiv:2205.09707 (2022)
34. Santhanam, K., Khattab, O., Saad-Falcon, J., Potts, C., Zaharia, M.: ColBERTv2: effective and efficient retrieval via lightweight late interaction. In: Proceedings of the 2022 Conference of the North American Chapter of the Association for Computational Linguistics: Human Language Technologies, pp. 3715–3734, Association for Computational Linguistics, Seattle, United States (2022)
35. Thakur, N., Reimers, N., Lin, J.: Domain adaptation for memory-efficient dense retrieval. arXiv preprint arXiv:2205.11498 (2022)
36. Thakur, N., Reimers, N., Rücklé, A., Srivastava, A., Gurevych, I.: BEIR: a heterogeneous benchmark for zero-shot evaluation of information retrieval models. In: Thirty-Fifth Conference on Neural Information Processing Systems Datasets and Benchmarks Track (Round 2) (2021)
37. Wang, K., Thakur, N., Reimers, N., Gurevych, I.: GPL: generative pseudo labeling for unsupervised domain adaptation of dense retrieval. arXiv preprint arXiv:2112.07577 (2021)
38. Wang, L., et al.: Text embeddings by weakly-supervised contrastive pre-training. arXiv preprint arXiv:2212.03533 (2022)
39. Willis, C., Sherman, G., Efron, M.: What makes a query temporally sensitive? In: Proceedings of the 39th International ACM SIGIR Conference on Research and Development in Information Retrieval, pp. 1065–1068 (2016)

40. Yang, E., Lawrie, D., Mayfield, J., Oard, D.W., Miller, S.: Translate-distill: learning cross-language dense retrieval by translation and distillation. In: Advances in Information Retrieval: 46th European Conference on IR Research, ECIR 2024 (2024)
41. Zhuang, S., Zhuang, H., Koopman, B., Zuccon, G.: A setwise approach for effective and highly efficient zero-shot ranking with large language models. In: Proceedings of the 47th International ACM SIGIR Conference on Research and Development in Information Retrieval, pp. 38–47 (2024)

Token Pruning Optimization for Efficient Multi-vector Dense Retrieval

Shanxiu He[1(✉)], Mutasem Al-Darabsah[2], Suraj Nair[2], Jonathan May[2,3], Tarun Agarwal[2], Tao Yang[1], and Choon Hui Teo[2]

[1] University of California, Santa Barbara, USA
{shanxiuhe,tyang}@cs.ucsb.edu
[2] Amazon, Palo Alto, USA
{mutasema,srjnair,tagar,choonhui}@amazon.com, jonmay@isi.edu
[3] University of Southern California, Los Angeles, USA

Abstract. Multi-vector dense retrieval with ColBERT has been shown to be effective in striking a good relevance and efficiency tradeoff for both in-domain and out-of-domain datasets through late interaction between queries and documents. However, the efficiency of ColBERT for a large-scale retrieval dataset is still constrained by its large memory footprint, as one embedding is stored per token; thus, previous work has studied static pruning of less significant tokens to enhance efficiency. To improve the adaptivity of prior work in zero-shot retrieval settings, this paper proposes a neural classification method that learns pruning decisions with Gumbel-Softmax, and provides an extension to adjust pruning decisions and meet memory space reduction requirements. We evaluate the effectiveness of our proposed method against several baseline approaches on out-of-domain datasets LoTTE and BEIR, and the in-domain MS MARCO passage dataset.

Keywords: Multi-vector neural representations · Late-interaction dense retrieval · Pruning · Space and time efficiency

1 Introduction

A transformer-based cross-encoder for document ranking delivers impressive performance but comes with extremely high computational costs during inference. To reduce complexity, many dense retrieval methods adopt a single-vector paradigm to represent queries and documents. However, deploying such single-vector models for large search datasets with diverse content domains and applications, such as product search involving a large number of entities, is challenging. As pointed out in the previous work [16,25,26], the single-vector representation paradigms have limited expressive power when handling out-of-domain datasets with scarce training data (including zero-shot retrieval) and in answering entity-centric questions. ColBERT [12] with a multi-vector representation can be viewed as a middle-ground between the single-vector bi-encoder and the cross-encoder

C. Hauff et al. (Eds.): ECIR 2025, LNCS 15572, pp. 101–115, 2025.
https://doi.org/10.1007/978-3-031-88708-6_7

design. Previous studies demonstrate that, compared to sparse representations like SPLADE [7,8], ColBERTv2 can achieve comparable performance on the MS MARCO and BEIR datasets [23,28], while also offering a relevance advantage when searching an out-of-domain data collection called LoTTE [23]. As a result, multi-vector dense representations can be desirable in certain search applications, especially on GPU-rich platforms where sparse retrieval cannot fully leverage GPU resources.

The expressiveness advantage of ColBERT over single-vector representations comes at a significant cost, namely, an order of magnitude increase in space and computational complexity. Reducing space complexity is important as the memory size of GPUs is much more limited than CPUs and maintaining high memory bandwidth becomes challenging as GPU memory capacity expands. Several efficiency optimization approaches have been developed recently for ColBERT, including offline techniques for space reduction such as quantization to compress the vector size [23], and token pruning [1,14]. Approaches of online inference acceleration include DESSERT [6], EMVB [20], CITADEL [16], and XTR [15].

Our paper focuses on token pruning optimization to minimize the number of tokens representing each document, thereby reducing space usage and accelerating inference time. Existing techniques like Top-k token pruning [14] face limitations due to the static nature of parameter k, which is not learned, making them less effective for out-of-domain search. Similarly, IDF-based pruning [1], while effective in some cases, can be unsafe for certain domains as removing frequently occurring tokens might inadvertently discard critical information. Increasing the IDF method parameter often leads to a significant relevance drop under tight space budgets. These challenges highlight the need for learnable context-aware pruning to improve adaptivity.

This paper develops a context-aware neural module that decides the pruning of tokens for each document in ColBERT-like dense retrieval. Our method, **LeapMV** (**Lea**rnable **P**runing for **M**ulti-**V**ector representations), is trained to minimize the multi-vector dense retrieval loss for one domain and is applicable to zero-shot retrieval in searching out-of-domain datasets. LeapMV can be extended to adjust its pruning aggressiveness to meet different space reduction requirements. Our evaluation shows that LeapMV improves the adaptivity of the previously proposed static pruning techniques in meeting different space reduction targets, which results in latency saving proportionally. Under the same relevance accuracy budget for LoTTE, LeapMV consumes up to 3.3× less space and is up-to 2.8× faster than the top-k token pruning and IDF methods.

2 Background and Design Considerations

The multi-vector representation in ColBERT encodes the tokens in each document (and a query) using embeddings and its index is built based on a set of token embedding vectors for each document. During inference time, ColBERT calculates the similarity between a document d and the given query q via the following "MaxSim" operations,

$$S_{q,d} = \sum_{i=1}^{N} \max_{j=1}^{M} Q_i \cdot D_j^T \tag{1}$$

where Q is the encoded token embedding set for a given query and D the token embedding sequence for each document d. N and M are the token sequence length for a query and a document respectively.

The required ColBERT online memory size for hosting the compressed multi-vector document representation is approximately proportional to the product of the average document length, the token embedding dimension, and the number of documents. This can easily flood out memory space for hosting a large search data collection with a long language-model embedding length. Furthermore, the online inference time complexity is proportional to the average number of tokens per document. Thus reducing the number of tokens per document can lead to near-linear decreases in the memory usage and search inference time.

Top-k Token Pruning. Static token pruning in [14] keeps top-k tokens per document, and several heuristics have been proposed to select such k items. Although the ColBERT model is optimized through training based on the given k hyperparameter, the trained ColBERT model based on one domain (e.g. MS MARCO) with a fixed k value can lose its advantage when this model is directly applied in a zero-shot manner to other search domains which have different document characteristics including lengths and IDF distributions.

IDF-based Token Pruning. The study in [1] compares several IDF-based static pruning methods for ColBERT with MS MARCO and finds the best method is called IDF uniform which removes top-τ BERT tokens (including stopwords) globally in all documents. Uniformly pruning popular words in all documents can inadvertently remove meaningful tokens in some contexts. For example, "the who" is the name of a well-known rock band, while "who" can stand for the World Health Organization–both of which could be pruned by the above IDF uniform method. When a larger space reduction is required, increasing the IDF filtering parameter can incur a significant relevance loss as demonstrated in our evaluation. Thus we seek a learned neural classifier for context-aware token pruning, applicable for zero-shot retrieval.

Text Chunking. We have also considered using text chunking, which originated in NLP research [22]. Chunking divides text into syntactically related, non-overlapping groups of words and reduces the number of representations per document by considering each chunk as a merged token. We have applied text chunking such as sentence chunking and word chunking for ColBERT and have found significant relevance degradations of such methods compared to other baseline methods.

Quantization and Online Inference Optimization. ColBERTv2 [23] and its highly optimized implementation PLAID [24] quantize every dimension of the encoded vector and represent the document as a summation from the cluster centroid and the residuals. Residual compression can further reduce the index size while preserving retrieval quality. ColBERTv2 and PLAID also use an inverted index with clustering to improve online search speed. Our evaluation has built LeapMV on top of PLAID, taking advantage of its highly optimized implementation. DESSERT [6] proposes to speed up online query processing using a randomized algorithm for LSH-based vector set queries. DESSERT does not reduce storage space, and it uses much more space than ColBERTv2 or PLAID, though its online inference can be fairly fast. CITADEL [16] proposes to route each token vector to the predicted lexical "keys" such that a query token vector only interacts with document token vectors routed to the same key. While CITADEL is faster than PLAID and intends to use quantization, it has been reported to use more space than PLAID when desiring competitive relevance. EMVB [20] proposes an efficient query processing framework for multi-vector dense retrieval with centroid-based document pre-filtering and space compression with product quantization. XTR [15] approximates document rank scoring of multi-vector dense retrieval by retrieving the most important query-specific document tokens first and accounting for the contribution of the missing tokens to the overall score with missing similarity imputation.

LeapMV is orthogonal and complementary to the above online techniques because online optimization can be applied after offline processing with LeapMV's static token pruning of each document.

3 Learnable Pruning for Multi-vector Representations

We now describe the LeapMV module, as illustrated in Fig. 1. Let D be the sequence of M tokens representing each document d: $D = (D_1, D_2, \cdots, D_M)$, where $D_j \in \mathbb{R}^h$. To decide if the j-th token can be pruned or not, we let $D'_j \in \mathbb{R}^{(2c+1)\cdot h}$ represent the concatenation of a contextualized window with a window size of c before and after the j-th token. Namely, $D'_j = (D_{j-c} \circ \cdots \circ D_j \circ \cdots \circ D_{j+c})$, which forms a representation of total size $(2c+1)\cdot h$. As shown in Fig. 1, the above representation is forwarded to one fully-connected neural network layer with dropout as a token-level classifier to obtain:

$$Y_j = W(D'_j \circ \xi) + b, \text{ with } j \in \{1, 2, \cdots, M\} \tag{2}$$

where W is a weight matrix transforming the inputs into classifier logits. $\xi \in \mathbb{R}^h$, is a vector of independent noise, drawn from a Bernoulli distribution with probability $1-p$ with p as dropout rate, and b is a bias term. Logits $Y_j \in \mathbb{R}^2$ denote classification scores for j-th token, i.e. dimension 0 is the unnormalized probability for keeping the token (decision [KEEP]) and dimension 1 the logit for dropping (decision [DROP]).

To obtain discrete decisions for pruning or keeping a token, we apply Straight-through Gumbel-Softmax [9,10,18] on Y_j, i.e.,

$$(P_j)_i = \frac{\exp(\log((Y_j)_i + \epsilon_i)/\tau)}{\sum_{i'=0}^{1} \exp(\log((Y_j)_{i'} + \epsilon_{i'})/\tau)}. \tag{3}$$

$$u_j = \underset{i \in \{0,1\}}{\arg\max}\, (P_j)_i. \tag{4}$$

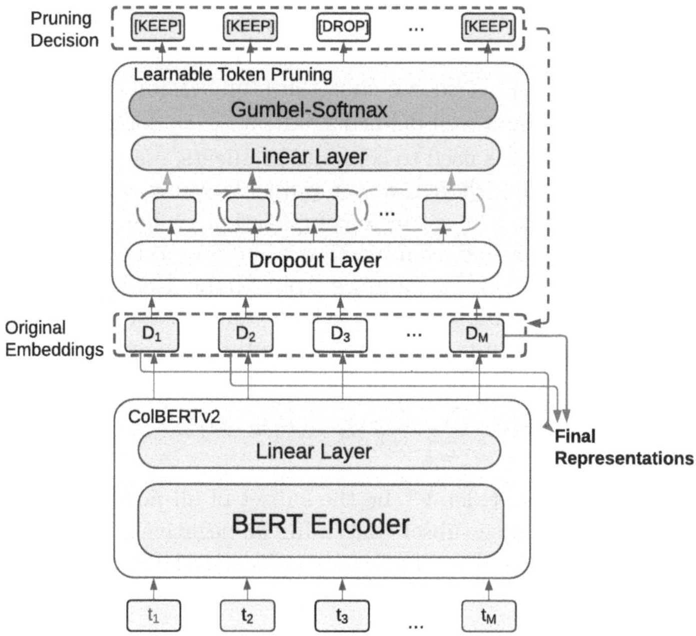

Fig. 1. LeapMV model architecture for pruning document tokens.

Here P_j is a vector with normalized probability, where $(P_j)_i$ is the i-th entry of vector P_j, and $(Y_j)_i$ is the i-th entry of vector Y_j. The hyperparameter τ is the softmax temperature, and ϵ_0, ϵ_1 are i.i.d samples from the Gumbel distribution $-\log(-\log(\text{Uniform}[0,1]))$. We obtain $u_j \in \{0,1\}$ as the prediction for the jth-token [KEEP] or [DROP] decision for document d.

Online Inference. For scoring shown in Formula (1), only the token predicted as [KEEP] would be included for MaxSim computation. With token pruning under the predicted value u_j as 0 or 1 for j-token of a document, we compute the rank score of each document with a MaxSim operation as:

$$S_{q,d} = \sum_{i=1}^{N} \max_{j=1}^{M} Q_i \cdot ((1-u_j) \cdot D_j^T). \tag{5}$$

In the above expression, tokens predicted as [DROP] are not used for calculating matching scores between queries and documents. The online inference complexity is proportional to the average number of kept tokens instead of the original number of tokens.

Training. During model training, our scheme performs token-level classification, uses the classification results to perform modified late interaction, and receives feedback similar to the original training of ColBERTv2, which includes a knowledge distillation loss from the cross-encoder teacher as well as in-batch negative loss. During training with backpropagation, the Gumbel-SoftMax function provides a differentiable approximation to the argmax function that prunes tokens if they are not selected. It is known that the Gumbel-Softmax distribution is smooth for $\tau > 0$, yielding a well-defined gradient $\frac{\partial (P_j)_i}{\partial (I_j)_i}$ for the predicted softmax probability. Thus this is used to compute gradients, essentially by replacing discretely classified pruning choices with Gumbel-Softmax samples.

For low temperatures, the softmax computation of Formula (3) smoothly approaches the discrete argmax computation while preserving the relative order of the Gumbels $(\log(Y_j + \epsilon)/\tau)$. Samples from Gumbel Softmax distributions become near identical to samples from a [KEEP] and [DROP] class distribution. Thus the MaxSim scoring in training loss computation is approximated as

$$\tilde{S}_{q,d} = \sum_{i=1}^{N} \max_{j=1}^{M} Q_i \cdot ((P_j)_0 \cdot D_j^T). \tag{6}$$

Given a training dataset, let $\mathcal{V}^+$ be the subset of all positive documents for training query q, and $\mathcal{V}^-$ be a subset containing all negative documents for query q. The top one probability distribution of positive or negative document d_i is computed as:

$$\frac{\exp(\tilde{S}_{q,d_i})}{\sum_{d_j \in \mathcal{V}^+ \cup \mathcal{V}^-} \exp(\tilde{S}_{q,d_j})}.$$

To train the ColBERT model with the above token pruning classifier, we use the cross-entropy rank loss with KL divergence for knowledge distillation [23].

Notice that there is a difference from the original scoring Formula (1) to the actual online scoring Formula (5) due to token pruning, and there is a training approximation from Formula (1) to Formula (6) for pruning optimization through SGD learning. Our evaluation shows such an approximation leads to a limited relevance drop while gaining a large advantage for space and time-saving.

Adjustment with a Space Reduction Target. Our evaluation finds that the above LeapMV module in Fig. 1 yields approximately 50% token pruning. We can modify the Gumbel-Softmax probability outcome for each document towards a more or less aggressive token pruning target based on a desired space reduction goal. Let α be the target percentage of tokens to prune in this document.

Given the outcome of Formula (3) for each token in a document, we sort $(P_j)_0$ in a non-descending order where $1 \leq j \leq M$ for this document with M tokens. Without loss of generality, assume after such sorting, $(P_j)_0 \leq (P_{j+1})_0$ where $1 \leq j \leq M - 1$. Pruning tokens from the sorted position 1 to x while keeping all other tokens from position $x + 1$ meets the target pruning ratio α. Then we set

$$\delta = 0.5 - 0.5((P_x)_0 + (P_{x+1})_0).$$

We can show that adding offset δ to the first dimension of the vectors (P_j) of this document will allow LeapMV to prune α percentage of tokens approximately. This is because the above offset δ satisfies

$$(P_j)_0 + \delta \leq (P_j)_1 - \delta \text{ where } 1 \leq j \leq \text{ x}$$

and

$$(P_j)_0 + \delta \geq (P_j)_1 - \delta \text{ where } x+1 \leq j \leq M.$$

With the randomization introduced in Formula (3), the chance of being equal in the above inequalities is small. Then tokens from the sorted positions 1 to x are pruned approximately when applying Formula (4).

Thus to adjust the chance of being pruned towards target ratio α, we modify the vectors (P_j) for each document as follows after computation of Formula (3) and before computing Formula (4):

$$(P_j)_0 = (P_j)_0 + \delta \text{ and } (P_j)_1 = (P_j)_1 - \delta \text{ where } 1 \leq j \leq M.$$

The evaluation in Sect. 4 finds that while LeapMV cuts about 50% of tokens in its default setting without α adjustment, choosing α betwen 25% and 75% gives a range towards the desired space reduction target while incurring a small or modest relevance loss.

4 Evaluation

Datasets. To evaluate LeapMV for its out-of-domain retrieval quality, we perform experiments on two out-of-domain search data collections:

LoTTE [23], a collection of 12 test sets, focuses on information-seeking user queries and answers on diverse long-tail topics from StackExchange forums. Topics for each dataset include writing, recreation, science, technology, and lifestyle. Each topic has 400–2100 queries and 100k-2M passages. The pooled set with 2.8M passages contains the passages and queries aggregated across all topics to form a more diverse corpus. LoTTE also provides search and forum queries, gathered from filtered Google search queries [11] and StackExchange communities respectively. The performance is reported separately for each group.

BEIR collection [26] consists of 13 publicly available datasets on fact-checking, citation prediction, duplicate question retrieval, argument retrieval, news retrieval, question answering, tweet retrieval, bio-medical IR, and entity retrieval tasks. The size of these data sets varies from a few thousand to 5.4M.

Model training uses MS MARCO data [3] with 8.8M passages. We also report in-domain retrieval evaluations with the development set with 6980 queries.

Implementation. LeapMV is trained under the same setting as the ColBERTv2, such as the starting checkpoint, learning rates, the number of training steps, and training data, with the exception of using a step accumulation of 2 to reach a batch size of 32. To be more specific, we initialize LeapMV with ColBERTv1.9 [4], the starting checkpoint released by ColBERT official repository, and train using four 40GB A100 GPUs for 400k steps. We train the model using 64-way mined hard negatives provided by ColBERT, consisting of a query, a highly-ranked passage (or gold labels), and one or more negative passages with the same cross-encoder teacher [19]. For indexing, we follow the default PLAID setting of hyper-parameters when the search depth is 1000 unless otherwise mentioned. We adopt a dropout rate of $p = 0.1$ for Formula (2) and temperature $\tau = 2/3$ for Formula (3), similar as [5,18].

Metrics. We follow the standard practice in the literature, reporting, success@5 (recall@5) for LoTTE, NDCG@10 for BEIR, and mean reciprocal rank at 10 (MRR@10) and recall at 1000 for MS MARCO Dev set.

For efficiency, we report the total amount of in-memory space in gigabytes needed to host the search data and the mean latency to execute a test query in milliseconds with a retrieval depth of 1000. For GPU latency, we measure on one NVIDIA A40 GPU and allow full access to 16 threads available. For CPU latency, we measure on an AMD EPYC 7763 (AMD Milan) running with 8 threads or 1 thread. On average, the 8-thread CPU time is about 3.7x faster than the 1-thread CPU. We report the 8-thread CPU time.

Baselines. We compare several baselines discussed in Sect. 2, which directly reduces the number of tokens in the ColBERT multi-vector representation.

- **Text Chunking**. We will compare two text chunking methods: phrase-based and sentence-based chunking. Both methods aggregate the embeddings for each phrase or each sentence by mean pooling. We use NLTK-tagger-chunker as labels of boundaries for each phrase and use NLTK sentence tokenization package to obtain sentence partitions [2].
- **Static Pruning**. We have implemented the first-k and attention-k methods of top k pruning [14], where we keep the first k tokens of a document or the top k tokens that receive the most amount of attention in the last layer of the document encoder. We also implemented IDF-uniform with parameter IDF threshold τ [1], which is the recommended technique from the paper. IDF parameter τ means to prune the top-τ tokens with the highest IDF value in MS MARCO corpus uniformly in all documents.

4.1 Zero-Shot Retrieval Performance in Out-of-Domain Datasets

We evaluate LeapMV outside the training domain to analyze its zero-shot effectiveness against pruning baselines and some state-of-the-art dense retrieval models.

Table 1 presents, under the same relevance accuracy constraints, the index space and latency requirements for different pruning methods in LoTTE pooled dataset. Accuracy 98% indicates that the method retains at least 98% of Success@5 performance compared to PLAID ColBERTv2. Each table entry lists the minimum space size (in GB) and GPU latency (in ms) for each method after pruning, based the PLAID framework, to meet or exceed the specified accuracy.

To meet the desired accuracy, we adjust k for the first-k and attention-k methods, IDF threshold τ in IDF-uniform, and α pruning ratio in LeapMV. The entry '-' means that such a method did not deliver the corresponding accuracy under the tested settings. Table 1 demonstrates that, under the same relevance constraint, LeapMV requires significantly less space and operates much faster.

Table 1. Space and latency for LoTTE pooled dataset under different accuracy constraints. PLAID ColBERTv2 takes 19.6 ms for the search queries and 24.8 ms for the forum queries with an index size of 13 GB.

Accuracy (%)	98% Space		98% Latency		97% Space		97% Latency		95% Space		95% Latency		93% Space		93% Latency	
	Search test queries															
First	-		-		12.0	(2.3×)	11.9	(1.5×)	9.2	(2.6×)	10.3	(1.5×)	6.9	(2.5×)	9.0	(1.5×)
Attn	10.0	(1.5×)	10.6	(1.2×)	8.4	(1.6×)	9.8	(1.2×)	6.8	(1.9×)	8.8	(1.3×)	5.9	(2.1×)	8.0	(1.3×)
IDF	-		-		8.8	(1.7×)	20.7	(2.6×)	7.5	(2.1×)	17.8	(2.7×)	6.2	(2.2×)	16.8	(2.7×)
LeapMV	**6.5**		**8.5**		**5.2**		**8.0**		**3.5**		**6.7**		**2.8**		**6.2**	
	Forum test queries															
First	-		-		-		-		11.0	(3.1×)	13.3	(1.8×)	9.2	(3.3×)	12.2	(1.8×)
Attn	11.0	(1.7×)	13.8	(1.4×)	10.0	(1.7×)	13.0	(1.4×)	7.6	(2.2×)	10.7	(1.4×)	5.9	(2.1×)	8.8	(1.3×)
IDF	8.8	(1.4×)	24.0	(2.4×)	8.0	(1.4×)	21.7	(2.3×)	7.2	(2.1×)	20.8	(2.8×)	6.2	(2.2×)	18.3	(2.7×)
LeapMV	**6.5**		**10.1**		**5.9**		**9.4**		**3.5**		**7.4**		**2.8**		**6.7**	

For example, with less than 2% loss (98% accuracy), LeapMV achieves 2× reduction in space and 2.5× speedup on GPU compared to PLAID. To achieve 97% of accuracy, LeapMV requires 5.2GB for search test queries, while IDF-uniform uses 1.7× more space and takes 2.6× longer.

Table 2. Relevance for LoTTE pooled dataset under space budgets (percentage of the PLAID space size). Entries are best Success@5 one method achieves under a space constraint.

Space budget	75%	50%	40%	32.5%	25%	17.5%
LoTTE search queries	Maximum Success@5					
First	68.3	65.3	63.1	61.1	57.8	53.8
Attn	70.6	67.9	66.6	65.1	60.5	49.8
IDF-uniform	70.3	67.5	62.6	60.7	59.5	-
LeapMV	**70.7**	**70.6**	**70.0**	**69.4**	**67.8**	**66.1**
LoTTE forum queries	Maximum Success@5					
First	58.9	56.3	55.2	53.1	50.3	45.6
Attn	61.8	59.1	58.1	56.0	51.8	43.0
IDF-uniform	62.5	59.9	56.3	54.0	51.8	48.8
LeapMV	**62.7**	**62.2**	**61.4**	**60.9**	**59.8**	**57.5**

Table 2 lists the retrieval quality, measured with Success@5 in LoTTE, under the same space budget constraint. Space size 75% means to have the index of at most about 75% of original ColBERTv2 PLAID, which is approximately 13GB for LoTTE pooled dataset. For example, using about 50% of the space ColBERTv2 PLAID needs (about 6.5GB), LeapMV can deliver Success@5 of 70.6

Table 3. Zero-shot retrieval performance on LoTTE test set benchmark.

BEIR Datasets	Rocket –QAv2	Retro –MAE	PLAID	First k=50	Attn k=50	IDF τ=100	IDF τ=300	LeapMV
Space	-	-	13GB	5.1GB	5.1GB	8GB	6.2GB	6.5GB
	LoTTE search test queries (Success@5)							
Writing	78.0	-	80.6	74.2	76.3	76.5	74.3	76.2
Recreation	72.1	-	72.3	64.1	67.2	72.4	71.4	72.2
Science	55.3	-	56.7	49.8	53.2	55.4	54.3	55.3
Technology	63.4	-	66.1	56.5	59.4	63.6	62.6	67.1
Lifestyle	82.1	-	84.3	73.1	80.9	83.4	82.5	84.7
Pooled	69.8	66.8	72.1	63.1	66.6	69.3	67.5	70.2
% Loss	-	-	0%	12.5%	7.6%	3.9%	6.4%	2.6%
	LoTTE forum test queries (Success@5)							
Writing	71.5	-	76.2	70.9	73.8	74.0	72.0	75.8
Recreation	65.7	-	71.4	60.4	66.5	70.9	69.3	69.6
Science	38.0	-	47.2	39.0	39.9	45.9	44.1	44.4
Technology	47.3	-	53.7	48.3	46.8	52.1	49.9	53.3
Lifestyle	73.7	-	76.9	68.0	72.7	76.5	75.2	77.1
Pooled	57.7	58.5	63.5	55.2	58.1	62.1	59.9	62.1
% Loss	-	-	0%	13.1%	8.5%	2.2%	5.7%	2.2%

while other methods can deliver at most 67.9. When adhering to a tight space budget, such as 17.5% of the original index, the baseline methods experience a decrease in performance from 25–31% compared to PLAID in search queries. In contrast, LeapMV only shows an 8.3% decline. Overall, LeapMV consistently delivers better relevance under the same space budget and is robust under more aggressive compression ratio.

To assess the performance for each subtopics of LoTTE, Table 3 presents the Success@5 (Recall@5) of different methods with several parameter settings with index sizes comparable to our default LeapMV. As a point of reference, we also include the average NDCG@10 of two recent single-vector dense retrievers RetroMAE [17] and RocketQAv2 [21]. The result from Table 3 shows that LeapMV can deliver about 2× reduction in space usage while incurring only 2.6% and 2.2% relevance loss. In comparison, IDF-uniform with τ=100 can deliver visibly smaller 1.6× space reduction while having a larger average loss of relevance (3.9% and 2.2%). These findings are consistent with the results discussed in Table 1 and Table 2.

Table 4 shows the zero-shot retrieval performance, measured in NDCG@10, across 13 BEIR datasets for out-of-domain search. For relevance comparison, we also list NDCG@10 of single-vector dense retrievers RetroMAE and RocketQAv2. The targeted space reduction is about 50% of the PLAID ColBERTv2

Table 4. Zero-shot retrieval performance on 13 BEIR datasets.

BEIR Datasets	Rocket- QAv2	Retro-MAE	PLAID	First-k	Attn-k	IDF-τ	LeapMV
% of PLAID space	-	-	100%	54%	54%	49%	49%
SciFact	56.8	65.3	69.3	54.9	47.5	62.9	68.9
NFCorpus	29.3	30.8	33.8	30.7	26.5	31.9	34.1
ArguAna	45.1	43.3	46.3	44.5	45.1	41.4	44.9
SCIDOCS	13.1	15.0	15.4	14.8	13.9	14.6	15.2
Touche-2020	24.7	23.7	26.3	22.8	22.1	26.2	24.0
FiQA	30.2	31.6	35.6	26.2	30.0	30.3	34.6
T-COVID	67.5	77.2	73.8	68.5	55.2	61.7	75.2
NQ	50.5	51.8	56.1	48.5	48.4	48.8	54.5
DBPedia	35.6	39.0	44.6	42.5	42.6	41.6	42.5
HotpotQA	53.3	63.5	66.7	57.8	58.5	62.6	59.6
FEVER	67.6	77.4	78.5	63.3	56.5	74.8	74.1
C-FEVER	18.0	23.2	17.6	13.7	11.5	17.6	18.1
Quora	74.9	84.7	85.2	85.4	85.4	76.9	82.5
Average NDCG@10	43.6	48.2	50.0	44.1	41.8	45.5	48.3
% Loss	-	-	0%	11.69%	16.35%	8.98%	3.28%

size, achieved by selecting $k = 50$ for top-k pruning and τ=1000 for IDF. By default, LeapMV reduces space usage by about 2×, with only a 3.28% relevance loss compared to ColBERTv2 PLAID. In comparison, the first-k, attention-k, and IDF uniform have an NDCG@10 loss varying from 8.98% to 16%.

4.2 In-Domain Search with MS MARCO

Table 5 lists the performance of LeapMV for MS MARCO passage ranking in comparison with the original PLAID ColBERTv2, phrase chunking, sentence chunking, top-k static pruning, and IDF-uniform. The results from Table 5 show the relevance of LeapMV is fairly close to static top-k pruning which uses a fixed top k parameter. LeapMV incurs a relevance loss of 2.7% for MRR@10 and 1.6% for Recall@1K compared to the original PLAID implementation of ColBERTv2, while achieving approximately 2× space reduction and 1.8× faster GPU latency. In contrast, IDF uniform with $\tau = 100$ uses 33% more space than LeapMV default while their Recall@1K difference is within 1%. When increasing τ to 3000, IDF's space cost 5.4GB approaches LeapMV with $\alpha = 80\%$, but results in a rapid decline in relevance, making it substantially less effective than LeapMV. Similar conclusions can be drawn with the first k and the attention k method, demonstrating the robustness of LeapMV. The two text chunking methods are not competitive due to their significant losses in retrieval performance.

As a reference, we also list the relevance performance of single-vector dense retrieval RetroMAE and RocketQAv2 [21]. The released RetroMAE checkpoint

Table 5. End-to-end in-domain retrieval performance for MS MARCO passages.

Methods	Relevance		Latency ms (Speedup)				Space (Ratio)	
	MRR@10	R@1K	GPU		8-thr. CPU			
RocketQA	38.8	98.1	-	-	-	-	27	-
RetroMAE (no title anno.)	36.0	97.7	-	-	-	-	27	-
PLAID ColBERTv2	39.4	97.6	18.9	(1×)	69.1	(1×)	22	(1×)
Phase Chunking	30.0	89.6	11.5	(1.6×)	56.5	(1.2×)	12	(1.8×)
Sentence Chunking	22.0	69.4	5.6	(3.4×)	16.3	(4.2×)	2.1	(10.5×)
First k=15	31.7	84.7	5.9	(3.2×)	30.6	(2.3×)	4.9	(4.5×)
First k=50	38.0	95.8	8.7	(2.2×)	59.8	(1.2×)	16	(1.4×)
Attn k=15	28.2	86.4	5.7	(3.3×)	29.1	(2.4×)	4.8	(4.6×)
Attn k=50	38.6	97.1	8.6	(2.2×)	62.2	(1.1×)	16	(1.4×)
IDF τ=3000	27.3	86.1	8.9	(2.1×)	27.3	(2.5×)	5.4	(4.1×)
IDF τ=100	38.5	97.0	13.0	(1.5×)	54.6	(1.3×)	15	(1.5×)
LeapMV, α=0.8	35.8	93.6	6.0	(3.2×)	26.1	(2.6×)	4.9	(4.5×)
LeapMV, default	38.4	96.0	8.0	(2.4×)	49.9	(1.4×)	12	(1.8×)
LeapMV, α=0.25	38.8	96.2	9.5	(2.0×)	59.5	(1.2×)	16	(1.4×)

gives 0.359 MRR@10 using the standard MS MARCO, and this is below 0.416 reported in [17]. The reason was explained in [13] that, following RocketQAv2 [21], the paper evaluates the modified MS MARCO dataset with title annotation, which is not fair. Since the original dataset released does not utilize title information, all experiments in ColBERT and our evaluation follows the standard approach to use the original MS MARCO without title annotation.

4.3 Ablation Studies

We present two ablation studies regarding model architecture design.

Impact of Window Size Choices. Table 6 shows the impact of the window size c in Formula (2) on the MS MARCO passage, Dev set. Due to the cost of model training, we train for 300k steps instead of 400k steps for this study. The result shows that imposing a wider context window does not yield much performance gain. We suspect that document token encodings already contain sufficient contextualized information in MS MARCO. For simplicity, all evaluations including zero-shot retrieval use $c = 0$.

Table 6. Impact of window size choice c on MS MARCO dev set.

c = 0		c = 2		c = 5	
MRR@10	Recall@1K	MRR@10	Recall@1K	MRR@10	Recall@1K
38.3	96.0	38.2	96.2	38.2	95.8

Linear Layer Versus Transformer in LeapMV. Figure 1 uses a linear layer to compute the pruning decision. Table 7 shows the impact of replacing a single linear layer with a more complex transformer encoder [27]. For the transformer encoder, hidden dimension d is equal to the default ColBERT encoding of 128, 8 attention heads, and feed-forward layer dimension of $4d$. Despite integrating this more complex module, both modules yield a similar relevance. For better indexing efficiency, we choose the linear layer with fewer parameters for LeapMV.

Table 7. Choice of linear layer versus transformer network on MS MARCO dev set for computing pruning decisions in LeapMV architecture.

Linear Layer		Transformer	
MRR@10	Recall@1K	MRR@10	Recall@1K
38.4	96.0	38.4	95.9

5 Concluding Remarks

We introduce LeapMV, a learned token pruning scheme designed for efficient multi-vector dense retrieval with a moderate relevance tradeoff to enhance efficiency. LeapMV makes a context-aware pruning decision for each document and is more adaptive than existing baselines. The degree of pruning in LeapMV can be adjusted with parameter α to meet the targeted space reduction goals, resulting in a near-linear reduction in both GPU and CPU latency.

Our evaluation shows the advantage of LeapMV over the previous baselines in adapting to various space reduction goals. For instance, Table 1 shows that under the same relevance accuracy budget, LeapMV consumes up to 3.3× less space and is up to 2.8× faster for LoTTE than top-k token pruning and IDF methods. Compared to the optimized PLAID ColBERTv2, LeapMV achieves 2× space saving and 2.5× GPU latency speedup while incurring less than 2% relevance loss on LoTTE forum queries. LeapMV yields more space and time saving when accuracy can be relaxed further while maintaining performance significantly better than its competitors under tight space budgets. Our current evaluation follows the default parameter setting in PLAID with a retrieval depth of 1000, and a future study is to investigate the impact of varying these settings.

Acknowledgments. We thank anonymous referees for their valuable comments. The major portion of this work was supported by Amazon. It was partially supported by U.S. NSF IIS-2225942 and its ACCESS program in using its computing resource. Any opinions, findings, conclusions or recommendations expressed in this material are those of the authors and do not necessarily reflect the views of Amazon or the NSF.

References

1. Acquavia, A., Macdonald, C., Tonellotto, N.: Static pruning for multi-representation dense retrieval. In: Proceedings of the ACM Symposium on Document Engineering 2023. DocEng '23, Association for Computing Machinery, New York, NY, USA (2023). https://doi.org/10.1145/3573128.3604896
2. Bird, S., Loper, E.: NLTK: the natural language toolkit. In: Proceedings of the ACL Interactive Poster and Demonstration Sessions, pp. 214–217. Association for Computational Linguistics, Barcelona, Spain (2004). https://aclanthology.org/P04-3031
3. Campos, D.F., et al.: MS MARCO: a human generated machine reading comprehension dataset. ArXiv abs/1611.09268 (2016)
4. colbertv1.9: https://huggingface.co/colbert-ir/colbertv1.9 (2022)
5. Devlin, J., Chang, M.W., Lee, K., Toutanova, K.: BERT: pre-training of deep bidirectional transformers for language understanding. In: NAACL (2019)
6. Engels, J., Coleman, B., Lakshman, V., Shrivastava, A.: DESSERT: an efficient algorithm for vector set search with vector set queries. In: Thirty-seventh Conference on Neural Information Processing Systems (2023). https://openreview.net/forum?id=kXfrlWXLwH
7. Formal, T., Lassance, C., Piwowarski, B., Clinchant, S.: From distillation to hard negative sampling: making sparse neural IR models more effective. SIGIR (2022)
8. Formal, T., Piwowarski, B., Clinchant, S.: SPLADE: sparse lexical and expansion model for first stage ranking. SIGIR (2021)
9. Gumbel, E.: Statistical Theory of Extreme Values and Some Practical Applications: A Series of Lectures. U.S. Government Printing Office, Applied mathematics series (1954)
10. Jang, E., Gu, S.S., Poole, B.: Categorical reparameterization with Gumbel-Softmax. ICLR (2017)
11. Khashabi, D., Ng, A., Khot, T., Sabharwal, A., Hajishirzi, H., Callison-Burch, C.: GooAQ: open question answering with diverse answer types (2021). https://arxiv.org/abs/2104.08727
12. Khattab, O., Zaharia, M.A.: ColBERT: efficient and effective passage search via contextualized late interaction over BERT. SIGIR (2020)
13. Lassance, C., Clinchant, S.: The tale of two MSMARCO - and their unfair comparisons. In: Proceedings of the 46th International ACM SIGIR Conference on Research and Development in Information Retrieval, pp. 2431–2435. SIGIR '23, ACM, New York, NY, USA (2023)
14. Lassance, C., Maachou, M., Park, J., Clinchant, S.: Learned token pruning in contextualized late interaction over BERT (ColBERT). In: Proceedings of the 45th International ACM SIGIR Conference on Research and Development in Information Retrieval, pp. 2232–2236. SIGIR '22, Association for Computing Machinery, New York, NY, USA (2022). https://doi.org/10.1145/3477495.3531835

15. Lee, J., et al.: Rethinking the role of token retrieval in multi-vector retrieval. In: Thirty-seventh Conference on Neural Information Processing Systems (2023). https://openreview.net/forum?id=ZQzm0Z47jz
16. Li, M., et al.: CITADEL: conditional token interaction via dynamic lexical routing for efficient and effective multi-vector retrieval, pp. 11891–11907 (2023)
17. Liu, Z., Xiao, S., Shao, Y., Cao, Z.: RetroMAE-2: duplex masked auto-encoder for pre-training retrieval-oriented language models. In: Rogers, A., Boyd-Graber, J., Okazaki, N. (eds.) Proceedings of the 61st Annual Meeting of the Association for Computational Linguistics (Volume 1: Long Papers), pp. 2635–2648. Association for Computational Linguistics, Toronto, Canada (2023)
18. Maddison, C.J., Mnih, A., Teh, Y.W.: The concrete distribution: a continuous relaxation of discrete random variables. ICLR (2017)
19. MiniLM-L-6-v2: https://huggingface.co/cross-encoder/ms-marco-minilm-l-6-v2 (2022)
20. Nardini, F.M., Rulli, C., Venturini, R.: Efficient multi-vector dense retrieval with bit vectors. In: Goharian, N., Tonellotto, N., He, Y., Lipani, A., McDonald, G., Macdonald, C., Ounis, I. (eds.) Advances in Information Retrieval (ECIR 2024), pp. 3–17. Springer Nature Switzerland, Cham (2024)
21. Ren, R., et al.: RocketQAv2: a joint training method for dense passage retrieval and passage re-ranking. In: Proceedings of the 2021 Conference on Empirical Methods in Natural Language Processing, pp. 2825–2835. ACM, Online and Punta Cana, Dominican Republic (2021)
22. Sang, E.F.T.K., Buchholz, S.: Introduction to the CoNLL-2000 shared task: Chunking (2000). https://arxiv.org/abs/cs/0009008
23. Santhanam, K., Khattab, O., Saad-Falcon, J., Potts, C., Zaharia, M.A.: ColBERTv2: effective and efficient retrieval via lightweight late interaction. ArXiv abs/2112.01488 (2021)
24. Santhanam, K., Khattab, O., Potts, C., Zaharia, M.: PLAID: an efficient engine for late interaction retrieval. In: Proceedings of the 31st ACM International Conference on Information & Knowledge Management, pp. 1747–1756. CIKM '22 (2022)
25. Sciavolino, C., Zhong, Z., Lee, J., Chen, D.: Simple entity-centric questions challenge dense retrievers. In: Proceedings of the 2021 Conference on Empirical Methods in Natural Language Processing, pp. 6138–6148. ACL (2021)
26. Thakur, N., Reimers, N., Rücklé, A., Srivastava, A., Gurevych, I.: BEIR: a heterogeneous benchmark for zero-shot evaluation of information retrieval models. In: NeurIPS (2021)
27. Vaswani, A., et al.: Attention is all you need (2023). https://arxiv.org/abs/1706.03762
28. Yang, Y., Qiao, Y., He, S., Yang, T.: Weighted KL-divergence for document ranking model refinement. arxiv and SIGIR 2024 (2024)

Advancing Math Formula Search Using Diverse Structural and Symbolic Representations

Sumedh Vemuganti(✉), Ayu Seiya, and Nickvash Kani

University of Illinois at Urbana Champaign, Champaign, IL 61820, USA
{sumedh2,aseiya2,kani}@illinois.edu
https://mlpgroup.xyz

Abstract. We present a comprehensive analysis of search engines from past MathIR competitions (ARQMath, NTCIR) and investigate the semantic representations required for effective formula retrieval. After evaluating these search engines' methodologies and performance, we uncover a diverse range of equation-parsing and feature-extraction strategies. Inspired by these approaches, we present an ensemble-based retrieval system, integrating symbolic, tree-based, and textual representations of formulas. A key innovation of our work is the Leaf-to-Leaf (L2L) feature extraction method, which captures structural dependencies in formula trees by identifying the shortest paths of operators between pairs of leaf nodes. Our system combines L2L features with symbol-based and LaTeX-based rankings, surpassing the top performing ARQMath-3 search engines in nDCG′ and mAP′ metrics, while matching or exceeding the top NTCIR-12 performers in PR′@{15, 20} and P′@{5, 10, 15, 20} measures. These results demonstrate the effectiveness of leveraging multiple feature spaces to enhance mathematical formula retrieval.

Keywords: mathematical information retrieval · set-based retrieval · query-by-expression

1 Introduction

Mathematical Information Retrieval (MathIR) has emerged as a critical area of research that encompasses: searching for information using mathematical notation [13], matching query formulas to those in existing repositories, [16,33], and ranking results based on their relevance to mathematical queries [24]. Historically, text-based retrieval systems have not accounted for the structural and symbolic complexity of formulas [11], while transformer-based models primarily detect shallow features, such as variable overlap and operator presence, neglecting deeper relationships, such as operator precedence. [24]; fine-tuning these models reinforces reliance on surface-level patterns. In contrast, top systems in previous MathIR competitions (ARQMath and NTCIR) improve retrieval accuracy by leveraging hierarchical and syntactic representations [35,37].

C. Hauff et al. (Eds.): ECIR 2025, LNCS 15572, pp. 116–131, 2025.
https://doi.org/10.1007/978-3-031-88708-6_8

An evaluation of the successes and shortcomings of past MathIR systems highlights the need to: (1) index features beyond plain text, (2) capture structure from trees, and (3) use robust back-end systems. Based on these insights, we introduce L2L, an improved IR system for retrieving and ranking mathematical expressions. L2L converts Content MathML into operator trees for feature extraction, applies variable grouping for symbol matching, and extracts shortest paths between leaf nodes to capture tree-based relationships. Integrating symbolic and LaTeX similarity, L2L utilizes information from multiple representation schemes in both indexing and ranking. L2L leverages Neo4J-a mature back-end system for rapid indexing and efficient query processing.

L2L outperforms all prior submissions to ARQMath-3 in nDCG′, MAP′ and NTCIR-12 in P′@{5, 10, 15, 20}. An ablation study on feature-space weights reveals that ranking precision depends on the inclusion of all three feature spaces, rather than their relative weighting. We conclude that ensemble methods combining structural and visual features are key to advancing MathIR.

2 Background

MathIR distinguishes itself from textual IR by addressing the challenge of handling layered structures and relationships unique to mathematical notation [19,23,29]. Researchers have explored textual indexing, tree-based representations, and feature-driven systems. This section reviews these techniques, highlighting their strengths, limitations, and associated lessons.

2.1 Text-Based Approaches

An analysis of systems submitted to the early NTCIR-10 competition [1] reveals that solely text-based indexing, even as early as 2013, was ineffective, prompting the development of advanced, semantic-aware strategies.

The BRKLY [10] system represented math operators as string tags in its indexing structure. When formulas were queried and matched to candidate responses, the gathered document pool-used for formula extraction-was suboptimal, leading to a low competition standing [1]; purely lexical indexing hindered performance.

In the same competition, the high-performing MCAT system [27] took a more sophisticated approach by capturing both tree structure and formula-symbols in its indexing scheme. This enabled relevant results to be retrieved even when minor structures were altered or variables were renamed [1]. By going beyond raw string-based indexing, MCAT delivered a stronger retrieval performance.

Moreover, none of the MathIR engines submitted to subsequent iterations of the same competition [3,21] relied solely on textual indexing. Each subsequent system leveraged the Mathematical Markup Language (MathML) [18] annotation scheme to represent mathematical expressions, capturing visual layout with Presentation MathML and underlying structure with Content MathML.

Lesson 1: Text-based indexing alone falls short in MathIR; advanced techniques integrating mathematical information are essential to handle formula variability and remain resilient to symbolic/structural changes.

2.2 Tree-Based Approaches

Tree-based representations of formulas capture structural relationships between operators and variables, as well as the relative spatial positioning of symbols. Incorporating information from trees into the indexing and search processes bolsters performance in MathIR tasks [1,2,21]. Moreover, even advanced encoder models perform better when combined with tree-based approaches [14,22].

The MCAT [9] and Tangent-S [5] systems leveraged information from tree representations to achieve competitive performance in the NTCIR-12 tasks [21]. MCAT encoded hierarchical information through root-to-node paths. Local context was captured through horizontal connections in sibling relationships. Similarly, Tangent-S scored formulas using relative paths between symbol pairs, encoding symbol-operator relationships. Modeling spatial relationships from trees is precisely what drove these systems' strong performance in NTCIR-12.

Tree-based paths have evolved into embedding schemes like Tangent-CTF [14], which encodes paths between symbols as tuples, embeds them using n-gram techniques, and averages the vectors to represent the tree. These embeddings reduce reliance on specific paths, improving retrieval for queries with unseen paths. However, combining Tangent-CTF with a tree-based search engine yields stronger results. This implies that raw tree paths capture critical information, beyond what is expressible through advanced embedding schemes [14].

A similar trend is observed in neural embedding methods such as MathBERT [22], which uses Masked Substructure Prediction (MSP) to learn representations of formulas. However, augmenting MathBERT with tree-based retrieval improves performance; this suggests that tree paths capture crucial structural information that even a machine learning embedding scheme cannot.

Lesson 2: Tree-based features capture structure and meaning that text alone cannot. Paths between tree nodes act as effective retrieval structures.

2.3 Feature-Based Approaches

The success of feature-based systems is driven by mature search engine software, such as Lucene [28], with built-in support for ranking and inverted indexing.

MIaS [26] leverages Lucene's backend to overcome three core complexities in its implementation: (1) MIaS requires indexing of sub-trees, which is facilitated by Lucene's tokenizer and rapid inverted indexing; (2) MIaS needs a modified scoring formula to account for formula-specific weights, which Lucene accommodates with its built-in scoring system; and (3) MIaS supports querying of both text and formulas together, a capability enabled by Lucene's multimodal query support. Lucene's mature, versatile features provide the foundation necessary for building the MIaS system. Tangent-L [31] relies on Lucene's rapid back-end to retrieve responses from large candidate pools and leverages Lucene's in-built BM25+ for swift ranking. Lucene provides the foundational backbone that enables Tangent-L to unify structure-aware matching with efficient ranking.

Lesson 3: Mature and scalable back-ends are crucial for feature-based MathIR, enabling efficient indexing, retrieval, and ranking of formulas.

3 Methods

This section outlines the methodology and motivation behind L2L's indexing, retrieval, and ranking schemes. We discuss how parallelism is used in runtime optimization and motivate the inclusion of hyperparameters in our ranking function for an exploratory post hoc analysis of the importance of feature spaces.

3.1 Indexing

Our indexing methodology, illustrated in Fig. 1, builds on the lessons outlined above (2) and distinguishes itself through refined granularity in node grouping, a balanced approach to incorporating root-based information in tree features, and the adoption of a robust Neo4J [6] back-end.

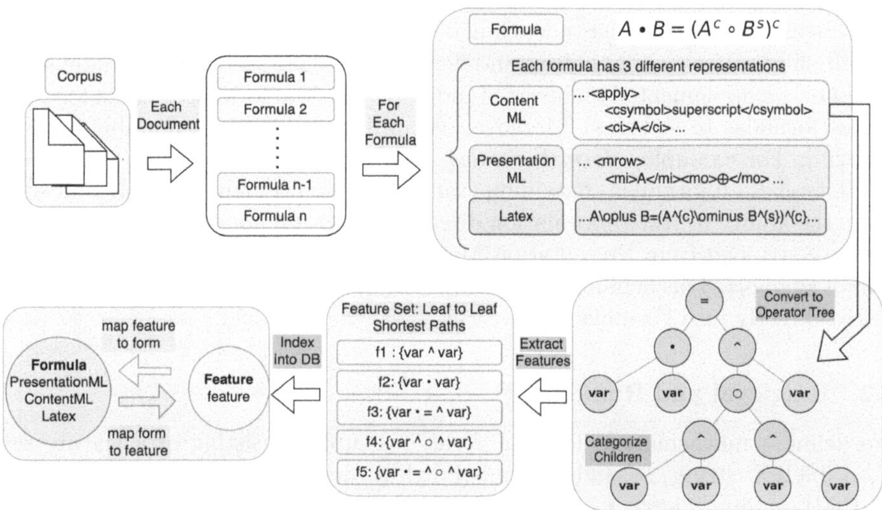

Fig. 1. Overview of the equation indexing process. Equations are extracted from the dataset in three forms: Content MathML, Presentation MathML, and LaTeX. For leaf-to-leaf (L2L) features, the ContentML string is converted into an operator tree, and shortest paths between variables are extracted as features. Both forward and inverted indices are constructed for efficient retrieval.

To combat text-based IR systems' inadequacies in handling notational variability (Lesson 1), our system transforms Content MathML strings into structured operator trees, grouping leaf nodes into three categories: "var" for alphabetic characters, "num" for numeric values, and "ascii" for ASCII symbols. We use fewer degrees of granularity than Tangent-S [34], which employs eight distinct node types (variables, fractions, text fragments, matrices, radicals, parenthesized expressions, commutative/non-commutative operators), but more than the broader granularity levels in MCAT [27], which categorizes nodes into only three levels: math, paragraph, and document. Hyper-specificity in leaf node categorization may inhibit the retrieval of relevant results, whereas excessive looseness can retrieve too many irrelevant results. We believe this approach to canonicalization strikes a balance, providing enough specificity to filter out irrelevant results while maintaining enough generality to capture relevant matches.

Guided by tree-based paths as retrieval units (Lesson 2), our system extracts the shortest paths between leaf nodes as features, encapsulating structural relationships. Our approach differs from former methods such as Tangent-S [34], which relied on full root-to-leaf paths. When the same authors built Tangent-CFT [14], they fully omitted root-to-leaf paths to prevent unnecessary information from clobbering embeddings. This tells us that root based information has some value, but if overused, can distract from other critical aspects of retrieval. We believe that our approach, which naturally includes the root in longer feature paths, but excludes the root in shorter feature paths balances the relevance of root-related information.

Lesson 3 prompted the adoption of a mature Neo4J [17] back-end with inbuilt support for custom indices and flexibility in data-storage. To handle large datasets, we implemented an inverted index (features to formulas) and a forward index (formulas to features). Moreover, GraphIR provides interpretable natural querying. For example, *(form:Formula) -> [form_to_ftr] -> (ftr: (var, times, var))*, fetches all formulas containing (var, times, var). Furthermore, Neo4J supports the ability to store information directly into graph nodes. During indexing, tags are stripped from Presentation MathML, reducing it to an ordered collection of symbols. This symbol set, in conjunction with the raw LaTeX strings are stored directly into Formula nodes.

3.2 Retrieval and Ranking

We define a mathematical formula X as a 3-tuple consisting of a feature set F, a LaTeX string L, and a symbol list S derived from the Presentation MathML string. Thus, $X = (F, L, S)$. Let the query formula be represented as $Q = (F_Q, L_Q, S_Q)$ and the response formula as $R = (F_R, L_R, S_R)$. During the query step, the Content MathML representation is processed to construct an operator tree. Leaf nodes in the tree are grouped in the same manner as they are during indexing. From this tree, all pairs of shortest paths between leaf nodes are extracted to generate the feature set F_Q. For each feature $f \in F_Q$, a graph query is performed to retrieve all information for formulas containing this feature (Fig. 2).

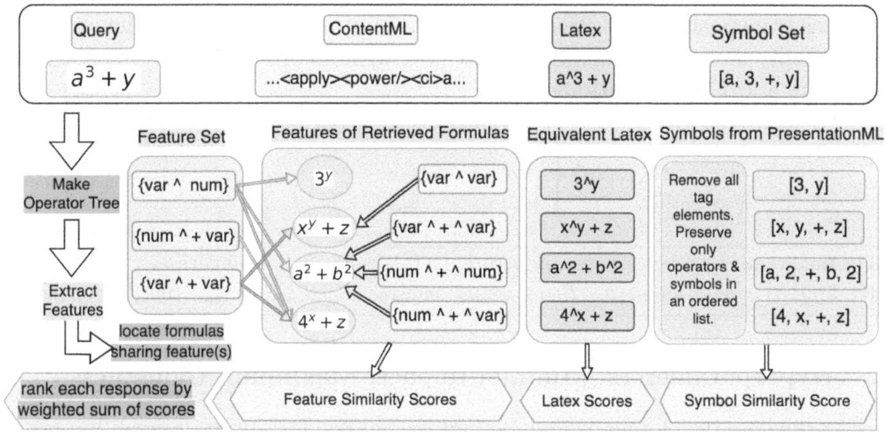

Fig. 2. Overview of the equation retrieval process: The Content MathML string is converted into an operator tree to extract feature paths. Formulas sharing at least one feature with the query are retrieved, and their features are extracted. Three scores are computed for each response: feature set similarity, LaTeX string similarity, and symbol similarity from Presentation MathML. The final rank is a weighted sum of these scores.

From this collection of formulas–sharing at least one feature with the query–the similarity between the query Q and each response R is computed in three ways: (1) a feature set similarity score between F_Q and F_R, (2) a LaTeX string similarity score between the LaTeX representations L_Q and L_R, and (3) a symbol similarity score between symbol sets S_Q and S_R. Finally, a weighted sum of these three similarity scores is computed for ranking the responses.

The feature similarity formula sim_{ftr} was designed to ensure that queries and responses that share more features are ranked higher. The intersection over the union [7] ranking metric was chosen such that equations are evaluated not just on the quantity of shared features, but also on the proportional relevance of those features to both the query and response:

$$\text{sim}_{\text{ftr}}(F_Q, F_R) = \frac{|F_Q \cap F_R|}{|F_Q \cup F_R|}$$

LaTeX strings have precise textual representations, where small changes can result in entirely different expressions. To measure similarity, we calculate the minimum number of edits required to convert one string into another-fewer edits indicate higher similarity. $E(L_Q, L_R)$ represents the edit distance [12] with unit-cost insertion and deletion (no transposition). The similarity is given by:

$$\text{sim}_{\text{latex}}(L_Q, L_R) = \frac{1}{1 + E(L_Q, L_R)}$$

After removing tags, Presentation MathML strings resemble ordered collections of variable, numeric, and operator symbols. Jaro-Winkler distance [8] penalizes transpositions less than edit distance but more that than order-agnostic set-similarity metrics, balancing sensitivity to symbol order. With m mismatched characters and t transpositions:

$$\text{sim}_{\text{sym}}(S_Q, S_R) \begin{cases} 0 & \text{if } m = 0, \\ \frac{1}{3}\left(\frac{m}{|S_Q|} + \frac{m}{|S_R|} + \frac{m-t}{m}\right) & \text{if } m > 0 \end{cases}$$

The final rank is a weighted sum of feature, LaTeX, and symbol similarities.
$\text{Rank}(Q, R) = w_{\text{ftr}} \cdot \text{sim}_{\text{ftr}}(F_Q, F_R) + w_{\text{latex}} \cdot \text{sim}_{\text{latex}}(L_Q, L_R) + w_{\text{sym}} \cdot \text{sim}_{\text{sym}}(S_Q, S_R)$

3.3 Hyperparameters: Minimal Impact, Maximum Robustness

During the core development and evaluation phases, L2L used uniform hyperparameter weights ($w_{\text{ftr}} = 1$, $w_{\text{latex}} = 1$, $w_{\text{sym}} = 1$). This design choice ensured an unbiased baseline, avoiding any tuning on the test set that could risk methodological invalidity or overfitting, as our data set of 20 queries for NTCIR-12 was too small to perform model cross-validation. After establishing these baseline results, we conducted an exploratory post hoc analysis to investigate the effects of varying the weights of the systems and to understand the contributions of the ranking components to the overall effectiveness of the system. This analysis, detailed in Sect. 4, revealed that the system's performance is robust to hyperparameter variations, provided that all components are included.

3.4 A Note on Runtime

Serial runtimes in Table 1 are accelerated with parallelized document processing during feature extraction, custom index utilization during retrieval, and concurrent scoring during ranking.

Table 1. Runtimes for indexing, retrieval, and ranking sub-processes with n variable nodes, m internal nodes, and r shared features. For retrieval and ranking, n and m refer to the larger of the two formulas being compared.

Process	Runtime
Operator Tree Construction	$O(n + m)$
Feature Extraction	$O(n^2 \log m)$
Retrieval and Ranking	$O(r \cdot n^2 \log m)$

When indexing a corpus, multiple documents are processed simultaneously. Within a document, feature-extraction is performed in parallel for all formulas. During retrieval, Neo4J's custom indices accelerate the process of gathering formulas containing a specific feature. During ranking, string similarity computations are limited to formulas that share at least one feature with the query, removing needless quadratic time computations. All retrieved results are ranked in parallel. Moreover, for a fixed response, each similarity metric is computed concurrently, ensuring rapid ranking. Parallelization, targeted-computations, and custom indices enable system efficiency for large, complex mathematical datasets.

4 System Results and Feature Space Ablation Study

The L2L MathIR Engine demonstrates strong performance across two major formula retrieval benchmarks: the ARQMath-3 formula retrieval task (Task 2) and the non-wildcard subset (Queries 1–20) of the Wikipedia formula retrieval task from NTCIR-12. Experimentation with feature-space parameters, combined with an analysis of queries most impacted by these changes, highlight the importance of all three representations for retrieval accuracy.

4.1 ARQ-3 and NTCIR-12 Performance

For ARQMath-3, L2L achieves the highest nDCG′, and mAP′, and second highest P′@10 score. P′@10 measures the proportion of the top 10 retrieved responses that are fully relevant. With uniform weights, our system achieves a P′@10 score of 0.6800, 0.008 below the best system, retrieving 0.08 fewer fully relevant responses in the top 10. mAP′ evaluates the average proportion of relevant responses across the top 1, top 2, top 3, ..., top k results. L2L achieves the highest mAP′ score of 0.6197, consistently ranking relevant responses effectively across retrieval levels. nDCG′ calculates the sum of relevance scores, weighted to emphasize ranking quality at all levels using a logarithmic adjustment. L2L achieves the maximum nDCG′ score (0.8488), preserving clarity in deeper ranks (Table 2).

For the NTCIR-12 competition, L2L fetched strong fully-relevant matches across all ranks and robust partial matches in deeper retrieved responses. P′@k measures the proportion of the top k retrieved results deemed fully relevant to the query, while PR′@k captures the proportion of the top k results marked as either partially or fully relevant. L2L matched the top system in P′@5 and outperformed it in P′@{10, 15, 20}, showcasing robust full-relevance matches at all retrieval depths. While PR'{5, 10} were slightly lower, indicating fewer partially relevant results at the top ranks, L2L compensated by delivering fully relevant results at these levels and strong partial matches at deeper ranks, where PR′@{15, 20} notably exceeded the top systems. This aligns with user expectations-highly relevant matches are critical in the top ranks, while partially relevant results at deeper levels are valuable for providing broader context.

Due to the small sample size (20 queries) and limited comparison teams (2 for NTCIR-12, 5 for ARQMath-3), statistical significance testing was not conducted; doing so would violate key assumptions such as normality. Instead, we report effect sizes, such as mean differences, to provide a more meaningful interpretation of performance differences in this context.

Table 2. PR′ and P′ Scores for NTCIR-12 non-wildcard queries on Wikipedia Formula Systems. Scores for the non-wildcard queries were computed using RITUW and MCAT [30] competition run files. Since the competition is over, the P′@k variant, which considers only scored sources, was used for comparison [36]. While we aimed to compare our results with embedding systems like MathBERT [22] and Tangent-CFT [14], these were optimized for BPREF@1000, a metric focusing on much deeper ranks, making comparisons to top-20 relevance metrics unsuitable.

System	PR'@5	PR'@10	PR'@15	PR'@20	P'@5	P'@10	P'@15	P'@20
L2L	0.460	0.4375	**0.4183**	**0.4050**	**0.265**	**0.2075**	**0.1783**	**0.1513**
RITUW_run1	0.470	**0.4425**	0.4133	0.3975	0.240	0.1775	0.1450	0.1188
RITUW_run2	0.460	0.4275	0.4000	0.3850	0.210	0.1650	0.1333	0.1150
RITUW_run3	0.455	0.4300	0.4067	0.3875	0.260	0.1750	0.1467	0.1250
RITUW_run4	0.455	0.4125	0.4033	0.3850	**0.265**	0.1850	0.1583	0.1388
RITUW_run5	**0.475**	0.4325	0.4167	**0.4050**	0.240	0.1775	0.1433	0.1225
MCAT_lr_unif	0.405	0.3900	0.3750	0.3550	0.230	0.1650	0.1283	0.1130
MCAT_lr	0.405	0.3825	0.3683	0.3500	0.225	0.1650	0.1283	0.1112
MCAT_unif	**0.475**	0.4325	0.4167	**0.4050**	0.240	0.1775	0.1433	0.1225

4.2 Experiment 1: Ranking Function Hyperparameter Sweep

To investigate whether L2L's performance was influenced by the coordinated unit-weight hyperparameter selection and to assess the system's robustness under varying conditions, we conducted a hyperparameter sweep. This experiment focused on P′@10, as it was the lowest-performing metric with unit weights, making it the most sensitive indicator of potential deficits or hyperparameter dependencies in the system′s design. During the sweep, we fixed w_{sym} at 1 and adjusted the relative contributions of the other two ranking components, measuring the system's performance on P′@10. The results are shown in Fig. 3.

P′@10 ranged from a minimum of 0.616 to a maximum of 0.688, matching the best system. This narrow range of variation-7.2% between the lowest and highest precision-indicates that the system consistently retrieves a high proportion of relevant results regardless of the weight configuration. By maintaining stable performance across weight shifts, L2L demonstrates resilience to hyperparameter changes, highlighting its robustness. The stability observed in P′@10, despite being the lowest-performing metric, suggests that it is the inclusion of all three feature spaces, rather than the reliance on fine-tuned parameters that drives L2L′s success.

4.3 Experiment 2: Suppression of Feature Space Component(s)

To understand the individual contributions of each ranking component and their combined impact on retrieval accuracy, we conducted two experiments. (1) We assess all seven possible combinations of the three components-symbol rank,

Table 3. Performance Comparison of formula retrieval systems in ARQMath-3 Task 2 against the novel system novel system L2L for: nDCG′, MAP′, and P′@10 Metrics.

Run	$nDCG'$	MAP'	P'@10
Baselines [32]			
Tangent-S	0.540	0.336	0.511
approach0 [37]			
fusion_alph05	**0.720**	**0.568**	**0.688**
fusion_alph03	0.720	0.565	0.665
fusion_alph02	0.715	0.558	0.659
a0	0.639	0.501	0.615
fusion02_ctx	0.631	0.490	0.611
DPRL [15]			
TangentCFT2ED	0.694	0.480	0.611
TangentCFT2	0.641	0.419	0.534
T-CFT2TED+MathAMR	0.640	0.388	0.478
LTR	0.575	0.377	0.566
MathAMR	0.316	0.160	0.253
MathDowsers [20]			
latex_L8_a040	0.640	0.451	0.549
latex_L8_a035	0.640	0.450	0.549
L8	0.633	0.445	0.549
XYPhoc [4]			
xy7o4	0.472	0.309	0.563
xy5	0.369	0.211	0.518
xy5IDF	0.322	0.180	0.461
JU_NITS [25]			
formulaL	0.161	0.059	0.125
formulaO	0.016	0.008	0.001
formulaS	0.000	0.000	0.000
L2L (Our System)			
Feature Only	0.8308	0.5997	0.6480
Latex Only	0.8171	0.5495	0.5760
Symbol Only	0.8266	0.5624	0.6000
Feature + Latex	0.8449	0.6160	0.6533
Feature + Symbol	0.8439	0.6088	0.6747
Latex + Symbol	0.8284	0.5732	0.6173
All (Unweighted)	**0.8488**	**0.6197**	0.6800
All (Retrospective)	0.8469	0.6128	**0.6880**

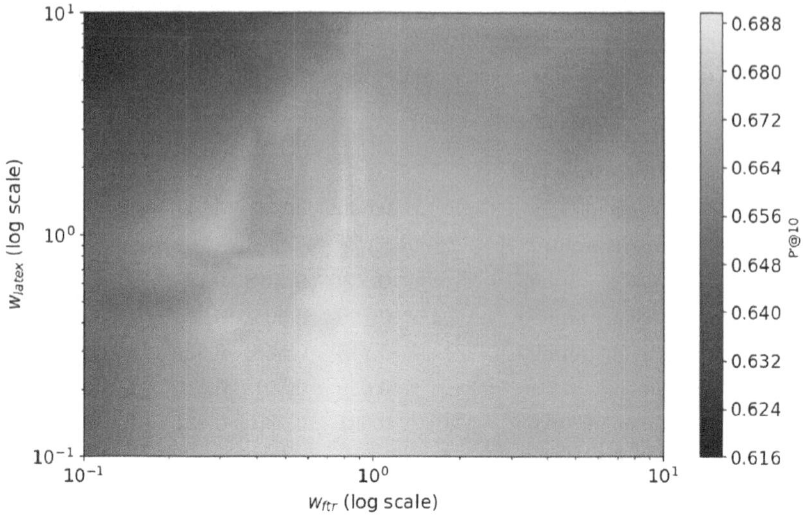

Fig. 3. The $P'@10$ values obtained by the L2L MathIR System on the ARQ3 corpus, by varying the weights w_{ftr} and w_{latex} while keeping w_{sym} fixed at 1.0.

feature rank, and LaTeX rank-on the system's overall P′@10 performance, as shown in the bottom rows of Table 3. (2) We observe the impact of suppressing each feature space on a query-by-query basis, as detailed in Fig. 4.

The last eight rows of Table 3 illustrate that systems excluding one or more ranking components consistently perform worse than those utilizing all three. Notably, the "All (Unweighted)" and "All (Retrospective)" configurations achieve the highest P′@10, nDCG′, and MAP′ scores. Moreover Fig. 4 illustrates that the exclusion of a single feature space impacts queries differently, with some exhibiting sharp declines when symbol rank is removed, indicating dependencies on specific formula variables, while others drop most with the removal of feature rank, reflecting a reliance on structural understanding. The variation in query performance to weight suppression highlights the complementary roles of these representations in achieving optimal retrieval accuracy. Moreover, these findings reinforce the diverse nature of mathematical queries and the necessity of all three components to ensure reliable performance across the query spectrum.

4.4 Qualitative Insights into Feature Space Suppression on Queries

Table 4 lists the most severely impacted queries when a specific ranking component is suppressed. This analysis evaluates whether the most impacted queries reflect the expected role of the suppressed ranking component, confirming that each feature space contributes as designed to the retrieval process.

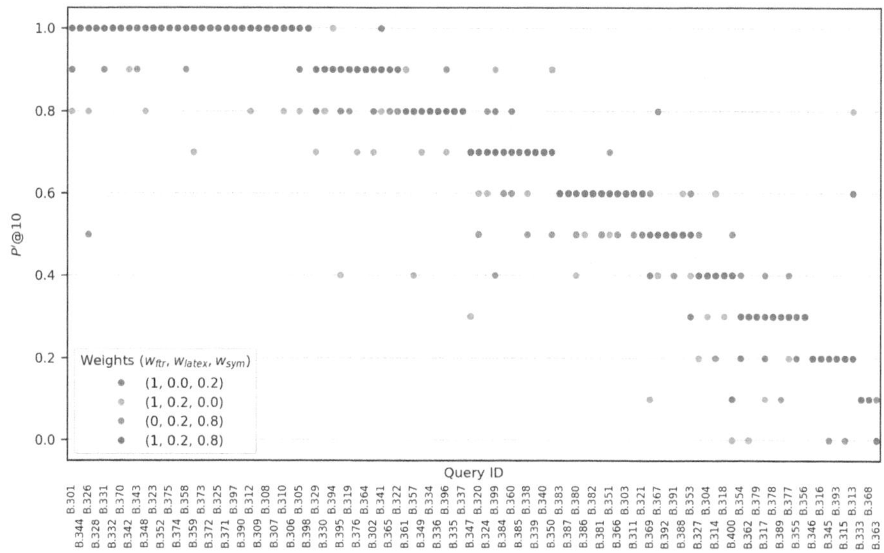

Fig. 4. P′@10 values obtained by the L2L MathIR System on the ARQMath-3 corpus, by suppressing one of the three weights w_{ftr}, w_{latex}, and w_{sym} on a query by query basis. Missing colors result from overlapping points.

Symbol Rank: Stripping tags from a Presentation MathML string reveals an ordered set of symbols. When this symbol set is omitted, the remaining representation, captured from tree paths, reduces specific symbols to generic group labels (num, var, ascii). This abstraction erases distinctions between individual symbols, such as $\mathbb{R}$ in $\mathbb{R}^{n+1} - \mathbb{R}^n$ and $\mathcal{A}$ in $\mathcal{A} = \{\{a\}\}$, both of which are replaced with the generic ascii label during L2L feature extraction. Without the symbol set, the system is forced to match formulas using a degraded representation of mathematics, resulting in poor matches for queries involving complex symbols.

LaTeX Rank is crucial for capturing the exact textual structure of queries, ensuring accurate differentiation between visually similar but mathematically distinct expressions. For example, $Cov(x, y) = 0$ requires the correct text-based representation to differentiate it from unrelated expressions, such as $Var(x) = 0$ or $Corr(x, y) = 0$. In matrices, even minor textual changes like swapping two elements can result in completely different eigenspaces. Without LaTeX rank, the system risks conflating distinct queries that depend on precise textual matches.

Feature Rank is needed for capturing operations that are not explicitly shown in textual representations, but are evident in tree-based features. For instance, in $\gamma'(t) = \lambda(t)\gamma(t)$, the implicit multiplication operator between $\lambda(t)$ and $\gamma(t)$ is identified through an L2L path. Similarly, in $\|A\|_2 = \sqrt{\rho(A^T A)}$, feature paths uncover the underlying matrix-matrix multiplication and normalization, which

Table 4. Queries with reduced P′@10 scores due to ranking component suppression.

Suppressed Component	Queries Impacted
Symbol Rank	$A_1 \subseteq A_2 \subseteq \ldots \subseteq A_n \subseteq A_{n+1} \subseteq \ldots$
	$\mathcal{A} = \{\{a\}\}$
	$\mathbb{R}^{n+1} - \mathbb{R}^n$
LaTeX Rank	$\begin{vmatrix} 1 & 1 & 1 \\ x & y & z \\ x^2 & y^2 & z^2 \end{vmatrix}$
	$Cov(x, y) = 0$
	$f(f(x)^2 + f(y)) = xf(x) + y$
Feature Rank	$\|A\|_2 = \sqrt{\rho(A^T A)}$
	$F = P \oplus T$
	$\gamma'(t) = \lambda(t)\gamma(t)$

could easily be conflated as a subscript in textual representations. Without L2L features, operator-operand relationships are insufficiently captured, leading to poor matching based on incomplete representations of mathematical formulas.

Omitting any component-symbol, LaTeX, or feature rank-results in a loss of query information, highlighting their unified roles in capturing mathematics.

5 Conclusion

Leveraging multiple feature spaces improves retrieval and ranking in mathematical formula search. An analysis of past MathIR engines reveals the inadequacy of directly applying text-only approaches to MathIR, and the need for tree and feature-based approaches. From this analysis, we present L2L, which combines three key components: leaf-to-leaf shortest paths to capture structural relationships, symbolic matching to reflect meaning stored in mathematical symbols, and LaTeX similarity to assess textual similarity. L2L achieves impressive results in ARQMath-3 and NTCIR-12, demonstrating its versatility. Moreover, altering feature-space weights demonstrate that even the weakest performing metric, P'@10, remains stable across various weight configurations, indicating that feature-space inclusion matters more than its relative contributions. Individual query performance degrades differently when individual feature-spaces are suppressed, while overall system performance degrades with any suppression, demonstrating the complementary roles of these representations across the query spectrum. A qualitative analysis of formulas most affected by the exclusion of a particular representation aligns with the representation's role in capturing specific query properties. Together, these findings demonstrate that an ensemble method is crucial for mathematical formula retrieval.

5.1 Future Work

Future work will focus on dynamic query adaptation and optimizing L2L for real-time applications. While this work highlights the importance of including all ranking components, future research will explore dynamic query-specific weighting schemes. For instance, symbolic-heavy queries could prioritize symbolic matching, while textual driven queries might emphasize LaTeX similarity. This approach has the potential to refine retrieval accuracy by tailoring rankings to the specific needs of individual query types. Additionally, the reliance on parallelism makes L2L efficient for batch retrieval, but real-time systems require lower latency. Strategies such as caching frequently queried features or employing approximate matching techniques will be explored to achieve the necessary speed without sacrificing accuracy.

5.2 Implications for the Information Retrieval Field

This research invites a reexamination of how features are conceptualized, weighted, and optimized in IR for non-textual data. Techniques like tree-based indexing, feature-driven ranking, and multi-space representations challenge conventional IR assumptions by showing that domain-specific representations are not just beneficial but necessary. The results suggest that a "one-size-fits-all" IR system may not be sufficient for many domains. Instead, task-specific retrieval models that integrate multiple feature spaces may be be more effective. This opens a dialogue within the IR community on how to design modular and adaptive systems for domain-specific tasks while maintaining generalizability.

References

1. Sakai, T., Oard, D.W., Kando, N. (eds.): Evaluating Information Retrieval and Access Tasks: NTCIR's Legacy of Research Impact. Springer Singapore, Singapore (2021)
2. Aizawa, A., et al.: NTCIR-11 math-2 task overview. In: Proceedings of the 11th NTCIR Conference on Evaluation of Information Access Technologies. National Institute of Informatics. Tokyo, Japan, (2014)
3. Aizawa, A., et al.: NTCIR-11 math-2 task overview. In: NII Testbeds and Community for Information Access Research (NTCIR). Tokyo, Japan (2014)
4. Avenoso, R., et al.: XY-PHOC symbol location embeddings for math formula retrieval and autocompletion. In: Proceedings of CLEF 2021 (CEUR Working Notes) (2021)
5. Davila, K., Zanibbi, R.: Layout and semantics: combining representations for mathematical formula search. In: Proceedings of the 40th International ACM SIGIR Conference on Research and Development in Information Retrieval. ACM (2017). https://doi.org/10.1145/3077136.3080748
6. Fernandes, D., Bernardino, J.: Graph Databases Comparison: AllegroGraph, ArangoDB, InfiniteGraph, Neo4J, and OrientDB. In: Proceedings of the 7th International Conference on Data Science, Technology and Applications (DATA). SciTePress (2018). https://doi.org/10.5220/0006910203730380

7. Jaccard, P.: Distribution de la flore alpine dans le Bassin des Dranses et dans quelques régions voisines. In: Bulletin de la Société Vaudoise des Sciences Naturelles (1901). https://doi.org/10.5169/seals-266440
8. Jaro, M.A.: Advances in record-linkage methodology as applied to matching the 1985 census of Tampa, Florida. J. American Stat. Assoc. **84**(406), 414–420 (1989). https://doi.org/10.1080/01621459.1989.10478785
9. Kristianto, G.Y., et al.: MCAT math retrieval system for NTCIR- 12 MathIR Task. In: Proceedings of the 12th NTCIR Conference on Evaluation of Information Access Technologies. Tokyo, Japan, June (2016)
10. Larson, R.R., et al.: The abject failure of keyword IR for mathematics search: Berkeley at NTCIR-10 Math. In: Proceedings of the NTCIR-10 Conference. National Institute of Informatics. Tokyo, Japan (2013)
11. Larson, R.R., et al.: the abject failure of keyword IR for mathematics search: Berkeley at NTCIR-10 math. In: NTCIR (2013)
12. Levenshtein, V.I.: Binary codes capable of correcting deletions, insertions, and reversals. In: Soviet Physics Doklady (1965)
13. Barrón-cedeño, A., et al. (eds.): Experimental IR Meets Multilinguality, Multimodality, and Interaction: 13th International Conference of the CLEF Association, CLEF 2022, Bologna, Italy, September 5–8, 2022, Proceedings. Springer International Publishing, Cham (2022)
14. Mansouri, B., et al.: Tangent-CFT: An Embedding Model for Mathematical Formulas. In: Proceedings of the 2019 ACM SIGIR International Conference on Theory of Information Retrieval. ACM. (2019)
15. Mansouri, B., et al.: DPRL systems in the CLEF 2021 ARQMath Lab: sentence-BERT for answer retrieval, learning-to-rank for formula retrieval. In: Proceedings of CLEF 2021 (CEUR Working Notes) (2021)
16. Barrón-cedeõn, A., et al. (eds.): Experimental IR Meets Multilinguality, Multimodality, and Interaction: 13th International Conference of the CLEF Association, CLEF 2022, Bologna, Italy, September 5–8, 2022, Proceedings. Springer International Publishing, Cham (2022)
17. Miller, J.J.: Graph database applications and concepts with Neo4j. In: Proceedings of the southern association for information systems conference, Atlanta, GA, USA (2013)
18. Miner, R.: The importance of MathML to mathematics communication. In: Notices of the AMS (2005)
19. MIR@MU. Mathematical Information Retrieval at Masaryk University. Accessed Oct 06 2024 2021
20. Ng, Y.K., et al.: Dowsing for Answers to Math Questions: Ongoing Viability of Traditional MathIR. In: (2021)
21. NTCIR-12 MathIR - Data. (Visited on 06/20/2023)
22. Peng, S., et al.: MathBERT: A Pre-trained Model for Mathematical Formula Understanding. In: arXiv preprint arXiv:2105.00377 (2021). https://doi.org/10.48550/arXiv.2105.00377
23. Centre de Recherches Mathématiques. Centre de Recherches Mathématiques. 2021. (Visited on 10/06/2024)
24. Goharian, N., et al. (eds.): Advances in Information Retrieval: 46th European Conference on Information Retrieval, ECIR 2024, Glasgow, UK, March 24–28, 2024, Proceedings, Part I. Springer Nature Switzerland, Cham (2024)
25. Sarkar, S., et al.: Formula Retrieval Using Structural Similarity. In: CLEF (Working Notes) (2022)

26. Sojka, P., et al.: MIaS: math-aware retrieval in digital mathematical libraries. In: Proceedings of the 27th ACM International Conference on Information and Knowledge Management. Torino Italy: ACM, Oct. 2018. https://doi.org/10.1145/3269206.3269233
27. Topic, G., et al.: The MCAT math retrieval system for NTCIR-10 math track. In: Proceedings of the 10th NTCIR Conference. National Institute of Informatics. Tokyo, Japan (2013)
28. Yang, P., et al.: Anserini: enabling the use of lucene for information retrieval research. In: Proceedings of the 40th International ACM SIGIR Conference on Research and Development in Information Retrieval. SIGIR'17. Shinjuku, Tokyo, Japan: Association for Computing Machinery (2017). https://doi.org/10.1145/3077136.3080721
29. Ell Zanibbi. DPRL - Document Processing Research Lab. Accessed Oct 07 2024
30. Hiemstra, D., Moens, M.-F., Mothe, J., Perego, R., Potthast, M., Sebastiani, F. (eds.): Advances in Information Retrieval: 43rd European Conference on IR Research, ECIR 2021, Virtual Event, March 28 – April 1, 2021, Proceedings, Part II. Springer International Publishing, Cham (2021)
31. Zanibbi, R., Davila, K.: Tangent-L: combining tree-based and feature-based approaches for math formula search. In: Journal of Information Retrieval (2019)
32. Zanibbi, R., et al.: The tangent search engine: improved similarity metrics and scalability for math formula search. In: arXiv:1507.06235 [cs] (July 2015). arXiv: 1507.06235
33. Zanibbi, R., et al.: NTCIR-12 MathIR Task Overview. In: NII Testbeds and Community for Information Access Research (NTCIR). Tokyo, Japan (2016)
34. Zanibbi, R., et al.: The tangent search engine: improved similarity metrics and scalability for math formula search. In: Proceedings of the 39th International ACM SIGIR Conference on Research and Development in Information Retrieval. ACM (2016). https://doi.org/10.48550/arXiv.1507.06235
35. Mansouri, B., Agarwal, A., Oard, D.W., Zanibbi, R.: Advancing math-aware search: the ARQMath-2 Lab at CLEF 2021. In: Hiemstra, D., Moens, M.-F., Mothe, J., Perego, R., Potthast, M., Sebastiani, F. (eds.) ECIR 2021. LNCS, vol. 12657, pp. 631–638. Springer, Cham (2021). https://doi.org/10.1007/978-3-030-72240-1_74
36. Zanibbi, R., et al.: Mathematical Information Retrieval: Search and Question Answering. In: arXiv preprint arXiv:2408.11646 (2024). https://doi.org/10.48550/arXiv.2408.11646
37. Zhong, W., et al.: Approach Zero and Anserini at the CLEF-2021 ARQMath track: applying substructure search and BM25 on operator tree path tokens. In: Proceedings of CLEF 2021 (CEUR Working Notes). Virtual Event, Sept (2021)

Ragnarök: A Reusable RAG Framework and Baselines for TREC 2024 Retrieval-Augmented Generation Track

Ronak Pradeep[1(✉)], Nandan Thakur[1], Sahel Sharifymoghaddam[1], Eric Zhang[1], Ryan Nguyen[1], Daniel Campos[2], Nick Craswell[3], and Jimmy Lin[1]

[1] University of Waterloo, Waterloo, Canada
rpradeep@uwaterloo.ca
[2] Snowflake Inc., New York, USA
[3] Microsoft, Seattle, USA

Abstract. Have you tried the new Bing Search? Or maybe you fiddled around with Google AI Overviews? These might sound familiar because the modern-day search stack has evolved to include retrieval-augmented generation (RAG) systems. They allow searching and incorporating real-time data into large language models (LLMs) to provide a well-informed, attributed, concise summary, in contrast to the traditional search paradigm that relies on displaying a ranked list of documents. Therefore, given these recent advancements, it is crucial to have an arena to build, test, visualize, and systematically evaluate RAG-based search systems. With this in mind, we propose TREC RAG to foster innovation in evaluating RAG systems. In our work, we lay out the steps we have taken towards making this track a reality—we describe the details of our reusable framework, Ragnarök, explain the curation of the new MS MARCO V2.1 collection, release the development topics, some relevance judgments and baselines for the track and standardize the I/O definitions which assist the end user. Next, using Ragnarök, we identify and provide key proprietary and open-source baselines such as OpenAI's GPT-4o, Cohere's Command R+, and Meta's LLaMA3.1-70B. Further, we introduce a web-based user interface for an interactive arena allowing benchmarking pairwise RAG systems by crowdsourcing. We open-source Ragnarök and baselines to achieve a unified standard for future RAG systems.

Keywords: Retrieval-Augmented Generation · Large Language Models · Ad Hoc Retrieval

1 Introduction

Retrieval-Augmented Generation (RAG) [9,21,23,33] has emerged as a popular technique to augment large language model (LLM) generation for knowledge-intensive tasks such as open-domain question answering or fact verification [51].

R. Pradeep and N. Thakur—Both authors contributed equally to this research.

C. Hauff et al. (Eds.): ECIR 2025, LNCS 15572, pp. 132–148, 2025.
https://doi.org/10.1007/978-3-031-88708-6_9

Using the top-k retrieved segments from a suitable retrieval system, RAG systems output an answer summary grounded on the relevant context. RAG systems mitigate factual inconsistencies in LLM outputs [19,27,33,39], and enhance interpretability [21] and generalization [20], thus facilitating a wider LLM adoption across several domains like Medicine [63] and Finance [24].

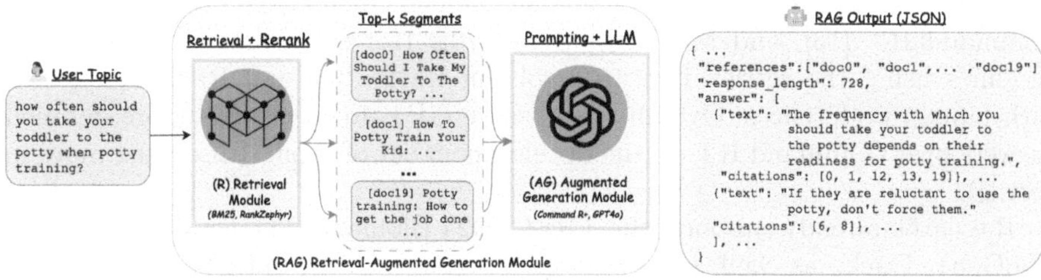

Fig. 1. Schematic diagram of the Ragnarök. Given a user topic (left), the process consists of two steps: (1) (R) retrieval (+ rerank), where the topic yields the top-k relevant segments from our document collection (e.g., potty training articles); and (2) (AG) augmented-generation, where the retrieved segments with a suitable prompt template are fed to the large language model (LLM) to generate the post-processed answer response (JSON) containing individual sentence-level citations.

Several companies provide end-to-end RAG systems such as Bing Search [45], or Google AI Overviews. Most of these systems are either proprietary or offer limited user customization. Likewise, the absence of a standardized framework makes implementing RAG at a large scale challenging. Implementing atop existing frameworks requires custom code for multiple steps including retrieval, reranking, and generation. To promote wider adoption of RAG in academia, we develop Ragnarök, a user-friendly, reusable, end-to-end RAG framework offering code to customize retrievers, rerankers, and generation models.

Ragnarök comprises two key modules: (R) Retrieval and (AG) Augmented Generation. The retrieval module incorporates both the retrieval and reranking stages to yield the top-k segments for a given user topic. Next, the augmented generation module uses both the user topic and retrieved segments as input to produce a RAG answer, formatted into individual sentences, citing the relevant information from the top-k retrieved and reranked segments. Ragnarök is deeply integrated with existing Python frameworks, such as `Pyserini` [36] and `rank_llm` [53,54] and can be easily installed via PyPI using "`pip install pyragnarok`". This framework offers REST APIs and an integrated WebUI to enhance user-friendliness and improve the human evaluation experience.

Ragnarök is used to provide baselines in the TREC 2024 Retrieval-Augmented Generation Track. An ideal framework should contain a sufficiently large document collection covering diverse information and non-factoid, decompositional topics requiring long-form answers. In our work, we deduplicate the existing MS MARCO V2 document collection. In addition, we provide a "segment" collection using a sliding-window chunking technique (discussed more in

Sect. 4). Further, we release two sets of development topics: (i) TREC-RAGgy 2024: a filtered subset of topics with long-form answers from TREC Deep Learning 2021–23 [14–16]; and (ii) TREC-Researchy 2024: a subset of the Researchy Questions introduced in Rosset et al. [57].

Ragnarök supports a head-to-head RAG battle arena for answer evaluation, heavily inspired by recent work such as the Chatbot Arena [12,66]. We include key proprietary and open-source baselines such as OpenAI GPT-4o [48], Cohere Command R+ [13], and Meta LLaMA3.1-70B [17] and evaluate some of the baselines using the retrieval setup involving BM25 [55] and RankZephyr [54] with human preferences. Overall, we observe GPT-4o to provide more detailed answers over Command R+ on the development set of topics (discussed more in Sect. 6).

Ragnarök already supports the TREC 2024 Biomedical Generative Retrieval (BioGen) Track, as well as report generation for the TREC 2024 Neural Cross-Language Information Retrieval (NeuCLIR) Track. This demonstrates the framework's adaptability across diverse domains, each characterized by distinct information needs. The flexibility and domain-general capabilities highlight Ragnarök and its potential to serve as a versatile tool in specialized RAG applications.

Finally, with the growing need for standardized RAG frameworks, we have made Ragnarök publicly available.[1] Moving forward, we plan to expand our retrieval corpora, incorporate additional baseline systems, and continuously enhance our framework to meet our new goals.

2 Related Work

RAG Frameworks. Existing RAG systems are primarily proprietary and difficult to reproduce. Open-source frameworks such as LangChain [11] and LlamaIndex [38], while available, are not research-friendly and lack proper evaluation and benchmarking. FlashRAG [25], a concurrent work, is a similarly motivated toolkit to improve the RAG experience for researchers. While the framework is extensive and designed for pipeline flexibility, Ragnarök offers a few additional capabilities—a WebUI serving a RAG battle arena, REST APIs, a standardized I/O definition working with sentence-level citations, and a tight integration with popular retrieval and reranking frameworks like Pyserini [36] and RankLLM.

Collection Selection. Current RAG datasets are constructed using the English Wikipedia as the document collection, However, their scale is limited to provide rich and comprehensive information to support RAG systems. A prominent option, ClueWeb22 [49] offers an extensive collection of 22 billion curated web pages, previously utilized in parts by TREC tracks such as the TREC Conversational Assistance Track (CAsT) [50] and the TREC Interactive Knowledge Assistance Track (iKAT) [3]. However, ClueWeb22 remains gated and serves as a significant hurdle for researchers with limited computational resources or those

[1] Ragnarök code repository: https://github.com/castorini/ragnarok.

unable to obtain access credentials. Another alternative is the MS MARCO V2 document collection, which was used primarily in the TREC Deep Learning (DL) track. The open availability of MS MARCO makes it an attractive option for researchers seeking to develop and evaluate RAG systems at scale.

Topic Selection. Recently, there has been a surge in datasets providing topics with long-form answers for evaluating RAG systems. ASQA [58], ELI5 [18], and QAMPARI [4] were utilized for evaluation in the Automatic LLMs' Citation Evaluation (ALCE) framework [19]. Similarly, related long-form QA datasets include AquaMuse [28], ExpertQA [42], and TruthfulQA [37]. Another recently introduced dataset is ClapNQ [56], created from the subset of Natural Questions (NQ) [29] and HAGRID [26] built on a subset of MS MARCO Dev [7]. Almost all previous datasets are built on English Wikipedia. In contrast, our work deliberately avoids English Wikipedia to prevent the overfitting seen in existing retrieval benchmarks [46,61]. In our work, we re-utilize topics from previous TREC tracks such as the Deep Learning (DL) track, because human judgments are available on the MS MARCO V2 corpora and Researchy Questions [57] as it covers a wide range of topics with multi-faceted information needs.

3 Our Framework

Ragnarök is an open-source, reproducible, and reusable framework implementing an end-to-end retrieval-augmented generation (RAG) pipeline, comprising two modules applied sequentially: (1) (R) retrieval and (2) (AG) augmented generation. Through Ragnarök, we provide several baselines to all participants in TREC RAG. An overview of the framework is provided in Fig. 1. We first describe both modules and expand on the I/O specifications in our framework.

Retrieval Module. This module retrieves the relevant segments for a user topic as the input. It supports (i) first-stage lexical retrieval models such as BM25 [55] and (ii) reranking models such as RankZephyr [54]. The retrieval system searches for relevant segments in the document collection and retrieves the top-100 segments further reranked by the reranker model to filter out the top-20 relevant segments for the next stage.

Augmented Generation Module. This module takes in the user topic and the top-20 retrieved segments (from the retrieval module) as the input and a prompting strategy to the LLM to generate the answer response with in-context citations for the topic. The answer response is divided into individual sentences, each sentence within the answer contains text and is grounded on retrieved documents provided as references to the LLM (if possible).

3.1 RAG Input/Output Definitions

RAG Input. The input specifications are straightforward as the user can formulate any question they wish to ask, provide the user topic as input, and call Ragnarök.

RAG Output. The user receives a JSON output in response to their topic from Ragnarök. The first key in the output JSON schema, `references`, provides a list of segment IDs that are referenced in the answer. These segments are selected from among the top-20 results returned by the retrieval module. Next, `answer`, provides the LLM-generated RAG answer to the user topic, presented as a top-to-bottom list of sentence-level texts with corresponding segment citations. All citations are zero-based indexed, indicating the exact position of the segment ID from the `references` list. Finally, `response_length`, provides the total count of the whitespace-separated words present in the output RAG answer.

Table 1. Comparison of document and segment counts between versions V2 and V2.1 (our version after removing near-duplicates) of the MS MARCO collection.

Collection	Version V2	Version V2.1 (Ours)
MS MARCO Document	11,959,635	10,960,555
MS MARCO Segment	124,131,414	113,520,750

4 Document Collection

The MS MARCO V2 document collection, previously used in the TREC-DL tracks, contains a substantial overlap of near-duplicates (documents with sufficiently similar text information) [15,16]. When left intact, these near-duplicates degrade the downstream retrieval accuracy and reduce the diversity of the collected documents, potentially impacting the effectiveness of RAG systems.

Documents in the existing collection tend to be verbose, containing extensive information about individual topics. Chunking, which breaks down a long verbose document into smaller compact representations is a key challenge, as the retrieved chunk representations correlate with the RAG answer quality [39].

MS MARCO V2.1 Document Collection. We conduct a deduplication strategy in the MS MARCO V2 document collection to avoid near-duplicates in two stages. In the first stage, we establish an equivalence class of the documents using Locality Sensitive Hashing (LSH) with MinHash [10] and 9-gram shingles. Next, we select a representative document for each equivalence class for our refined MS MARCO V2.1 document collection, reducing the duplicates in the original MS MARCO V2 document collection by 8.35% as shown in Table 1.

MS MARCO V2.1 Segment Collection. We segment the MS MARCO V2.1 document collection into overlapping segments (or chunks) and develop the MS MARCO V2.1 segment collection, with more than 113 million text segments (Table 1). We utilize a sliding window technique to generate the segments, by fixing the sliding window size of 10 sentences and a stride of 5 sentences to create each segment, roughly on average, between 500–1000 characters long. To easily map each segment back to the document, every segment contains the document ID within the segment ID. Further, two new fields: `start_char` and `end_char`, which indicate the start (inclusive) and the end position character (exclusive) of where the segment begins and ends in the mapped MS MARCO V2.1 document collection, respectively.

5 Topic Collection

Topics, i.e., user queries, are crucial for robust evaluation of RAG systems. Traditionally, popular retrieval and traditional QA benchmarks primarily consist of factoid queries, where answers are typically found within a single sentence or paragraph. However, these topics lack complexity, leading to short answers that can be easily memorized by LLMs. For instance, MS MARCO [7] surprisingly contains up to 55% factoid queries [8,57]. To avoid short-form answers in RAG, we utilize two collections containing non-factoid topics covering diverse information and requiring long-form answers. We describe these collections below:

Table 2. TREC-RAGgy 2024 and TREC-Researchy 2024 topic distribution. The table shows the top-5 categories in topic classification for TREC-RAGgy 2024 (each topic classified into a single category), intrinsic attributes for TREC-Researchy 2024, and the first word in all topics.

TREC-RAGgy 2024				TREC-Researchy 2024			
Topic Category	%	First Word	%	Intrinsic Attributes	%	First Word	%
Aggregation	24.2	What	37.5	Knowledge-Intensive	79.8	How	41.0
Simple w/ cond.	23.3	How	27.5	Multi-Faceted	75.7	Why	25.5
Set	20.8	Why	3.3	Reasoning-Intensive	75.5	What	15.0
Simple	10.0	Is	2.5	Subjective	48.5	Is	5.2
Comparison	6.7	When	1.7	Assumptive	25.7	Should	2.2

TREC-RAGgy 2024. We develop TREC-RAGgy 2024, a collection with topics filtered from TREC Deep Learning 2021–2023 tracks [14–16], based on topic category and generated-answer classification. We classify each available topic into seven categories and filter out a subset of topics that either have a long-form answer or require information aggregation across multiple sources of information. Out of the 210 original topics available, we filter and include 120 topics (57.1%) in the TREC-RAGgy 2024 collection. From Table 2, we observe 24.2% of the topics included are "aggregation", indicating RAG systems should aggregate information from multiple segments to generate an accurate long-form answer. Similarly, 65% of the topics start with "what" or "how" questions.

The TREC-RAGgy 2024 collection includes mapped document-level relevance judgments from previous TREC Deep Learning tracks. These relevance judgments, originally associated with the full documents, have been mapped to the new deduplicated corpus, enabling effective evaluation of the retrieval systems. However, since most of our work operates at the segment level, prior to the evaluation, we perform a MaxP step—leveraging the score of the most relevant segment as the representative score of the document.

TREC-Researchy 2024. Researchy Questions, introduced in Rosset et al. [57], contains 102K non-factoid topics with long-form answers. These topics were curated from Bing Search logs and evaluated by GPT-4 on a scale of 0–10 based on eight intrinsic attributes, such as subjectivity and multifacetedness. Notably, unlike TREC-RAGgy 2024, these queries lack relevance judgments. To curate a smaller development subset for a faster evaluation of RAG systems, we employ a sampler designed to maximize diversity based on the eight intrinsic attributes. This is achieved by iteratively selecting the query with the highest l_1 norm in the intrinsic attribute space (of all eight dimensions) relative to the already-sampled set. We refer to the resultant topic set as TREC-Researchy 2024. From Table 2, about 80% of the topics are Knowledge-Intensive, and about 76% are Multi-Faceted, highlighting the need for effective RAG systems. Additionally, 66.5% of topics start with "how" or "why", emphasizing explanatory questions. These distributions suggest that TREC-Researchy 2024 prioritizes complex and multi-dimensional topics.

For the TREC RAG test topics, we released a new and fresh scrape of topics close to the submission period. This approach compiled a fresh and newer set of topics, similar to Rosset et al. [57], thereby minimizing the risk of data leakage and ensuring a fair evaluation with existing commercially available LLMs.

6 TREC RAG Baselines

6.1 Retrieval

Our retrieval module integrates both first-stage retrievers and rerankers. For traditional dual encoders, we employ BM25 and BM25 + Rocchio, available in Anserini [65] and retrieve the top 3000 segments for a given topic. BM25 has proven effective as a first-stage retriever due to its capability to capture lexical overlap, with Rocchio introducing relevance feedback to further refine the search results based on the initial BM25 retrieval.

We further extend our retrieval evaluation by incorporating GTE-L, a dense dual encoder model from Alibaba-NLP [34]. GTE-L is their larger variant comprising 434M parameters, with embeddings of dimensionality 1024, and support for a maximum sequence length of 8196 tokens. This model is particularly suited for long-context retrieval tasks, enabling it to effectively capture relationships across larger text segments. GTE-L is a strong baseline demonstrating state-of-the-art effectiveness in the MTEB [46] benchmark, when it was introduced.

Additionally, we evaluate two other dense dual encoders from Snowflake: ArcticEmbed-M and ArcticEmbed-L [44]. ArcticEmbed-M is a 137M parameter model, generating embeddings with a dimensionality of 768, while ArcticEmbed-L is a larger model with 335M parameters and an embedding dimensionality of 1024. Both models have shown state-of-the-art performance in retrieval tasks, particularly on the BEIR benchmark [61], where they excel in diverse retrieval settings. ArcticEmbed-M and ArcticEmbed-L provide efficient and scalable solutions for semantic retrieval, complementing our first-stage retrieval suite.

For each retrieval method (BM25, BM25 + Rocchio, GTE-L, ArcticEmbed-M, and ArcticEmbed-L), we retrieve the top 3000 segments. To leverage the strengths of multiple retrieval strategies, we employ Reciprocal Rank Fusion (RRF) as a hybrid approach. The RRF technique fuses the retrieval results from BM25 + Rocchio, GTE-L, ArcticEmbed-M, and ArcticEmbed-L, with the aim of producing a better candidate set for reranking, as hybrid approaches have been shown to improve retrieval effectiveness in multiple TREC tracks [32].

Following retrieval, we explore neural reranking models for first-stage reranking. We leverage monoT5-3B, a pointwise reranker with 3B parameters, that demonstrates strong effectiveness in reranking by assigning relevance scores to individual document-query pairs [47,52]. Note that we fuse these results with the results from the prior stage.

Next, we incorporate RankZephyr, a highly effective listwise reranker in Pradeep et al. [54], which takes as input a query and a list of passages and outputs a reordered list based on relevance. RankZephyr is particularly effective in scenarios where contextual relationships between documents are useful for ideal reranking. In our pipeline, we use RankZephyr to rerank the top 100 resultant candidates from monoT5-3B, ensuring a more precise final ranking. Note that we fuse the RankZephyr results with the results from the prior stage.

Additionally, we evaluate RankZephyr$_\rho$, which adopts a progressive reranking strategy. RankZephyr$_\rho$ iteratively refines the ranking of candidate documents over three passes, progressively improving the precision of the final ranked list. This iterative refinement has been shown to enhance retrieval effectiveness, particularly in cases where high-precision ranking is required.

Both monoT5-3B and RankZephyr and also other rerankers such as LRL [41], RankGPT [59] and LiT5 [60], are available through the `rank_llm` package, which we leverage as part of our reranking pipeline in Ragnarök. While budget constraints prevent us from testing every reranking model, support for this functionality continues through Ragnarök. Some of these retrieved results in the case of TREC-RAGgy 2024 can be evaluated after running the MaxP operation to get the representative score for the document. Finally, the top-20 reranked segments are passed on to the next stage, i.e., augmented generation.

Results. Table 3 presents the nDCG@10, MAP@100, and recall@100 for the primary retrieval baselines under consideration. The traditional lexical retrievers BM25 (1a) and BM25+Rocchio (1b) demonstrate comparable scores with only marginal differences. Dense dual encoders (1c, 1d, 1e) exhibit superior effectiveness compared to traditional lexical methods (1a, 1b). Among the dense models, the ArcticEmbed variants (1d, 1e) achieve slightly higher effectiveness metrics than GTE-L (1c). Notably, the application of Reciprocal Rank Fusion across

Table 3. Results on the Document Ranking Task of the TREC-RAGgy 2024.

Model	nDCG@10	MAP@100	Recall@100
Lexical & Dual Encoders			
(1a) BM25	0.4227	0.1561	0.2807
(1b) BM25+Rocchio	0.4188	0.1818	0.3141
(1c) GTE-L	0.5682	0.2162	0.3512
(1d) ArcticEmbed-M	0.5749	0.2349	0.3692
(1e) ArcticEmbed-L	0.5776	0.2277	0.3623
(1f) RRF(1b,1c,1d,1e)	0.6064	0.2592	0.3990
Rerankers			
(2a) RRF(1f, monoT5-3B)	0.6175	0.2708	0.4208
(2b) RRF(2a, RankZephyr)	0.6357	0.2770	0.4208
(2c) RRF(2a, RankZephyr$_\rho$)	0.6317	0.2771	0.4208

these models (1f) yields substantial improvements compared to any individual retriever (1a–e).

The integration of the pointwise reranker monoT5-3B (2a) further enhances retrieval effectiveness across all metrics. Subsequent application of listwise reranking models (2b, 2c) yields additional effectiveness gains. However, the progressive reranking approach does not demonstrate meaningful improvements in high-precision metrics for this particular collection, as evidenced between the multiple and single-pass RankZephyr variants (2c vs. 2b).

6.2 Augmented Generation

Our generation module is designed to support a wide range of both open-source and proprietary models, enabling flexible integration for various use cases. At the core of our open-source support is vLLM [30], an optimized framework for efficient large language model (LLM) inference. vLLM excels in handling resource-intensive models by leveraging tensor parallelism and dynamic batching, ensuring scalability and performance across distributed systems. This makes it well-suited for open-source LLMs, where efficient resource management is critical. In addition, vLLM provides a wide support for text-based LLMs[2] including custom models fine-tuned with LoRA [22].

Our focus is on three popular LLMs: (i) GPT-4o is the latest GPT version from OpenAI [48]; (ii) Command R+ is Cohere's open-source instruction following LLM developed for complex RAG pipelines [13]; (iii) LLaMA-3.1-70B-Instruct is Meta's flagship open-source instruction tuned generative model optimized for multilingual dialogue use cases [17].

Given that Command R+ cites in a span level, we map the citations to their parent sentences. For GPT-4o and LLaMA-3.1-70B, we follow the ChatQA

[2] https://docs.vllm.ai/en/latest/models/supported_models.html.

```
System: This is a chat between a user and an artificial intelligence
assistant. The assistant gives helpful, detailed, and polite answers
to the user's questions based on the context. The assistant should
also indicate when the answer cannot be found in the context.

INSTRUCTION: Please give a complete answer to the question. Cite
each context document that supports your answer within brackets []
using the IEEE format.

QUESTION: {query}

CONTEXTS:
[1] {Passage title}: {Passage text}
[2] {Passage title}: {Passage text}
...
[20] {Passage title}: {Passage text}

INSTRUCTION: Please give a complete answer to the question. Cite
each context document that supports your answer within brackets []
using the IEEE format.
```

Fig. 2. ChatQA prompt template [40] used for RAG generation with in-text citations with LLaMA3.1-70B and GPT-4o in our Ragnarök framework.

prompt template [40] and cite relevant segments within the text (in-line) using the IEEE format. An example of the prompt template is shown in Fig. 2.

Additionally, Ragnarök incorporates a variety of refined ChatQA prompts specifically tailored for biomedical RAG tasks, particularly within the context of the TREC 2024 Biomedical Generative REtrieval (BioGen) track. These refinements enhance the system's ability to address domain-specific queries effectively in the expected manner and steer the model with expected word counts and sentence structures (each sentence is an independent assertion). Furthermore, the foundational ChatQA template was repurposed for the TREC 2024 Neural Cross-Language Information Retrieval (NeuCLIR) track's report generation task. This adaptability underscores Ragnarök's versatility in prompt management, allowing users to modify prompts with minimal intervention, typically requiring only a single addition within the prompt template file. This streamlined approach ensures that users can efficiently adapt our framework across various domains.

6.3 RAG-Bench Evaluation

Evaluating different RAG answers is challenging as multiple factors within the output response are crucial for effectiveness evaluation. To combat this, recent works rely on an LLM-as-a-judge setup [66], where strong LLM assessors judge the RAG-generated output in a pairwise evaluation style (side-by-side) in a head-on tournament. In our work, we briefly overview our baseline techniques using human evaluators. The Command R+ baseline outputs shorter answers and cites more relevant segments, whereas, the GPT-4o baseline outputs longer and more detailed answers and cites fewer segments. Therefore, for topics in both TREC-RAGgy 2024 and TREC-Researchy 2024, GPT-4o intuitively is the better choice for RAG answer generation. We found that the LLaMA3.1-70B model produces long and detailed answers but sometimes generates (from model parameters) a separate references section, from which it cites. We leave it for future work, to empirically compute the win rates (in %) between our baselines in the RAG-bench evaluation.

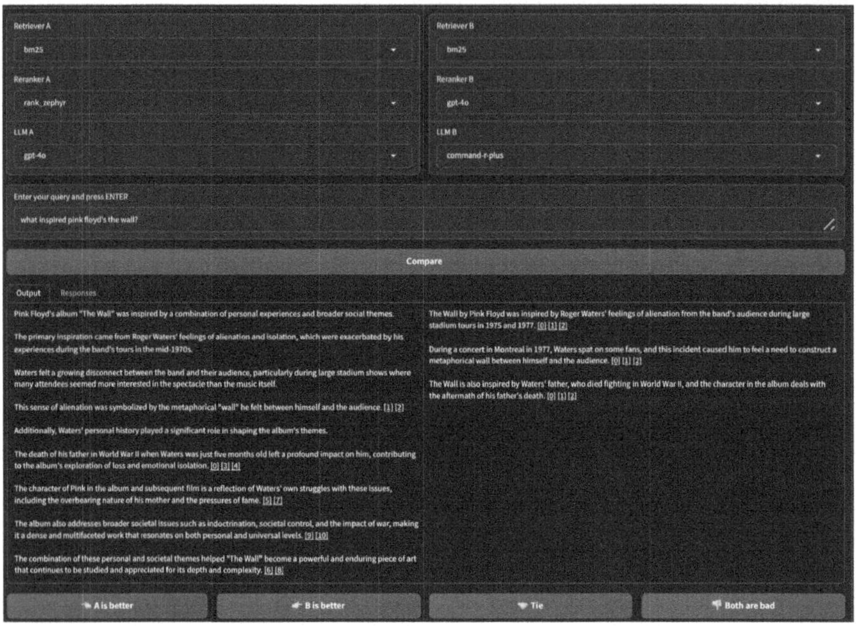

Fig. 3. WebUI showcasing the Ragnarök System Arena and the user query, "what inspired pink floyd's the wall?", with answers from two pipelines side-by-side comparing GPT-4o (left) and Command R+ (right).

7 Ragnarök System Arena

Heavily inspired by the success of Chatbot Arena [12,66], a crowdsourcing benchmark WebUI featuring anonymous battles, we extend the concept to multi-stage configurable RAG pipelines with Ragnarök. In the arena, users interact with two unblinded/blinded RAG systems simultaneously, issuing the same topic to both. The participants evaluate and select the pipeline that delivers their most preferred response, with the identities of the modules in the end-to-end pipeline revealed after the voting process in the blinded case. We leverage Gradio [1] to build the WebUI for Ragnarök. Each step of the pipeline uses REST APIs for intercommunication, enabling easy module switching within the pipeline. This modular design simplifies the integration of different retrieval and LLM configurations, enhancing scalability and maintainability.

Figure 3 illustrates an example topic "what inspired pink floyd's the wall?" processed by two different pipelines: Pipeline A, comprising BM25 → RankZephyr → GPT-4o (left), and Pipeline B, comprising BM25 → RankGPT-4o → Command R+ (right) in the unblinded tab. The outputs generated by each pipeline are compared, allowing users to discern which system provided a more satisfactory answer. Note that when the user hovers the mouse over a citation, they can preview the cited segment. We also provide a similar blinded pairwise evaluation interface along with the ability to view the responses in the JSON form (a tab) in the WebUI for Ragnarök.

After reading both outputs, the user has to pick one of the four choices for preference scoring: (i) A is better (ii) B is better (iii) Tie, or (iv) Both are bad. These results from preference scoring are organized into three separate ELO leaderboards, leveraging an SQLite database to track module effectiveness. These leaderboards evaluate language models (LLMs) for augmented generation-only (AG), retrieval models (R), and retrieval-augmented generation pipelines (RAG), providing a comprehensive view of their relative contributions.

8 Ongoing Work

Ragnarök is the first step for the ongoing work in the TREC RAG, by releasing the document collections, development topics, and baseline strategies for participants. We will continue to update the pipelines to include more diverse retrieval models including other sparse dual encoders such as SPLADE-v3 [31] and effective pairwise rerankers [52]. It is also in our critical path to evaluate the retrieved results from submissions to TREC-RAGgy 2024, after pooling, both with NIST annotators and LLM judges [62]. We plan to add additional support for more advanced RAG techniques like SelfRAG [6] and CRAG [64].

The next phase of our efforts will focus on finalizing the evaluation methodology of the generated answer in the RAG output. We are planning to include three evaluation metrics: nugget recall, support, and fluency. Automatic nugget-based evaluation is gaining popularity [2,5,43] and becoming the de facto strategy for RAG evaluation. Building on earlier work by Lin et al. [35], we plan to build a list of nuggets for each question recursively passing through the pooled results. We evaluate each generated answer based on precision and recall metrics. Next, for support evaluation, we plan to measure whether the cited documents sufficiently ground the information present in the answer. Lastly, for fluency evaluation, we measure whether the RAG output is coherent and fluent without any grammatical mistakes.

9 Conclusion

The emergence of retrieval-augmented generation (RAG) has revolutionized modern search systems by allowing real-time data incorporation into large language models (LLMs). Our work presents Ragnarök, a reusable and open-source end-to-end framework designed to provide reproducible baselines and a WebUI serving a RAG battle arena for retriever, reranker, and generation models.

In addition to introducing the MS MARCO V2.1 collection, we carefully curated topics from TREC-DL 2021–2023 and Researchy Questions. The TREC-RAGgy 2024 subset and the corpus can be collectively viewed as an ad-hoc retrieval collection on which we evaluated several state-of-the-art dual encoders and rerankers. Our work also defines I/O specifications to assist users in the RAG paradigm. By tightly integrating with popular retrieval frameworks such as Pyserini and `rank_llm`, Ragnarök ensures seamless usage.

We identify key baselines from industry, including OpenAI's GPT-4o, Cohere's Command R+ and Meta's LLaMA3.1-70B, and provide a qualitative analysis of these baselines on the development topics. Our framework provides a WebUI for head-to-head RAG system comparisons, inspired by recent work such as Chatbot Arena, to facilitate user evaluation and preference scoring.

Finally, by open-sourcing Ragnarök, we aim to standardize RAG applications in preparation for TREC RAG, promoting wider adoption and fostering innovation in the RAG research community. We plan to continuously update Ragnarök to include more advanced retrieval models and RAG techniques, ensuring it remains a valuable tool for researchers and practitioners alike.

Acknowledgments. We thank Ian Soboroff for his support with MS MARCO V2 document collection deduplication, Cohere for providing us with the necessary credits to evaluate Command-R+, and Microsoft for providing Azure credits to evaluate GPT-4o. Additionally, we thank Corby Rosset for the discussions surrounding Researchy Questions [57]. Additional funding is provided by Snowflake.

References

1. Abid, A., Abdalla, A., Abid, A., Khan, D., Alfozan, A., Zou, J.Y.: Gradio: hassle-free sharing and testing of ML models in the wild. CoRR arxiv:1906.02569 (2019)
2. Alaofi, M., Arabzadeh, N., Clarke, C.L.A., Sanderson, M.: Generative information retrieval evaluation. CoRR arxiv:2404.08137 (2024)
3. Aliannejadi, M., Abbasiantaeb, Z., Chatterjee, S., Dalton, J., Azzopardi, L.: TREC ikat 2023: the interactive knowledge assistance track overview. CoRR arxiv:2401.01330 (2024)
4. Amouyal, S.J., Rubin, O., Yoran, O., Wolfson, T., Herzig, J., Berant, J.: QAMPARI: an open-domain question answering benchmark for questions with many answers from multiple paragraphs. CoRR arxiv:2205.12665 (2022)
5. Arabzadeh, N., Clarke, C.L.A.: A comparison of methods for evaluating generative IR. CoRR arxiv:2404.04044 (2024)
6. Asai, A., Wu, Z., Wang, Y., Sil, A., Hajishirzi, H.: Self-RAG: learning to retrieve, generate, and critique through self-reflection. CoRR arxiv:2310.11511 (2023)

7. Bajaj, P., et al.: MS MARCO: a human generated machine reading comprehension dataset. CoRR arxiv:1611.09268 (2016)
8. Bolotova, V., Blinov, V., Scholer, F., Croft, W.B., Sanderson, M.: A non-factoid question-answering taxonomy. In: Amigó, E., Castells, P., Gonzalo, J., Carterette, B., Culpepper, J.S., Kazai, G. (eds.) SIGIR '22: The 45th International ACM SIGIR Conference on Research and Development in Information Retrieval, Madrid, Spain, 11–15 July 2022, pp. 1196–1207. ACM (2022)
9. Borgeaud, S., et al.: Improving language models by retrieving from trillions of tokens. In: Chaudhuri, K., Jegelka, S., Song, L., Szepesvári, C., Niu, G., Sabato, S. (eds.) International Conference on Machine Learning, ICML 2022, Baltimore, Maryland, USA, 17–23 July 2022. Proceedings of Machine Learning Research, vol. 162, pp. 2206–2240. PMLR (2022)
10. Broder, A.Z.: On the resemblance and containment of documents. In: Carpentieri, B., Santis, A.D., Vaccaro, U., Storer, J.A. (eds.) Compression and Complexity of SEQUENCES 1997, Positano, Amalfitan Coast, Salerno, Italy, 11–13 June 1997, Proceedings, pp. 21–29. IEEE (1997)
11. Chase, H.: Langchain (2022)
12. Chiang, W., et al.: Chatbot arena: an open platform for evaluating LLMs by human preference. CoRR arxiv:2403.04132 (2024)
13. Cohere: Introducing Command R+: a scalable llm built for business (2024)
14. Craswell, N., Mitra, B., Yilmaz, E., Campos, D., Lin, J.: Overview of the TREC 2021 deep learning track. In: Soboroff, I., Ellis, A. (eds.) Proceedings of the Thirtieth Text REtrieval Conference, TREC 2021, online, 15–19 November 2021. NIST Special Publication, vol. 500-335. National Institute of Standards and Technology (NIST) (2021)
15. Craswell, N., et al.: Overview of the TREC 2022 deep learning track. In: Soboroff, I., Ellis, A. (eds.) Proceedings of the Thirty-First Text REtrieval Conference, TREC 2022, online, 15–19 November 2022. NIST Special Publication, vol. 500-338. National Institute of Standards and Technology (NIST) (2022)
16. Craswell, N., et al.: Overview of the TREC 2023 Deep Learning Track. In: Text REtrieval Conference (TREC). NIST, TREC (2024)
17. Dubey, A., et al.: The Llama 3 herd of models. arXiv:2407.21783 (2024)
18. Fan, A., Jernite, Y., Perez, E., Grangier, D., Weston, J., Auli, M.: ELI5: long form question answering. In: Korhonen, A., Traum, D.R., Màrquez, L. (eds.) Proceedings of the 57th Conference of the Association for Computational Linguistics, ACL 2019, Florence, Italy, 28 July–2 August 2019, vol. 1: Long Papers, pp. 3558–3567. Association for Computational Linguistics (2019)
19. Gao, T., Yen, H., Yu, J., Chen, D.: Enabling large language models to generate text with citations. In: Bouamor, H., Pino, J., Bali, K. (eds.) Proceedings of the 2023 Conference on Empirical Methods in Natural Language Processing, EMNLP 2023, Singapore, 6–10 December 2023, pp. 6465–6488. Association for Computational Linguistics (2023)
20. Gao, Y., et al.: Retrieval-augmented generation for large language models: a survey. CoRR arxiv:2312.10997 (2023)
21. Guu, K., Lee, K., Tung, Z., Pasupat, P., Chang, M.: Retrieval augmented language model pre-training. In: Proceedings of the 37th International Conference on Machine Learning, ICML 2020, 13–18 July 2020, Virtual Event. Proceedings of Machine Learning Research, vol. 119, pp. 3929–3938. PMLR (2020)
22. Hu, E.J., et al.: Lora: low-rank adaptation of large language models. In: The Tenth International Conference on Learning Representations, ICLR 2022, Virtual Event, 25–29 April 2022. OpenReview.net (2022)

23. Izacard, G., Grave, E.: Leveraging passage retrieval with generative models for open domain question answering. In: Merlo, P., Tiedemann, J., Tsarfaty, R. (eds.) Proceedings of the 16th Conference of the European Chapter of the Association for Computational Linguistics: Main Volume, EACL 2021, Online, 19–23 April 2021, pp. 874–880. Association for Computational Linguistics (2021)
24. Jimeno-Yepes, A., You, Y., Milczek, J., Laverde, S., Li, R.: Financial report chunking for effective retrieval augmented generation. CoRR arxiv:2402.05131 (2024)
25. Jin, J., Zhu, Y., Yang, X., Zhang, C., Dou, Z.: FlashRAG: a modular toolkit for efficient retrieval-augmented generation research. CoRR arxiv:2405.13576 (2024)
26. Kamalloo, E., Jafari, A., Zhang, X., Thakur, N., Lin, J.: HAGRID: a human-LLM collaborative dataset for generative information-seeking with attribution. CoRR arxiv:2307.16883 (2023)
27. Khandelwal, U., Levy, O., Jurafsky, D., Zettlemoyer, L., Lewis, M.: Generalization through memorization: nearest neighbor language models. In: 8th International Conference on Learning Representations, ICLR 2020, Addis Ababa, Ethiopia, 26–30 April 2020. OpenReview.net (2020)
28. Kulkarni, S., Chammas, S., Zhu, W., Sha, F., Ie, E.: Aquamuse: automatically generating datasets for query-based multi-document summarization. CoRR arxiv:2010.12694 (2020)
29. Kwiatkowski, T., et al.: Natural Questions: a benchmark for question answering research. Trans. Assoc. Comput. Linguist. **7**, 452–466 (2019)
30. Kwon, W., et al.: Efficient memory management for large language model serving with pagedattention. In: Proceedings of the ACM SIGOPS 29th Symposium on Operating Systems Principles (2023)
31. Lassance, C., Déjean, H., Formal, T., Clinchant, S.: SPLADE-v3: new baselines for SPLADE. arXiv:2403.06789 (2024)
32. Lassance, C., Pradeep, R., Lin, J.: naverloo @ TREC deep learning and NeuCLIR 2023: as easy as zero, one, two, three—cascading dual encoders, mono, duo, and listo for ad-hoc retrieval. In: Text REtrieval Conference (TREC). NIST (2024)
33. Lewis, P.S.H., et al.: Retrieval-augmented generation for knowledge-intensive NLP tasks. In: Larochelle, H., Ranzato, M., Hadsell, R., Balcan, M., Lin, H. (eds.) Advances in Neural Information Processing Systems 33: Annual Conference on Neural Information Processing Systems 2020, NeurIPS 2020, 6–12 December 2020, virtual (2020)
34. Li, Z., Zhang, X., Zhang, Y., Long, D., Xie, P., Zhang, M.: Towards general text embeddings with multi-stage contrastive learning. arXiv:2308.03281 (2023)
35. Lin, J., Demner-Fushman, D.: Methods for automatically evaluating answers to complex questions. Inf. Retr. **9**(5), 565–587 (2006)
36. Lin, J., Ma, X., Lin, S.C., Yang, J.H., Pradeep, R., Nogueira, R.: Pyserini: a python toolkit for reproducible information retrieval research with sparse and dense representations. In: Proceedings of the 44th Annual International ACM SIGIR Conference on Research and Development in Information Retrieval (SIGIR 2021), pp. 2356–2362 (2021)
37. Lin, S., Hilton, J., Evans, O.: Truthfulqa: measuring how models mimic human falsehoods. In: Muresan, S., Nakov, P., Villavicencio, A. (eds.) Proceedings of the 60th Annual Meeting of the Association for Computational Linguistics, ACL 2022, Dublin, Ireland, 22–27 May 2022, vol. 1: Long Papers, pp. 3214–3252. Association for Computational Linguistics (2022)
38. Liu, J.: Llamaindex (2022)

39. Liu, N.F., Lin, K., Hewitt, J., Paranjape, A., Bevilacqua, M., Petroni, F., Liang, P.: Lost in the middle: how language models use long contexts. Trans. Assoc. Comput. Linguist. **12**, 157–173 (2024)
40. Liu, Z., et al.: Chatqa: building GPT-4 level conversational QA models. CoRR arxiv:2401.10225 (2024)
41. Ma, X., Zhang, X., Pradeep, R., Lin, J.: Zero-shot listwise document reranking with a large language model. arXiv:2305.02156 (2023)
42. Malaviya, C., Lee, S., Chen, S., Sieber, E., Yatskar, M., Roth, D.: Expertqa: expert-curated questions and attributed answers. CoRR arxiv:2309.07852 (2023)
43. Mayfield, J., et al.: On the evaluation of machine-generated reports. In: Proceedings of the 47th International ACM SIGIR Conference on Research and Development in Information Retrieval (2024)
44. Merrick, L., Xu, D., Nuti, G., Campos, D.: Arctic-embed: scalable, efficient, and accurate text embedding models (2024)
45. Microsoft: Reinventing search with a new AI-powered Microsoft Bing and Edge, your copilot for the web (2023)
46. Muennighoff, N., Tazi, N., Magne, L., Reimers, N.: MTEB: massive text embedding benchmark. In: Vlachos, A., Augenstein, I. (eds.) Proceedings of the 17th Conference of the European Chapter of the Association for Computational Linguistics, EACL 2023, Dubrovnik, Croatia, 2–6 May 2023, pp. 2006–2029. Association for Computational Linguistics (2023)
47. Nogueira, R., Jiang, Z., Pradeep, R., Lin, J.: Document ranking with a pretrained sequence-to-sequence model. In: Findings of the Association for Computational Linguistics: EMNLP 2020, pp. 708–718 (2020)
48. OpenAI: Hello GPT-4o (2024)
49. Overwijk, A., Xiong, C., Callan, J.: Clueweb22: 10 billion web documents with rich information. In: Amigó, E., Castells, P., Gonzalo, J., Carterette, B., Culpepper, J.S., Kazai, G. (eds.) SIGIR '22: The 45th International ACM SIGIR Conference on Research and Development in Information Retrieval, Madrid, Spain, 11–15 July 2022, pp. 3360–3362. ACM (2022)
50. Owoicho, P., Dalton, J., Aliannejadi, M., Azzopardi, L., Trippas, J.R., Vakulenko, S.: TREC CAsT 2022: going beyond user ask and system retrieve with initiative and response generation. In: Soboroff, I., Ellis, A. (eds.) Proceedings of the Thirty-First Text REtrieval Conference, TREC 2022, online, 15–19 November 2022. NIST Special Publication, vol. 500-338. National Institute of Standards and Technology (NIST) (2022)
51. Petroni, F., et al.: KILT: a benchmark for knowledge intensive language tasks. In: Toutanova, K., et al. (eds.) Proceedings of the 2021 Conference of the North American Chapter of the Association for Computational Linguistics: Human Language Technologies, NAACL-HLT 2021, Online, 6–11 June 2021, pp. 2523–2544. Association for Computational Linguistics (2021)
52. Pradeep, R., Nogueira, R., Lin, J.: The expando-mono-duo design pattern for text ranking with pretrained sequence-to-sequence models. arXiv:2101.05667 (2021)
53. Pradeep, R., Sharifymoghaddam, S., Lin, J.: RankVicuna: zero-shot listwise document reranking with open-source large language models. CoRR arxiv:2309.15088 (2023)
54. Pradeep, R., Sharifymoghaddam, S., Lin, J.: RankZephyr: effective and robust zero-shot listwise reranking is a breeze! CoRR arxiv:2312.02724 (2023)
55. Robertson, S.E., Zaragoza, H.: The probabilistic relevance framework: BM25 and beyond. Found. Trends Inf. Retr. **3**(4), 333–389 (2009)

56. Rosenthal, S., Sil, A., Florian, R., Roukos, S.: CLAPNQ: cohesive long-form answers from passages in natural questions for RAG systems. CoRR arxiv:2404.02103 (2024)
57. Rosset, C., et al.: Researchy questions: a dataset of multi-perspective, decompositional questions for LLM web agents. CoRR arxiv:2402.17896 (2024)
58. Stelmakh, I., Luan, Y., Dhingra, B., Chang, M.: ASQA: factoid questions meet long-form answers. In: Goldberg, Y., Kozareva, Z., Zhang, Y. (eds.) Proceedings of the 2022 Conference on Empirical Methods in Natural Language Processing, EMNLP 2022, Abu Dhabi, United Arab Emirates, 7–11 December 2022, pp. 8273–8288. Association for Computational Linguistics (2022)
59. Sun, W., Yan, L., Ma, X., Ren, P., Yin, D., Ren, Z.: Is ChatGPT good at search? investigating large language models as re-ranking agent. arXiv:2304.09542 (2023)
60. Tamber, M.S., Pradeep, R., Lin, J.: Scaling down, LiTting up: efficient zero-shot listwise reranking with Seq2seq encoder-decoder models. arXiv:2312.16098 (2023)
61. Thakur, N., Reimers, N., Rücklé, A., Srivastava, A., Gurevych, I.: BEIR: a heterogeneous benchmark for zero-shot evaluation of information retrieval models. In: Vanschoren, J., Yeung, S. (eds.) Proceedings of the Neural Information Processing Systems Track on Datasets and Benchmarks 1, NeurIPS Datasets and Benchmarks 2021, December 2021, virtual (2021)
62. Upadhyay, S., Pradeep, R., Thakur, N., Craswell, N., Lin, J.: UMBRELA: UMbrela is the (open-source reproduction of the) bing RELevance assessor. arXiv:2406.06519 (2024)
63. Xiong, G., Jin, Q., Lu, Z., Zhang, A.: Benchmarking retrieval-augmented generation for medicine. CoRR arxiv:2402.13178 (2024)
64. Yan, S., Gu, J., Zhu, Y., Ling, Z.: Corrective retrieval augmented generation. CoRR arxiv:2401.15884 (2024)
65. Yang, P., Fang, H., Lin, J.: Anserini: enabling the use of lucene for information retrieval research. In: International Conference on Research and Development in Information Retrieval (SIGIR) (2017)
66. Zheng, L., et al.: Judging LLM-as-a-judge with MT-bench and chatbot arena. In: Oh, A., Naumann, T., Globerson, A., Saenko, K., Hardt, M., Levine, S. (eds.) Advances in Neural Information Processing Systems 36: Annual Conference on Neural Information Processing Systems 2023, NeurIPS 2023, New Orleans, LA, USA, 10–16 December 2023 (2023)

Retrieve, Annotate, Evaluate, Repeat: Leveraging Multimodal LLMs for Large-Scale Product Retrieval Evaluation

Kasra Hosseini(✉), Thomas Kober, Josip Krapac, Roland Vollgraf, Weiwei Cheng, and Ana Peleteiro Ramallo

Zalando SE, Berlin, Germany
k.hosseinizad@gmail.com

Abstract. Evaluating production-level retrieval systems at scale is a crucial yet challenging task due to the limited availability of a large pool of well-trained human annotators. Large Language Models (LLMs) have the potential to address this scaling issue and offer a viable alternative to humans for the bulk of annotation tasks. In this paper, we propose a framework for assessing the product search engines in a large-scale e-commerce setting, leveraging multimodal LLMs for (i) generating *tailored annotation guidelines* for individual queries, and (ii) conducting the subsequent *annotation task*. Our method, validated on a large e-commerce platform, achieves comparable quality to human annotations, significantly reduces time and cost, enables rapid problem discovery, and provides an effective solution for scalable, production-level quality control.

Keywords: Relevance Assessment · Large Language Models · LLM · Information Retrieval · System Evaluation · E-commerce Search · NLP Applications

1 Introduction

Search functionality is a fundamental component of e-commerce platforms, with the objective of finding the most relevant products in a dynamic product database. Customers using search often exhibit a higher intent to find specific products [16], leading to greater engagement and conversion rates. However, they may struggle to articulate their needs in a search query. Even if they do express their intent clearly, information retrieval (IR) systems might fail to interpret it correctly, resulting in irrelevant search results [28].

Evaluating product retrieval systems on a large scale—especially in multilingual settings and for a diverse range of customer queries—is an intricate but essential task for maintaining a high-quality user experience and driving business success. A prerequisite for this evaluation is the availability of a large enough pool of query-product relevance labels [11,27], which indicate whether a retrieved product is semantically relevant to the query. Semantic relevance depends solely on the query and the product, without considering other contextual factors, such as individual customer preferences.

C. Hauff et al. (Eds.): ECIR 2025, LNCS 15572, pp. 149–163, 2025.
https://doi.org/10.1007/978-3-031-88708-6_10

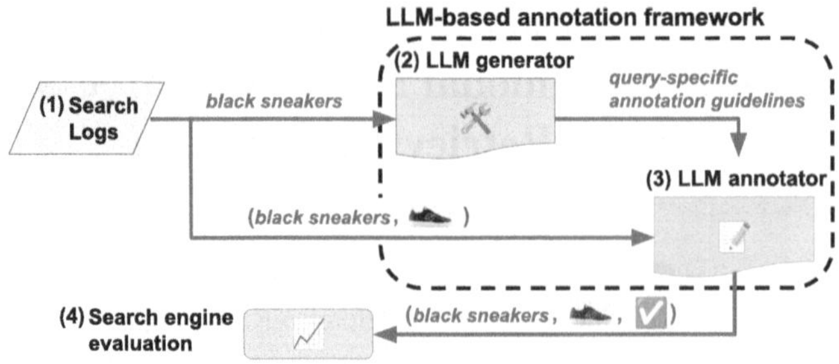

Fig. 1. Our proposed framework works by extracting a query-product pair from our search query-click logs (1). The query (e.g. *black sneakers*) is then passed on to the **LLM generator** (2) which creates *tailored annotation guidelines* for the given query. The query-specific annotation guidelines and the query-product pair (e.g. *black sneakers* and the retrieved product) are provided as input to the **LLM annotator** (3). Lastly, the annotated query-product pair is forwarded to the search engine evaluation module (4). (Color figure online)

Creating annotation guidelines to codify semantic relevance is a complex task [23]. It involves crafting guidelines that are digestible, concise, and precise, along with curating illustrative examples of varying difficulty. Even with well-defined guidelines and well-trained human annotators, manual annotation is slow and costly.

The advent of crowd-sourcing platforms has increased scalability [2,4,6,13,15], allowing for a trade-off between speed and cost. However, increasing the number of annotators can lead to inconsistencies, as even the same annotator may provide contradictory annotations for similar query-product pairs, let alone multiple annotators. Consistency can be improved by using more annotators per pair (see, e.g., [10]), but this results in increased cost. In large e-commerce systems, the volume of data that needs to be annotated leads to prohibitively high costs when using crowd-sourcing platforms that rely on human annotators.

While the rate of manual relevance judgement varies depending on the task [7,21,22,27], in our use case, we estimate a throughput of 2–3 query-product pairs per minute. As an example, 50,000 queries[1] and 20 products per query results in one million query-product pairs, which takes 5,500–8,500 h of human labour, assuming one annotation per pair. Moreover, evaluation is not a one-off practice; ideally, search engines are continuously assessed to ensure their effectiveness over time. This is particularly important because user preferences and behaviours evolve, queries may change seasonally or in response to new trends, sales events, or shifting consumer needs. Additionally, product catalogues in e-commerce are highly dynamic, and updates to search algorithms or ranking models can inadvertently introduce new biases or degrade performance.

The sheer volume of required annotations in multiple languages, along with the need for continuous evaluation, makes human-generated relevance judgements the primary

[1] A modest number for the evaluation of large IR systems.

bottleneck in creating product retrieval evaluation datasets. To overcome these challenges, there has been growing interest in leveraging LLMs [3,8,9,14,20,22,24–26].[2]

In this study, we propose a framework that leverages the capabilities of Multimodal Large Language Models (MLLMs) for assessing the semantic relevance of query-product pairs (Fig. 1). Our method draws on the strengths of LLMs and MLLMs in understanding natural language queries across various languages and processing both textual and visual features of products. Unlike traditional per-task annotation guidelines, such as those discussed by [22], we employ LLMs to generate annotation guidelines *specific to each query*. Additionally, our pipeline's modular design allows for caching and parallel processing, which is crucial for scaling up to larger systems. This framework has enabled daily evaluations of our product retrieval systems. It has also facilitated the comparison of different search models, increasing our confidence in offline evaluations and complementing our online evaluation techniques, such as A/B testing and other controlled online experiments [12]. Moreover, we have used the relevance assessments' outputs to train and evaluate other components of our search and ranking systems.

Our results show that while human-human and human-LLM agreement scores are on par, humans and LLMs tend to make very different types of annotation errors. These findings suggest that LLMs are very effective for the bulk work of annotations whereas human expertise is better leveraged for more complex cases.

In summary, our contributions are as follows:

- We introduce a multimodal LLM-based evaluation framework for large-scale product retrieval systems. We propose *query-specific* annotation guidelines and utilise multimodal inputs (text and images) for relevance assessment.
- We evaluate the performance of our framework against 20,000 human annotations on real-world production search queries in a multilingual setting and analyse the different types of errors that humans and LLMs tend to make.
- We demonstrate the cost-effectiveness and efficiency of our approach for conducting large-scale evaluations. We also compare the performance of different types of LLMs [1,5,19] for relevance assessment.

2 Multimodal LLM-Based Relevance Assessment

Our framework, shown in Fig. 2, leverages (M)LLMs for efficient evaluation of large-scale product retrieval systems and consists of six main steps:

(1) For a given query and its context (e.g., selected gender and market), an LLM generates a query requirement list and a *query-specific* annotation guideline. The query requirement list captures the relevant pieces of information in the user's query and their level of importance. For example, for the query *Nike red shoes*, the query requirement list includes the brand (*Nike*), colour (*red*) and product category (*shoe*). The query-specific annotation guideline is generated by the LLM based on the query and its requirement list. It outlines criteria for each predefined label. In our experiments, we

[2] While LLMs are being increasingly explored for relevance labelling, several studies have raised concerns about their trustworthiness, biases, and limitations [8,29].

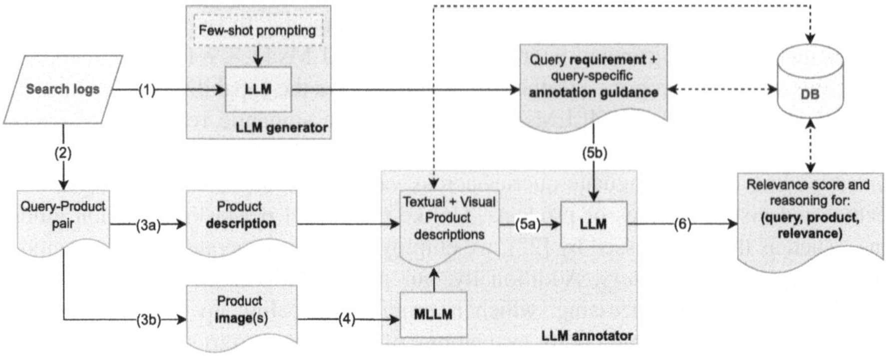

Fig. 2. Our proposed multimodal LLM-powered framework enables offline evaluation of large-scale product retrieval systems and presents significant time and cost reductions compared to existing evaluation techniques. The orange rectangle indicates where a "one-step" Multimodal LLM (MLLM) could be utilised, instead of employing one MLLM to create a textual description for image inputs (Step 4) followed by an LLM (Step 5a). In the one-step MLLM, both textual descriptions and the product image are directly fed into the LLM annotator, along with query requirements and query-specific annotation guidelines. The depiction of the pipeline is simplified for readability. (Color figure online)

defined three relevance labels for a query-product pair: "irrelevant", "acceptable substitute" and "highly relevant".[3]

(2) The query and its context are sent to the search engine, which retrieves a set of products. For simplicity, we illustrate this process using a single query-product pair. However, in practice, we work with multiple query-product pairs and may utilise two or more retrieval systems, particularly when comparing their performance.

(3a,b) For each retrieved product, we have access to its textual description and image.

(4) A multimodal LLM uses the product image to generate descriptive text. This step is optional, as a "one-step" MLLM could be utilised to directly feed both the textual descriptions and product image into the LLM annotator.

(5a,b) The combined text and visual product descriptions, along with the Step 1 outputs (query requirements and query-specific annotation guidelines), are sent to an LLM.

(6) The LLM assigns a relevance score to the query-product pair using a set of predefined labels. In its simplest form, the output is a database with one row for each `(query, product, relevance score)`.

In Steps 1 and 6, we utilise chain-of-thought (CoT) prompting [17,30] to enhance the quality of (M)LLM outputs and facilitate debugging.

As illustrated with dashed lines in Fig. 2, all outputs and intermediate steps are stored in a database. This caching serves two key functions in our pipeline. Firstly, it facilitates efficient retrieval and reuse. When evaluating a new search engine configura-

[3] See Appendix A for details.

tion (or a variation of existing ones), the database is queried to retrieve relevant pieces of information, including the query requirement list, query-specific annotation guidance, textual and visual product descriptions, and relevance scores. We only compute the missing pieces of information. Secondly, it ensures consistent evaluation across different search engines, as intermediate steps (such as query-specific annotation guidelines) are computed only once and then used to evaluate various search engines.

3 Experiments and Results

3.1 Dataset

Data Collection. As a starting point for data collection, we used one year's worth of production search query traffic.[4] We then performed stratified sampling along the following dimensions: a) search engine, b) activated gender filter on the website, c) query frequency, and d) query length in tokens.

After sampling and manual curation,[5] we obtained 500 queries in German and English each. For each query, we then used our search engine to retrieve 20 products, selecting 15 products from the top of the retrieved results and randomly sampling the remaining 5 from rank 500 onwards. This resulted in 10,000 unique query-product pairs for German and English each. Table 1 outlines the statistics of the collected dataset.

Table 1. Dataset statistics.

Language	Unique pairs	Unique queries	Avg. tokens per query	Unique products
German	10,000	500	3.68	8,076
English	10,000	500	3.99	8,652

Data Annotation. We contracted an external agency to facilitate crowd-sourcing annotations based on the guidelines we provided.[6] Our requirements specified that only native speakers (German and English, see Table 1) annotate the data. Each query-product pair was to be annotated by two human annotators, with an additional third annotation as a tiebreaker in cases where the two original annotators disagreed. The data annotation process was done in three phases: (i) a pilot phase to onboard annotators and resolve outstanding loopholes and questions in the annotation guidelines; (ii) the main annotation phase; (iii) the tiebreaker phase.

[4] Our data collection process complies with the regulations defined in the GDPR and other existing regulatory frameworks around data privacy and safety in the European Union.

[5] We manually checked all sampled queries to ensure we cover a diverse and broad spectrum. For example, we would manually replace *yet another* sneakers-related query with a boots-related query of the same length.

[6] See Appendix A for an overview of our annotation guidelines.

In total, the data annotation process—from the handover of the initial version of the annotation guidelines by us to the final delivery of annotated data by the external agency—took approximately 8 weeks,[7] with roughly 3 weeks needed for completing the main annotation and tiebreaker phases. The total cost of data annotation was € 30,000.

3.2 LLM Versus Human Annotators

Using our proposed framework, we assessed the relevance quality of the 20,000 unique query-product pairs. Table 2 summaries the results of our experiments using few-shot prompting, where we incorporated examples into the system prompts of the (M)LLMs. Here, we provided the LLMs with example customer queries, their corresponding requirement lists, and quality labels, but not complete product descriptions or images.

Initially, we randomly sampled 100 examples from the English dataset and examined the relevance labels assigned by both LLMs and human annotators. We used the results of this step to adjust the few-shot examples in the system prompt.

We compare the performance of different versions of our pipeline with human annotations. In Table 2, these versions are labelled as "LLM-text", "MLLM-text", and "MLLM-multi".

"LLM-text" is the simplest version where only product descriptions in textual form are used, without incorporating product images. In "MLLM-text", we employ a vision model to generate textual descriptions of product images (Step 4 in the pipeline, see

Table 2. Agreements between (M)LLM and human annotator groups (A1 and A2). We compare agreements based on: (i) matches with either A1 *or* A2, and (ii) inter-annotator agreement between human annotators (A1 vs. A2), as well as between the (M)LLM and the human majority vote. For the A1 *or* A2 column, we use the same majority vote methodology to measure agreements among the human annotators. Results are reported separately for English and German. For human annotations, we report the total time and cost. We use GPT-4o in all steps of our LLM annotation pipeline (Fig. 2). Refer to Table 3 for a more detailed comparison between human annotator groups (A1, A2, and tiebreaker) and different versions of our LLM-powered framework.

Annotator	Agreement with A1 OR A2 (%)		Inter Annotator Agreement (%)		Total time		Cost	
	EN	DE	EN	DE	EN	DE	EN	DE
A1	76.6	75.9	60.2	60.5	3 weeks		€15,000	
A2	75.8	76.1	60.2	60.5	3 weeks		€15,000	
LLM-text	75.4	75.6	62.8	61.4	8 min	7 min	€70	€70
MLLM-text	76.6	77.0	64.3	63.1	12 min	13 min	€92	€93
MLLM-multi	**78.1**	**78.4**	**65.6**	**64.7**	10 min	10 min	€72	€72
MLLM-multi (gpt-4-turbo)	75.4	75.6	62.9	62.0	30 min	23 min	€155	€156
Without query-specific annotation guidance								
LLM-text	71.2 (-4.2)	67.0 (-8.6)	57.6 (-5.2)	52.2 (-9.2)	7 min	7 min	€45	€45
MLLM-text	72.5 (-4.1)	68.0 (-9.0)	58.7 (-5.6)	53.4 (-9.7)	11 min	11 min	€66	€67
MLLM-multi	74.1 (-4.0)	71.9 (-6.5)	61.6 (-4.0)	59.9 (-4.8)	4 min	4 min	€25	€24

[7] This excludes the effort needed to write the annotation guidelines and to sample the query-product candidate pairs.

Fig. 2). The generated textual description of the product image is then concatenated with the product description itself (Step 5a). "MLLM-multi" utilises the same textual input as "LLM-text", while also incorporating the product image as an additional input.[8] Comparing "MLLM-multi" and "LLM-text" highlights the impact of multimodal inputs on our task.

In all cases, the (M)LLM uses product information (in different modalities, depending on the version), query requirements, and query-specific annotation guidance to assign relevance labels.

Overall, Table 2 demonstrates that the agreement scores between human-human annotators and human-LLM annotators are comparable, supporting the scalability of LLM-based annotation for production-level traffic.

Table 2 also shows the results of an ablation study that removed the query-specific annotation guideline (Step 1 in Fig. 2). The inclusion of such guidelines improved agreements by approximately 4–10%. More importantly, this component in our framework is essential for enhancing the interpretability and debugging of LLM-based decisions. However, as expected, incorporating query-specific annotation guidelines and chain-of-thought reasoning increased the evaluation costs.

We also tested the impact of different (M)LLM architectures in our pipeline.[9] The results shown in Table 2 are based on GPT-4o [18] except for one row, labelled "MLLM-multi (gpt-4-turbo)". In the case of GPT-4 Turbo, the agreement with human annotators consistently fell below that of GPT-4o, while its costs and evaluation times exceeded those of all other architectures.

In Table 4, we repeated the experiments using GPT-3.5 Turbo. As expected, the results were significantly worse compared to GPT-4o or GPT-4 Turbo. However, the cost and time required for GPT-3.5 Turbo were lower than for the other architectures.

4 Discussion

Agreement Between LLM and Human Annotators. The agreements between "MLLM-multi" and the human majority vote—65.6% for EN and 64.7% for DE, as shown in Table 2—are in line with the human inter-annotator agreement, which is 60.2% for EN and 60.5% for DE.

To better identify discrepancies between LLM and human annotations, we focused our analysis on hard disagreements between the two. We consider a hard disagreement to be when, for example, the LLM considers a product to be "highly relevant" for a given query, whereas the human majority judgement would be "irrelevant", and vice versa. In total, we found that approximately 15% of annotations[10] in our dataset were hard disagreements. For manual analysis, we sampled 20% of the hard disagreements and found that in 50% of the cases, the human annotation was wrong, in 31% the LLM

[8] See the orange rectangle in Fig. 2 where both the textual product description and its image are fed into an MLLM.

[9] Refer to Table 3 and 4 in Appendix B for the results of similar experiments conducted with different LLM types.

[10] 2,971 out of 20,000 query-product pairs ($\approx$15%) had hard disagreements between the LLM and human majority vote.

was wrong, and in 19% of cases, both the LLM and the humans provided a wrong annotation.

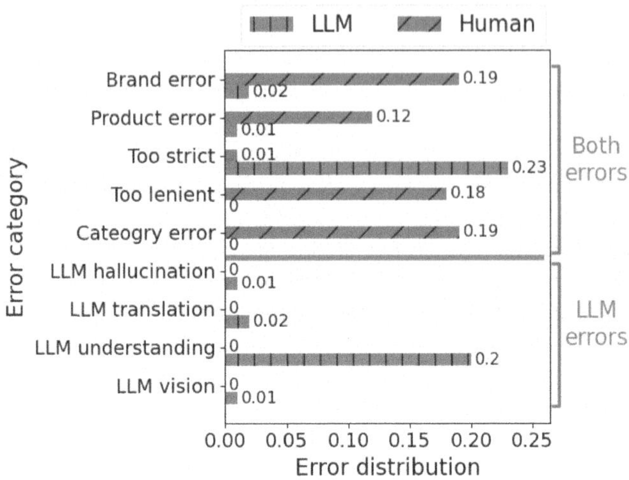

Fig. 3. Distribution of errors for LLMs and humans on hard disagreements (50% human errors, 31% LLM errors, and 19% cases where both made an error). The upper part ("Both errors") focuses on errors that either LLMs or humans could make. It highlights that LLMs and humans make very different types of errors. In addition, the lower part ("LLM errors") shows the distribution of errors that only an LLM would make. Predominantly these are misunderstandings of a part of the search query.

We further categorised the hard disagreements into nine error classes and found that LLMs and humans tend to make very different kinds of errors.[11] For example, as shown in Fig. 3, the main errors made by the LLM are either being *too strict* in their judgement (e.g., considering a product as "irrelevant" where "acceptable substitute" would have been more appropriate), or misunderstanding a part of the query (e.g., interpreting *On Vacation* in its literal sense rather than the fashion brand). On the other hand, humans would tend to be *too lenient* when LLMs were *too strict* (e.g., considering a product as "highly relevant" when "acceptable substitute" would have been more appropriate). Furthermore, human annotations frequently exhibited brand errors (e.g., considering a pair of *Lee* jeans as "highly relevant" for a query requesting *Levi's jeans*), product errors (e.g., considering an *Adidas Samba* sneaker as "highly relevant" for a query requesting an *Adidas Stan Smith* sneaker), or category errors (e.g., considering a pair of *Nike shirts* as "highly relevant" for a query requesting *Nike shoes*), which we barely ever observed for LLMs. We hypothesise that the latter three kinds of human errors are primarily due to annotation fatigue as specifically these cases have been prominently and unambiguously featured in the annotation guidelines.

[11] See Appendix C for the definition of these nine classes and examples of errors we observed.

These findings suggest that LLMs might be a more reliable source for the bulk of annotations, freeing human labour to focus on trickier cases.[12] In the human-machine collaboration spectrum introduced by [8], our approach can be classified as a "Human Verification" (or human-in-the-loop) approach.

Subjective Nature of Relevance Judgements. We found that human disagreements were dominated by two main factors: (i) human errors due to annotation fatigue as described above,[13] and (ii) the inherent subjective nature of the task. For the latter, we attribute the source of disagreements to either the ambiguity in the annotation guidelines (even comprehensive guidelines cannot cover all possible cases), or to the subjective judgement of the annotator.[14] Ideally, the annotation guidelines should make the task as objective as possible; however, in practice, there is always a level of subjectivity.

Annotation Time and Cost. (M)LLMs are approximately 100 to 1,000 times cheaper than human annotators, and the time required to complete all 20,000 annotations using (M)LLMs is significantly smaller (around 20 min for (M)LLMs compared to about 3 weeks for human annotators). Note that several human annotators worked in each group ("A1" and "A2" in Table 2), and the total time reported in Table 2 is for annotating all query-product pairs. This excludes the time spent on scoping and onboarding human annotators. For (M)LLMs, the reported time excludes the pipeline development time and only includes the actual annotation time.

We anticipate that both cost and time will decrease even further as LLMs and their APIs become more efficient. Moreover, new approaches, such as batch processing and prompt caching[15], can further reduce costs (e.g., OpenAI's new batch processing is half the price of non-batch queries[16]). Indeed, in production, we use batch processing to assess query-product pairs across markets on a nightly basis.

Relevance Assessment in Production. High relevance is a necessary, but not sufficient, condition for high customer engagement, as other factors such as personal preferences, product availability, and price expectations also play a significant role. In this paper, we focus on semantic relevance; however, in production, we rank retrieved products based on various features to account for both query relevance and personal customer preferences.

We currently employ the LLM-powered evaluation framework discussed in this paper in production to continuously assess relevance at scale. Our framework primarily monitors the performance of high-volume queries. Additionally, we evaluate retrieval

[12] This typically includes new trending terms or styles that LLMs have not observed yet, but also ambiguous queries such as *old money* or *dark academia*, that specify a fashion style.

[13] In fact, we found that human-human hard disagreements, which made up approximately 15% of the annotated data, were primarily due to brand errors, product errors, and category errors—the same types of errors we observed in LLM-human hard disagreements.

[14] See Appendix D for examples.

[15] https://platform.openai.com/docs/guides/prompt-caching (accessed on 2025-01-17).

[16] https://platform.openai.com/docs/guides/batch (accessed on 2025-01-17).

performance for queries identified as low-performing. These queries are flagged based on signals such as low engagement with the results and high customer friction, including a high reformulation rate[17] or high exit rates. This approach enables us to reduce costs and enhance customer experience more efficiently by prioritizing queries that need the most attention, thereby optimizing resources accordingly.

5 Conclusion

Our novel evaluation method, which utilises multimodal LLMs, offers a highly efficient approach to assessing large-scale IR systems in product retrieval. By introducing query-specific annotation guidelines and leveraging the multimodal capabilities of foundation models, we improve the quality and consistency of relevance assessments at scale. Our LLM-powered framework, enhanced with caching and parallel processing, achieves significant reductions in both time and cost. The method's scalability, ability to handle multilingual queries and products, and support for continuous offline evaluations make it well-suited for large IR systems operating in diverse markets. Experimental validation against 20,000 human annotations confirms its effectiveness and efficiency. We are currently using this framework in production to continuously perform relevance assessments at scale and maintain a high-quality user experience, while also leveraging the outputs of these assessments to train and evaluate other components of our search and ranking systems.

6 Ethics Statement

Our data collection process strictly adheres to the General Data Protection Regulation (GDPR) and other relevant data privacy and safety laws within the European Union. We ensure that all data utilised, including human evaluation data, is anonymised to safeguard against the disclosure of any personally identifiable information.

We do not suggest replacing human annotators with large language models (LLMs). Instead, we focus on leveraging the strengths of both. Our analysis indicates that human annotators may make errors due to annotation fatigue or lack of domain knowledge-errors not observed with LLMs. Therefore, we recommend using LLMs for bulk annotation work while reserving human expertise for more complex cases.

We are committed to advancing responsible and unbiased AI technologies and welcome any inquiries regarding the ethical aspects of our work.

Disclosure of Interests. The authors have no competing interests to declare that are relevant to the content of this article.

[17] The reformulation rate refers to the percentage of queries modified and resubmitted within the same search session, indicating an initial failure to meet the user's intent and a subsequent attempt to refine the search.

A Human Annotation Guidelines

For human annotators, we focused on three classes:

- **highly relevant**: The retrieved product satisfies all the specifications in the query.
- **acceptable substitute**: The item fulfils some, but not all, aspects of the query, and can be used as a functional substitute for the intended item.
- **irrelevant**: A central aspect of the query is not fulfilled (e.g. wrong brand, wrong category, wrong product).

We decided against a more granular annotation scale to reduce mental load on annotators and to (hopefully) harness higher agreement scores among annotators.

Our annotation guidelines also reflect requirements that are more business-specific than content-specific. For example, annotators have been explicitly briefed that if a query requests a specific brand (e.g. *Polo Ralph Lauren jumpers*), any retrieved item that is not from the requested brand is to be regarded as "irrelevant".

Another business-specific rule was that if a query requests a particular product (e.g. *The North Face 1996 retro nuptse jacket*), any retrieved product that is not that particular type of The North Face jacket is to be regarded as "irrelevant".

Despite explicit mentions of these rules, numerous examples provided across product categories, and an additional briefing session after the annotation pilot phase, brand and product errors remained among the most common human errors (see Fig. 3).

B Experiments with LLM Types: GPT-3.5, GPT-4, and GPT-4o

Table 3. Agreements between (M)LLM and the human annotator groups (A1, A2 and tiebreaker). Similar to Table 2, but with additional columns showing the agreements of A1, A2, and the tiebreaker groups with other annotators. We report agreements separately for English and German. For human annotations, we report the total time and cost. We use GPT-4o in all steps of our LLM-powered framework (Fig. 2).

Annotator	Agreement with A1 OR A2 (%)		Agreement with A1 (%)		Agreement with A2 (%)		Agreement with Tiebreaker (%)		Inter Annotator Agreement (%)		Total time		Cost	
	EN	DE	EN	DE	EN	DE	EN	DE	EN	DE	EN	DE	EN	DE
A1	76.6	75.9	-	-	**60.2**	60.5	40.0	37.5	60.2	60.5	3 weeks		€15,000	
A2	75.8	76.1	60.2	60.5	-	-	37.5	36.6	60.2	60.5	3 weeks		€15,000	
LLM-text	75.4	75.6	63.9	60.8	55.8	59.3	41.9	33.9	62.8	61.4	8 min	7 min	€70	€70
MLLM-text	76.6	77.0	64.9	62.0	56.9	60.7	43.2	35.5	64.3	63.1	12 min	13 min	€92	€93
MLLM-multi	**78.1**	**78.4**	**66.0**	**63.1**	58.2	**62.0**	**43.9**	37.4	**65.6**	**64.7**	10 min	10 min	€72	€72
MLLM-multi (gpt-4-turbo)	75.4	75.6	63.9	61.1	55.8	59.3	41.7	34.8	62.9	62.0	30 min	23 min	€155	€156
Without query-specific annotation guidance														
LLM-text	71.2	67.0	57.5	52.9	54.4	51.4	39.0	34.4	57.6	52.2	7 min	7 min	€45	€45
MLLM-text	72.5	68.0	58.8	53.8	55.4	52.4	39.7	34.1	58.7	53.4	11 min	11 min	€66	€67
MLLM-multi	74.1	71.9	61.1	57.1	56.3	56.4	41.9	**42.3**	61.6	59.9	4 min	4 min	€25	€24

In this section, we compare human annotator groups with (M)LLMs using different architectures. The results in Table 3 are primarily based on GPT-4o [18], except for the row labelled "MLLM-multi (gpt-4-turbo)". For GPT-4 Turbo, the agreement with

Table 4. Like Table 3, except we use GPT-3.5 Turbo (specifically, "gpt-3.5-turbo-0125") for text inputs and GPT-4o for generating textual descriptions for image inputs (Step 4 in Fig. 2). Here, we do not have "MLLM-multi" as GPT-3.5 Turbo does not accept multimodal (text and image) inputs.

Annotator	Agreement with A1 OR A2 (%)		Agreement with A1 (%)		Agreement with A2 (%)		Agreement with Tiebreaker (%)		Inter Annotator Agreement (%)		Total time		Cost	
	EN	DE	EN	DE	EN	DE	EN	DE	EN	DE	EN	DE	EN	DE
A1	**76.6**	75.9	-	-	**60.2**	**60.5**	40.0	37.5	**60.2**	**60.5**	3 weeks		€15,000	
A2	75.8	**76.1**	**60.2**	**60.5**	-	-	37.5	36.6	**60.2**	**60.5**	3 weeks		€15,000	
LLM-text	65.2	63.4	51.3	48.5	49.6	48.7	40.1	37.7	52.1	49.3	6 min	6 min	€6	€6
MLLM-text	66.1	62.6	51.6	48.2	50.9	48.1	40.8	39.2	53.1	49.6	11 min	11 min	€24	€25
Without query-specific annotation guidance														
LLM-text	59.1	57.0	43.9	43.0	46.5	42.8	41.1	47.9	47.6	46.3	4 min	4 min	€4	€4
MLLM-text	60.1	57.2	45.0	42.9	47.4	43.4	**41.9**	**50.2**	48.9	47.4	9 min	10 min	€21	€22

human annotators was consistently lower than that of GPT-4o, while its costs and evaluation times exceeded those of all other architectures. In Table 4, we repeated the experiments using GPT-3.5 Turbo. As expected, its results were significantly worse compared to GPT-4o and GPT-4 Turbo, but its cost and time requirements were lower than those of the other architectures.

C LLM Versus Human Error Types

After manually inspecting a sample of hard disagreements,[18] we defined the following nine error classes, some of which are applicable to LLMs and humans, and some to LLMs only:

1. Brand error. When a user specifies a brand name in the search query, e.g. *Lee jeans*, *Nike sneakers*, or *Mascara dresses*, we consider any retrieved item as "irrelevant" if it is not from the requested brand. This is independent of whether the retrieved item would be visually similar to the requested one. This requirement has been covered in the LLM prompt as well as the human annotation guidelines. Predominantly, this error has been made by human annotators (see Fig. 3).

2. Product error. When a user specifies a specific product in the search query, e.g. *Levis 501* or *Adidas Stan Smith*, we consider any retrieved item that is not exactly the requested item as "irrelevant". This requirement has been covered in the LLM prompt as well as the human annotation guidelines. Predominantly, this error has been made by human annotators (see Fig. 3).

3. Too strict. This error happened when a product was judged as "irrelevant" for a given query despite fulfilling *almost* all the requirements of the query. This error has been predominantly made by LLMs (Fig. 3), for example, when a query requested *black Levis jeans with holes*, but the retrieved product was a grey pair of Levis jeans with holes. In such cases, the LLM would typically annotate the retrieved product as "irrelevant."

[18] We consider a hard disagreement to be when, for example, the LLM considers a product to be "highly relevant" for a given query, whereas the human majority judgement would be "irrelevant", and vice versa. Refer to Sect. 4 for details.

4. Too lenient. This error happened when a product was judged as "highly relevant" for a given query, despite not meeting all the specified requirements of the query. This error has been exclusively made by human annotators (see Fig. 3), for example, where for a query like *Nike Air Force One high-top*, humans annotated a *Nike Air Force One low-top* sneaker as "highly relevant".

5. Category error. When a user specifies the category of a fashion item in the search query, e.g. *dress*, *sneakers*, *belts*, we consider any retrieved item that does not match the category as "irrelevant". This requirement has been covered in the LLM prompt as well as the human annotation guidelines. Predominantly, this error has been made by human annotators (see Fig. 3).

6. LLM hallucination error. We rarely observed hallucinations as a source of error. Interestingly, when hallucinations did occur, they were exclusively related to size queries, such as *t-shirt xxl*. In such cases, the LLM would hallucinate various available sizes for a given retrieved product and make a relevancy judgement on the basis of its hallucinations.

7. LLM translation error. Since our dataset contained German and English queries, the LLM was prompted to translate a German query into English before starting its reasoning process. This sometimes resulted in translation errors that subsequently led to incorrect relevancy judgements. For example, it happened for queries containing the term *Unterziehhose*, meaning some sports leggings one can wear underneath sports shorts, which the LLM incorrectly translated as *underpants*.

8. LLM understanding error. This error category is somewhat broader. We would categorise an LLM error as *understanding error* whenever the LLM misinterpreted a part of the query or the product. For example, this error occurred when the LLM would misinterpret a query for *Nike Tech Fleece* to be focused on the material whereas *Tech Fleece* typically refers to a particular series of Nike sports clothing. Another example is the misinterpretation of brand names, such as for *On Vacation* (interpreted in its literal meaning), or for *Evry Jewels* (where *Evry* would be interpreted to mean *Every*). To our amusement during error analysis, we also observed a brand misinterpretation for the query *miniature winter jackets for kids*, where *Mini A Ture* is a kids' clothing brand. The LLM interpreted *miniature* in its literal sense and reasoning that *[...] the sizes available are for kids, which fits the 'miniature' requirement.*

9. LLM vision error. Some of our models included the visual interpretation of a product image in its relevancy assessment.[19] We only rarely observed LLM vision errors. If they did occur, it was typically when the product image was taken at a slight angle-for instance, with a pair of sneakers where the LLM erroneously identified them as high-top due to the photo angle. Errors were also more likely when the image included a human model, which acted as a distractor.

D Subjective Nature of Relevance Judgements

The difficulty in judging query-product relevancy can vary widely. For example, for queries such as *Nike Air Max 95* or *Paul Smith long sleeve polo shirt*, there is barely any

[19] MLLM-text in Tables 2, 3 and 4.

room for subjectivity—the retrieved products are either matches or not. And indeed, this is reflected in the human-human inter-annotator agreement (95% and 82%, respectively) and the LLM-human agreement (98% and 89%, respectively), for these two examples.

However, there are numerous queries that are much more open to subjective judgement. One such example is the query *smart casual shoes*, where the human-human agreement was only 12% and the LLM-human agreement was 24%. The range of suitable products for this query spans various types of shoes, and whether or not a particular shoe can be categorised as *smart casual* is typically not included in the product data. In these cases, humans and LLMs would draw on their prior knowledge for making a relevance judgement. LLMs would generally be a stricter judge and consider anything that resembles a sneaker too closely, or is not in a shade of black or brown, as "irrelevant". Human strictness in relevance judgments for this query ranged from very formal to (loosely speaking) "anything goes" extremes.

References

1. Achiam, J., et al.: Gpt-4 technical report. arXiv preprint arXiv:2303.08774 (2023)
2. Alonso, O., Mizzaro, S.: Using crowdsourcing for trec relevance assessment. Inf. Process. Manag. **48**(6), 1053–1066 (2012)
3. Bergum, J.K.: Improving retrieval with llm-as-a-judge (2024). https://blog.vespa.ai/improving-retrieval-with-llm-as-a-judge/. Accessed 10 June 2024
4. Blanco, R., et al.: Repeatable and reliable search system evaluation using crowdsourcing. In: Proceedings of the 34th International ACM SIGIR Conference on Research and Development in Information Retrieval, pp. 923–932 (2011)
5. Brown, T., et al.: Language models are few-shot learners. Adv. Neural. Inf. Process. Syst. **33**, 1877–1901 (2020)
6. Chen, K.T., Alonso, O., Larson, M., King, I.: Introduction to the special issue on crowd in intelligent systems (2016)
7. Chen, Y., Liu, S., Liu, Z., Sun, W., Baltrunas, L., Schroeder, B.: WANDS: dataset for product search relevance assessment. In: Hagen, M., et al. (eds.) ECIR 2022. LNCS, vol. 13185, pp. 128–141. Springer, Cham (2022). https://doi.org/10.1007/978-3-030-99736-6_9
8. Faggioli, G., et al.: Perspectives on large language models for relevance judgment. In: Proceedings of the 2023 ACM SIGIR International Conference on Theory of Information Retrieval, ICTIR '23, pp. 39–50. Association for Computing Machinery, New York (2023). https://doi.org/10.1145/3578337.3605136
9. Faggioli, G., et al.: Who determines what is relevant? Humans or ai? Why not both? Commun. ACM **67**(4), 31–34 (2024)
10. Ferrante, M., Ferro, N., Maistro, M.: Aware: exploiting evaluation measures to combine multiple assessors. ACM Trans. Inf. Syst. (TOIS) **36**(2), 1–38 (2017)
11. Halvey, M., Villa, R., Clough, P.D.: Sigir 2014: workshop on gathering efficient assessments of relevance (gear). SIGIR Forum **49**(1), 16–19 (2015). https://doi.org/10.1145/2795403.2795409
12. Kohavi, R., Longbotham, R., Sommerfield, D., Henne, R.M.: Controlled experiments on the web: survey and practical guide. Data Min. Knowl. Disc. **18**, 140–181 (2009)
13. Lease, M., Yilmaz, E.: Crowdsourcing for information retrieval: introduction to the special issue. Inf. Retr. **16**, 91–100 (2013)

14. MacAvaney, S., Soldaini, L.: One-shot labeling for automatic relevance estimation. In: Proceedings of the 46th International ACM SIGIR Conference on Research and Development in Information Retrieval, SIGIR '23, pp. 2230–2235. Association for Computing Machinery, New York (2023).https://doi.org/10.1145/3539618.3592032
15. Marcus, A., Parameswaran, A., et al.: Crowdsourced data management: industry and academic perspectives. Found. Trends® Databases **6**(1-2), 1–161 (2015)
16. Moe, W.W.: Buying, searching, or browsing: differentiating between online shoppers using in-store navigational clickstream. J. Consum. Psychol. **13**(1–2), 29–39 (2003). https://doi.org/10.1207/S15327663JCP13-1&2_03
17. Nye, M., et al.: Show your work: scratchpads for intermediate computation with language models. arXiv preprint arXiv:2112.00114 (2021)
18. OpenAI: Hello gpt-4o (2024). https://openai.com/index/hello-gpt-4o/. Accessed 17 July 2024
19. Radford, A., Wu, J., Child, R., Luan, D., Amodei, D., Sutskever, I., et al.: Language models are unsupervised multitask learners. OpenAI blog **1**(8), 9 (2019)
20. Rahmani, H.A., Craswell, N., Yilmaz, E., Mitra, B., Campos, D.: Synthetic test collections for retrieval evaluation (2024)
21. Sanderson, M., et al.: Test collection based evaluation of information retrieval systems. Found. Trends® Inf. Retr. **4**(4), 247–375 (2010)
22. Soviero, B., Kuhn, D., Salle, A., Moreira, V.P.: Chatgpt goes shopping: LLMs can predict relevance in ecommerce search. In: Goharian, N., et al. (eds.) Advances in Information Retrieval, pp. 3–11. Springer, Cham (2024). https://doi.org/10.1007/978-3-031-56066-8_1
23. Spark-Jones, K.: Report on the need for and provision of an'ideal'information retrieval test collection. Computer Laboratory (1975)
24. Sun, W., et al.: Is ChatGPT good at search? investigating large language models as re-ranking agents. In: Bouamor, H., Pino, J., Bali, K. (eds.) Proceedings of the 2023 Conference on Empirical Methods in Natural Language Processing, pp. 14918–14937. Association for Computational Linguistics, Singapore (2023). https://doi.org/10.18653/v1/2023.emnlp-main.923. https://aclanthology.org/2023.emnlp-main.923/
25. Thomas, P., Spielman, S., Craswell, N., Mitra, B.: Large language models can accurately predict searcher preferences. arXiv preprint arXiv:2309.10621 (2023)
26. Upadhyay, S., Kamalloo, E., Lin, J.: LLMs can patch up missing relevance judgments in evaluation. arXiv preprint arXiv:2405.04727 (2024)
27. Voorhees, E.M.: The philosophy of information retrieval evaluation. In: Peters, C., Braschler, M., Gonzalo, J., Kluck, M. (eds.) CLEF 2001. LNCS, vol. 2406, pp. 355–370. Springer, Heidelberg (2002). https://doi.org/10.1007/3-540-45691-0_34
28. Wang, H., Na, T.: Rethinking e-commerce search. SIGIR Forum **57**(2) (2024). https://doi.org/10.1145/3642979.3643007
29. Wang, P., et al.: Large language models are not fair evaluators. In: Ku, L.W., Martins, A., Srikumar, V. (eds.) Proceedings of the 62nd Annual Meeting of the Association for Computational Linguistics, vol. 1: Long Papers, pp. 9440–9450. Association for Computational Linguistics, Bangkok (2024). https://doi.org/10.18653/v1/2024.acl-long.511. https://aclanthology.org/2024.acl-long.511/
30. Wei, J., et al.: Chain-of-thought prompting elicits reasoning in large language models. Adv. Neural. Inf. Process. Syst. **35**, 24824–24837 (2022)

Graph Representation of Tables+Text and Compact Subgraph Retrieval for QA Tasks

Vishwajeet Kumar[1(✉)], Jaydeep Sen[1], Bhawna Chelani[2], and Soumen Chakrabarti[2]

[1] IBM Research, Gurgaon, India
{vishk024,jaydesen}@in.ibm.com
[2] IIT Bombay, Mumbai, India
soumen@cse.iitb.ac.in

Abstract. Large language models (LLMs) are increasingly used for table+text question answering (QA), but there is insufficient focus on unified table and text representation needed for effective retrieval guidance. In response, we present TabSegNet, comprised of (1) a fine-tuned LLM to decompose the question into segments that access individual tables;(2) a unified graph representation of tables+text, where retrieval amounts to identifying a compact node subset collectively covering the question segments; and (3) a novel graph neural network (GNN) whose messages are informed by the question and its segmentation. Experiments with existing and newly-created data sets demonstrate the promise of our approach, compared to sparse and dense nearest neighbor search, or using an LLM for retrieval.

1 Introduction

Passage-based QA was among the earliest successful applications of large language models (LLMs), particularly when assisted with query-time retrieval. To satisfy the needs of many QA applications that benefit from tabular data, many systems converted tables into linear text, appended the question, encoded this sequence using an LLM, which then decoded the response [4, 5, 13]. Early LLMs were unable to support large tables, given their input length limits and quadratic attention structure. Even recent LLMs can get overwhelmed if the answer has to be inferred across multiple large tables.

A compelling enterprise scenario involves long documents (e.g., financial reports), with many embedded tables with captions [6, 14, 28]. Tables and their elements are referenced from text. Cells may have textual and numeric content. A complex question is asked in the context of one or few such documents, without specifying which tables are relevant. Elements or slices from relevant table(s) need to be accessed to answer the question, together with other text elements. These whole documents may be too large to present to even recent LLMs. Moreover, indiscriminate table linearization may lose vital structural clues.

Graphs provide a uniform representation of such multimodal objects. They have already been used to represent tables in isolation [22, 26]. However, this

C. Hauff et al. (Eds.): ECIR 2025, LNCS 15572, pp. 164–180, 2025.
https://doi.org/10.1007/978-3-031-88708-6_11

In the table below, the aggregate contractual principal amount of loans on nonaccrual status and/or more than 90 days past due (which excludes loans carried at zero fair value and considered uncollectible) exceeds the related fair value primarily because the firm regularly purchases loans, such as distressed loans, at values significantly below the contractual principal amounts:

($ in millions)	As of December	
	2016	2015
Performing loans and long-term receivables		
Aggregate contractual principal in excess of fair value	$ 478	$ 1,330
Loans on nonaccrual status and/or more than 90 days past due		
Aggregate contractual principal in excess of fair value	8,101	9,600
Aggregate fair value on loans on nonaccrual status and/or more than 90 days past due	2,138	2,391

The table below presents information about our funding sources.

	As of December			
$ in millions	**2018**		**2017**	
Deposits	$158,257	25%	$138,604	23%
Collateralized financings:				
Repurchase aggrements	78,723	13%	84,718	14%
Securities loaned	11,808	2%	14,793	2%
Other secured financings	21,433	3%	24,788	2%
Total collateralized financings	111,964	18%	124,299	20%
Unsecured short-term borrowings	40,502	7%	46,922	8%
Unsecured long-term borrowings	224,149	36%	217,687	36%
Total shareholders' equity	90,185	14%	82,243	13%
Total funding sources	$625,057	100%	$609,755	100%

Question: What is the sum of securities loaned in 2017 and aggregate contractual principal in excess of fair value in 2015 (in millions)?

Fig. 1. Example instance showing the need for matching collectively between the question, tables and linear text. Various colored spans in the question are best matched to correspondingly colored spans in the document, to extract the values (orange borders) needed to answer the question. Our approach to convert the document to a query-modulated graph network is expected to focus on these matches while tuning down other possible spurious matches (not shown). (Color figure online)

is not sufficient to handle situations where fragments of the query must match various sites in text and table elements that are collectively proximate, in some non-trivial sense, to the answer. In Fig. 1, the span `securities loaned` matches a row header and `2017` matches a column header in the lower table. The span `aggregate contractual principal in excess of fair value` matches a row header and `2015` matches a column header in the upper table. Not highlighted are partial matches between query spans and the unstructured text spans in the document (which may be useful or spurious). The spatial relations between these match locations have semantics that help us retrieve values from the two cells with orange borders, and then compute the response, possibly using a calculator tool or by invoking code. In this paper, we focus on the first *retrieval* step.

In this work, we model a document as a sequence of sentences, interspersed with, or connected with, tables with possibly complex layouts. Earlier work [22, 26] provides an idiom for converting standalone tables to graphs. We extend the graph by adding a node for each unstructured sentence, and connecting them up in a chain, passing through or linked to nodes that are placeholders for whole tables and connecting to its headers and cells. Edges in this graph are richly typed, depending on what kind of nodes they connect. The nodes contain text from table cells or sentences, as the case may be. Another innovation is that the nodes and edges of this graph are featurized via early interaction with the question, or segments of the question inferred by an LLM.

Given a question, the table+text retrieval problem thus reduces to selecting a subset of nodes from the graph induced from tables and associated texts. Given the graph may have thousands to tens of thousands of nodes, this initial filtering or retrieval step must be fast and have high recall. Subsequent steps may refine these candidates down to the answer node(s) via computationally expensive operations such as cross-interaction LLMs.

Contributions: We present a high-recall retrieval/filtration stage for text+table QA with the following salient combination of innovations.

- **`Question segmentation`** —We use LLMs' vast pretrained world knowledge to segment the question into portions that target single tables, or a single connection between a pair of tables or sentences.
- **`Segment-informed GNN construction`** —The question and its segments are used in conjunction with the uniform graph representation described above, to define a novel hybrid early-interaction graph neural network (GNN).
- **`Compact subgraph retrieval`** —The GNN is trained to identify a subset of nodes that collectively cover all segments, and are connected by edges strongly triggered by the question.

The result is TabSegNet, a modular graph filter that achieves high recall of ground-truth answer-containing nodes, as evidenced by extensive experiments over several table+text QA benchmarks.

2 Related Work

End-to-End LLMs for QA: Question answering over multi-modal documents including table+text presents a very practical use-case and therefore, has garnered much interest in the retrieval community. A common approach in many of the recent works on this topic has been to use LLMs to solve it as an end-to-end QA task. Typically these approaches [1,30] propose to linearize tables along with the question as inputs to an encoder, followed by a causal decoder to generate the answer. Although the maximum input context length of LLMs has rapidly increased, LLMs cannot yet exploit long contexts reliably [16,17,24].

Therefore, selecting relevant sentences and table slices becomes a key factor in ensuring LLMs efficacy, and yet, there is insufficient scrutiny of this critically important first step. Existing work focuses more on end-to-end QA accuracy than this filtering or retrieval stage, which is fraught with two serious dangers:(1) LLM-based end-to-end QA systems may elicit correct answers from pre-trained knowledge rather than deriving it from the retrieved table(s) and text, thus being vulnerable to make mistakes if the relevant text or tables change or gets updated, which is quite common in enterprise settings, especially with streaming data repository. (2) If retrieved supporting evidence is noisy, the LLM runs a greater risk of hallucinating incorrect responses.

Table to Graph Representation: Given the heterogeneous nature of multimodal context, several recent systems [9,27] have recognized the versatility of graphs for developing a uniform representation for tables and text [23,26]. GTR [23] proposed a way to convert tables (even with nested row/column hierarchies) to graphs with typed edges, added edge-based biases [21] to a transformer network to define a GNN, and showed that table embeddings obtained from pooled node embeddings from this GNN are useful for whole-table retrieval. From this starting point, we make some critical enhancements. (1) We extend the representation to add unstuctured text nodes and connect them to nodes representing table elements. (2) Our GNN is defined as an early-interaction function of the question (and its decomposed segments) as well as the graph, not the graph alone. (3) We support fine-grained node filtering and retrieval, not whole-graph retrieval.

3 Baseline (Sub)graph Retrieval Approaches

We list baseline approaches for table+text QA and highlight potential drawbacks to justify TabSegNet, presented in Sect. 4. Some of these baselines will be used in our experiments.

LLM as Retriever (LLM-R): For a fair comparison across retrieval techniques, we pass linearized tables and text, interspersed with unique, opaque[1] IDs (corresponding to graph nodes) into the LLM and prompt it to generate a ranked list of the most relevant IDs [20], which is used to evaluate recall@K, MAP, etc. Some fraction of instances overflowed the context capacity, and it took some prompt tweaking to reduce hallucination and elicit enough useful IDs from the LLM.

Sparse and Dense Single-Vector Retrieval: Classic sparse and dense retrieval provide more baselines. For sparse retrieval we use vector space models **TFIDF** and **BM25** [18]. For dense passage retrieval (**DPR**), we encode the question and each node's text into single dense vectors and use the popular DiskANN package [19] with cosine similarity to rank graph nodes. Failure analysis shows that single-vector search is inadequate; evidence collection must happen across multiple table slices and text segments, also informed by graph proximity of matched nodes. TFIDF, BM25 and DPR all score nodes *in competition with* other nodes, failing to capture this collective scoring requirement.

ColBERT Multi-vector Dense Retrieval: [15] avoid premature aggregation of question and node texts into fixed-size vectors, while preserving some contextualization. ColBERT is known to work better than single-vector DPR. Each question word finds its best 'partner' word in the text of a candidate node. Our system uses ColBERT's scoring function, but, used by itself, it may also suffer from not modeling *collective* satisfaction of query words by a candidate subgraph.

[1] I.e., unlikely to collide with meaningful tokens in the LLM's vocabulary.

4 Proposed Method: TabSegNet

Preliminaries: We are given a query $\boldsymbol{q}$ which is a sequence of tokens. Along with the query, we are given a document D containing spans of text interleaved with tables $T_1, \ldots, T_N$. The document text is segmented into sentence-sized spans. Table T_n has a caption text. It is composed of cell elements $T_n[i, j]$. Our goal is to retrieve a ranked list of most relevant table elements and sentences.

In the following sections, we describe the essential components of TabSegNet: (1) question segmentation using a tuned LLM, (2) construction of a feature-rich graph from the question segments and the document, and (3) formulation of a GNN-based subgraph selection method.

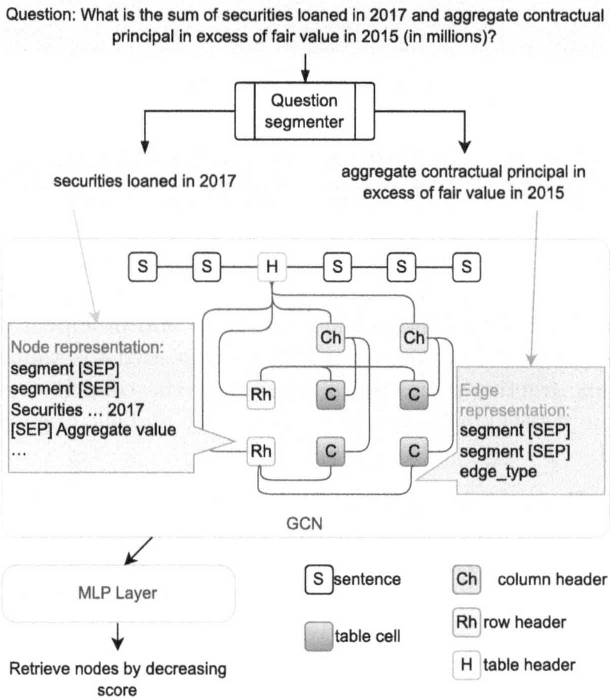

Fig. 2. TabSegNet block diagram. A fine-tuned LLM segments the query. Text interspersed with tables are converted into a uniform graph representation. Nodes and edges are featurized based on their edge types and the query segments. Then a graph convolutional network (GCN/GNN) followed by a MLP layer score nodes for retrieval. The parameters in these stages are trained. Some cell-to-cell edges are elided for clarity.

4.1 Question Segmentation

Many simplistic retrieval systems score and rank individual nodes against the whole question $\boldsymbol{q}$. As established in Fig. 1, if the question entails combining information across multiple nodes, it stands to reason that various spans of the question match portions of text in these nodes, but no single node 'covers' the whole question.

```
Instruction: Find segments of the given question which are needed to answer
it. (separate segments by SEP)
Query: What's the difference of Securities America, Inc.-3(4) between 2010
and 2009 for Actual Capital? (in million)
Predicted Segments: Securities America, Inc.-3(4) Actual Capital 2010 (in
millions) SEP Securities America, Inc.-3(4) Actual Capital 2009 (in
millions)
```

Fig. 3. Segmenting a question. Black text shows encoder input and teal text shows expected decoder output.

To address this problem, we employ an encoder-decoder LLM to generate *segments* $\{s_1, \ldots, s_M\}$ from the question. Graph nodes will compete to 'cover' each segment, but, in the end, we will choose a *set* of nodes to cover all segments well. We fine-tune a pre-trained LLM for the task of query segmentation. An example input and output are shown in Fig. 3. For fine-tuning, ideally, we need manually segmented queries, but such data is hard to come by. Instead, we use the available supervision data at the sentence and table cells to derive the desired segments of the question. To finetune, we concatenated the instruction, the question, then the text of each segment separated by **SEP**. Because we fine-tune, no in-context few-shot examples are used. Note that, when the LLM segmenter is deployed, our segments are arbitrary text generated by the LLM, not necessarily contiguous token spans from the question, or text from table or corpus text elements.

Recently, [2] used an LLM to identify, from a natural language question, tables and columns in a relational schema that should appear in a SQL translation. They do not *segment* the question in our sense, or match these segments against unstructured text. The items ranked are whole tables and their columns, not individual cells.

4.2 Table+Text Graph Construction

Given a document D containing tables passages, we create a **hybrid graph** (V, E) by extending the paradigm of [22,26], and others. Figure 2 shows the typical topology of the graph obtained by conversion from text and tables. Each table, table element and table caption becomes a node in the graph. Some of the table elements are headers. Table-derived nodes are connected similar to GTR [23] (however, GTR provides only whole-table retrieval and not subsets of table elements and text spans like us). Each sentence in the document becomes a node in the graph. Successive sentence nodes are connected with an edge. Documents vary with regard to meaningful placement of tables. If the tables are placed deliberately and manually, we connect the table header node to adjacent sentence nodes. Some typesetting tools may migrate tables and figures for improved layout. In such cases, we can detect references from the text to a table, and insert edges in the text chain from those reference sites. This is left for future work. If (u, v) is an edge, $t(u, v) \in [1, T]$ will denote its type index. We list in Fig. 4 some of the edge types we use in our table+text to graph representation. Depending on the application, we can add make every edge bidirected, adding distinct edge types in the two directions.

passage to another passage	table caption to previous passage
passage to table caption	table caption to next table caption
table caption to row header	row header to table caption
table caption to column header	column header to table caption
table caption to all table cells	table cell to table caption
table cell to its row header	table cell to its column header
table cell to next cell in same row	table cell to previous cell in same row
table cell to next cell in same column	table cell to previous cell in same column

Fig. 4. Sample edge types introduced into the unified graph.

4.3 Subgraph Selection via GNN

In a broad departure from earlier table QA methods, we bring together the question, its segments, and the hybrid graph (in a process of early interaction) to define a GNN, which is then used to select a subgraph as retrieved for the question.

CLS vectors output by a pretrained BERT [8] model is used to obtain the base node and query representations. Suppose $\text{text}(u)$ is the token sequence associated with node $u \in V$ in the graph. Then we write

$$\boldsymbol{q} = \text{BERT}(\boldsymbol{q})[\mathsf{CLS}] \text{ and} \tag{1}$$

$$\boldsymbol{u} = \text{BERT}(\text{text}(u))[\mathsf{CLS}]. \tag{2}$$

Similarly, if $\boldsymbol{s}_m$ is a segment, we write

$$\boldsymbol{s}_m = \text{BERT}(\boldsymbol{s}_m)[\mathsf{CLS}]. \tag{3}$$

At this point, most GNN-based QA systems would initialize the first layer node embeddings as $\boldsymbol{x}^{(0)}(u) \leftarrow \boldsymbol{u}$ and then iterate

$$\boldsymbol{x}^{(\ell+1)}(u) \leftarrow \text{comb}\Big(\boldsymbol{x}^{(\ell)}(u), \sigma\big(\textstyle\bigoplus_{(u,v)\in E} \boldsymbol{W}\boldsymbol{x}^{(\ell)}(v)\big)\Big), \tag{4}$$

where $\bigoplus$ is a symmetric aggregator and $\sigma(\cdot)$ is some activation function. After the last Lth layer, they would obtain node representations $\boldsymbol{x}^{(L)}(u)$ and compute the score of node u as the similarity $\text{sim}(\boldsymbol{x}^{(L)}(u), \boldsymbol{q})$. The intuition behind such use of a GNN is that, if nodes u and v collectively cover the question well, their last-layer embeddings will resemble $\boldsymbol{q}$ only if they are nearby in the graph.

Question-Driven Edge Modulation: The above intuition might work if all edges convey relations of comparable importance, which is definitely not our situation. Initial experiments were unimpressive, leading to the insight that although E contains many types of edges, only a few of them are important from the point of view of a given question; the other edges might well not exist.

To implement this intuition, we provide an embedding vector for each edge (relation) type in the form of matrix $\boldsymbol{R} \in \mathbb{R}^{T \times d'}$, with row $\boldsymbol{R}[t(u,v), \star]$ being the base embedding of edge $(u,v) \in E$. The question-dependent importance or 'strength' of edge (u,v) is computed as the scalar value

$$s((u,v)|\boldsymbol{q}) = \boldsymbol{q} \cdot \tanh\left(\boldsymbol{R}[t(u,v), \star]\boldsymbol{A} + \boldsymbol{b}\right) \tag{5}$$

where $\boldsymbol{A} \in \mathbb{R}^{d' \times dim\_in\_channel}, \boldsymbol{b} \in \mathbb{R}^{dim\_in\_channel}$ are global trainable model weights, with $dim\_in\_channel$ being the input channels dimension of GNN

Question-Informed Node Initialization: As with edges, a question-independent node initialization fails to capture early the affinity of the question to various nodes. To rectify this, we modify node initialization to

$$\boldsymbol{x}^{(0)}(u) \leftarrow \tanh\big([\boldsymbol{u},\, \boldsymbol{q}]\boldsymbol{C} + \boldsymbol{d}\big), \tag{6}$$

where $[\boldsymbol{u},\, \boldsymbol{q}]$ is the concatenation of node and question embeddings, and $\boldsymbol{C}, \boldsymbol{d}$ are more global mode weights to learn.

Segment-Informed Node Initialization: Ideally, we wish to inform node initialization from not only the whole query but also the segment best covering the node. To reflect this, we modify $[\boldsymbol{u},\, \boldsymbol{q}]$ to also append the embedding $\boldsymbol{s}^*(u)$ of the segment $\boldsymbol{s}^*(u)$ that best covers node u, writing this as $[\boldsymbol{u},\, \boldsymbol{q},\, \boldsymbol{s}^*(u)]$, and compute

$$\boldsymbol{x}^{(0)}(u) \leftarrow \tanh\big([\boldsymbol{u},\, \boldsymbol{q},\, \boldsymbol{s}^*(u)]\boldsymbol{C} + \boldsymbol{d}\big), \tag{7}$$

adjusting the shape of $\boldsymbol{C}$ suitably.

GNN: At this point we are ready to launch the modified GNN, with layer updates

$$\boldsymbol{x}^{(\ell+1)}(u) \leftarrow \text{comb}\Big(\boldsymbol{x}^{(\ell)}(u), \sigma\big(\textstyle\bigoplus_{(u,v)\in E} s((u,v)|\boldsymbol{q})\, \boldsymbol{x}^{(\ell)}(v)\big)\Big), \tag{8}$$

where $\boldsymbol{W}$ has been replaced with edge importance $s((u,v)|\boldsymbol{q})$. The score of node u is calculated as

$$\text{score}(u|D, \boldsymbol{q}) = \text{MLP}_\theta\left(\boldsymbol{x}^{(L)}(u)\right), \tag{9}$$

where MLP_θ is a single layer neural network with parameters θ. We sort the nodes by decreasing $\text{score}(u|D, \boldsymbol{q})$.

Training Protocol: Training instances provide $\boldsymbol{q}$ and D, together with the set of 'gold' nodes $\{u_\oplus\}$. Let all other nodes form the set $\{u_\ominus\} = V \setminus \{u_\oplus\}$ of 'worse' nodes. We use a standard pairwise hinge loss [12] to encourage the scores of gold nodes to exceed the scores of worse nodes:

$$\sum_{u_\oplus, u_\ominus} \text{ReLU}\left(\delta + \text{score}(u_\ominus | D, \boldsymbol{q}) - \text{score}(u_\oplus | D, \boldsymbol{q})\right), \qquad (10)$$

where δ is a margin hyperparameter. The trainable weights of TabSegNet consist of $\theta, \boldsymbol{A}, \boldsymbol{b}, \boldsymbol{C}, \boldsymbol{d}$, which makes for a fairly frugal model with fast training speeds.

4.4 Training the Retriever

We use a sentence transformer as the encoder to get initial representations for both node and query representations. For each node in the graph, we extract its text and use encoder to get its representation. Similarly, we use encoder to get representations for the query and its segments. After assigning initial representations for node, query and its segments, the representations are updated via the GCN layer during each iteration of training. Each iteration refines the node's representation by combining it with the representations of its neighboring nodes in the graph. At the end of training we get refined node representations for all the nodes and we score the nodes based on their similarity between node and query representations. The basic idea is that nodes close to each other in the graph and relevant to the query will have similar representations.

Node Representation Each node in the graph is represented by concatenating the node's data embedding with the embeddings of the query and all its segments. This combined representation is then compressed using a linear layer followed by a tanh activation function. This approach allows the node representation to capture interactions with the query and segments, enabling the model to learn the relationship between the query and nodes and score the nodes based on their relevance to the query. To better capture the relevance of the query to different nodes, we modify the initial node embeddings. Instead of using only the node's encoder representation, we also include the query's representation. This helps the GCN to start with an understanding of how each node relates to the query.

Edge Type and Edge Representation Each edge type in the graph is represented as the element-wise dot product of the query embedding and the relation type embedding (similar to a base word embedding matrix). This design allows the edge representation to adapt to the query, learning the relationship between the query and different edge types. For example, for an aggregate query, the model learns to emphasize edges connecting column headers to cells, which are crucial for aggregating information. We assign an embedding to each type of edge and compute its importance using the query. This ensures that only the most relevant edges are emphasized during the GCN's propagation process.

The GNN processes the node representations and aggregates information from their neighbors based on the graph structure. The output from the GCN is fed into a single Multi-Layer Perceptron (MLP) layer. This MLP layer is responsible for scoring each node, determining its relevance to the query. Training involves using a set of training instances where each instance includes a query, a document, and the set of 'gold' nodes (the most relevant nodes). We train the model to ensure that the scores of these gold nodes are higher than those of other nodes. We use a hinge loss function to achieve this, which encourages the model to maximize the difference in scores between relevant and irrelevant nodes. The model parameters are adjusted during training to improve its performance, leading to an efficient model that trains quickly.

5 Experimental Setup

In this section we detail our experimental setup including the benchmarks, baselines, implementation details and finally a thorough analysis of results and discuss key takeaways.

5.1 Benchmarks

Although our problem setting is well-motivated, public domain data sets exactly satisfying our requirements are not plentiful. Our requirements are that, to answer a question successfully, the system must collate evidence from a combination of tables and passages from a document having several passages interspersed with tables. Other popular table-QA data sets like HybridQA [5], TaT-QA [31], OTT-QA [3] or AIT-QA [14] do not satisfy all our requirements.

We use the MultiHierTT [28] data set as the latest and among the most challenging data sets in the table+text QA space. We also extend the FinQA [6] data set to fit our requirements. For both data sets, our focus will be on retrieving relevant evidence, rather than inferring the final response. We will make our extended data sets publicly available.

MultiHierTT: The MultiHierTT [29] data set consists of 7830 training instances and 1044 dev instances. Each instance contains (1) a context document consisting of passages interspersed with structured tables with complex layouts, (2) a natural language question that can be answered using the context document, (3) manual annotation of ground-truth evidences (table elements and text spans) required to answer the query, and (4) the ground-truth answer. The context documents contain 70 passages and 4 tables on average. In general, evidences from multiple passages and tables are required to answer the questions. The manual annotations are used in two ways: (1) text from ground-truth evidence and answer are used to fine-tune our LLM segmenter, and (2) ground-truth evidence and answer are mapped to gold nodes $\{u_\oplus\}$, used to supervise the GNN (10).

FinQA: The FinQA[2] [6] data set consists of 6251 training instances and 883 dev instances. On average, in this data set, the context for every query contains

[2] Processed data and scripts are available here https://github.com/vishwajeetkumar93/tabsegnet_data_scripts.

23 passages and a single table. To stress the ability of TabSegNet to retrieve the correct table out of competitive choices in a more realistic setting, we add a set of distractor [25] tables to the context. We first tokenize all tables in the data set and prepare a sparse index. Then, for each instance, we tokenize the table containing the ground-truth answer, tokenize it, and look it up in the index using the BM25 similarity score. The three most similar tables are artificially introduced into the context document at randomly selected positions.

5.2 Baselines

We provide the configuration details for the baselines as mentioned in Sect. 3 .

TFIDF: Scikit implementation with default settings.

BM25: Okapi BM25 library with default settings.

DPR: Facebook DPR library with default settings in zero-shot mode. Our benchmarks have been publicly available for years, so fine-tuning is unlikely to result in large improvements. Tables are linearized and table element texts are concatenated before passing into DPR's encoder.

ColBERT: We used the code and model checkpoint from existing public code repository. Tables are linearized similar to DPR. No fine-tuning is done for the same reason as in DPR.

Splade: We also use learned sparse retriever Splade_v2 [10] as a baseline here, reusing the code and model checkpoints publicly available from the authors.

5.3 Implementation Details

All experiments ran on a single NVIDIA A100_80GB GPU. Some implementation details of TabSegNet follow. For **query segmentation**, we fine-tuned FLAN-T5-XL [7] using AdamW with learning rate 2×10^{-5}. For **graph representation**, we used the NetworkX library. To derive initial **node and edge embeddings**, we used pre-trained SentenceTransformers to get a 768 dimensional vector for the `[CLS]` embedding to represent the query and each of the nodes. For **GNN computations**, we used GraphConv, with $L = 1$ layer ($L > 1$ made minimal difference) and max as the aggregator function $\bigoplus$ in Eqn. 8, 120 input and 150 output channels. MLP_θ has one layer. Model weights were initialized with Xavier Uniform [11] and were optimized using Adam, setting learning rate to 0.001. Models were trained for 20 epochs on the train set and the best checkpoint was selected based on the lowest loss on the dev set.

Table 1. TABSEGNET overall retrieval comparison with baselines.

Model	FinQA		MultiHierTT	
	R@10	MAP	R@10	MAP
BM25	0.610	0.47	0.472	0.35
TFIDF	0.700	0.56	0.535	0.40
DPR	0.640	0.52	0.491	0.36
ColBERT	0.714	0.60	0.566	0.44
SPLADE	0.732	**0.61**	0.591	0.46
TabNet_No_Etype	0.745	0.42	0.651	0.43
TabNet	0.798	0.46	0.634	0.41
TabSegNet	**0.836**	0.54	**0.675**	**0.47**

5.4 Evaluation Metrics

Our data sets identified ground-truth answer text spans and table elements. These are easily mapped to nodes in the graph we build, which we call 'gold' nodes. Any system being evaluated is expected to return a ranked list of nodes. From this ranking, we compute recall@10 (R@10) and mean average precision (MAP), which reduces to mean reciprocal rank (MRR) in case only one gold node exists. (The MultiHierTT leaderboard and FinQA leaderboard are online, but they report end-to-end answer accuracy, which includes arithmetic abilities beyond retrieval, which is outside our scope. Most chart-toppers focus, and are evaluated, on the correctness of the *program* computing the answer, not *retrieval.*)

6 Results and Analysis

We first describe key research questions that we want to address —**RQ1:** Does TABSEGNET improve retrieval accuracy for complex multimodal QA where a query needs multiple tablular and text segments? We also want to investigate the effect of key components in TABSEGNET design. **RQ2:** How important is the segmentation step for TABSEGNET performance? **RQ3:** How important are having edge type distinctions?

6.1 TABSEGNET vs. baselines

Table 1 shows that, in terms of R@10, TABSEGNET is better than all other retriever baselines. Among existing baselines, ColBERT does better than DPR or (untrained) sparse retrievers, which is expected given its ability to allow multi-vector interaction between query and document. The difference in performance between TABSEGNET and sparse retrievers is even higher, given that the sparse retrievers rely on keyword matches irrespective of their table/text positions.

SPLADE is the best baseline method. It relies on learned keyword expansions which acts like semantic synonyms of important terms in the document. Moreover, SPLADE and ColBERT also achieves better MAP scores because they work at token level matches for relevance scoring. However, for downstream applications using LLMs over the retrieved context, a better recall becomes the necessary criterion where TabSegNet improves on both ColBERT and SPLADE in R@10, supporting RQ1. Both TabSegNet and TabNet gets better recall than baselines. This validates the need for adopting a graph structure unifying tabular and text data, plus graph-enabled learning to score nodes.

6.2 Benefit of Segmentation

We also see in Table 1 that, without segmentation, there is a considerable drop in both recall and MAP. Question segmentation helps in finding focused localized matches and also differentiates between a compact sub-graph covering important question words, versus a set of unrelated nodes that match parts of the query. This result answers RQ2 in the affirmative, i.e., segmentation is indeed needed to boost accuracy of table+text retrieval. Summarizing, TabSegNet with question segmentation, graph representation, and learning to score nodes via GNN training can effectively induce the much-needed notion of compact subgraph retrieval.

6.3 Benefit from Edge Types and Weights

Table 1 further shows that if we remove distinct edge types, recall and MAP significantly decreases for FinQA, although not so much for MultiHierTT. With designated, informative edge types, the GNN can better learn associations between the question and valuable edges, as well as differentiate between promising versus unrelated neighbourhood nodes. Such learning translates to better evidence retrieval during inference. FinQA numbers answer our third research question RQ3 in the affirmative, i.e. edge types play a vital role.

Table 2. TabSegNet

Model	FinQA		MultiHierTT	
	Recall@10	MAP	Recall@10	MAP
TabSegNet_large	0.836	0.542	0.675	0.468
TabSegNet_base	0.790	0.462	0.648	0.440
TabNet_large	0.798	0.462	0.634	0.407
TabNet_base	0.731	0.396	0.542	0.347

6.4 Ablation on GNN Model Size

Results in Table 2 explores the effect of changing the number of GNN parameters. Here 'large' models have GNN input width of 128 as opposed to 120 in 'base' models and the output node representation with is 256 in 'large' models, as opposed to 150 in 'base'. Furthermore, the 'large' model uses an edge embedding size of 128 whereas it is 10 for 'base'.

As we can see in Table 2 the change in representation sizes in GCN nodes and edge embedding impacts recall and MAP in predictable ways. The 'large' configuration has more representation power and is able to learn better associations between the question and relevant graph nodes and edges, resulting in better performance. Also, the effects of having a larger variation of GNN params impacts both TabSegNet and TabNet, once again validating the fact that a robust GNN helps the overall system—with or without segmentation.

6.5 Discussion and Anecdotes

Segmenter Errors. Unlike MultiHierTT, which gold evidence nodes accompanying each instance, for FinQA we derive those using gold program annotations and hence call them silver segments. We do a quality check of these silver segments and perform necessary cleaning to keep FinQA data relevant for our use-case. We see two key failure cases - pertaining to segmenter and the silver segments.

Incorrect labeled data —Careful inspection revealed spurious program annotations in FinQA, which renders the silver segments spurious and uncorrelated to the question. Such error cases can not be recovered from any segmentation method. One example is shown:

Question
`what was jpmorgan chase & co's common equity tier 1 ( cet1 ) ratio in 2008?`

Silver segments
`2008 total tier 1 capital ( a ) [SEP] 2008 risk-weighted assets`

Inferred segments
`2008 common equity tier 1 ( cet1 ) [SEP]`
`2007 common equity tier 1 ( cet1 )`

Although the question specifically refers to `common equity tier 1 (cet1)`, the gold program annotation refers to a unrelated (incorrect) table with column header `total tier 1 capital`. This leads to the silver segments from this annotation include tokens such as `total`, which is not in the original query itself. Our trained segmenter does a satisfactory job of predicting `2008 common equity tier 1 ( cet1 )` and `2007 common equity tier 1 ( cet1 )` as possible query segments which seem plausible. We purposefully exclude such misleading instances to report on a sanitized FinQA.

Only textual evidence —There are many instances in FinQA where the answer is directly contained in a single text span. Such instances are not suitable for our use-case where the aim is to evaluate systems on their ability to discover *connections* between multiple tables and text spans for accurate evidence retrieval. In fact, for such examples, the silver segments are empty, again misleading the training of our LLM-based segmenter. There are a total of 235 instances where the derived silver segment is empty. We remove those from our evaluation.

7 Conclusion

We presented TabSegNet, a retrieval and filtering stage for table+text QA. In TabSegNet, we use a uniform graph representation for table+text, upon which we define a GNN whose node and edge weights are carefully informed by question segments, which are obtained from a fine-tuned LLM. Experiments show that TabSegNet has superior recall of ground-truth text spans and table elements compared to recent competitive dense retrieval methods.

References

1. Andrejczuk, E., Eisenschlos, J., Piccinno, F., Krichene, S., Altun, Y.: Table-to-text generation and pre-training with TabT5. In: Goldberg, Y., Kozareva, Z., Zhang, Y. (eds.) Findings of the Association for Computational Linguistics: EMNLP 2022, pp. 6758–6766. Association for Computational Linguistics, Abu Dhabi (2022). https://doi.org/10.18653/v1/2022.findings-emnlp.503. https://aclanthology.org/2022.findings-emnlp.503
2. Chen, P.B., Zhang, Y., Roth, D.: Is table retrieval a solved problem? exploring join-aware multi-table retrieval. In: ACL (2024). https://arxiv.org/abs/2404.09889
3. Chen, W., Chang, M.W., Schlinger, E., Wang, W., Cohen, W.W.: Open question answering over tables and text (2021)
4. Chen, W., et al.: Tabfact: a large-scale dataset for table-based fact verification. CoRR arxiv:1909.02164 (2019)
5. Chen, W., Zha, H., Chen, Z., Xiong, W., Wang, H., Wang, W.Y.: HybridQA: a dataset of multi-hop question answering over tabular and textual data. In: Cohn, T., He, Y., Liu, Y. (eds.) Findings of the Association for Computational Linguistics: EMNLP 2020, pp. 1026–1036. Association for Computational Linguistics, Online (2020). https://doi.org/10.18653/v1/2020.findings-emnlp.91. https://aclanthology.org/2020.findings-emnlp.91
6. Chen, Z., et al.: Finqa: a dataset of numerical reasoning over financial data (2022)
7. Chung, H.W., et al.: Scaling instruction-finetuned language models (2022)
8. Devlin, J., Chang, M.W., Lee, K., Toutanova, K.: BERT: pre-training of deep bidirectional transformers for language understanding. In: NAACL Conference (2019). https://www.aclweb.org/anthology/N19-1423.pdf
9. Du, L., et al.: TabularNet: a neural network architecture for understanding semantic structures of tabular data. In: Proceedings of the 27th ACM SIGKDD Conference. KDD 2021. ACM (2021). https://doi.org/10.1145/3447548.3467228

10. Formal, T., Piwowarski, B., Clinchant, S.: Splade: sparse lexical and expansion model for first stage ranking. In: Proceedings of the 44th International ACM SIGIR Conference on Research and Development in Information Retrieval, pp. 2288–2292 (2021)
11. Glorot, X., Bengio, Y.: Understanding the difficulty of training deep feedforward neural networks. In: Teh, Y.W., Titterington, M. (eds.) Proceedings of the Thirteenth International Conference on Artificial Intelligence and Statistics. Proceedings of Machine Learning Research, vol. 9, pp. 249–256. PMLR, Chia Laguna Resort (2010). https://proceedings.mlr.press/v9/glorot10a.html
12. Joachims, T.: Optimizing search engines using clickthrough data. In: SIGKDD Conference, pp. 133–142. ACM (2002). http://www.cs.cornell.edu/People/tj/publications/joachims_02c.pdf
13. Katsis, Y., et al.: AIT-QA: Question answering dataset over complex tables in the airline industry. In: Loukina, A., Gangadharaiah, R., Min, B. (eds.) Proceedings of the 2022 Conference of the North American Chapter of the Association for Computational Linguistics: Human Language Technologies: Industry Track, pp. 305–314. Association for Computational Linguistics, Hybrid: Seattle, Washington + Online (2022). https://doi.org/10.18653/v1/2022.naacl-industry.34. https://aclanthology.org/2022.naacl-industry.34
14. Katsis, Y., et al.: AIT-QA: question answering dataset over complex tables in the airline industry. In: NAACL Conference (2022). https://openreview.net/forum?id=VkY8WURpggN
15. Khattab, O., Zaharia, M.: Colbert: efficient and effective passage search via contextualized late interaction over bert. In: Proceedings of the 43rd International ACM SIGIR Conference on Research and Development in Information Retrieval, pp. 39–48 (2020)
16. Li, T., Zhang, G., Do, Q.D., Yue, X., Chen, W.: Long-context llms struggle with long in-context learning (2024)
17. Liu, N.F., et al.: Lost in the middle: how language models use long contexts. Trans. Assoc. Comput. Linguist. **12**, 157–173 (2024). https://doi.org/10.1162/tacl_a_00638
18. Manning, C.D., Schütze, H.: Foundations of Statistical Natural Language Processing. MIT Press, Cambridge (1999). https://nlp.stanford.edu/fsnlp/
19. Simhadri, H.V., et al.: DiskANN: graph-structured indices for scalable, fast, fresh and filtered approximate nearest neighbor search (2023). https://github.com/Microsoft/DiskANN
20. Tay, Y., et al.: Transformer memory as a differentiable search index. In: Oh, A.H., Agarwal, A., Belgrave, D., Cho, K. (eds.) Advances in Neural Information Processing Systems (2022). https://openreview.net/forum?id=Vu-B0clPfq
21. Wang, B., Shin, R., Liu, X., Polozov, O., Richardson, M.: Rat-sql: relation-aware schema encoding and linking for text-to-sql parsers (2021)
22. Wang, F., Sun, K., Chen, M., Pujara, J., Szekely, P.: Retrieving complex tables with multi-granular graph representation learning. In: SIGIR Conference, pp. 1472–1482 (2021). https://dl.acm.org/doi/pdf/10.1145/3404835.3462909
23. Wang, F., Sun, K., Chen, M., Pujara, J., Szekely, P.A.: Retrieving complex tables with multi-granular graph representation learning. In: Proceedings of the 44th International ACM SIGIR Conference on Research and Development in Information Retrieval (2021). https://api.semanticscholar.org/CorpusID:233739808
24. Xiang, Y., Yan, H., Gui, L., He, Y.: Addressing order sensitivity of in-context demonstration examples in causal language models. ACL Findings (2024). https://arxiv.org/abs/2402.15637

25. Xiao, Y., et al.: Dynamically fused graph network for multi-hop reasoning. arXiv preprint arXiv:1905.06933 (2019)
26. Zayats, V., Toutanova, K., Ostendorf, M.: Representations for question answering from documents with tables and text. In: EACL Conference (2021). https://www.aclweb.org/anthology/2021.eacl-main.253.pdf
27. Zhang, X., Shou, L., Pei, J., Gong, M., Wen, L., Jiang, D.: A graph representation of semi-structured data for web question answering. In: COLING (2020). https://arxiv.org/abs/2010.06801
28. Zhao, Y., Li, Y., Li, C., Zhang, R.: Multihiertt: numerical reasoning over multi hierarchical tabular and textual data. arXiv preprint arXiv:2206.01347 (2022)
29. Zhao, Y., Li, Y., Li, C., Zhang, R.: MultiHiertt: Numerical reasoning over multi hierarchical tabular and textual data. In: Muresan, S., Nakov, P., Villavicencio, A. (eds.) Proceedings of the 60th Annual Meeting of the Association for Computational Linguistics, vol. 1: Long Papers, pp. 6588–6600. Association for Computational Linguistics, Dublin (20). https://doi.org/10.18653/v1/2022.acl-long.454. https://aclanthology.org/2022.acl-long.454
30. Zhou, Y., Bao, J., Duan, C., Wu, Y., He, X., Zhao, T.: UniRPG: unified discrete reasoning over table and text as program generation. In: Goldberg, Y., Kozareva, Z., Zhang, Y. (eds.) Proceedings of the 2022 Conference on Empirical Methods in Natural Language Processing, pp. 7494–7507. Association for Computational Linguistics, Abu Dhabi (2022). https://doi.org/10.18653/v1/2022.emnlp-main.508. https://aclanthology.org/2022.emnlp-main.508
31. Zhu, F., et al.: TAT-QA: a question answering benchmark on a hybrid of tabular and textual content in finance. In: Zong, C., Xia, F., Li, W., Navigli, R. (eds.) Proceedings of the 59th Annual Meeting of the Association for Computational Linguistics and the 11th International Joint Conference on Natural Language Processing, vol. 1: Long Papers, pp. 3277–3287. Association for Computational Linguistics, Online (2021). https://doi.org/10.18653/v1/2021.acl-long.254. https://aclanthology.org/2021.acl-long.254

Higher Order Knowledge Graph Embeddings

Giuseppe Pirrò(✉)

University of Calabria, Rende, CS, Italy
giuseppe.pirro@unical.it

Abstract. Nested knowledge graphs (NKGs), where subjects or objects can themselves be triples, encode higher-order relationships that traditional knowledge graph embedding (KGE) methods struggle to capture. These nested structures introduce complexity in reasoning tasks like link prediction, as they require handling relationships between triples. We propose HOKE, a hybrid embedding model designed specifically for NKGs to address this. HOKE utilizes a three-layer architecture: *local embeddings* capture the semantics of original KG triples; *structural embeddings* refine these by emphasizing shared elements between triples via a line graph representation; and *nested embeddings* encode higher-order relationships using a nested line graph, where triples are treated as nodes. This approach preserves both local semantic meaning and global structural dependencies. Additionally, HOKE introduces a task-driven subgraph extraction mechanism that enhances reasoning efficiency by focusing only on relevant portions of the NKG. HOKE bridges the gap in existing KGE models, providing a scalable solution to handle the complex relational structures inherent in large-scale NKGs.

Keywords: Knowledge Graphs · Query subgraphs · Embeddings

1 Introduction

Knowledge graphs (KGs) represent structured information about entities and their relationships. One of the most common ways to perform knowledge graph reasoning is through *knowledge graph embeddings* (KGE), which involves mapping entities and relations into continuous vector spaces [3]. This enables learning models to evaluate the plausibility of unseen or incomplete triples by calculating the compatibility between embeddings of entities and relations. KGE has been adopted in various applications, from link prediction to question answering [19]. In many KGs, relationships are not limited to simple triples. Instead,

We acknowledge financial support from the PNRR MUR project FAIR - Future AI Research (PE00000013), Spoke 9 WP 9.2 - Green-aware AI, funded by the Next Generation EU.

C. Hauff et al. (Eds.): ECIR 2025, LNCS 15572, pp. 181–195, 2025.
https://doi.org/10.1007/978-3-031-88708-6_12

we often encounter *nested triples*, where the head, the tail, or both can themselves be triples. For example, a nested triple may represent a temporal relationship between two other triples, such as *Alice worked at CompanyX while Bob managed TeamA*. In these nested knowledge graphs (NKGs), the relationship's subject and object are complex structures rather than single entities. Recent efforts focused on both the standardization (e.g., RDF-star [11]) and the construction of NKGs [1]. In NKGs, traditional tasks like link prediction become more challenging, as they involve predicting links between triples rather than entities. Existing KGE methods have mostly overlooked the NKG scenario.

Related Work. KGE approaches are broadly categorized into two types. *Semantic models* embed triples by representing the head, relation, and tail independently in a vector space, using a scoring function to predict triple validity. Examples include TransE [2], ConvE [6], and RotatE [17]. While efficient, these models overlook broader graph structure, capturing only direct interactions between triples. Recently, BiVe [4] and NestE [20] extended semantic models to handle relations between entire triples in NKGs. On the other hand, *structural models* leverage the graph structure, propagating information across entities and neighbors using *Graph Neural Networks (GNNs)*. Examples include RGCN [16] and MINERVA [5]. While these methods capture multi-hop dependencies, they struggle with scalability and do not address nested triples. Recent methods have adopted sampling strategies to mitigate computational costs. Subgraph-wise sampling, like GraIL [18], extracts subgraphs around the head entity, while layer-wise sampling methods like GraphSAGE [9] selects subgraphs at each model layer. Techniques like Personalized PageRank [23] further reduce complexity by focusing on relevant graph portions for link prediction.

Novelty and Contributions. While existing models like NestE and BiVe advance reasoning in NKGs, they often overlook crucial structural context, such as the neighborhood of triples, which provides rich semantic information. These models focus on triples individually, missing how inter-triple relationships contribute to deeper understanding. We propose HOKE, a novel framework that integrates semantic and structural reasoning in a layered approach. HOKE represents Nested Graph Knowledge (NGK) hierarchically, progressively refining embeddings to capture both local and global relational information. Additionally, HOKE introduces a *task-driven subgraph extraction* mechanism that significantly reduces computational overhead by focusing on semantically relevant subgraphs for each query, rather than processing the entire graph. The key contributions of HOKE are: (i) a hybrid model combining semantic and structural reasoning, effectively handling nested triples while capturing local and global relationships; and (ii) a query-driven subgraph sampling strategy, which improves efficiency by extracting subgraphs relevant to each query, maintaining high accuracy while reducing computational cost.

HOKE provides a robust framework for complex scenarios in Information Retrieval (IR) and Recommender Systems (RecSys). In IR, it enables semantic search and query expansion by embedding intricate relationships, such as temporal or contextual qualifiers, improving result ranking based on subtle dependencies. In RecSys, HOKE facilitates context-aware and cross-domain recommendations, encoding higher-order interactions like user preferences and nested relationships.

The paper is organized as follows. We start with some background (Sect. 2), then introduce the subgraph sampling strategy (Sect. 3.1) and the reasoning framework (Sect. 4). We discuss the experimental evaluation (Sect. 5) and conclude (Sect. 6).

2 Preliminaries

A Knowledge Graph ($\mathcal{G}$) is a labeled directed multigraph $\mathcal{G} = (\mathcal{E}, \mathcal{R}, \mathcal{T}_s)$ where $\mathcal{E} = \{e_1, \ldots, e_n\}$ is a finite set of *entities* and $\mathcal{R} = \{r_1, r_2, \ldots, r_m\}$ is a finite set of *relations*. $\mathcal{T}_s$ is a set of triples of the form (h,r,t) where $h \in \mathcal{E}$ is the *head entity*, $r \in \mathcal{R}$ the *relation*, and $t \in \mathcal{E}$ the *tail entity*. We now extend the definition of KG to include *nested triples*, that is, triples where at least one between the subject and the object is another triple.

Definition 1 (Nested Knowledge Graph). *A nested knowledge graph $\mathcal{G}_n$ is a tuple $\mathcal{G}_n = (\mathcal{E}, \mathcal{R}, \mathcal{R}_m, \mathcal{T}_s, \mathcal{T}_n)$ that extends a KG with a finite set of* meta-relations *$\mathcal{R}_m = \{m_1, m_2, \ldots, m_d\}$, that is, higher-order relationships that connect entire triples or entities. $\mathcal{T}_n$: $\langle h^n, r^n, t^n \rangle$ is a set of nested triples, with $r^n \in \mathcal{R}_m$ and at least one between h^n and t^n belongs to $\mathcal{T}_s$ while the other can be an entity belonging to $\mathcal{E}$.*

2.1 Problem Statement

Despite their value, real-world (nested) knowledge graphs are often incomplete or contain noisy data, making predictive tasks such as link prediction and triple classification essential for inferring missing knowledge and validating existing information [12]. Although these tasks have been extensively studied in standard KGs, there has been limited exploration in the context of NKGs. We now formalize two key predictive tasks: triple prediction and conditional link prediction that build upon and extend previous definitions [20]. For simplicity, in the definitions, we consider that the prediction always occurs in the tail position (i.e., *?t*). In the experiments, we considered both tail and head positions as missing (i.e., predicting either *?h* or *?t*) depending on the query.

Definition 2 (Triple Prediction). *Given $\mathcal{G}_n = (\mathcal{E}, \mathcal{R}, \mathcal{R}_m, \mathcal{T}_s, \mathcal{T}_n)$, the triple prediction problem is to answer a query $\langle T_i, r^n, ?t \rangle$, $r^n \in \mathcal{R}_m$ and $?t$ must be resolved to an entity in $\mathcal{E}$ or triple in $\mathcal{T}_s$.*

Definition 3 (Conditional Link Prediction). *Given $\mathcal{G}_n = (\mathcal{E}, \mathcal{R}, \mathcal{R}_m, \mathcal{T}_s, \mathcal{T}_n)$, let $T_i = (h_i, r_i, t_i)$. The conditional link prediction problem involves queries $\langle T_i, r^n, (h_j, r_j, ?t) \rangle$ or $\langle T_i, r^n, (?h, r_j, t_j) \rangle$, where variables need to be bound to entities in $\mathcal{E}$.*

3 Capturing Relevant Information

We now introduce the HOKE framework starting from a data modeling perspective. We represent an NKG from a three-layer perspective, separating base elements (entities and relations) from complex ones (triple and meta-relations). This hierarchical view of a NKG allows to capture both local and global structural information as well as higher-order semantic relationships. Figure 1 provides an example of NKG represented from the HOKE perspective.

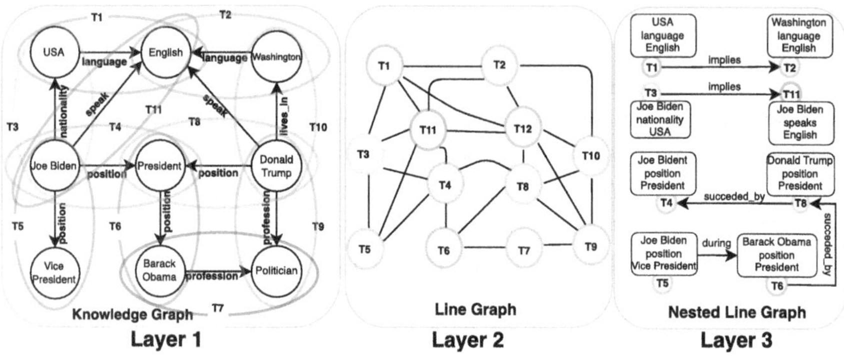

Fig. 1. Multi-layer view of a nested knowledge graph.

Layer 1: The knowledge graph $\mathcal{G}$ directly models the basic relations between entities. It provides a detailed, fine-grained view of how entities are connected, which is critical for understanding individual relationships.

Layer 2: The *line graph* $\mathcal{L}_t = (V_t, E_t)$ is constructed from the knowledge graph $\mathcal{G} = (\mathcal{E}, \mathcal{R}, \mathcal{T})$ at *Layer 1*. Nodes V_t of $\mathcal{L}_t$ represent triples and edges E_t connect nodes if their corresponding triples share at least one entity. The line graph captures structural dependencies between triples based on shared entities.

Layer 3: The *nested line graph* (NLG) $\mathcal{G}_l = (V, \mathcal{R}_{\mathrm{m}}, \mathcal{T}_n)$ includes nodes $V = \mathcal{T}_s \cup \mathcal{E}_n$ containing standard triples ($\mathcal{T}_s$) and $\mathcal{E}_n \subseteq \mathcal{E} \in \mathcal{G}$ appearing in nested triples. $\mathcal{R}_{\mathrm{m}}$ is the set of higher-order relations between nested triple elements.

By focusing on how triples interact with each other beyond just shared entities, this layer can capture more complex, higher-order logical connections between facts (e.g., Joe Biden succeeded Donald Trump as President) shifting from individual entities and relations to structural and meta-relational dependencies.

3.1 Extracting Query-Focused Subgraphs

In large-scale NKGs, reasoning over the entire graph is computationally prohibitive and often introduces unnecessary noise for a specific query. To address this, we define a *top-down approach* for extracting query-specific subgraphs by sampling relevant entities, relations, and triples while accounting for the query's semantics. Our multi-layer strategy begins with *Layer 3*, which focuses on higher-order semantic relations; then it moves at *Layer 2* by leveraging relatedness to capture neighboring triples and then to *Layer 1* to refine the context by focusing on entity-level facts. This approach ensures that the most relevant information is captured in the resulting *query-focused subgraphs.*

Definition 4 (Query-Focused Subgraph). *Given a query q and a nested knowledge graph $\mathcal{G}_n$, the* query-focused subgraph *is $\mathcal{S}^q = (\mathcal{E}^q, \mathcal{R}^q, \mathcal{R}^q_m, \mathcal{T}^q_s, \mathcal{T}^q_n)$, where each set represents a subset of the corresponding entities, relations, meta-relations, standard triples, and nested triples in $\mathcal{G}_n$ that are relevant to q.*

The query-focused subgraph ensures that only the most essential parts of the graph are considered during reasoning, thus minimizing both noise and computational load. The subgraph extraction algorithm for *Triple Prediction* is outlined in Algorithm 1. It employs a *relatedness-aware* BFS strategy with a bounded depth. The algorithm first identifies the relevant nested triples from *Layer 3* based on: (i) the relatedness between triple meta-relations and the query's meta-relation; (ii) the subject or object of the nested triple is directly connected to the query triple. Meta-relation relatedness is computed based on meta-relation co-occurrences in nested triples on $\mathcal{G}_n$ [14].

Algorithm 1. Query-Focused Subgraph Extraction: Triple Prediction

1: **Input:** Query $q = \langle (h_q, r_q, t_q), r^n, ?t \rangle$, KG $\mathcal{G}$, Line Graph $\mathcal{L}_t = (V_t, E_t)$, Nested KG $\mathcal{G}_n$, BFS depth $maxDepth$, Relation Relatedness Matrix $MRel$ and *Meta* Relation Relatedness Matrix $MNRel$, Relatedness threshold $ThRel$, Meta Relatedness Threshold $MThRel$
2: **Output:** Query-focused subgraph $\mathcal{S}^q$
3: Let $\tau \in V_t$ be the node in the line graph corresponding to the triple (h_q, r_q, t_q)
4: Initialize $\mathcal{S}^q$, priority queue Q_2 with $(\tau, 0)$, visited set $V = \emptyset$
▷ Step 1: **Select Relevant Nested Triples** (*Layer 3*)
5: **for all** nested triple $(T_i, \hat{r}, O)$ in $\mathcal{G}_n$ **do**
6: Compute $rel = MNRel(r^n, \hat{r})$
7: **if** $rel > MThRel$ **and** T_i or O shares entities with $\langle h_q, r_q, t_q \rangle$ **then**
8: Add $(T_i, \hat{r}, O)$ to $\mathcal{T}^q$
▷ Step 2: **Relatedness-driven BFS on Line Graph** (*Layer 2*)
9: **while** Q_2 is not empty **do**
10: Pop $(current\_node, depth)$ from Q_2
11: **if** $depth > maxDepth$ or $current\_node \in V$ **then continue**
12: Add the triple corresponding to the $current\_node$ to $\mathcal{T}^q_s$
13: **for all** $neighbor$ of $current\_node$ **do**
14: Extract relation $r_{neighbor}$ in the triple corresponding to $neighbor$
15: Compute relatedness $rel = MRel(r_q, r_{neighbor})$
16: **if** $rel > ThRel$ **then**
17: Push $(neighbor, depth + 1)$ to Q_2
▷ Step 3: **Lower-Level Triple Materialization** (*Layer 1*)
18: **for all** $T_i \in \mathcal{T}^q_s$ **do**
19: Map T_i to Layer 1, extract (h, r, t), and expand $\mathcal{E}^q = \mathcal{E}^q \cup \{h, t\}$, $\mathcal{R}^q = \mathcal{R}^q \cup r$
return $\mathcal{S}^q$

The algorithm maps the head triple (h_q, r_q, t_q) in the query triple q to a node in the line graph; then it progressively builds the subgraph at *Layer 2* (lines 6–12). Neighboring nodes in the line graph are selected based on the relatedness of their predicates to the query predicate r_q, measured via another precomputed *relation* relatedness matrix this time computed on $\mathcal{G}$. After traversing *Layer 2*, the relevant triples are mapped back to *Layer 1* for further materialization (lines 14–15). The subgraph extraction algorithm adapts to the specific task. For *Triple Prediction*, when the missing element *?t* is an entity, the algorithm could prioritize *Layer 1* to focus on entity-related triples further expanding the set of triples $\mathcal{T}_s^q$. However, if the missing element is a triple, the focus shifts to *Layer 2*, which captures structural relationships between triples. With a slight adaptation Algorithm 1 can extract subgraphs for conditional link prediction tasks like $\langle (h_w, r_q, t_q), r^n, (h_j, r_j, ?t) \rangle$. The difference is that besides the line graph relatedness-aware traversal (line 6–12) starting from the node T_i in the line graph corresponding to (h_w, r_q, t_q) and considering r_q, an *additional* traversal is performed at *Layer 1* starting from h_i and considering r_j as reference relation. By sampling starting from *Layer 3* and progressively downward, the algorithm ensures that the hierarchical relationships are maintained, while important contextual information is preserved.

4 Higher Order Embeddings

HOKE operates on a query-focused subgraph $\mathcal{S}^q = (\mathcal{E}^q, \mathcal{R}^q, \mathcal{R}_{\mathrm{m}}^q, \mathcal{T}_s^q, \mathcal{T}_n^q)$ instead of the initial nested knowledge graph. This time HOKE proceeds in a *bottom-up* manner, progressing from entity and relation embeddings to nested triple embeddings. This hierarchical approach ensures that detailed information at the lower levels feeds into the more abstract representations, allowing the model to capture both local details and global structures effectively.

Layer 1: Learning Entity and Relation Embeddings. At the foundational layer, HOKE learns embeddings for entities and relations to construct local embeddings for standard triples. Given the set of triples $\mathcal{T}_s^q = \{(h_i, r_i, t_i)\}$ in $\mathcal{S}^q$, where $h_i, t_i \in \mathcal{E}^q$ are entities and $r_i \in \mathcal{R}^q$ are relations, HOKE learns embeddings $\mathbf{e}_{h_i}, \mathbf{e}_{t_i} \in \mathbb{R}^d$ for entities and $\mathbf{e}_{r_i} \in \mathbb{R}^d$ for relations. They are combined to form local triple embeddings: $\mathbf{z}_T = [\mathbf{e}_h; \mathbf{e}_r; \mathbf{e}_t]$. Embeddings are learned using a scoring function f_{local} that captures the relational structure within the triple. For example, using TransE, $f_{\text{TransE}}(\mathbf{e}_{h_i}, \mathbf{e}_{r_i}, \mathbf{e}_{t_i}) = \|\mathbf{e}_{h_i} + \mathbf{e}_{r_i} - \mathbf{e}_{t_i}\|$. *Layer 1* establishes the foundation by learning low-level embeddings for entities and relations, which are critical for constructing more complex triple embeddings in later layers.

Layer 2: Refinement with Structural Information. Using the entity embeddings from *Layer 1*, HOKE computes embeddings for triples within the line graph $\mathcal{L}_t^q$ obtained from the standard triples $\mathcal{T}_s^q$ in $\mathcal{S}^q$. The initial embedding for each node in $\mathcal{L}_t^q$ is formed by concatenating the embeddings from *Layer 1*: $\mathbf{z}_{(h,r,t)}^{(0)} = [\mathbf{e}_h; \mathbf{e}_r; \mathbf{e}_t]$. HOKE refines these embeddings *Layer 2* using a Graph Neural Network (GNN) that iteratively updates the embeddings by aggregating information from neighboring triples in $\mathcal{L}_t^q$:

$$h_T^{\ell+1} = \text{DROPOUT}\big(\text{ACT}\big(\text{AGG}\big(\text{MESS}(h_X^\ell, h_T^\ell) | X \in \mathcal{N}(T)\big)\big)\big),$$

where h_T^ℓ is the embedding of triple T at layer ℓ, $\mathcal{N}(T)$ denotes the neighbors of T in $\mathcal{L}_t^q$, and MESS, AGG, ACT, and DROPOUT are the message passing, aggregation, activation, and dropout functions, respectively.

Layer 3: Refining Embeddings with Meta-Relations. At *Layer 3*, HOKE utilizes the nested triples $\mathcal{T}_n^q$ in $\mathcal{S}^q$ to capture higher-order semantic relationships. Starting with the *refined* triple embeddings $\mathbf{z}_T^L = h_T^l$ from *Layer 2*, where L is the final GNN layer in *Layer 2*, and entity embeddings from *Layer 1*, HOKE computes nested triple embeddings. Given a nested triple $T^n = (T_i, r^n, O)$, where T_i is a triple, r^n is a meta-relation, and O is either a triple T_j or an entity e_j, HOKE computes the final embedding $\mathbf{z}_{T^n}^{\text{final}}$ as: $\mathbf{z}_{T^n}^{\text{final}} = f_{\text{nested}}\left(\mathbf{z}_{T_i}^L,\ \mathbf{q}_{r^n},\ \mathbf{z}_O\right)$,where: if $O = T_j$, then $\mathbf{z}_O = \mathbf{z}_{T_j}^L$ or if $O = e_j$, then $\mathbf{z}_O = \mathbf{e}_{e_j}$. However, due to the selective nature of $\mathcal{S}^q$, it is possible that the head or tail elements of a nested triple (whether a triple or an entity) may be missing from the query-focused subgraph. If the missing element is a triple, HOKE can compute its embedding in an inductive manner using the structure of the line graph. By leveraging the known triples and their relationships in *Layer 2*, the system can infer an approximate embedding for the missing triple aggregating the embeddings of neighbor triples, ensuring the overall consistency of the nested triple embedding like: $\mathbf{z}_{T_{\text{missing}}}^{\text{ind}} = \text{AGG}\left(\left\{z_T^{(L)} | T \in \mathcal{N}(T_{\text{missing}})\right\}\right)$, where AGG is the aggregation function. If the missing components cannot be retrieved through backtracking to the embeddings in *Layer 2* or *Layer 1*, HOKE falls back on approximate embeddings. In such cases, missing components can be initialized using average or related embeddings based on the meta-relation r^n, allowing the computation to continue. The function f_{nested} combines the embeddings to capture semantic relationships among the nested triple's components. For example, using quaternion multiplication: $\mathbf{z}_{T^n}^{\text{final}} = \mathbf{z}_{T_i}^L \cdot \mathbf{q}_{r^n} \cdot \mathbf{z}_O$ where $\cdot$ denotes quaternion multiplication, and $\mathbf{q}_{r^n} \in \mathbb{R}^{4d}$ is the quaternion embedding of r^n. This formulation captures interactions between triples and meta-relations.

4.1 Overall Learning Objective

HOKE learns embeddings at multiple levels, refining them across the layers. We define loss functions for each layer and combine them to form the overall loss function: $\mathcal{L} = \mathcal{L}_{\text{base}} + \lambda_1 \mathcal{L}_{\text{line}} + \lambda_2 \mathcal{L}_{\text{nested}}$ where λ_1 and λ_2 are hyperparameters controlling the importance of each component selected based on validation performance. At each layer, a different aspect of the graph structure is captured, and the overall objective is to combine these representations to form a coherent final embedding. For standard triples, we use:

$$\mathcal{L}_{\text{base}} = \sum_{(h,r,t) \in \mathcal{T}_{\text{train}}} g(-\phi(\mathbf{e}_h, \mathbf{e}_r, \mathbf{e}_t)) + \sum_{(h',r',t') \in \mathcal{T}'_{\text{neg}}} g(\phi(\mathbf{e}_{h'}, \mathbf{e}_{r'}, \mathbf{e}_{t'})), \tag{1}$$

with $g(x) = \log(1 + \exp(x))$, ϕ is the scoring function (e.g., $\phi(\mathbf{e}_h, \mathbf{e}_r, \mathbf{e}_t) = \|\mathbf{e}_h + \mathbf{e}_r - \mathbf{e}_t\|$) and $\mathcal{T}'_{neg}$ are negative examples obtained by corrupting the head or tail entities from existing triples. We define the similarity function $\psi(v_i, v_j)$ between nodes v_i and v_j in the line graph as: $\psi(v_i, v_j) = \mathbf{z}_{v_i}^\top \mathbf{z}_{v_j}$, where $\mathbf{z}_{v_i}$ and $\mathbf{z}_{v_j}$ are the embeddings of nodes v_i and v_j. The loss function for the line graph embeddings is:

$$\mathcal{L}_{\text{line}} = -\sum_{(v_i,v_j)\in\mathcal{E}_{\text{line}}} \log \sigma\left(\psi(v_i, v_j)\right) - \sum_{(v_i,v_k)\notin\mathcal{E}_{\text{line}}} \log\left(1 - \sigma\left(\psi(v_i, v_k)\right)\right), \quad (2)$$

where $\sigma(x) = \frac{1}{1+\exp(-x)}$ is the sigmoid function, $\mathcal{E}_{\text{line}}$ is the set of edges in the line graph, and negative samples (v_i, v_k) are pairs of nodes not connected in the line graph. The loss function encourages the model to assign higher similarity scores to connected nodes and lower scores to unconnected nodes, thereby capturing the structural relationships in the line graph. We define the scoring function ρ for a nested triple (T_i, r^n, O) as: $\rho(\mathbf{z}_{T_i}^L, \mathbf{q}_{r^n}, \mathbf{z}_O) = \left\| \mathbf{z}_{T_i}^{(L)} \cdot \mathbf{q}_{r^n} \cdot \mathbf{z}_O^{-1} \right\|$, where: $\mathbf{z}_{T_i}^L$ is the embedding of triple T_i from *Layer 2*, $\mathbf{q}_{r^n}$ is the embedding of the meta-relation r^n, and $\mathbf{z}_O^{-1}$ is quaternion inverse of $\mathbf{z}_O$, which is $\mathbf{z}_{T_j}^L$ if $O = T_j$, or $\mathbf{e}_{e_j}$ if $O = e_j$. The loss for nested triples is:

$$\mathcal{L}_{\text{nested}} = \sum_{(T_i,r^n,O)\in\mathcal{T}_{\text{nested}}} g\left(-\rho(\mathbf{z}_{T_i}^{(L)}, \mathbf{q}_{r^n}, \mathbf{z}_O)\right) + \sum_{(T_i',r^{n'},O')\in\mathcal{T}_{\text{neg}}'} g\left(\rho(\mathbf{z}_{T_i'}^{(L)}, \mathbf{q}_{r^{n'}}, \mathbf{z}_{O'})\right) \quad (3)$$

Algorithm 2 provides an overview of the training phase.

Algorithm 2. Multilayer Embedding Learning on Query-Focused Subgraphs

1: **Input:** Batch of query-focused subgraphs $\mathcal{S}^q$
2: **Output:** Learned embeddings and predictions
3: **for** each subgraph $S_i^q \in \mathcal{S}^q$ **do**
4: **Layer 1:** Compute entity and relation embeddings using f_{local}
5: **Layer 2:** Refine triple embeddings via GNN over the Line Graph
6: **Layer 3:** Compute nested triple embeddings using meta-relations and f_{nested}
7: **Compute Overall Loss** $\mathcal{L} = \mathcal{L}_{\text{base}} + \lambda_1 \mathcal{L}_{\text{line}} + \lambda_2 \mathcal{L}_{\text{nested}}$
8: **Update Model Parameters** by minimizing $\mathcal{L}$
9: **Return:** Learned embeddings

Predictions and Inductive Learning. The refined embeddings are utilized for prediction tasks given a query q. We emphasize that predictions do not involve the original nested knowledge graph but the query-focused subgraph $\mathcal{S}^q$, which is much smaller in general. For *Triple Prediction*, given q=$(T_i, r^n, ?T)$, where T_i is a known triple, r^n is a relation, and $?T$ represents the unknown triple/entity to be predicted, we score candidates using: $S_{\text{tp}}(T_i, r^n, T_j) = \langle \mathbf{q}_{r^n} \otimes \mathbf{z}_{T_i}, \mathbf{z}_{T_j} \rangle$, where $\mathbf{z}_{T_i}$ is the embeddings of the triple T_i, $\mathbf{z}_{T_j}$ is the entity/triple embedding of T_j, and $\mathbf{q}_{r^n}$ is the embedding of the meta-relation r^n. Here, $\otimes$ denotes element-wise multiplication and $\langle \cdot, \cdot \rangle$ the dot product. For *Conditional Link Prediction*, given q=$\langle T_i, r^n, (h_j, r_j, ?t) \rangle$, we aim to predict the tail entity $x \in \mathcal{E}^q$ that bounds to the variable $?t$. The score is computed as: $S_{\text{clp}}(x) = S_b(h_j, r_j, x) + \gamma \cdot S_l(T_i, r^n, \mathbf{z}_{(h_j,r_j,x)}^{\text{ind}})$, where $S_b(h_j, r_j, x)$ is a base scoring function (e.g., TransE) that evaluates the plausibility of the triple (h_j, r_j, x), $S_l(T_i, r^n, \mathbf{z}^{\text{ind}}_{(h_j,r_j,x)})$ is a nested scoring function (e.g., Quaternion-based), and γ is a hyperparameter controlling the contribution of the inductive term. Here, $\mathbf{z}^{\text{ind}}_{(h_j,r_j,x)}$ is an *inductively* learned embedding for the triple (h_j, r_j, x) that leverages the line graph structure captured in *Layer 2*, regardless of whether the triple exists in the underlying knowledge graph. It is defined as: $\mathbf{z}^{\text{ind}}_{(h_j,r_j,x)} = \text{AGG}\left(\{\mathbf{z}^{\text{line}}_T \mid T \in \mathcal{N}(h_j, r_j, x)\}\right)$ where $\mathcal{N}(h_j, r_j, x)$ represents the set of neighboring triples in the line graph that share at least one element (entity or relation) with (h_j, r_j, x). Moreover, $\mathbf{z}^{\text{line}}_T$ is the embedding of triple T from the line graph representations (*Layer 2*) and AGG the aggregation function. By leveraging the query-focused subgraph, HOKE keeps the number of candidate entities and the computational cost of inductive learning manageable. This enables the model to effectively perform conditional link prediction with inductive embeddings, enhancing its ability to predict links involving unseen or non-existent triples without significant overhead.

5 Experimental Evaluation

The primary goal was to evaluate the HOKE's performance on the novel *nested* knowledge graph tasks of triple classification and conditional link prediction.

Datasets and Metrics. We used the state-of-the-art nested knowledge graphs: FBH, FBHE, and DBHE, which contain nested facts and have been constructed and manually curated [4]. FBH and FBHE are derived from FB15K-237 (Freebase), while DBHE is based on DB15K (DBpedia). FBH consists only of nested facts inferred from triple facts, such as *implies*, whereas FBHE and DBHE also include externally sourced knowledge such as *transfers-to*. Dataset statistics are presented in Table 1. We used the same dataset splits of NestE available online[1]

Baselines. We consider BiVE [4] as well as NestE [20] as our primary baselines since they are specifically designed for NKGs and have shown significant improvements over methods that were not designed to deal with nested triples. For each such methods we report the variants with the best performance [20]. We also rule-based approaches, including Neural-LP [21], DRUM [15], and AnyBURL [13], which indirectly account for relations between facts using first-order logic-like expressions. Additionally, we include QuatE [22] and BiQUE [8] as state-of-the-art triple-based methods based on 4D hypercomplex numbers, although they do not directly handle nested facts. To accommodate the datasets to the usage with HOKE we explicitly create the nested line graph, where atomic facts are converted into entities, and nested facts into triples [20]. We precomputed the line graphs for the datasets considering base triples as nodes [7]. The relatedness matrices were computed by counting predicate co-occurrences in triples [14].

Our Approach. We implement three variants of HOKE. HOKE, which uses all three layers; HOKE-N, which skips the Line Graph and computes triple embeddings directly from *Layer 1* concatenating the head, relation, and tail embeddings; HOKE-B, which skips *Layer 1* and begins with structural refinement at

[1] https://github.com/xiongbo010/NestE.

Table 1. NKGs statistics. $\mathcal{G}_n = (\mathcal{E}, \mathcal{R}, \mathcal{R}_{\mathrm{m}}, \mathcal{T}_s, \mathcal{T}_n)$. $|T'|$ represents the number of atomic triples involved in the nested triples.

Dataset	$\|\mathcal{E}\|$	$\|\mathcal{R}\|$	$\|\mathcal{T}_s\|$	$\|\mathcal{R}_{\mathrm{m}}\|$	$\|\mathcal{T}_n\|$	$\|T'\|$
FBH	14,541	237	310,117	6	27,062	33,157
FBHE	14,541	237	310,117	10	34,941	33,719
DBHE	12,440	87	68,296	8	6,717	8,206

Layer 2 (Line Graph); here, triple embeddings are initialized randomly. In all variants, *nested* triple embeddings are learned at *Layer 3* through higher-order reasoning based on meta-relations. The hyperparameters were tuned via grid search and validation. A learning rate of 0.001 provided stable convergence without overfitting, a max. BFS depth $maxD$=3 balanced accuracy and computational cost. We set relation ($ThRel$) and metarelation ($MThRel$) relatedness thresholds to 0.5 to retain relevant triples while filtering noise. As for the structural approach (*Layer 2*) we used multi-head attention and ReLU activation gave the best generalization, while a 0.3 dropout prevented overfitting. Using 2 layers was optimal, as more layers added complexity without significant performance gains. Moreover, for both base embeddings (*Layer 1*) and nested embeddings (*Layer 3*) we adopted QuatE [22]. Embedding dimensionality was set to $d = 200$. The contributions of λ_1, nested loss λ_2, and the inductive component γ for conditional link prediction were all set to 0.5.

Metrics. Filtered Mean Rank (FMR) is a metric used in KG evaluations to measure the average rank of correct answers in link prediction tasks, after filtering out invalid candidates (e.g., other valid triples in the graph). Unlike metrics such as Mean Reciprocal Rank (MRR) or Hits@10, FMR is unbounded and reflects the overall rank distribution, including long-tail predictions. This complements MRR and Hits@10 by providing insight into how consistently a model ranks correct answers, rather than just focusing on top-ranked results. High FMR values alongside good MRR or Hits@10 scores indicate that while the model excels at ranking some answers highly, it may struggle with consistency across the dataset. In what follows, the mean performance is reported to be over 10 random seeds for each method, with relatively small standard deviations omitted.

Implementation. HOKE has been implemented based on OpenKE [10], Apple MLX and Pytorch; the hardware configuration was a Mac Studio with 192GB of shared memory. For the competitors, we reused the best-performing parameters without performing additional searches[2]. The code is available online[3].

[2] As reported at https://github.com/xiongbo010/NestE.
[3] https://github.com/giuseppepirro/HOKE.

5.1 Results on Triple Prediction

Most of the baseline methods considered cannot handle triple prediction and conditional link prediction due to their inability to work with NGKs. Therefore, to compare with these baselines, we adopt the approach of BiVE and NestE and separate nested triples from standard triples, and create a novel standard KG where the entities are triples. We also note that NestE and BiVe address both standard and nested triples simultaneously. The results of our evaluation, shown in Table 2, suggest that HOKE effectively bridges the gap between older and newer models, delivering precise, context-aware triple predictions while efficiently handling complex relational structures. HOKE's performance highlights the benefits of its *multi-layered approach*, which combines both *semantic* understanding and *structural* refinement via the line graph. This architecture allows the model to capture both broad relational patterns and higher-order dependencies. The consistently lower MR and higher MRR scores suggest that HOKE balances ranking efficiency with prediction accuracy, particularly in the more challenging higher-order relational contexts found in datasets like DBHE. By refining candidate triples through its layered approach, HOKE demonstrates competitive performance, outperforming both traditional methods and more recent models like NestE-SB in triple prediction tasks.

Table 2. Triple Prediction comparison. The best scores are boldfaced. * denotes results taking from [20].

	FBH			FBHE			DBHE		
	MR (↓)	MRR (↑)	Hit@10 (↑)	MR (↓)	MRR (↑)	Hit@10 (↑)	MR (↓)	MRR (↑)	Hit@10 (↑)
QuatE*	145603.8	0.103	0.114	94684.4	0.101	0.209	26485.0	0.157	0.179
BiQUE*	81687.5	0.104	0.115	61015.2	0.135	0.205	19079.4	0.163	0.185
Neural-LP*	115016.6	0.070	0.073	90000.4	0.238	0.274	21130.5	0.170	0.209
DRUM*	115016.6	0.069	0.073	90000.3	0.261	0.274	21130.5	0.166	0.209
AnyBURL*	108079.6	0.096	0.108	83136.8	0.191	0.252	20530.8	0.177	0.214
BiVE-B*	8.63	0.833	0.924	9.53	0.705	0.860	4.66	0.718	0.945
NestE-SB*	3.34	0.922	0.982	3.05	0.851	0.962	2.07	0.862	0.984
HOKE-B	4.83	0.718	0.766	4.44	0.676	0.757	2.03	0.678	0.778
HOKE-N	3.06	0.719	0.777	2.82	0.674	0.764	2.10	0.674	0.777
HOKE	**2.34**	**0.945**	**0.987**	**2.13**	**0.896**	**0.973**	**1.45**	**0.893**	**0.989**

5.2 Results on Conditional Link Prediction

We compared HOKE with the state-of-the-art systems on conditional link prediction. We recall that conditional link prediction tasks in *nested* knowledge graphs are more challenging than standard link prediction because they involve predicting missing links conditioned on additional triples or context. Results are reported in Table 3. The table shows that HOKE outperforms other models in almost all scenarios. It achieves the lowest MR and highest MRR and Hits@10 but in FBH. While NestE-SB performs competitively, particularly in FBHE, HOKE consistently dominates across metrics, demonstrating its advantage in handling higher-order relationships. Older models such as QuatE and BiQUE fall behind significantly, especially in MR, showing their limitations in capturing complex conditional relationships in higher-order scenarios. We hypothesize that the HOKE's layered approach, which progressively refines candidate entities, contributes to learning robust base embeddings that capture broad relational patterns and help efficiently narrow down potential candidates. Notably, in *Layer 2*, the use of focused sampling, guided by predicate relatedness, extracts query-relevant subgraphs, allowing for more precise refinement of these candidates. Additionally, line graph refinement in this layer captures higher-order dependencies between triples, which is crucial for understanding complex relational contexts. This combination enhances candidate identification and ranking, contributing to HOKE's effectiveness in conditional link prediction tasks.

Table 3. Performance on conditional link prediction. The best scores are boldfaced. * denotes results taken from [20].

	FBH			FBHE			DBHE		
	MR (↓)	MRR (↑)	Hit@10 (↑)	MR (↓)	MRR (↑)	Hit@10 (↑)	MR (↓)	MRR (↑)	Hit@10 (↑)
QuatE*	163.7	0.346	0.494	1546.4	0.124	0.189	551.6	0.208	0.309
BiQUE*	111.0	0.423	0.641	90.1	0.387	0.617	29.5	0.378	0.677
Neural-LP*	185.9	0.433	0.648	146.2	0.466	0.716	32.2	0.517	0.756
DRUM*	262.7	0.394	0.555	207.6	0.413	0.620	49.0	0.470	0.732
AnyBURL*	228.5	0.380	0.563	166.0	0.418	0.607	81.7	0.403	0.594
BiVE-Q*	4.33	0.826	0.948	6.56	0.761	0.886	2.69	0.852	0.971
NestE-SB*	1.52	0.934	**0.991**	2.61	0.867	0.951	1.72	0.919	0.990
HOKE-B	4.83	0.718	0.896	2.64	0.771	0.857	2.03	0.778	0.838
HOKE-N	3.06	0.719	0.977	2.52	0.874	0.864	2.10	0.874	0.877
HOKE	**1.34**	**0.945**	0.988	**2.13**	**0.896**	**0.973**	**1.45**	**0.943**	**0.993**

5.3 Ablation Analysis of Query-Focused Subgraph Extraction

We conducted a series of ablation experiments to assess the impact of different components in the query-focused subgraph extraction algorithm. The goal of the analysis was to isolate the contribution of *Layer 2*, *Layer 1*, and the biased BFS using predicate relatedness. We defined several variants of the extraction algorithm: (i) $\text{HOKE}_{\overline{Rel}}$: the biased BFS at *Layer 2* does not use predicate relatedness, effectively treating all neighbors equally; (ii) $\text{HOKE}_{\overline{L1}}$: this variant omits the BFS expansion in *Layer 1*, materializing triples from *Layer 2* but not exploring further entities in *Layer 1*; (iii) $\text{HOKE}_{\overline{L2}}$: this variant omits the line graph refinement at *Layer 2*; (iv) HOKE: this is the complete query-focused subgraph extraction algorithm with all steps included, starting from *Layer 3*, biased BFS in *Layer 2*, and materialization with expansion in *Layer 1*. Table 4 reports the results of the ablation study for conditional link prediction. The variant $\text{HOKE}_{\overline{Rel}}$, where predicate relatedness is omitted in the biased BFS of *Layer 2*, shows a marked drop in performance, with MRR in FBH significantly decreasing. This highlights the importance of predicate-relatedness in guiding BFS towards relevant triples. Similarly, removing *Layer 1* expansion ($\text{HOKE}_{\overline{L1}}$) leads to a moderate decline, suggesting that *Layer 1* is crucial for refining candidate entities early on. The largest performance drop is seen in $\text{HOKE}_{\overline{L2}}$, where omitting line graph refinement results in a notable decrease in Hits@10, especially in DBHE. This confirms the role of line graph refinement in capturing higher-order dependencies. The full model (HOKE) consistently outperforms the ablated versions, showing the effectiveness of combining predicate-relatedness, *Layer 1* expansion, and line graph refinement for conditional link prediction.

Table 4. Ablation results for conditional link prediction.

Variant	FBH			FBHE			DBHE		
	MR (↓)	MRR (↑)	Hit@10 (↑)	MR (↓)	MRR (↑)	Hit@10 (↑)	MR (↓)	MRR (↑)	Hit@10 (↑)
$\text{HOKE}_{\overline{Rel}}$	0.315	0.472	0.339	0.310	0.450	0.320	0.300	0.460	0.340
$\text{HOKE}_{\overline{L1}}$	2.04	0.911	0.901	2.78	0.882	0.917	2.05	0.923	0.947
$\text{HOKE}_{\overline{L2}}$	3.15	0.871	0.877	3.35	0.851	0.789	3.65	0.875	0.734
HOKE	**1.34**	**0.945**	**0.988**	**2.13**	**0.896**	**0.973**	**1.45**	**0.943**	**0.993**

5.4 Effects of the Relatedness-Aware Subgraph Construction

We empirically evaluate to what extent plain BFS and Rel-BFS can identify the results relevant to triple prediction queries in a query-focused subgraph $\mathcal{S}^q$. For both BFS and Rel-BFS we tested max depth (maxD) values of 2, 3, and 4. In particular, we computed $\text{RC} = \frac{1}{|\mathcal{Q}^{\text{test}}|} \sum_{q \in \mathcal{Q}^{\text{test}}} \mathbf{I}\{\text{target}(q) \in \mathcal{T}_s^q\}$ where $\mathcal{Q}^{\text{test}}$ is the set of triple prediction queries, target(q) refers to the missing element in the query (i.e., a triple), and $\mathcal{T}_s^q$ are the (standard) triples in the query-focused subgraph. The results in Table 5 show that Rel-BFS consistently outperforms plain BFS across all datasets and depths, demonstrating the importance of using predicate-relatedness in query-focused subgraph extraction. Increasing the search depth improves result coverage for both methods, but Rel-BFS achieves higher coverage even at smaller depths. These results highlight Rel-BFS's efficiency in capturing relevant triples with fewer graph explorations.

6 Concluding Remarks and Future Work

We introduced HOKE, a novel approach that effectively bridges semantic and structural reasoning for *nested* knowledge graphs. By combining layered embeddings with query-driven subgraph extraction, HOKE captures both local and

Table 5. Result Coverage for BFS and Rel-BFS at different depths.

Strategy	FBH			FBHE			DBHE		
	maxD = 2	maxD = 3	maxD = 4	maxD = 2	maxD = 3	D = 4	maxD = 2	maxD = 3	maxD = 4
BFS	0.718	0.849	0.858	0.472	0.612	0.735	0.328	0.628	0.860
Rel-BFS	**0.745**	**0.898**	**0.993**	**0.737**	**0.835**	**0.991**	**0.775**	**0.943**	**0.998**

global relational patterns, enabling precise triple and conditional link prediction even in complex, large-scale knowledge graphs. Our results hint that HOKE consistently outperforms state-of-the-art models, highlighting the importance of considering both nested relationships and structural dependencies in knowledge graph reasoning. Looking forward, several promising directions can further enhance HOKE's capabilities. First, extending the model to *dynamic* knowledge graphs could unlock new applications in areas such as real-time knowledge updating. Incorporating contextual data, such as temporal and probabilistic, could allow HOKE to handle even more advanced reasoning tasks.

References

1. Arenas-Guerrero, J., Iglesias-Molina, A., Chaves-Fraga, D., Garijo, D., Corcho, O., Dimou, A.: Declarative generation of RDF-star graphs from heterogeneous data. Semantic Web 1–19 (2024, preprint)
2. Bordes, A., Usunier, N., García-Durán, A., Weston, J., Yakhnenko, O.: Translating embeddings for modeling multi-relational data. In: Burges, C.J.C., Bottou, L., Ghahramani, Z., Weinberger, K.Q. (eds.) Advances in Neural Information Processing Systems 26: Proceedings of the 27th Annual Conference on Neural Information Processing Systems, Lake Tahoe, Nevada, USA, pp. 2787–2795 (2013)
3. Cai, H., Zheng, V.W., Chang, K.: A comprehensive survey of graph embedding: problems, techniques, and applications. Trans. Knowl. Data Eng. **30**(9), 1616–1637 (2018)
4. Chung, C., Whang, J.J.: Learning representations of bi-level knowledge graphs for reasoning beyond link prediction. In: Proceedings of the AAAI Conference on Artificial Intelligence, pp. 4208–4216 (2023)
5. Das, R., et al.: Go for a walk and arrive at the answer: reasoning over paths in knowledge bases using reinforcement learning. In: Proceedings of the 6th International Conference on Learning Representations (ICLR) (2018)
6. Dettmers, T., Minervini, P., Stenetorp, P., Riedel, S.: Convolutional 2D knowledge graph embeddings. In: Proceedings of the AAAI Conference on Artificial Intelligence, pp. 1811–1818 (2018)
7. Fionda, V., Pirrò, G.: Learning triple embeddings from knowledge graphs. In: Proceedings of the AAAI Conference on Artificial Intelligence, pp. 3874–3881 (2020)
8. Guo, J., Kok, S.: BiQUE: biquaternionic embeddings of knowledge graphs. In: Moens, M.F., Huang, X., Specia, L., Yih, S.W. (eds.) Proceedings of the 2021 Conference on Empirical Methods in Natural Language Processing (2021)
9. Hamilton, W.L., Ying, Z., Leskovec, J.: Inductive representation learning on large graphs. In: Advances in Neural Information Processing Systems, vol. 30, pp. 1024–1034 (2017)

10. Han, X., et al.: OpenKE: an open toolkit for knowledge embedding. In: Proceedings of the EMNLP (2018)
11. Hartig, O.: RDF* and SPARQL*: an alternative approach to annotate statements in RDF. In: ISWC (Posters, Demos & Industry Tracks), pp. 1–4 (2017)
12. Kazemi, S.M., Poole, D.: Simple embedding for link prediction in knowledge graphs. In: Advances in Neural Information Processing Systems, vol. 31 (2018)
13. Meilicke, C., Chekol, M.W., Betz, P., Fink, M., Stuckenschmidt, H.: Anytime bottom-up rule learning for large-scale knowledge graph completion. VLDB J. **33**(1), 131–161 (2024)
14. Pirrò, G.: Building relatedness explanations from knowledge graphs. Semantic Web **10**(6), 963–990 (2019)
15. Sadeghian, A., Armandpour, M., Ding, P., Wang, D.Z.: DRUM: end-to-end differentiable rule mining on knowledge graphs. In: Advances in Neural Information Processing Systems, vol. 32 (2019)
16. Schlichtkrull, M., Kipf, T.N., Bloem, P., van den Berg, R., Titov, I., Welling, M.: Modeling relational data with graph convolutional networks. In: Gangemi, A., et al. (eds.) ESWC 2018. LNCS, vol. 10843, pp. 593–607. Springer, Cham (2018). https://doi.org/10.1007/978-3-319-93417-4_38
17. Sun, Z., Deng, Z.H., Nie, J.Y., Tang, J.: RotatE: knowledge graph embedding by relational rotation in complex space. In: Proceedings of the International Conference on Learning Representations (2019)
18. Teru, K., Denis, M.A., Hamilton, W.L.: Inductive relation prediction by subgraph reasoning. In: Proceedings of the International Conference on Machine Learning, pp. 9448–9457. PMLR (2020)
19. Wang, Q., Mao, Z., Wang, B., Guo, L.: Knowledge graph embedding: a survey of approaches and applications. Trans. Knowl. Data Eng. **29**(12), 2724–2743 (2017)
20. Xiong, B., Nayyeri, M., Luo, L., Wang, Z., Pan, S., Staab, S.: NestE: modeling nested relational structures for knowledge graph reasoning. In: Proceedings of the AAAI Conference on Artificial Intelligence, pp. 9205–9213 (2024)
21. Yang, F., Yang, Z., Cohen, W.W.: Differentiable learning of logical rules for knowledge base reasoning. In: Advances in Neural Information Processing Systems, vol. 30 (2017)
22. Zhang, S., Tay, Y., Yao, L., Liu, Q.: Quaternion knowledge graph embedding. arXiv preprint arXiv:1904.10281 (2019)
23. Zhou, Z., Zhang, Y., Yao, J., Yao, Q., Han, B.: Less is more: one-shot subgraph reasoning on large-scale knowledge graphs. In: Proceedings of the Twelfth International Conference on Learning Representations (ICLR) (2024)

Improving the Reusability of Conversational Search Test Collections

Zahra Abbasiantaeb[1(✉)], Chuan Meng[1], Leif Azzopardi[2], and Mohammad Aliannejadi[1]

[1] University of Amsterdam, Amsterdam, The Netherlands
z.abbasiantaeb@uva.nl
[2] University of Strathclyde, Glasgow, UK

Abstract. Incomplete relevance judgments limit the reusability of test collections. When new systems are compared to previous systems that contributed to the pool, they often face a disadvantage. This is due to pockets of unjudged documents (called "holes") in the test collection that the new systems return. The very nature of Conversational Search (CS) means that these holes are potentially larger and more problematic when evaluating systems. In this paper, we aim to extend CS test collections by employing Large Language Models (LLMs) to fill holes by leveraging existing judgments. We explore this problem using TREC iKAT 23 and TREC CAsT 22 collections, where information needs are highly dynamic and the responses are much more varied – leaving bigger holes to fill. Our experiments reveal that CS collections show a trend towards less reusability in deeper turns. Also, fine-tuning the Llama-3.1 model leads to high agreement with human assessors, while few-shot prompting the GPT-3.5 results in low agreement with humans. Consequently, filling the holes of a new system using GPT-3.5 leads to higher change in the location of the new system. While regenerating the assessment pool with few-shot prompting the GPT-3.5 model and using it for evaluation achieves a high rank correlation with human-assessed pools. We show that filling the holes using few-shot training the Llama-3.1 model enables a fairer comparison between the new system and the systems contributed to the pool. Our hole-filling model based on few-shot training of the Llama-3.1 model can improve the reusability of test collections.

Keywords: Conversational search · Large language models · Judgments

1 Introduction and Background

Building reusable test collections in a cost-efficient manner to evaluate current and future systems has been a long-standing challenge in the field of Information Retrieval (IR) [13,43]. The predominant strategy for creating such collections has been through the use of pooling [36,43] – where the top-k documents from each contributing system are assessed for relevance. This approach is considered

C. Hauff et al. (Eds.): ECIR 2025, LNCS 15572, pp. 196–213, 2025.
https://doi.org/10.1007/978-3-031-88708-6_13

an acceptable compromise, moving away from the "*ideal test collection*" with complete relevance judgments, which is neither feasible nor practical. While the pooling strategy is fairly robust [9,14,21,34], it leads to various evaluation biases (e.g., [7,9]) where systems that did not contribute to the pool, can be significantly disadvantaged. This is because documents that haven't been judged are considered irrelevant. Therefore, the fewer judged documents returned in a ranking, the lower the retrieval performance ceiling [8]. However, the fewer the judgments required to compare systems the cheaper the test collection. Researchers have tried to address these trade-offs by proposing new metrics and methodologies for compensating for the unjudged documents (e.g., [5,8,9]), and/or developing new pooling and judgment strategies (e.g., [10,20,28]). Either way, "holes" in the pools still remain [22,42]. Various studies have proven that the level of missing judgment does not affect the final ranking significantly [15,41]. However, with new advanced ranking models, nowadays, it is more likely that the majority of the top-ranked passages are unjudged.

However, with the advances in the development of powerful Large Language Models (LLMs) and other neural-based models, new opportunities have emerged for creating scalable, robust, and reusable test collections at a lower cost. LLMs provide the possibility to: (1) assess large volumes of documents reasonably cheaply, especially compared to human judgments [35], (2) consistent yet potentially biased evaluations [3], (3) collect judgments under the same conditions at different times, if the LLM and prompts are fixed and shared. (4) typically achieve higher quality than "typical" crowd workers [2].

Indeed, recent studies [18,19,25,33,37,39,40] have shown the effectiveness of using LLMs to generate relevance judgments in the scenario of ad-hoc search automatically. These works demonstrate that LLM-based judgments exhibit a high correlation with human judgments. Existing works [18,37] prompted commercial LLMs (e.g., GPT-3.5/4) to generate relevance judgments. However, commercial LLMs come with limitations like non-reproducibility, non-deterministic outputs, and potential data leakage between training and evaluation data, which impend their utility in scientific research [31]. [22] and [19] prompted an Open Source (OS) LLM, Flan-T5 [11], for generating relevance judgments. While OS LLMs are less effective, they do offer the potential for the development of reproducible and reusable test collections at scale. This led to efforts by [25], who fine-tuned the Llama [38] model to better condition the LLM for performing the task of assigning relevance judgments. They found that fine-tuned Llama-3 leads to better agreement with human annotators than GPT-3.5 [18]. Further suggesting that more complete test collections could be produced, with high quality, for lower cost. [40] introduced UMbrela, an OS toolkit designed for relevance assessment leveraging OpenAI models. UMbrela is designed based on the research by [37] and is employed for a large-scale relevance evaluation in the TREC RAG competition [39], demonstrating a strong correlation with human assessors.

While this prospect is very appealing, it is fraught with new, unexplored challenges. For example, [3] demonstrated that randomly inserting query words into passages can mislead LLMs, causing them to incorrectly assess irrelevant passages as relevant. Of interest, in this work is the notion of grounding. Training systems on the judgments of LLMs, and then evaluating those systems on subsequent test collections, based on judgments from LLMs, creates a potentially dangerous cycle that may amplify and re-enforce existing biases inherent in LLMs [12]. Grounding the LLM-based judgments given the human judgments provides a mechanism to condition the LLMs to be more aligned with human annotators – serving as a means to avoid the hypothesized AI doom loop [30]. To this end, [22] focus on a setting where the LLM is provided with one relevant example in the prompt to help ground the subsequent judgments. In this paper, we draw upon this direction in the context of Conversational Search (CS) where we aim to build and augment test collections with grounded LLM-based judgments.

CS is defined as responding to the user's information needs in the context of the conversation [6,23,24,26,27,32]. In CS the user's information need depends on the query, the context of the conversation, and the user's personal preferences [4]. This results in highly dynamic, non-linear conversational trajectories, where a user's information need can be addressed differently depending on the system's interpretation, as those needs evolve and change in response to the presented information. Therefore, to address such dynamic information needs, systems can follow many different paths. As a result, they retrieve a much larger array of documents, which leads to more unjudged documents. This creates 'bigger holes' in the evaluation of new or future systems, severely limiting the reusability of test collections [1]. We propose that LLMs can effectively and efficiently augment and extend CS test collections, thereby improving their reusability.

In this paper, we leverage both commercial and OS LLMs in zero-shot, few-shot, and fine-tuning settings to automatically generate relevance judgments for the CS scenario. We compare the generated relevance labels with the official relevance labels by human using various metrics. Specifically, for commercial LLMs, we use the GPT-3.5 model with different prompts, including zero-shot, one-shot, and two-shot prompts. For OS LLMs, we consider two setups: (1) directly prompting the Llama-3.1 [17] model in a zero-shot manner, and (2) fine-tuning the Llama-3.1 model on partial official relevance labels from CS collections and testing it on the remaining relevance labels. We evaluate our approach using the TREC iKAT 23 [4] and TREC CAsT 22 [29] CS collections, where TREC iKAT 23 serves as a personalized CS collection. In this work, we answer the following research questions:

RQ1 How can we measure the reusability of CS collections? And how does it change at different conversation depths?
RQ2 How do LLM-generated judgments compare to human-generated judgments?
RQ3 Can we make the CS collections more reusable using automatic relevance judgments?

To answer the RQs, we conduct a set of analyses and experiments. For RQ1, we report the number of missing relevant judgments at different conversation depths as a proxy for the collection's reusability. For RQ2, we create a training and test set of relevance labels and compare Llama-3.1 and GPT-3.5 in zero-shot, few-shot, and fine-tuning settings. We use GPT-3.5 and Llama-3.1 to generate relevance labels on the official pool and use it to rank the official TREC runs. We compare the relevance labels generated by LLM with human labels in both absolute label prediction and relative ranking of retrieval systems. To answer RQ3, we conduct multiple experiments where at each experiment, we remove the unique relevance labels of each run from the pool[1], simulating the case where a new model is being assessed using the original pool. We then generate the relevance labels using LLMs and use those labels to fill the holes.

2 Methodology

In this section, first, we define our method to evaluate the reusability of CS. Then we explain our approach for using LLMs as relevance assessors. Lastly, we explain our approach to use LLMs for enhancing the reusability of CS collections. To simulate the case where a system needs to be compared against systems in the original pool, we propose the following experimental setups.

- **Leave-one-model-out**: In this scenario, our goal is to compare a new system that did not contribute to the pool against the systems contributed to the pool. To do so, we use the retrieval systems contributed to the pool of the existing collections and remove one system at a time from the pool. We remove the relevance judgment of top-k documents retrieved by a given system (R) from the original pool (P) to create a modified pool, called P_{hole}. Hence P_{hole} is a subset of the original pool. Among the top-k documents retrieved by system R, we remove the documents that have only been retrieved by system R and did not appear in the list of top-k documents retrieved by other systems.
- **Leave-one-team-out**: The retrieval systems submitted by the same team at TREC are often very similar, so removing just one system while keeping the others from the same team in the pool may not accurately simulate having a significantly different system. To address this, instead of removing a single system's judgments at a time, we remove all judgments for all systems submitted by a given team. We then follow the same procedure as in the leave-one-model-out scenario to evaluate the impact.

2.1 Conversational Reusability

To establish the extent of the problem (i.e., how big the judgment holes are) while answering RQ1 on how reusable existing CS collections are, we analyze two major CS test collections. One of the simplest ways to estimate the reusability of a test collection is to test whether the basic concept of pooling can be verified. In top-k pooling, the general idea is that the majority of relevant documents appear

[1] We do not remove the documents that are in common with other runs.

and get assessed while the pool is created from a sufficiently large and diverse set of retrieval systems. Therefore, as a new system gets evaluated against the pool, the lower the number of unjudged documents is, the more reusable the collection is. As a proxy to reusability, in this section, we report two metrics per dataset, namely, missing judgments (ϕ) and missing relevant judgments (ϕ^+) for a new system or a set of new systems by one team, that did not contribute to the pooling.

Missing judgments (ϕ) denotes the number of judgment holes for a new system (or a set of systems by one team). These judgment holes can be relevant or irrelevant based on human judgment.

Missing relevant judgments (ϕ^+) denotes the number of judgment holes for a new system (or a set of systems by one team) that are relevant based on human judgment. Note that ϕ^+ is a subset of ϕ, denoting only the missing judgments of relevant documents. This is particularly important as it demonstrates how much the new system's assessment could be negatively affected by the missing judgments.

2.2 LLMs as Relevance Assessors

Models. To answer RQ2 we describe our model setting to use LLMs as relevance assessors. The task of relevance assessment is defined as assigning a relevance score called s to the given passage (d_i), given the user's information need and prior conversation. Therefore unlike ad-hoc retrieval, where the information need is captured in a single, self-contained query, CS requires understanding the context of the ongoing conversation to assess relevance.

Formally, a conversation contains several turns of user-system interactions where user utterance and system response at turn i are indicated by u_i and r_i. The context of the conversation at turn t is defined as $c_t = \{(u_1, t_1) \ldots (u_{t-1}, r_{t-1})\}$. The user utterance ($u_t$) is not representative of the user's full information need. Hence, to better understand the information need at turn i, the context of the conversation up to this turn is required. Some existing CS datasets include the resolved utterance by human, called u'_t, which clarifies co-references and dependencies in the user utterance based on previous turns. We define function f as follows:

$$s = f(u'_t, c_t, d_i) \ , \tag{1}$$

which given the resolved utterance (u'_t), context of the conversation (c_t), and document (d_i) predicts the relevance score (s). The scale of the score s varies from one collection to another collection. We use both few-shot prompting and fine-tuning techniques to use LLMs for relevance assessment.

- **Few-shot prompting**: We design three different prompts for few-shot training. Our designed prompts are inspired by the prompts used for relevance assessment in ad-hoc search by Thomas et al. [37] and MacAvaney and Soldaini [22]. We use zero-shot, one-shot, and two-shot prompts where in the

one-shot one we pass the canonical response as a positive example with maximum relevance score. The canonical response is the perfect system response to the user utterance, provided by the TREC organizers. In our two-shot prompt, we randomly sample one relevant and one irrelevant document from the collection. We design two versions of the one-shot prompt based on using 1) both the context and resolved utterance and 2) only using the resolved utterance.[2] We use both OS and commercial LLMs and do the relevance assessment task over all query-document pairs from the existing pool. We call the new pool generated by LLM as relevance assessors P_{LLM}.
- **Fine-tuning**: For fine-tuning we split the judgments from the existing collections into train, test, and validation sets. We fine-tune the models on the train set and use the fine-tuned models to assess the relevance of query-document pairs from the test set. In addition, we sample the judgments of few-shot prompting the LLMs on the test set for comparison with fine-tuned models. We call the new judgment set generated by LLMs on the test set $P^{\text{test}}_{\text{LLM}}$.

LLMs vs. Human Assessors. To further answer RQ2, we compare the LLM and the human judgments. Given a new judgment set P_{LLM} (or $P^{\text{test}}_{\text{LLM}}$) generated by LLM, we report the kappa agreement between labels by LLMs and labels by human on the same set. We report the agreement using both graded labels (0–4) and binary labels. To compute the agreement considering binary labels, we convert scores lower than two into the irrelevant and scores higher than one into the relevant classes. In addition, we use the judgments by LLM (P_{LLM} or $P^{\text{test}}_{\text{LLM}}$) and the judgments by human on the same set to evaluate the retrieval systems. We rank the systems based on the value of nDCG@5 metric and compare two rankings using Kendall's tau (τ) rank correlation metric.

2.3 LLMs for Reusability

To study the impact of judgment holes on the evaluation of the new retrieval systems and answer RQ3, we simulate the leave-one-model-out scenario. After removing system R in the leave-on-model-out scenario, we calculate the average number of holes among the top-k documents retrieved by R, across all user queries in the dataset. This metric is referred to as *Unjudged@k*. We repeat this process for all systems contributed to the pooling. Also, after removing system R, we fill the holes using LLM for relevance assessment and form the new assessment set called P_{filled}. This assessment pool is a mix of judgments by human and LLM. We evaluate and rank the systems using both the original pool (P) and P_{filled}. We compute the absolute distance between the position of a the system R in the two ranking lists and call it D. For example, if the system R is ranked 3rd using the original pool (P), and ranked 5th using the P_{filled}, then $D = 2$.

To study the application of LLMs to enhance the reusability of existing collections (RQ3), we test them on leave-one-team-out scenarios. We repeat the

[2] We provide our prompts in our GitHub: https://github.com/ZahraAbbasiantaeb/LLMs-for-relevance-judgment-assessment/.

same process mentioned above and form the P_{filled}. After removing all retrieval systems submitted by one team, we fill the holes using judgments by LLM and form the new assessment set (P_{filled}). We evaluate and rank all retrieval systems using the original pool (P) and P_{filled} over turns with different depth from the conversation. Then, we compute the rank correlation (τ) and rank change (D) using (P) and P_{filled} for ranking the retrieval systems.

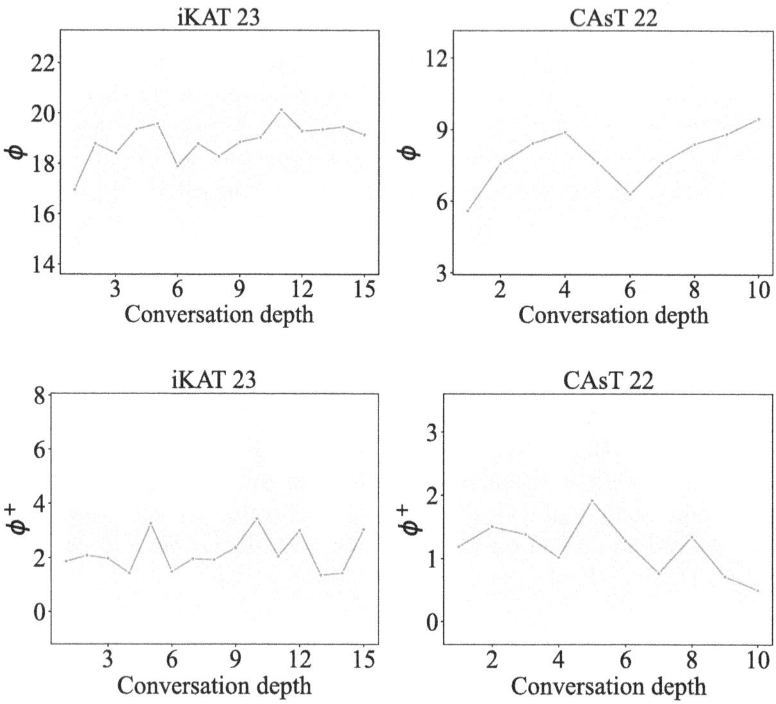

Fig. 1. Distribution of average and standard deviation of the count of unique documents (ϕ) and relevant unique documents (ϕ^+) retrieved by systems per depth of the conversation. These plot are based on leave-one-team-out scenario.

3 Experimental Setup

The Choice of LLMs. As a commercial proprietary LLM, we consider the GPT-3.5 (`gpt-3.5-turbo-0125`) model with `temperature=0` and `top_p=1`. As an OS LLM, we experiment with Llama-3.1 (`llama-3.1-instruct` (8B)) [17] by both few-shot prompting and fine-tuning. For fine-tuning the Llama-3.1 model, we follow Meng et al. [25] to use parameter-efficient fine-tuning (PEFT) method, 4-bit QLoRA [16]. Also, we fine-tune the Flan-T5 (`base`) [11] model with a maximum sequence length of 512 and a learning rate of 2e-5.

Datasets. We conduct our experiments on the TREC iKAT 23 [4] and TREC CAsT 22 [29] collections. In these benchmarks, the relevance of each query-document pair is assessed on the scale of 0 to 4. The TREC CAsT 22 and iKAT 23 collections include the assessment of 42,196 and 26,159 query-document pairs, respectively. For fine-tuning, we divide the assessments into train, test, and validation sets. First, we randomly remove a portion of irrelevant documents (those with a score < 2) from the pool to balance the number of relevant (score $>= 2$) and irrelevant documents per query. Next, we randomly distribute the documents for each query into the train, test, and validation sets, maintaining a 70%-15%-15% split. We ensure that all user queries appear in the train set, with at least one positive and one negative document per query. The size of train, test, and validation sets of TREC iKAT 23 are 3,582, 917, and 709, respectively. For TREC CAsT 22, the train, test, and validation sets include 8,995, 2,287, and 1,827 samples, respectively.

Retrieval Systems. To evaluate the correlation of the generated assessments, we use the baselines and runs submitted to the TREC iKAT 23 and CAsT 22 collections, ensuring that all models are equally treated while being human-assessed. In total, there are 28 and 38 systems submitted to TREC iKAT 23 and TREC CAsT 22, respectively. We utilize the retrieval outputs from these submissions, as provided by the organizers.

4 Results

In this section, we report the results of our designed experiments and answer our research questions.

4.1 Conversational Reusability

Missing Judgments. We simulate the leave-one-team-out setup and plot the distribution of ϕ and ϕ^+ for all retrieval systems over different depths of the conversation in Fig. 1. As can be seen, by increasing the depth of the conversation, the average of ϕ is increasing and it is not consistent over different depths of the conversation. Hence, the pool becomes more diverse by increasing the depth of the conversation in these CS collections. In addition, the average value of ϕ is around 18.55 for iKAT 23 collection while it is around 7.61 for CAsT 22 collection. We argue that a lower value for ϕ represents the higher reusability of the test collections which can be the result of 1) using a larger number of retrieval systems for top-k pooling, 2) larger value of k in top-k pooling, 3) covering a diverse set of systems in the pooling, or 4) differences in creating the test collections. The iKAT 23 dataset is a personalized test collection while CAsT is not personalized. The personalized attribute of the iKAT 23 collection affects the assessment and creation of the topics in iKAT 23. Moreover, fewer systems are used for pooling in iKAT 23 compared to CAsT 22. It is noteworthy that iKAT 23 also features a number of LLM-based systems, which could lead to a more diverse set of documents in the pool, hence having a higher ϕ value. However, we cannot verify this effect and leave a deeper analysis of it for the future.

Missing Relevant Judgments. As for the relevant missing judgments (ϕ^+), we observe a different trend. By going deeper in the conversation, the number of relevant missing judgments is decreasing. As we go deeper in the conversation the systems are diverging more and number of irrelevant documents retrieved by them is increasing. As mentioned earlier, the missing relevant judgments are those that actually lead to an unfair assessment of systems. Therefore, even a small number of missing relevant documents is important. We see in the case of iKAT 23 that at depth 10, we even have cases where the missing relevant documents count up to 8 documents, which clearly puts a new system in an unfair comparison with the others. On CAsT 22, we see that the overall trend is much lower (hence better) than iKAT 23 with an average missing relevant document count equal or below 2. But also, if we see the std. dev. values, a high std. dev. indicates that there are still systems that are unfairly ranked because of the high number of missing relevant documents.

Upshot. We answer RQ1 by concluding that the CS collections show a trend towards less reusability in deeper turns. We think that this is because of more complexity of the user utterances and information needs as the conversation progresses, making it less likely for different systems to return a similar set of documents.

Table 1. Cohen's kappa agreement between human and LLM labels on complete set and test set of pools from TREC iKAT 23 and TREC CAsT 22 collections. Agreement over binary and graded labels are indicated by (B) and (G), respectively.

LLM	Prompt	Context	TREC iKAT 23				TREC CAsT 22			
			Complete		Test		Complete		Test	
			B	G	B	G	B	G	B	G
GPT-3.5	zero-shot	✗	20.7	4.1	48.9	11.7	21.7	8.7	37.1	10.0
	one-shot	✗	23.5	13.7	49.9	21.2	24.5	10.8	43.5	13.7
	one-shot	✓	**27.3**	**20.2**	45.3	26.8	**29.1**	**20.4**	43.0	21.5
	two-shot	✗	21.8	15.2	45.4	18.4	28.5	18.3	42.9	19.2
Llama-3.1	one-shot	✗	2.4	2.8	2.4	1.7	3.2	1.9	1.5	1.3
	one-shot	✓	5.5	3.9	2.1	3.8	4.8	2.8	4.0	2.8
	FT	✗	-	-	71.8	53.0	-	-	**64.9**	**47.1**
	FT	✓	-	-	**73.3**	**55.7**	-	-	62.2	45.4
FlanT5	FT	✗	-	-	51.7	34.7	-	-	46.1	32.1

Table 2. Comparison between the relative ranking of retrieval systems of TREC iKAT 23 and TREC CAsT 22 collections using LLM- and human-generated pools. The relative ranking is compared using Kendall's tau (τ) metric and retrieval systems are ranked based on nDCG@5 metric.

LLM	Prompt	Context	TREC iKAT 23		TREC CAsT 22	
			Complete	Test	Complete	Test
GPT-3.5	zero-shot	✗	0.852	0.778	0.892	0.656
	one-shot	✗	**0.862**	0.778	**0.900**	0.676
	one-shot	✓	0.630	0.624	0.883	0.670
	two-shot	✗	0.825	0.746	0.886	0.650
Llama-3.1	one-shot	✗	0.450	0.307	0.289	0.280
	one-shot	✓	0.550	0.471	0.710	0.616
	FT	✗	-	0.841	-	0.881
	FT	✓	-	0.751	-	0.878
FlanT5	FT	✗	-	**0.889**	-	**0.886**

4.2 LLMs as Relevance Assessors

Agreement with Human Labels. In Table 1, we report our proposed LLM assessors' agreement with human labels. The experiments reveal that we can improve the agreement by fine-tuning the Llama-3.1 model. As can be seen, the fine-tuned Llama-3.1 model achieves the agreement of 71.8 and 64.9 on the binary level over TREC iKAT 23 and TREC CAsT 22 collections, respectively. Moreover, the one-shot learning on the GPT-3.5 model has the highest agreement with human labels among zero- and two-shot learning techniques. In addition, using the conversation context improves the agreement when using GPT-3.5 over the both collections. However, for the fine-tuned Llama-3.1 model, using context improves the agreement only on the TREC iKAT 23 collection.

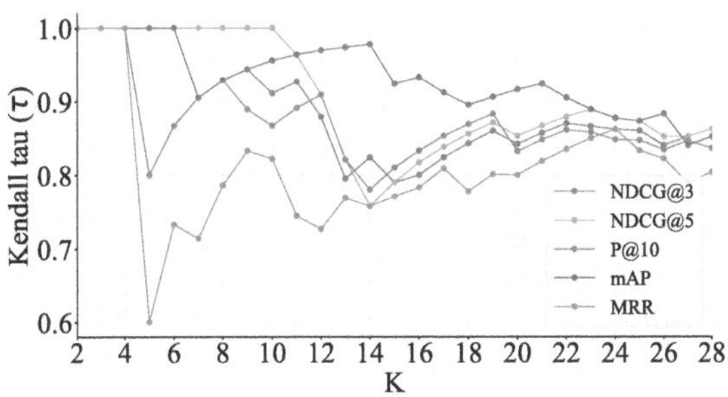

Fig. 2. Rank correlation between the K best-performing systems from TREC iKAT 23 using P and P_{llm}.

Rank Correlation of LLMs with Human. We report the correlation between the ranking of retrieval systems using P_{LLM} and P in Table 2. More correlation indicates the effectiveness of the LLM-generated pool for evaluating the retrieval systems. Surprisingly, the gap between the correlation of the fine-tuned Llama-3.1 and GPT-3.5 is highly lower on test data compared to the gap between their agreement with human labels. This could be due to the different labeling biases that the models have where GPT-3.5 labels could be more different from human labels in terms of absolute numbers (usually more positive [18]), but when we compare different documents they are more similar relatively. The FlanT5 model achieves comparable rank correlation compared to Llama-3.1 while it exhibits a lower agreement on generated labels compared to Llama-3.1. In addition, the GPT-3.5 model achieves the highest rank correlation using the one-shot prompt on the entire set from TREC iKAT 23 collection. Moreover, using the context in the prompt results in a lower rank correlation while it increases the agreement. We conclude that the high agreement between LLM labels and human labels does not necessarily result in a higher rank correlation.

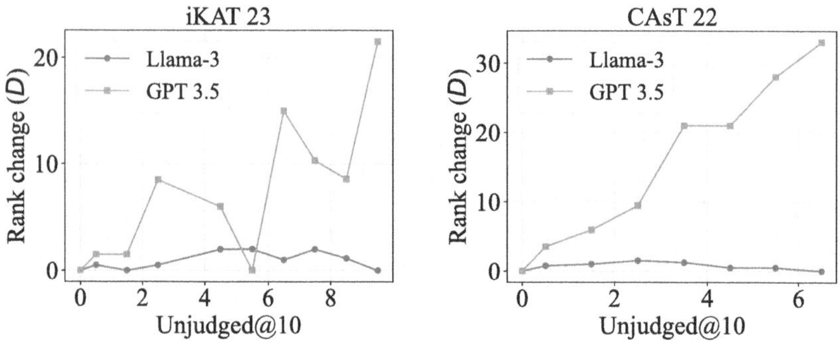

Fig. 3. Absolute distance between the location of a new system before and after filling the holes using GPT-3.5 and Llama-3.1 in leave-one-model-out setup.

Ranking of Top Systems. In Fig. 2, we show the correlation between the relative ranking of K best-performing systems using P_{LLM} and P. The relative ranking using P_{LLM} is the same as using the P over all ranking metrics for four best-performing systems ($K = 4$). The relative ranking of the top 10 best-performing systems based on nDCG@5 using the judgments by LLM, is the same as using human judgments ($\tau = 1$). According to Fig. 2, as the value of K increases (more systems are included in the comparison), the value of τ converges. This finding represents the reliability of LLM-generated judgments for evaluation.

Upshot. We answer RQ2, by concluding that fine-tuning the LLMs can lead to high agreement for absolute label prediction. However, a low agreement does not result in a low rank correlation when ranking retrieval systems. In addition, by one-shot prompting the GPT-3.5 we can achieve a high rank correlation.

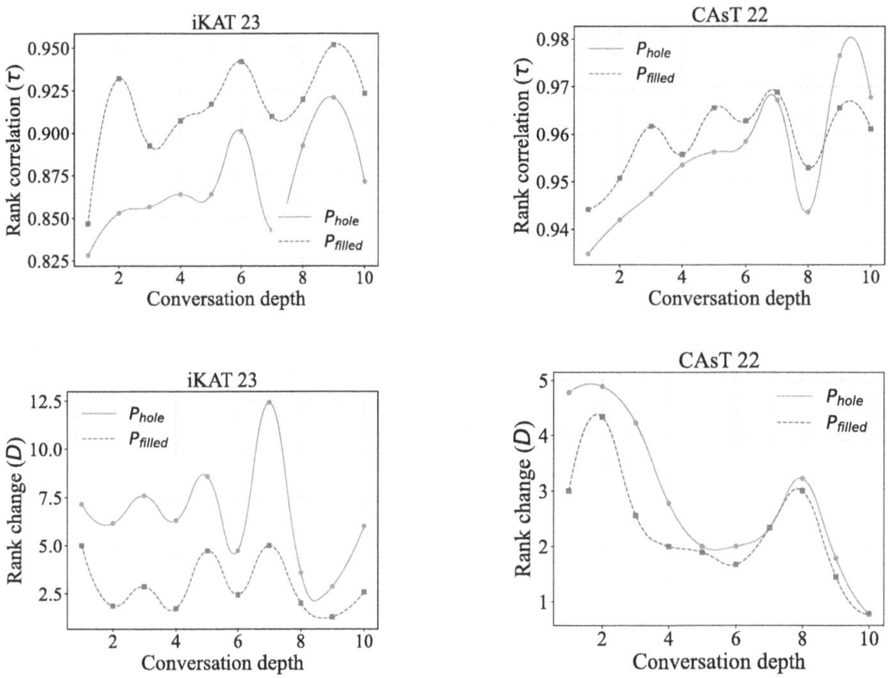

Fig. 4. Average rank correlation (τ) of systems and rank change (D) of a new system using P_{filled} or P_{hole} with original pool P over conversational turns with different depth. The P_{filled} is formed using the judgments by few-shot prompting the Llama-3.1 model and leave-one-team-out scenario.

4.3 LLMs for reusability

Reliability of LLM Labels. To answer RQ3, we simulate the leave-one-model-out scenario and use one-shot prompting the GPT-3.5 model with no context to form P_{filled}. The value of the absolute distance (D) in the two rankings based on the portion of the *Unjudged@10* documents for the new retrieval system is shown in Fig. 3. As can be seen, as the value of *Unjudged@10* increases, the absolute distance (D) increases. This makes sense as we know the GPT-3.5 model is biased to rate the documents with higher scores compared to human [18, 25]. We re-create the P_{filled} by one-shot prompting the Llama-3.1 model. Interestingly, we see that the results of Llama-3.1 exhibit a completely different trend where the number of judgment holes does not seem to matter. We see in the plot that Llama-3.1 can consistently rank the missing systems close to its original ranking and even achieves perfect ranking at some points.

To further answer RQ3, we do the leave-one-team-out analysis at the different depths of the conversation and use one-shot prompting the Llama-3.1 model to form the P_{filled}. We plot the average rank correlation between the ranking of systems and the average absolute distance between the position of the new systems using each of P_{filled} and P pools in Fig. 4. To better study the application of LLMs for enhancing the reusability of the test collections, we also plot the rank correlation and rank change between P_{hole} and P.

According to Fig. 1, the average number of unique documents (ϕ) retrieved by systems increases by increasing the depth and we have more judgment holes per deeper turns. Consequently, we expect the rank correlation between using P_{hole} and P to decrease by an increase in the depth. However, we observe a different pattern in Fig. 4, where the value of τ increases as we go deeper. Looking at the relevance label of unique retrieved documents in Fig. 1, we realize that even though the value of ϕ is increasing, most of these unique retrieved documents are irrelevant. Also, we observe a drop in the average performance of systems, indicating that the turns become generally "harder" and systems fail more. As the conversation progresses (higher depth), modeling the information need of the user becomes more complex and the retrieval systems diverge more.

According to Fig. 4, by filling judgment holes using the few-shot learned Llama-3.1 model, we can achieve a higher correlation with humans compared to using a pool with hole P_{hole}. Also, as can be seen in Fig. 3, using Llama-3.1 for filling the judgment holes helps to rank the new system closer to its real position in the ranking of systems compared to using the pool with hole P_{hole}.

Upshot. We answer RQ3 by concluding that given a new ranking model with a lot of missing judgments (a larger value for *Unjudged@10*), it is advisable to recreate the whole pool using one-shot prompting the GPT-3.5 model, rather than augmenting the existing human-created pool by judgments of the GPT-3.5 model. Because one-shot Llama-3.1 aligns better with human labels, it is more effective to use it for assessing the missing judgments, rather than re-assessing the whole pool. As our analyses reveal the CS collections become less reusable as the conversations progress (i.e., deeper turns). This is where we can take advantage of LLM-generated labels even more.

5 Conclusion

Our results show that ranking of the retrieval systems on CS test benchmarks using human- and LLM-generated are highly correlated, even though the LLM-generated and human judgments have a low agreement in binary- and graded-level labels, in line with the findings of [18]. In addition, the correlation of different IR metrics converges as we add more retrieval systems to the comparison pool, indicating that LLM-generated pools are reliable for comparing different retrieval models in CS. Moreover, we show that in the case of evaluating a new retrieval model, filling missing judgments with a zero-shot Llama-3.1 model results in less significant shifts in the ranking of the corresponding retrieval model compared to using one-shot GPT-3.5, perhaps due to higher agreement of the ratings, and because the one-shot GPT-3.5 model is biased to predict higher relevance scores [37]. In addition, we show that in CS datasets, the number of judgment holes increases and retrieval systems diverge more as the conversation progresses, making the existing collections less reusable for deeper turns.

Our experiments show less reusability of iKAT 23 collection compared to CAsT 22 collection. We leave further analysis for the reason behind this finding as future work. Interestingly, we find that at further turns in a conversation, the information need becomes more complex and the queries are more difficult since the average performance of models declines. This fact results in poorer comparisons and less re-usability for the new and more effective retrieval systems in the future – as the chance of having larger holes for future retrieval systems is greater. Our hole-filling approach using few-shot training the Llama-3.1 model ameliorates this problem and enables a fairer comparison between the systems that contributed to the pool and those that did not. In conclusion, by strategically leveraging LLMs, we can build more robust, reusable, and cost-effective CS test collections that facilitate more equitable comparisons between systems both present and future. In this study we relied on few-shot prompting the LLMs for filling the holes in leave-one-model-out scenario. We leave the fine-tuning technique in this scenario as feature work.

References

1. Abbasiantaeb, Z., Aliannejadi, M.: Generate then retrieve: conversational response retrieval using LLMs as answer and query generators. CoRR abs/2403.19302 (2024). https://doi.org/10.48550/arxiv.2403.19302
2. Alaofi, M., Arabzadeh, N., Clarke, C.L.A., Sanderson, M.: Generative information retrieval evaluation, pp. 135–159. Springer, Cham (2025). ISBN 978-3-031-73147-1. https://doi.org/10.1007/978-3-031-73147-1_6
3. Alaofi, M., Thomas, P., Scholer, F., Sanderson, M.: LLMs can be fooled into labelling a document as relevant: best café near me; this paper is perfectly relevant. In: Sakai, T., Ishita, E., Ohshima, H., Hasibi, F., Mao, J., Jose, J.M. (eds.) Proceedings of the 2024 Annual International ACM SIGIR Conference on Research and Development in Information Retrieval in the Asia Pacific Region, SIGIR-AP 2024, Tokyo, Japan, 9–12 December 2024, pp. 32–41. ACM (2024). https://doi.org/10.1145/3673791.3698431
4. Aliannejadi, M., Abbasiantaeb, Z., Chatterjee, S., Dalton, J., Azzopardi, L.: TREC ikat 2023: a test collection for evaluating conversational and interactive knowledge assistants. In: Yang, G.H., Wang, H., Han, S., Hauff, C., Zuccon, G., Zhang, Y. (eds.) Proceedings of the 47th International ACM SIGIR Conference on Research and Development in Information Retrieval, SIGIR 2024, Washington DC, USA, 14–18 July 2024, pp. 819–829. ACM (2024). https://doi.org/10.1145/3626772.3657860
5. Aslam, J.A., Yilmaz, E.: Inferring document relevance from incomplete information. In: Silva, M.J., Laender, A.H.F., Baeza-Yates, R.A., McGuinness, D.L., Olstad, B., Olsen, Ø.H., Falcão, A.O. (eds.) Proceedings of the Sixteenth ACM Conference on Information and Knowledge Management, CIKM 2007, Lisbon, Portugal, 6–10 November 2007, pp. 633–642. ACM (2007). https://doi.org/10.1145/1321440.1321529
6. Azzopardi, L., Dubiel, M., Halvey, M., Dalton, J.: A conceptual framework for conversational search and recommendation: conceptualizing agent-human interactions during the conversational search process. In: Proceedings of the CAIR'18: Second International Workshop on Conversational Approaches to Information Retrieval at SIGIR 2018 (2018)

7. Baillie, M., Azzopardi, L., Ruthven, I.: A retrieval evaluation methodology for incomplete relevance assessments. In: Amati, G., Carpineto, C., Romano, G. (eds.) ECIR 2007. LNCS, vol. 4425, pp. 271–282. Springer, Heidelberg (2007). https://doi.org/10.1007/978-3-540-71496-5_26
8. Baillie, M., Azzopardi, L., Ruthven, I.: Evaluating epistemic uncertainty under incomplete assessments. Inf. Process. Manag. **44**(2), 811–837 (2008)
9. Buckley, C., Voorhees, E.M.: Retrieval evaluation with incomplete information. In: Sanderson, M., Järvelin, K., Allan, J., Bruza, P. (eds.) SIGIR 2004: Proceedings of the 27th Annual International ACM SIGIR Conference on Research and Development in Information Retrieval, Sheffield, UK, 25–29 July 2004, pp. 25–32, ACM (2004). https://doi.org/10.1145/1008992.1009000
10. Carterette, B., Allan, J., Sitaraman, R.K.: Minimal test collections for retrieval evaluation. In: Efthimiadis, E.N., Dumais, S.T., Hawking, D., Järvelin, K. (eds.) SIGIR 2006: Proceedings of the 29th Annual International ACM SIGIR Conference on Research and Development in Information Retrieval, Seattle, Washington, USA, 6–11 August 2006, pp. 268–275. ACM (2006). https://doi.org/10.1145/1148170.1148219
11. Chung, H.W., et al.: Scaling instruction-finetuned language models. J. Mach. Learn. Res. **25**(70), 1–53 (2024)
12. Clarke, C.L., Dietz, L.: LLM-based relevance assessment still can't replace human relevance assessment. arXiv preprint arXiv:2412.17156 (2024)
13. Cleverdon, C.: The cranfield tests on index language devices. In: Readings in Information Retrieval, pp. 47–59. Morgan Kaufmann Publishers Inc., San Francisco (1997). ISBN 1558604545
14. Cormack, G.V., Palmer, C.R., Clarke, C.L.A.: Efficient construction of large test collections. In: Croft, W.B., Moffat, A., van Rijsbergen, C.J., Wilkinson, R., Zobel, J. (eds.) SIGIR 1998: Proceedings of the 21st Annual International ACM SIGIR Conference on Research and Development in Information Retrieval, 24–28 August 1998, Melbourne, Australia, pp. 282–289. ACM (1998). https://doi.org/10.1145/290941.291009
15. Craswell, N., Mitra, B., Yilmaz, E., Campos, D., Voorhees, E.M., Soboroff, I.: TREC deep learning track: reusable test collections in the large data regime. In: Diaz, F., Shah, C., Suel, T., Castells, P., Jones, R., Sakai, T. (eds.) SIGIR 2021: The 44th International ACM SIGIR Conference on Research and Development in Information Retrieval, Virtual Event, Canada, 11–15 July 2021, pp. 2369–2375. ACM (2021). https://doi.org/10.1145/3404835.3463249
16. Dettmers, T., Pagnoni, A., Holtzman, A., Zettlemoyer, L.: Qlora: efficient finetuning of quantized LLMs. In: Oh, A., Naumann, T., Globerson, A., Saenko, K., Hardt, M., Levine, S. (eds.) Advances in Neural Information Processing Systems 36: Annual Conference on Neural Information Processing Systems 2023, NeurIPS 2023, New Orleans, LA, USA, 10–16 December 2023 (2023). http://papers.nips.cc/paper_files/paper/2023/hash/1feb87871436031bdc0f2beaa62a049b-Abstract-Conference.html
17. Dubey, A., et al.: The llama 3 herd of models. CoRR abs/2407.21783 (2024). https://doi.org/10.48550/ARXIV.2407.21783
18. Faggioli, G., et al.: Perspectives on large language models for relevance judgment. In: Yoshioka, M., Kiseleva, J., Aliannejadi, M. (eds.) Proceedings of the 2023 ACM SIGIR International Conference on Theory of Information Retrieval, ICTIR 2023, Taipei, Taiwan, 23 July 2023, pp. 39–50. ACM (2023). https://doi.org/10.1145/3578337.3605136

19. Khramtsova, E., Zhuang, S., Baktashmotlagh, M., Zuccon, G.: Leveraging LLMs for unsupervised dense retriever ranking. In: Yang, G.H., Wang, H., Han, S., Hauff, C., Zuccon, G., Zhang, Y. (eds.) Proceedings of the 47th International ACM SIGIR Conference on Research and Development in Information Retrieval, SIGIR 2024, Washington DC, USA, 14–18 July 2024, pp. 1307–1317. ACM (2024). https://doi.org/10.1145/3626772.3657798
20. Lipani, A., Palotti, J., Lupu, M., Piroi, F., Zuccon, G., Hanbury, A.: Fixed-cost pooling strategies based on IR evaluation measures. In: Jose, J.M., et al. (eds.) ECIR 2017. LNCS, vol. 10193, pp. 357–368. Springer, Cham (2017). https://doi.org/10.1007/978-3-319-56608-5_28
21. Lu, X., Moffat, A., Culpepper, J.S.: The effect of pooling and evaluation depth on IR metrics. Inf. Retr. J. **19**(4), 416–445 (2016)
22. MacAvaney, S., Soldaini, L.: One-shot labeling for automatic relevance estimation. In: Chen, H., Duh, W.E., Huang, H., Kato, M.P., Mothe, J., Poblete, B. (eds.) Proceedings of the 46th International ACM SIGIR Conference on Research and Development in Information Retrieval, SIGIR 2023, Taipei, Taiwan, 23–27 July 2023, pp. 2230–2235. ACM (2023). https://doi.org/10.1145/3539618.3592032
23. Meng, C., Aliannejadi, M., de Rijke, M.: System initiative prediction for multi-turn conversational information seeking. In: Frommholz, I., Hopfgartner, F., Lee, M., Oakes, M., Lalmas, M., Zhang, M., Santos, R.L.T. (eds.) Proceedings of the 32nd ACM International Conference on Information and Knowledge Management, CIKM 2023, Birmingham, United Kingdom, 21–25 October 2023, pp. 1807–1817. ACM (2023). https://doi.org/10.1145/3583780.3615070
24. Meng, C., Arabzadeh, N., Aliannejadi, M., de Rijke, M.: Query performance prediction: from ad-hoc to conversational search. In: Chen, H., Duh, W.E., Huang, H., Kato, M.P., Mothe, J., Poblete, B. (eds.) Proceedings of the 46th International ACM SIGIR Conference on Research and Development in Information Retrieval, SIGIR 2023, Taipei, Taiwan, 23–27 July 2023, pp. 2583–2593. ACM (2023). https://doi.org/10.1145/3539618.3591919
25. Meng, C., Arabzadeh, N., Askari, A., Aliannejadi, M., de Rijke, M.: Query performance prediction using relevance judgments generated by large language models. CoRR abs/2404.01012 (2024). https://doi.org/10.48550/arxiv.2404.01012
26. Meng, C., Ren, P., Chen, Z., Ren, Z., Xi, T., de Rijke, M.: Initiative-aware self-supervised learning for knowledge-grounded conversations. In: Diaz, F., Shah, C., Suel, T., Castells, P., Jones, R., Sakai, T. (eds.) SIGIR 2021: The 44th International ACM SIGIR Conference on Research and Development in Information Retrieval, Virtual Event, Canada, 11–15 July 2021, pp. 522–532. ACM (2021). https://doi.org/10.1145/3404835.3462824
27. Meng, C., et al.: Dukenet: a dual knowledge interaction network for knowledge-grounded conversation. In: Huang, J.X., et al. (eds.) Proceedings of the 43rd International ACM SIGIR Conference on Research and Development in Information Retrieval, SIGIR 2020, Virtual Event, China, 25–30 July 2020, pp. 1151–1160. ACM (2020). https://doi.org/10.1145/3397271.3401097
28. Moffat, A., Webber, W., Zobel, J.: Strategic system comparisons via targeted relevance judgments. In: Kraaij, W., de Vries, A.P., Clarke, C.L.A., Fuhr, N., Kando, N. (eds.) SIGIR 2007: Proceedings of the 30th Annual International ACM SIGIR Conference on Research and Development in Information Retrieval, Amsterdam, The Netherlands, 23–27 July 2007, pp. 375–382. ACM (2007). https://doi.org/10.1145/1277741.1277806

29. Owoicho, P., Dalton, J., Aliannejadi, M., Azzopardi, L., Trippas, J.R., Vakulenko, S.: TREC cast 2022: going beyond user ask and system retrieve with initiative and response generation. In: Soboroff, I., Ellis, A. (eds.) Proceedings of the Thirty-First Text REtrieval Conference, TREC 2022, online, 15–19 November 2022, NIST Special Publication, vol. 500-338, National Institute of Standards and Technology (NIST) (2022). https://trec.nist.gov/pubs/trec31/papers/Overview_cast.pdf
30. Peterson, A.J.: AI and the problem of knowledge collapse. CoRR abs/2404.03502 (2024). https://doi.org/10.48550/arxiv.2404.03502
31. Pradeep, R., Sharifymoghaddam, S., Lin, J.: Rankzephyr: effective and robust zero-shot listwise reranking is a breeze! CoRR abs/2312.02724 (2023). https://doi.org/10.48550/arxiv.2312.02724
32. Radlinski, F., Craswell, N.: A theoretical framework for conversational search. In: Nordlie, R., Pharo, N., Freund, L., Larsen, B., Russel, D. (eds.) Proceedings of the 2017 Conference on Conference Human Information Interaction and Retrieval, CHIIR 2017, Oslo, Norway, 7–11 March 2017, pp. 117–126. ACM (2017). https://doi.org/10.1145/3020165.3020183
33. Rahmani, H.A., Wang, X., Yilmaz, E., Craswell, N., Mitra, B., Thomas, P.: Syndl: a large-scale synthetic test collection for passage retrieval. CoRR abs/2408.16312 (2024). https://doi.org/10.48550/arxiv.2408.16312
34. Sanderson, M., Zobel, J.: Information retrieval system evaluation: effort, sensitivity, and reliability. In: Baeza-Yates, R.A., Ziviani, N., Marchionini, G., Moffat, A., Tait, J. (eds.) SIGIR 2005: Proceedings of the 28th Annual International ACM SIGIR Conference on Research and Development in Information Retrieval, Salvador, Brazil, 15–19 August 2005, pp. 162–169. ACM (2005). https://doi.org/10.1145/1076034.1076064
35. Soviero, B., Kuhn, D., Salle, A., Moreira, V.P.: Chatgpt goes shopping: LLMs can predict relevance in ecommerce search. In: Goharian, N., et al. (eds.) Advances in Information Retrieval - 46th European Conference on Information Retrieval, ECIR 2024, Glasgow, UK, 24–28 March 2024, Proceedings, Part IV. Lecture Notes in Computer Science, vol. 14611, pp. 3–11. Springer, Cham (2024). https://doi.org/10.1007/978-3-031-56066-8_1
36. Sparck-Jones, K., Van Rijsbergen, C.J.: Report on the need for and provision of an 'ideal' information retrieval test collection. Technical report, British Library Research and Development Report No. 5266, Computer Laboratory, University of Cambridge (1975)
37. Thomas, P., Spielman, S., Craswell, N., Mitra, B.: Large language models can accurately predict searcher preferences. In: Yang, G.H., Wang, H., Han, S., Hauff, C., Zuccon, G., Zhang, Y. (eds.) Proceedings of the 47th International ACM SIGIR Conference on Research and Development in Information Retrieval, SIGIR 2024, Washington DC, USA, 14–18 July 2024, pp. 1930–1940. ACM (2024). https://doi.org/10.1145/3626772.3657707
38. Touvron, H., et al.: Llama: open and efficient foundation language models. CoRR abs/2302.13971 (2023). https://doi.org/10.48550/arxiv.2302.13971
39. Upadhyay, S., et al.: A large-scale study of relevance assessments with large language models: an initial look. CoRR abs/2411.08275 (2024). https://doi.org/10.48550/arxiv.2411.08275
40. Upadhyay, S., Pradeep, R., Thakur, N., Craswell, N., Lin, J.: UMBRELA: umbrela is the (open-source reproduction of the) bing relevance assessor. CoRR abs/2406.06519 (2024). https://doi.org/10.48550/arxiv.2406.06519

41. Voorhees, E.M.: On building fair and reusable test collections using bandit techniques. In: Cuzzocrea, A., et al. (eds.) Proceedings of the 27th ACM International Conference on Information and Knowledge Management, CIKM 2018, Torino, Italy, 22–26 October 2018, pp. 407–416. ACM (2018). https://doi.org/10.1145/3269206.3271766
42. Voorhees, E.M., Craswell, N., Lin, J.: Too many relevants: whither cranfield test collections? In: Amigó, E., Castells, P., Gonzalo, J., Carterette, B., Culpepper, J.S., Kazai, G. (eds.) SIGIR 2022: The 45th International ACM SIGIR Conference on Research and Development in Information Retrieval, Madrid, Spain, 11–15 July 2022, pp. 2970–2980. ACM (2022). https://doi.org/10.1145/3477495.3531728
43. Voorhees, E.M., Harman, D.K.: TREC: Experiment and Evaluation in Information Retrieval (Digital Libraries and Electronic Publishing). The MIT Press (2005). ISBN 0262220733

Repeat-Bias-Aware Optimization of Beyond-Accuracy Metrics for Next Basket Recommendation

Yuanna Liu(✉), Ming Li, Mohammad Aliannejadi, and Maarten de Rijke

University of Amsterdam, Amsterdam, The Netherlands
{y.liu8,m.li,m.aliannejadi,m.derijke}@uva.nl

Abstract. In next basket recommendation (NBR) a set of items is recommended to users based on their historical basket sequences. In many domains, the recommended baskets consist of both repeat items and explore items. Some state-of-the-art NBR methods are heavily biased to recommend repeat items so as to maximize utility. The evaluation and optimization of beyond-accuracy objectives for NBR, such as item fairness and diversity, has attracted increasing attention. How can such beyond-accuracy objectives be pursued in the presence of heavy repeat bias? We find that only optimizing diversity or item fairness without considering repeat bias may cause NBR algorithms to recommend more repeat items. To solve this problem, we propose a model-agnostic repeat-bias-aware optimization algorithm to post-process the recommended results obtained from NBR methods with the objective of mitigating repeat bias when optimizing diversity or item fairness. We consider multiple variations of our optimization algorithm to cater to multiple NBR methods. Experiments on three real-world grocery shopping datasets show that the proposed algorithms can effectively improve diversity and item fairness, and mitigate repeat bias at acceptable Recall loss.

Keywords: Next basket recommendation · Repeat bias · Beyond-accuracy metrics · Re-ranking

1 Introduction

In NBR a recommender system is meant to recommend a set of items at once [22]. In many e-commerce scenarios in which NBR are deployed users display *repetitive* consumption behavior (e.g., purchasing milk every week, or listening to the same song during workouts) as well as *exploratory* consumption behavior (e.g., buying Christmas gifts) at the same time. Hence, items in a recommended basket can be categorized into repeat items and explore items, depending on whether a user has consumed an item before.

Beyond-Accuracy Metrics and NBR. Many machine learning techniques have been applied to optimize the accuracy achieved on the NBR task, based on k-nearest neighbors (KNNs) [8,11,28], recurrent neural networks (RNNs) [36], or graph neural networks (GNNs) [37]. However, measuring and optimizing beyond-accuracy objectives for NBR remains largely unexplored. Of particular interest are *diversity*, to combat the problem of recommendation homogeneity and generate diversified baskets [15,33], and

C. Hauff et al. (Eds.): ECIR 2025, LNCS 15572, pp. 214–229, 2025.
https://doi.org/10.1007/978-3-031-88708-6_14

item fairness, to ensure a fair distribution of the exposure assigned to different groups of items [25]. Optimizing for beyond-accuracy metrics such as diversity and item fairness in the context of NBR is made more complex than in other recommendation scenarios due to the presence of repeat bias.

Repeat Bias in NBR. Experiments have shown that repeat items contribute most of the accuracy performance [16,18,19,21]. This is because repetition prediction is an easier task, typically with just dozens or hundreds of candidate items and explicit user feedback [20]. However, explore items hold great potential for long-term value in particular and for beyond-accuracy goals in general. E.g., exploring uncertain regions enhances exposure to new and long-tail content, reshaping the overall distribution of the contents, which ultimately improves long-term user experience [32]. Therefore, when curating sets that include both previously seen and new items, it is important to strike the right balance between the repeat and explore categories [2]. Importantly, finding a good balance between repeat and explore items in NBR is made more complex when beyond-accuracy metrics are considered [20]: if most utility is obtained from a relatively small number of items (the repeat items), there is little room to improve beyond-accuracy metrics without sacrificing utility.

Repeat-Bias-Aware Optimization for NBR. For the generic top-k recommendation task, re-ranking is a direct and effective method for improving beyond-accuracy performance, e.g., through greedy algorithms and constrained optimization [39]. The essence of re-ranking is to determine and exploit an effective trade-off between predicted relevance score and beyond-accuracy metrics. However, only optimizing diversity or item fairness may lead to increased repeat bias in some cases according to our experiments, i.e., to an increase in the deviation of the repeat ratio in recommended baskets vs. in ground truth baskets [35]. Even though previous research highlights the problem of repeat bias, no one has tried to optimize to mitigate it. To the best of our knowledge, we are the first to optimize repeat bias jointly with other metrics to seek a balance among (mitigating) repeat bias, accuracy, and beyond-accuracy metrics.

To solve this optimization problem, we propose a model-agnostic repeat-bias-aware diversity optimization (RADiv) algorithm and a repeat-bias-aware item fairness optimization (RAIF) algorithm based on *mixed-integer linear programming* (MILP) for NBR. The repeat ratio (or rather: reducing it) is one of the optimization objectives. The proposed algorithms optimize for predicted relevance, diversity (or item fairness), and repeat ratio simultaneously. We offer several flavors of our mixed-integer linear programming (MILP)-based optimization algorithm, to benefit from the peculiarities of different families of NBR methods. To the best of our knowledge, we are the first to apply a re-ranking algorithm to jointly optimize beyond-accuracy metrics and repeat bias in NBR.

Summarizing, the **main contributions** of the paper are:

- We propose repeat-bias-aware optimization algorithms (RADiv and RAIF) for NBR, which mitigate repeat bias while improving the diversity and item fairness of recommended baskets.
- We extend these algorithms to multiple NBR paradigms, including ones that merge and optimize items from separate repetition and exploration models.

– We conduct experiments on three retail datasets and find that the proposed RADiv and RAIF algorithms can significantly improve diversity, and item fairness and mitigate repeat bias with an acceptable loss in Recall.

2 Related Work

Beyond-Accuracy Objectives in Recommender System. In recommendation, optimizing only for accuracy measures is limiting and misguided [3]. There is a growing interest in beyond-accuracy metrics, which measure other recommendation qualities [30]. Kaminskas and Bridge [13] study the definitions and metrics of diversity, serendipity, novelty, and coverage. They implement re-ranking strategies for these beyond-accuracy metrics to investigate correlations between the two objectives.

Fairness of recommender systems has sparked much research in the community, considering multiple stakeholders of typical e-commerce platforms. From the user side, user fairness requires that a fair system should provide the same recommendation quality for different user groups [23]. From the item side, the goal is to measure the exposure assigned to each item, or each group, and evaluate this distribution to ensure fair principles, such as statistical parity or equal opportunity [29]. Usually, the exposure of an item in a ranked list is computed based on a user browsing model [6,26].

Beyond-Accuracy Objectives in NBR. In NBR, most deep-learning-based models predict the top-k relevant items via a user representation to form a basket. This paradigm leads to the problem of over-homogenization of the recommended baskets [33]. To address this problem, Sun et al. [33] apply an autoregressive item-level decoder to generate items one by one to ensure diversified baskets. Leng et al. [15] employ a deconvolutional network to generate diverse NBR results. Regarding item fairness, Liu et al. [25] reproduce a set of item fairness metrics to evaluate representative NBR methods. Li et al. [20] propose a framework to identify short-cuts in achieving better accuracy and beyond-accuracy performance and advocate fine-grained evaluations in NBR.

Apart from the work listed above, optimizing beyond-accuracy objectives in NBR remains unexplored. Regarding the imbalance between repeat and exploration recommendations, Li et al. [19] first define the repeat ratio of the recommended basket and point out the problem of repeat bias in NBR. Furthermore, Tran et al. [35] formulate the repeat bias as the deviation of the repeat ratio of the recommended baskets and ground truth baskets. We contribute by optimizing both diversity and item fairness through re-ranking, while also mitigating repeat bias to improve overall recommendation quality.

Fair/Diverse Re-ranking. Re-ranking algorithms are usually designed to adjust the ranked results obtained from information access systems considering beyond-accuracy objectives. Commonly used re-ranking algorithms are constrained optimization and maximal marginal relevance (MMR) [5]. For diverse re-ranking, Zhang and Hurley [38] summarize three classic patterns for improving the diversity of recommendation lists as constrained optimization problems: (i) maximize the diversity under the constraint of relevance tolerance; (ii) maximize relevance under the constraint of diversity tolerance; and (iii) maximize the weighted sum of relevance and diversity. In terms of fair re-ranking, Biega et al. [4] propose an online optimization approach that uses integer linear programming (ILP) for re-ranking based on accumulated attention and relevance scores while constraining according to a bound on the loss of ranking quality. Singh and Joachims [31] optimize a probabilistic ranking to maximize expected utility under three optional fairness constraints. For fair recommendation, Li et al. [23] set the user fairness metric as a constraint to reduce the recommendation quality gap between the advantaged and disadvantaged groups. CPFair [27] simultaneously optimizes consumer and producer fairness in the objective function for multiple rankings. Compared to previous work, which primarily focuses on fairness or diversity in isolation, we are the first to employ a MILP-based re-ranking algorithm to optimize diversity or item fairness, specifically accounting for repeat bias in NBR task.

3 Method

Our notation is summarized in Table 1.

3.1 Problem Formulation

In NBR, given a user set U and an item set I, for each user $u \in U$, the purchase history is represented as a sequence of baskets $[B_u^1, B_u^2, \dots, B_u^t]$, where each basket contains a set of items $B_u^t = \{i_1, i_2, \dots, i_m | i \in I\}$. In NBR the task is to predict the next basket B_u^{t+1} for each user. Following the common setting, the size of the recommended basket is fixed as K. Generally, NBR methods generate an item ranking $L(u)$ for each user based on the predicted relevance score S_{ui}, and then select the top-K relevant items to form the next basket, i.e., $B_u^{t+1} = L_K(u)$.

3.2 Model Overview

In the proposed re-ranking procedure, we select top-N relevant items $L_N(u)$ for each user as candidates. The relevance scores of candidate items of all users can be represented as $S = [S_{ui}]_{|U| \times N}$. A new basket $L'_K(u)$ is selected from the candidate list $L_N(u)$ taking into account the predicted relevance, as well as the diversity, item fairness, and repeat bias of the basket. $L'_K(u) \subset L_N(u), N > K$. The re-ranking procedure can be formalized as a MILP problem, which is a mathematical optimization and can be solved by heuristic algorithms and optimization solvers. In our MILP formulation, the objective function is designed to maximize relevance, diversity (or item fairness) and reduce repeat bias simultaneously. W is a binary decision matrix, $W = [W_{ui}]_{|U| \times N}$. The element $W_{ui} = 1$ indicates recommending item i to user u, and 0 otherwise. The proposed RADiv and RAIF algorithms are specifically adapted for different families of NBR methods: unified methods and combined methods.

3.3 Optimization Objectives

In this section, we introduce the definition of three optimization objectives that we aim to achieve.

Diversity. Diverse recommendation aims to recommend items of various and different categories to users. In this work, we use the diversity score (DS) [24] as the diversity objective, computed by dividing the number of recommended categories by the basket size K given the decision matrix W. Diversity objective $\mathrm{DS}(C, W)$ is the sum of DS among all users:

$$\mathrm{DS}(C, W) = \sum_{u \in U} \frac{(\#\text{recommended categories} \mid W)}{K}. \tag{1}$$

Item Fairness. We evaluate the item fairness between popular item group I_1 and unpopular group I_2. The recommended items will receive exposure related to the position. Demographic parity (DP) [31] defines fairness of exposure as equal average exposure between the two groups. The average exposure of a group I_k is computed as:

$$Exposure(I_k \mid W) = \frac{1}{|I_k|} \sum_{i \in I_k} Exposure(i \mid W). \tag{2}$$

Inspired by the construction in [27], the item fairness objective is designed as the difference of average exposure between group I_1 and I_2. The closer the IF value is to zero, the fairer a ranking is.

$$\mathrm{IF}(I_1, I_2, W) = Exposure(I_1 \mid W) - Exposure(I_2 \mid W). \tag{3}$$

Repeat Bias. Some NBR methods are either biased to recommending too many repeat items or explore items compared with the ground truth baskets. We aim to mitigate the repeat bias of recommended baskets. Since the repeat ratio of ground truth baskets $RepRatio_{gt}$ is unknown, we directly choose the repeat ratio of recommended baskets $RepRatio_{rec}$ as the optimization objective. Each user has a repeat item set I_u^{rep}, which contains all the items the user has bought before. Given the decision matrix W, the repeat ratio objective can be expressed as:

$$RepRatio_{rec}(I_u^{rep}, W) = \sum_{u \in U} \frac{(|L'_K(u) \cap I_u^{rep}| \mid W)}{K}. \tag{4}$$

3.4 Repeat-Bias-Aware Optimization for Unified NBR Methods

In this section, we post-process the recommendation lists of users generated by unified NBR methods, where a unified model generates the relevance scores of all items. We select top-N relevant items of each user as candidates. The relevance scores of these item candidates can be represented as $S = [S_{ui}]_{|U| \times N}$. We apply a MILP model to re-rank the top N candidates for each user. RADiv algorithm simultaneously optimizes diversity and repeat ratio as shown in Eq. 5:

$$\begin{aligned} &\max \frac{1}{K} \sum_{u \in U} \sum_{i=1}^{N} S_{ui} W_{ui} + \epsilon_1 \,\mathrm{DS}(C, W) - \lambda RepRatio_{rec}(I_u^{rep}, W) \\ &\text{such that } \sum_{i=1}^{N} W_{ui} = K, W_{ui} \in \{0, 1\}. \end{aligned} \tag{5}$$

Table 1. Notation used in the paper.

$u \in U$	Users	$i \in I$	Items
$RepRatio_{rec}$	Repeat ratio of recommendation	K	Basket size
$RepRatio_{gt}$	Repeat ratio of ground truth	N	Number of item candidates
W^r	Decision matrix of repeat items	C	Categories of items
W^e	Decision matrix of explore items	I_u^{rep}	Repeat item set of user u
S	Predicted relevance matrix	W	Decision matrix
S^r	Predicted relevance matrix of repeat items	I_1	Popular item group
S^e	Predicted relevance matrix of explore items	I_2	Unpopular item group
$L'_K(u)$	New basket of user u after re-ranking	$L_N(u)$	Item candidate list of user u
$H(\theta)$	Number of repeat items in combined baskets		

We design the objective function to maximize the sum of relevance scores, the diversity score $\mathrm{DS}(C, W)$, while minimizing the $RepRatio_{rec}(I_u^{rep}, W)$. Here, ϵ_1 and λ are the weighting parameters of the diversity term and repeat ratio term, respectively. W is a binary decision matrix determining whether to recommend item i to user u. The constraint indicates that the algorithm ultimately recommends K items to each user, which is the basket size. It is worth noting that the minus sign in front of $RepRatio_{rec}$ is designed for repeat-biased NBR methods with quite high $RepRatio_{rec}$. The minus implies that the algorithm aims to reduce the $RepRatio_{rec}$ to be closer to $RepRatio_{gt}$. In contrast, for explore-biased methods with quite low $RepRatio_{rec}$ values, this term should be set to a positive sign: $+\lambda RepRatio_{rec}(I_u^{rep}, W)$. The algorithm will increase the $RepRatio_{rec}$ so as to approach $RepRatio_{gt}$.

Similarly, RAIF simultaneously optimizes item fairness and repeat ratio by Eq. 6.

$$\begin{aligned} &\max \sum_{u \in U} \sum_{i=1}^{N} S_{ui} W_{ui} - \alpha_1 \, \mathrm{IF}(I_1, I_2, W) - \lambda RepRatio_{rec}(I_u^{rep}, W) \\ &\text{such that} \sum_{i=1}^{N} W_{ui} = K, W_{ui} \in \{0, 1\}. \end{aligned} \tag{6}$$

The objective function is designed to maximize the combination of relevance scores, item fairness $\mathrm{IF}(I_1, I_2, W)$, and minimize $RepRatio_{rec}(I_u^{rep}, W)$. Here, α_1 and λ are the weighting parameters of item fairness term and repeat ratio term, respectively.

3.5 Repeat-Bias-Aware Optimization for Combined NBR Methods

In the previous setting, the optimization algorithm is designed for unified NBR methods, where the relevance scores of repeat items and explore items are generated by the same model and are comparable. However, there is another NBR paradigm, such as two-step repetition-exploration (TREx) [20], where the repeat item list and explore item list are obtained from different models. This paradigm allows one to combine the strongest repetition model and the strongest exploration model flexibly, and adjust the proportion of these two parts. When we optimize diversity (or item fairness) and repeat bias of combined NBR methods, the challenge lies in the fact that the predicted relevance scores are not comparable between repeat items and explore items.

Inspired by the TREx framework [20], we use a threshold θ to filter the repeat item candidates and count the number of repeat items with a relevance score larger than θ. The numbers of repeat items for each user are saved as vector $H(\theta)_{\text{aux}}$. The final repeat position number for each user is $H(\theta) = \min(H(\theta)_{\text{aux}}, K)$. Here, θ indirectly controls $RepRatio_{rec}(\theta)$. The higher the θ becomes, the lower the number of repeat slots $H(\theta)$ and $RepRatio_{rec}(\theta)$ become. $H(\theta)$ is an important variable to avoid the comparison between repeat and explore relevance scores. In this way, repeat items are compared internally and $H(\theta)$ of them are finally selected. Explore items compete internally and $K - H(\theta)$ of them are chosen.

Algorithm 1: RADiv and RAIF for Combined NBR Methods

Input: Basket size K, Categories C, Number of item candidates N, Item group I_1, I_2, Parameters $\epsilon_2, \alpha_2, \theta$

Output: Recommendation matrix W^r, W^e

1. $S^r, S^e \leftarrow$ The top-N repeat and explore relevance scores of users.
2. Compare θ with S^r_{ui}, and obtain $H(\theta) = \min(H(\theta)_{\text{aux}}, K)$.
3. Solve the optimization problem following Eq. 7 (or 8).
4. **Return** W^r, W^e

The RADiv algorithm is adapted for combined NBR methods as shown in Eq. 7. $S^r = [S^r_{ui}]_{|U|\times N}$ and $S^e = [S^e_{ui}]_{|U|\times N}$ are predicted relevance score matrices of repeat items and explore items from different recommender systems, respectively. $W^r = [W^r_{ui}]_{|U|\times N}$, $W^e = [W^e_{ui}]_{|U|\times N}$ are corresponding binary decision matrices to determine the selection of repeat items and explore items. ϵ_2 is the weighting parameter of the diversity term. The constraints indicate the repeat slots for each user are $H(\theta)$, while the explore slots for each user are $K - H(\theta)$. Similarly, RAIF is adjusted to optimize item fairness $\text{IF}(I_1, I_2, W^r, W^e)$ and $RepRatio_{rec}(\theta)$ in Eq. 8. α_2 is the weighting parameter of item fairness term. Algorithm 1 summarizes our algorithms for combined NBR methods.

$$\begin{aligned}&\max \frac{1}{K}\sum_{u\in U}\sum_{i=1}^{N}(S^r_{ui}W^r_{ui}+S^e_{ui}W^e_{ui})+\epsilon_2\,\mathrm{DS}(C,W^r,W^e)\\&\text{such that}\sum_{i=1}^{N}W^r_{ui}=H(\theta),\sum_{i=1}^{N}W^e_{ui}=K-H(\theta),W^r_{ui},W^e_{ui}\in\{0,1\};\end{aligned} \tag{7}$$

$$\begin{aligned}&\max \sum_{u\in U}\sum_{i=1}^{N}(S^r_{ui}W^r_{ui}+S^e_{ui}W^e_{ui})-\alpha_2\,\mathrm{IF}(I_1,I_2,W^r,W^e)\\&\text{such that}\sum_{i=1}^{N}W^r_{ui}=H(\theta),\sum_{i=1}^{N}W^e_{ui}=K-H(\theta),W^r_{ui},W^e_{ui}\in\{0,1\}.\end{aligned} \tag{8}$$

4 Experimental Setup

NBR Methods. We select and investigate the following 5 representative NBR methods:

- **UP-CF@r**, which combines recency-aware user-wise popularity and collaborative filtering while considering the recent shopping behavior [8].
- **TIFUKNN**, which models the temporal dynamics of users' past baskets by using a KNN-based approach based on the personalized item frequency (PIF) [11].
- **Dream**, which forms basket representations using a pooling strategy and models sequential user behavior through an RNN [36].
- **DNNTSP** uses GNN and self-attention mechanisms to encode item-to-item relations across baskets and capture temporal dependencies [37].
- **TREx**, which has a repetition module considering item repurchase features and users' interests, and an exploration module targeted for beyond-accuracy metrics. Then, repeat items and explore items are combined to form the final basket [20].

We exclude NBR methods that only focus on the repetition recommendation (i.e., P-TopFreq, ReCANet [1], and NBRR [14]) or exploration recommendation (i.e., NNBR [17]), as we aim to investigate and balance repetition *and* exploration in NBR.

Datasets. Following previous NBR studies [17,19,20,25], we select three publicly available grocery shopping datasets: TaFeng [34], Dunnhumby [7], and Instacart [12]. They exhibit different characteristics in terms of repetition and exploration, which are critical for verifying the effectiveness of our proposed repeat-bias-aware optimization algorithms.

For each dataset, we remove users with fewer than three baskets and items purchased fewer than five times, as done in [1]. Due to the large size of the Instacart dataset, memory limitations occurred during the calculation of some methods. Following [28], we also randomly sampled 20,000 users from Instacart before applying any filtering. Table 2 provides the statistics of the three datasets after preprocessing. The average $RepRatio_{gt}$ refers to the average proportion of repeat items in the ground truth baskets, as defined in [19].

We split each dataset following the approach in [1,8,28]. The training set includes all baskets from each user except the last basket. For users with more than 50 baskets in the training data, we limit the training set to their most recent 50 baskets [19]. The last baskets of all users are then randomly split equally into a validation set (50%) and a test set (50%).

Table 2. Statistics of the datasets after preprocessing.

Dataset	#Users	#Items	#Baskets	Avg. #baskets/user	Avg. #items/basket	Avg. $RepRatio_{gt}$
Instacart	19,210	29,399	305,582	15.91	10.06	0.60
Dunnhumby	2,482	37,162	107,152	43.17	10.07	0.43
TaFeng	10,182	15,024	82,387	8.09	6.14	0.21

Evaluation Metrics. We use the following widely used metrics in our experiments to measure how the proposed algorithms balance multiple objectives. In terms of accuracy, Recall measures the system's ability to retrieve items that users will purchase in their next baskets. For item fairness, demographic parity (DP) measures the ratio of the average exposure of the popular group to the average exposure of the unpopular group. Following [29], we use logDP to deal with the empty-group case. The closer the logDP value is to zero, the fairer the recommendation is. The diversity score (DS) measures the number of categories within the recommended basket divided by basket size. Repeat bias is computed as: RepBias = $RepRatio_{rec} - RepRatio_{gt}$ [35]. Following [27], we use a comprehensive metric to evaluate the overall performance of item fairness and RepBias: mFR = $\omega|\text{logDP}| + (1-\omega)|\text{RepBias}|$. Similarly, the overall performance of diversity and RepBias is evaluated as mDR $= \omega\text{DS} - (1-\omega)|\text{RepBias}|$.

Implementation Details. In our experiments, we follow [11,20] and set basket size $K = 20$. The number of item candidates $N = 100$. We compute the popularity of all items and select the top 20% as popular I_1 and the rest belong to the unpopular group I_2. We apply grid search to select hyperparameters on the validation set. We select α_1 and α_2 in [0, 0.001, 0.01, 0.1, 1, 10, 20, 30, 40, 50, 60, 70, 80, 90, 100, 200]. ϵ_1 and select ϵ_2 from [0, 0.001, 0.01, 0.02, 0.04, 0.06, 0.08, 0.1, 0.12, 0.14, 0.16, 0.18, 0.2]. We choose λ in [0, 0.001, 0.01, 0.1, 0.2, 0.3, 0.4, 0.5, 0.6, 0.7, 0.8, 0.9, 1]. θ candidates are selected based on S^r.

We select the optimal hyperparameter combination according to these two conditions: (i) satisfy a 10% Recall drop tolerance; and (ii) maximize mDR or minimize mFR, where $\omega = 0.5$ following [27]. We use Gurobi,[1] which is an industrial optimization solver capable of delivering practical and effective feasible solutions. We release our code and hyperparameters at https://github.com/lynEcho/Repbias_NBR.

5 Experimental Results

We design experiments to answer three research questions: **RQ1** How does optimizing only for diversity or item fairness affect the repeat bias? **RQ2** What is the performance

[1] https://www.gurobi.com.

of proposed RADiv and RAIF algorithms on NBR methods? **RQ3** How does RADiv and RAIF algorithms strike a balance between utility and beyond-accuracy metrics?

We start by studying how optimizing for beyond-accuracy metrics, without taking the repeat bias, would affect the baskets in terms of repeat-explore items, answering RQ1. We apply a re-ranking algorithm without considering $RepRatio_{rec}$ to optimize the recommended baskets obtained from different NBR methods on Instacart, Dunnhumby, and TaFeng. For diverse re-ranking, the optimization objective is formulated as $\max \frac{1}{K} \sum_{u \in U} \sum_{i=1}^{N} S_{ui} W_{ui} + \epsilon \, \mathrm{DS}(C, W)$. In terms of fair re-ranking, the objective function is $\max \sum_{u \in U} \sum_{i=1}^{N} S_{ui} W_{ui} - \alpha \, \mathrm{IF}(I_1, I_2, W)$. The parameters ϵ and α are used to adjust the weight of the diversity and item fairness terms, respectively. Figure 1 shows the performance of diversity and item fairness optimization on Dunnhumby.[2]

We make the following observation: (i) Diversity and item fairness optimizations significantly improve the NBR methods in terms of logDP and DS. For item fairness optimization, as the parameter ϵ increases ($0 \rightarrow 200$), all NBR methods become fairer. Also, the DS of all NBR methods increase as α increases ($0 \rightarrow 0.2$). (ii) In most cases, Recall exhibits a declining trend as DS and logDP become better. In diversity optimization, the Recall of UP-CF@r, Dream, and DNNTSP improve slightly when $\alpha = 0.01$. (iii) In terms of repeat bias, which is measured as the gap between $RepRatio_{rec}$ and $RepRatio_{gt}$ (green dashed line in Fig. 1). We see that the repeat bias is likely to either intensify or diminish, showing no clear trend. In item fairness optimization, the repeat bias of TIFUKNN and Dream grows as ϵ increases, showing that it is important to consider repeat bias while optimizing beyond-accuracy metrics.

To answer RQ2, we apply RADiv and RAIF algorithms on five NBR methods. Of these, TIFUKNN, UP-CF@r, DNNTSP, and Dream are unified repeat-explore recommendation methods while TREx is a combined method. Since TREx does not provide predicted relevance scores for explore items, here we use the explore item list obtained from UP-CF@r to combine with repeat items of TREx. Table 3 reports the results on Instacart.[3] We also conduct an ablation study, i.e., optimizing only for diversity, item fairness, and repeat bias, respectively.

We arrive at the following conclusions: (i) The proposed RADiv algorithm (RD) achieves a significant improvement in DS, RepBias, and comprehensive metric mDR in comparison to the original (Ori.) results on all NBR methods. Surprisingly, the Recall of Dream increases after RADiv optimization ($0.0977 \rightarrow 0.1468$). (ii) The RAIF algorithm (RF) achieves better logDP, less RepBias, and better overall metric mFR than the original (Ori.) results, proving the effectiveness of RAIF algorithm. (iii) RADiv and RAIF algorithms can effectively adjust the $RepRatio_{rec}$ of methods to approach $RepRatio_{gt}$. Take Instacart as an example, RADiv and RAIF algorithms decrease the $RepRatio_{rec}$ of TIFUKNN, UP-CF@r, DNNTSP, and increase the $RepRatio_{rec}$ of Dream to make them closer to ground truth 0.6. (iv) In our ablation study, we find that only optimizing $RepRatio_{rec}$ (R) will decrease the diversity and item fairness of

[2] We observe a similar trend on the Instacart and TaFeng datasets. Because of space limitations, we report the results on Instacart and TaFeng in the anonymous repository.

[3] Similar patterns are observed on the Dunnhumby and TaFeng datasets. Due to space limitations, we report the results on Dunnhumby and TaFeng in the anonymous repository.

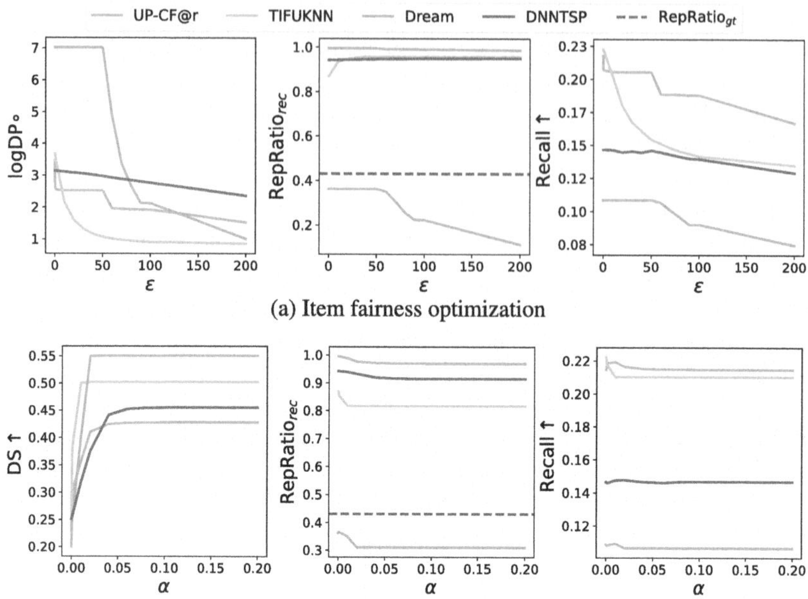

(a) Item fairness optimization

(b) Diversity optimization

Fig. 1. Performance of item fairness and diversity optimization without considering $RepRatio_{rec}$ on different NBR methods. logDP∘ means the closer logDP value is to zero, the fairer it is. (Color figure online)

TIFUKNN, UC-CF@r, DNNTSP and TREx. This means that it is necessary to optimize both $RepRatio_{rec}$ and diversity, both $RepRatio_{rec}$ and item fairness, as implemented by the proposed RADiv and RAIF algorithms.

To answer RQ3, we take UP-CF@r as an example to illustrate the trade-off between Recall and other metrics on Instacart. For the two parameters ϵ_1 and λ in the RADiv optimization, we change one parameter each time with the other one fixed. Similarly, we perform the same operation for α_1 and λ in RAIF optimization. The analysis is shown in Fig. 2.

We make the following observations: (i) In Fig. 2a, we first fix λ and change ϵ_1 from 0 to 0.2. We see that DS continuously increases at the cost of Recall. The red star indicates that the algorithm chooses $\epsilon_1 = 0.2$ as the optimal solution to achieve a balance between diversity and Recall. When we fix ϵ_1 and change λ from 0 to 1, $RepRatio_{rec}$ and Recall decline consistently. Since our algorithm optimizes $RepRatio_{rec}$ to be approaching $RepRatio_{gt}$, it selects $\lambda = 0.01$ as the optimal trade-off between Recall and repeat bias. This process reflects the balanced strategy of RADiv to maximize diversity and mitigate repeat bias within the tolerance of utility loss. (ii) In Fig. 2b, we fix λ and adjust α_1 from 0 to 200. logDP decreases indicating unpopular items are assigned with more exposure. The recommendation becomes fairer at the cost of a Recall drop. The algorithm chooses $\alpha_1 = 200$ to achieve a balance between logDP and Recall. When we change λ from 0 to 1 with a fixed α_1, $RepRatio_{rec}$ and Recall show a consistent

Table 3. The performance of RADiv and RAIF algorithms on Instacart ($RepRatio_{gt} = 0.6$). *Ori.* indicates the original baskets obtained from each method. *RD* indicates RADiv. *RF* refers to RAIF. In ablation study, *D*, *F*, *R* indicate optimizing only for diversity, item fairness, and repeat bias, respectively. RepR indicates $RepRatio_{rec}$ (best close to $RepRatio_{gt}$). RepBias and logDP (best close to zero).

Meth.	Diversity optimization						Item fairness optimization					
	Type	Recall↑	DS↑	RepR	RepBias	mDR↑	Type	Recall↑	logDP	RepR	RepBias	mFR↓
TIFUKNN	Ori.	**0.4559**	0.3615	0.9248	0.3248	0.0184	Ori.	**0.4559**	3.1252	0.9248	0.3248	1.7250
	D	0.4259	0.5897	0.8984	0.2984	0.1457	F	0.4269	2.3510	0.9252	0.3252	1.3381
	R	0.4537	0.3587	0.9100	0.3100	0.0244	R	0.4232	3.2760	**0.7939**	**0.1939**	1.7350
	RD	0.4245	**0.5898**	**0.8874**	**0.2874**	**0.1512**	RF	0.4098	**2.3271**	0.8718	0.2718	**1.2995**
UP-CF@r	Ori.	**0.4405**	0.3489	0.8905	0.2905	0.0292	Ori.	**0.4405**	3.3966	0.8905	0.2905	1.8436
	D	0.3983	**0.6375**	0.7896	0.1896	0.2239	F	0.4282	2.4860	0.8858	0.2858	1.3859
	R	0.4353	0.3424	0.8812	0.2812	0.0306	R	0.4373	3.3997	0.8791	0.2791	1.8394
	RD	0.3968	**0.6375**	**0.7837**	**0.1837**	**0.2269**	RF	0.4052	**2.1810**	**0.8019**	**0.2019**	**1.1915**
DNNTSP	Ori.	**0.4347**	0.3402	0.9133	0.3133	0.0135	Ori.	**0.4347**	3.2573	0.9133	0.3133	1.7853
	D	0.4046	0.6009	0.8660	0.2660	0.1675	F	0.4332	**2.9780**	0.9152	0.3152	1.6466
	R	0.4337	0.3388	0.9080	0.3080	0.0154	R	0.4256	3.3204	**0.8640**	**0.2640**	1.7922
	RD	0.4059	**0.6013**	**0.8623**	**0.2623**	**0.1695**	RF	0.4277	2.9901	0.8837	0.2837	**1.6369**
Dream	Ori.	0.0977	0.1000	0.1923	−0.4077	−0.1539	Ori.	0.0977	7.3111	0.1923	−0.4077	3.8594
	D	0.0723	**0.7000**	0.1267	−0.4733	0.1133	F	0.0709	**2.1018**	0.1226	−0.4774	**1.2896**
	R	**0.1540**	0.1243	**0.4202**	**−0.1798**	−0.0277	R	**0.1288**	7.3111	**0.3108**	**−0.2892**	3.8002
	RD	0.1468	0.5654	**0.4202**	**−0.1798**	**0.1928**	RF	0.0939	2.3893	0.2094	−0.3906	1.3900
TREx	Ori.	**0.4595**	0.3533	0.9265	0.3265	0.0134	Ori.	**0.4595**	3.2182	0.9265	0.3265	1.7724
	D	0.4455	0.5296	0.9265	0.3265	0.1016	F	0.4569	2.9052	0.9265	0.3265	1.6159
	R	0.4368	0.3280	**0.7973**	**0.1973**	0.0654	R	0.4277	3.4281	**0.7487**	**0.1487**	1.7884
	RD	0.4195	**0.5874**	**0.7973**	**0.1973**	**0.1951**	RF	0.4226	**2.6528**	**0.7487**	**0.1487**	**1.4008**

downward trend. The algorithm selects $\lambda = 0.9$ to balance repeat bias and Recall. Considering the inverse relationship between Recall and logDP, repeat bias, the RAIF algorithm decides to sacrifice a little utility in exchange for a greater return on item fairness and repeat bias.

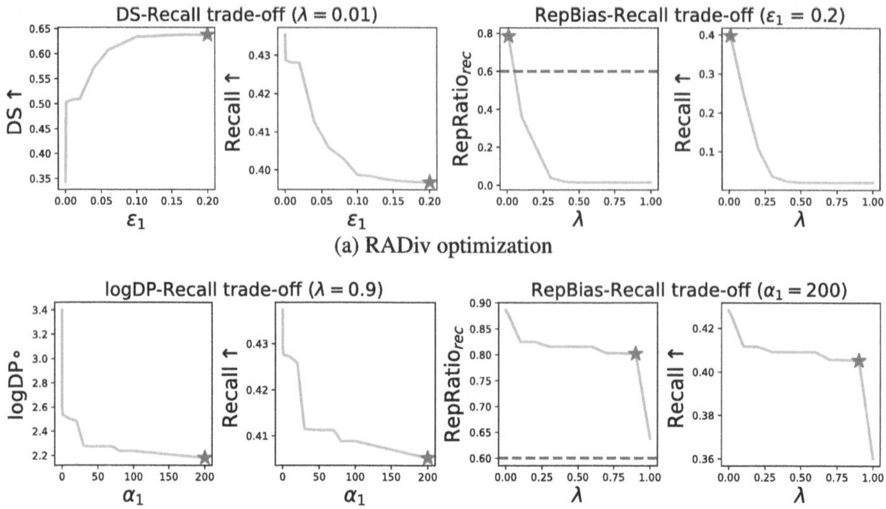

(b) RAIF optimization

Fig. 2. Trade-off between Recall and diversity, item fairness, repeat bias. Take UP-CF@r as an example on the Instacart dataset. The red star indicates the optimal solution chosen by the proposed algorithm. The green dashed line is value of $RepRatio_{gt}$. (Color figure online)

6 Conclusion

We have proposed repeat-bias-aware optimization algorithms for improving diversity and item fairness while mitigating repeat bias in NBR. Our RADiv and RAIF algorithms re-rank the preliminary recommended results obtained from various NBR baselines according to different beyond-accuracy objectives and seek a balance between repeat items and explore items simultaneously while optimizing for beyond-accuracy metrics. We have extended our approach to multiple NBR paradigms [20], in particular, one that fuses repeat item lists and explores item lists from separate models, allowing us to select tailored models for each list. We have conducted experiments on three real-world retail datasets. We find that only optimizing for diversity or item fairness will increase repeat bias in many cases, which can reduce user satisfaction. The proposed RADiv and RAIF algorithms can effectively optimize diversity and fairness while mitigating the repeat bias issue under an acceptable utility loss. Finally, we have investigated trade-offs between Recall and other beyond-accuracy metrics, including diversity, item fairness, and repeat bias.

Based on our experiments, we find that it is critical to evaluate the utility of NBR methods and measure their repeat bias at the same time. In many cases, a recommendation model can achieve the best performance in terms of Recall by recommending only repeat items (i.e., highest repeat bias), but this reduces the likelihood of novelty and serendipity in recommendations, leading to filter bubble and echo chamber issues [9, 10]. Re-ranking may be an effective solution to this problem, as we have shown that it can strike a balance between accuracy and beyond-accuracy metrics, while taking into account the repeat bias in the re-ranking process. However, in extreme cases where the original ranking does not include many explore items, re-ranking cannot really be effective. Therefore, we suggest to consider accuracy and beyond-accuracy objectives in the original ranking. In addition, the definition and quantification of repeat bias is also a direction worth exploring in the future.

Acknowledgments. This research was (partially) supported by the Dutch Research Council (NWO), under project numbers 024.004.022, NWA.1389.20.183, and KICH3.LTP.20.006, and the European Union's Horizon Europe program under grant agreement No 101070212.

All content represents the opinion of the authors, which is not necessarily shared or endorsed by their respective employers and/or sponsors.

References

1. Ariannezhad, M., Jullien, S., Li, M., Fang, M., Schelter, S., de Rijke, M.: ReCANet: a repeat consumption-aware neural network for next basket recommendation in grocery shopping. In: Proceedings of the 45th International ACM SIGIR Conference on Research and Development in Information Retrieval, pp. 1240–1250 (2022)
2. Ariannezhad, M., Li, M., Jullien, S., de Rijke, M.: Complex item set recommendation. In: Proceedings of the 46th International ACM SIGIR Conference on Research and Development in Information Retrieval, pp. 3444–3447 (2023)
3. Bauer, C., Bagchi, C., Hundogan, O.A., van Es, K.: Where are the values? A systematic literature review on news recommender systems. ACM Trans. Recomm. Syst. (2024)
4. Biega, A.J., Gummadi, K.P., Weikum, G.: Equity of attention: amortizing individual fairness in rankings. In: The 41st International ACM SIGIR Conference on Research & Development in Information Retrieval, pp. 405–414 (2018)
5. Carbonell, J., Goldstein, J.: The use of MMR, diversity-based reranking for reordering documents and producing summaries. In: Proceedings of the 21st Annual International ACM SIGIR Conference on Research and Development in Information Retrieval, pp. 335–336 (1998)
6. Chapelle, O., Metlzer, D., Zhang, Y., Grinspan, P.: Expected reciprocal rank for graded relevance. In: Proceedings of the 18th ACM Conference on Information and Knowledge Management, pp. 621–630 (2009)
7. Dunnhumby: Source files (2024). https://www.dunnhumby.com/source-files/
8. Faggioli, G., Polato, M., Aiolli, F.: Recency aware collaborative filtering for next basket recommendation. In: Proceedings of the 28th ACM Conference on User Modeling, Adaptation and Personalization, pp. 80–87 (2020)
9. Flaxman, S., Goel, S., Rao, J.M.: Filter bubbles, echo chambers, and online news consumption. Public Opin. Q. **80**, 298–320 (2016)
10. Ge, Y., et al.: Understanding echo chambers in e-commerce recommender systems. In: Proceedings of the 43rd International ACM SIGIR Conference on Research and Development in Information Retrieval, pp. 2261–2270 (2020)
11. Hu, H., He, X., Gao, J., Zhang, Z.L.: Modeling personalized item frequency information for next-basket recommendation. In: Proceedings of the 43rd International ACM SIGIR Conference on Research and Development in Information Retrieval, pp. 1071–1080 (2020)
12. Instacart: Market basket analysis (2017). https://www.kaggle.com/c/instacart-market-basket-analysis/data

13. Kaminskas, M., Bridge, D.: Diversity, serendipity, novelty, and coverage: a survey and empirical analysis of beyond-accuracy objectives in recommender systems. ACM Trans. Interact. Intell. Syst. (TiiS) **7**(1), 1–42 (2016)
14. Katz, O., Barkan, O., Koenigstein, N., Zabari, N.: Learning to ride a buy-cycle: a hyper-convolutional model for next basket repurchase recommendation. In: Proceedings of the 16th ACM Conference on Recommender Systems, pp. 316–326 (2022)
15. Leng, Y., Yu, L., Xiong, J., Xu, G.: Recurrent convolution basket map for diversity next-basket recommendation. In: International Conference on Database Systems for Advanced Applications, pp. 638–653. Springer, Cham (2020)
16. Li, M., Ariannezhad, M., Yates, A., De Rijke, M.: Who will purchase this item next? Reverse next period recommendation in grocery shopping. ACM Trans. Recomm. Syst. **1**(2), 1–32 (2023)
17. Li, M., Ariannezhad, M., Yates, A., de Rijke, M.: Masked and swapped sequence modeling for next novel basket recommendation in grocery shopping. In: Proceedings of the 17th ACM Conference on Recommender Systems, pp. 35–46 (2023)
18. Li, M., Huang, J., de Rijke, M.: Repetition and exploration in offline reinforcement learning-based recommendations. In: Proceedings of the 4th Workshop on Deep Reinforcement Learning for Information Retrieval at CIKM. ACM (2023)
19. Li, M., Jullien, S., Ariannezhad, M., de Rijke, M.: A next basket recommendation reality check. ACM Trans. Inf. Syst. **41**(4), 1–29 (2023)
20. Li, M., et al.: Are we really achieving better beyond-accuracy performance in next basket recommendation? In: Proceedings of the 47th International ACM SIGIR Conference on Research and Development in Information Retrieval, pp. 924–934 (2024)
21. Li, M., Vardasbi, A., Yates, A., de Rijke, M.: Repetition and exploration in sequential recommendation. In: Proceedings of the 46th International ACM SIGIR Conference on Research and Development in Information Retrieval, pp. 2532–2541. ACM (2023)
22. Li, R., Zhang, L., Liu, G., Wu, J.: Next basket recommendation with intent-aware hypergraph adversarial network. In: Proceedings of the 46th International ACM SIGIR Conference on Research and Development in Information Retrieval, pp. 1303–1312 (2023)
23. Li, Y., Chen, H., Fu, Z., Ge, Y., Zhang, Y.: User-oriented fairness in recommendation. In: Proceedings of the Web Conference 2021, pp. 624–632 (2021)
24. Liang, Y., Qian, T., Li, Q., Yin, H.: Enhancing domain-level and user-level adaptivity in diversified recommendation. In: Proceedings of the 44th International ACM SIGIR Conference on Research and Development in Information Retrieval, pp. 747–756 (2021)
25. Liu, Y., Li, M., Ariannezhad, M., Mansoury, M., Aliannejadi, M., de Rijke, M.: Measuring item fairness in next basket recommendation: a reproducibility study. In: European Conference on Information Retrieval, pp. 210–225. Springer, Cham (2024)
26. Moffat, A., Zobel, J.: Rank-biased precision for measurement of retrieval effectiveness. ACM Trans. Inf. Syst. (TOIS) **27**(1), 1–27 (2008)
27. Naghiaei, M., Rahmani, H.A., Deldjoo, Y.: CPFair: personalized consumer and producer fairness re-ranking for recommender systems. In: Proceedings of the 45th International ACM SIGIR Conference on Research and Development in Information Retrieval, pp. 770–779 (2022)
28. Naumov, S., Ananyeva, M., Lashinin, O., Kolesnikov, S., Ignatov, D.I.: Time-dependent next-basket recommendations. In: European Conference on Information Retrieval, pp. 502–511. Springer, Cham (2023)
29. Raj, A., Ekstrand, M.D.: Measuring fairness in ranked results: an analytical and empirical comparison. In: Proceedings of the 45th International ACM SIGIR Conference on Research and Development in Information Retrieval, pp. 726–736 (2022)
30. de Rijke, M.: Beyond accuracy goals, again. In: WSDM 2023: The Sixteenth International Conference on Web Search and Data Mining, pp. 2–3. ACM (2023)

31. Singh, A., Joachims, T.: Fairness of exposure in rankings. In: Proceedings of the 24th ACM SIGKDD International Conference on Knowledge Discovery & Data Mining, pp. 2219–2228 (2018)
32. Su, Y., et al.: Long-term value of exploration: measurements, findings and algorithms. In: Proceedings of the 17th ACM International Conference on Web Search and Data Mining, pp. 636–644 (2024)
33. Sun, W., Xie, R., Zhang, J., Zhao, W.X., Lin, L., Wen, J.R.: Generative next-basket recommendation. In: Proceedings of the 17th ACM Conference on Recommender Systems, pp. 737–743 (2023)
34. TaFeng: Grocery dataset (2001). https://www.kaggle.com/datasets/chiranjivdas09/ta-feng-grocery-dataset
35. Tran, V.A., Salha-Galvan, G., Sguerra, B., Hennequin, R.: Transformers meet ACT-R: repeat-aware and sequential listening session recommendation. In: RecSys2024 (2024)
36. Yu, F., Liu, Q., Wu, S., Wang, L., Tan, T.: A dynamic recurrent model for next basket recommendation. In: Proceedings of the 39th International ACM SIGIR Conference on Research and Development in Information Retrieval, pp. 729–732 (2016)
37. Yu, L., Sun, L., Du, B., Liu, C., Xiong, H., Lv, W.: Predicting temporal sets with deep neural networks. In: Proceedings of the 26th ACM SIGKDD International Conference on Knowledge Discovery & Data Mining, pp. 1083–1091 (2020)
38. Zhang, M., Hurley, N.: Avoiding monotony: improving the diversity of recommendation lists. In: Proceedings of the 2008 ACM Conference on Recommender Systems, pp. 123–130 (2008)
39. Zhao, Y., Wang, Y., Liu, Y., Cheng, X., Aggarwal, C.C., Derr, T.: Fairness and diversity in recommender systems: a survey. ACM Trans. Intell. Syst. Technol. (2024)

Guiding Retrieval Using LLM-Based Listwise Rankers

Mandeep Rathee[1(✉)], Sean MacAvaney[2], and Avishek Anand[3]

[1] L3S Research Center, Hannover, Germany
rathee@l3s.de
[2] University of Glasgow, Glasgow, UK
sean.macavaney@glasgow.ac.uk
[3] Delft University of Technology, Delft, The Netherlands
avishek.anand@tudelft.nl

Abstract. Large Language Models (LLMs) have shown strong promise as rerankers, especially in "listwise" settings where an LLM is prompted to rerank several search results at once. However, this "cascading" retrieve-and-rerank approach is limited by the bounded recall problem: relevant documents not retrieved initially are permanently excluded from the final ranking. Adaptive retrieval techniques address this problem, but do not work with listwise rerankers because they assume a document's score is computed independently from other documents. In this paper, we propose an adaptation of an existing adaptive retrieval method that supports the listwise setting and helps guide the retrieval process itself (thereby overcoming the bounded recall problem for LLM rerankers). Specifically, our proposed algorithm merges results both from the initial ranking and feedback documents provided by the most relevant documents seen up to that point. Through extensive experiments across diverse LLM rerankers, first stage retrievers, and feedback sources, we demonstrate that our method can improve nDCG@10 by up to 13.23% and recall by 28.02%–all while keeping the total number of LLM inferences constant and overheads due to the adaptive process minimal. The work opens the door to leveraging LLM-based search in settings where the initial pool of results is limited, e.g., by legacy systems, or by the cost of deploying a semantic first-stage.

Keywords: Reranking · Adaptive Retrieval · LLM

1 Introduction

One of the most prevalent approaches in document search systems is the two-stage retrieve-and-rerank ("cascading") pipeline, where an initial retrieval phase is followed by a more involved reranker [18]. In this paradigm, the results from a retrieval model are re-ordered by a reranker—a slower but more capable relevance model—based on their estimated relevance to the query. For example, LLM-based rankers have recently gained prominence due to their ability to

C. Hauff et al. (Eds.): ECIR 2025, LNCS 15572, pp. 230–246, 2025.
https://doi.org/10.1007/978-3-031-88708-6_15

understand and process complex language patterns, often performing effectively in listwise reranking settings where the goal is to produce an ordered list of documents without explicit relevance scores for individual documents [10,26,27,34]. However, a significant limitation of reranking systems is the *bounded recall problem*. Specifically, if a relevant document fails to be retrieved in the initial retrieval phase, it is irrevocably excluded from the final ranked list, regardless of its potential relevance. This constraint limits the overall effectiveness of cascading systems, as the presence of relevant documents in the retrieved set is a prerequisite for successful reranking.

Recently, Adaptive Retrieval (AR) has emerged as a promising solution to mitigate the bounded recall problem [12]. Unlike conventional reranking, which solely reorders the initially retrieved documents, AR dynamically expands the retrieval set by leveraging the reranker's assessments during the reranking process itself. Instead of reranking all retrieved documents indiscriminately, AR applies the Clustering Hypothesis, which suggests that similar documents tend to be relevant to the same query [5], to selectively identify unretrieved documents that are similar to those identified as highly relevant by the reranker. AR is a form of Pseudo-Relevance Feedback (PRF), though unlike traditional PRF that applies query rewriting, the query remains unchanged and the index is not re-queried. AR facilitates this property by performing lookups in a corpus graph that encodes the top-k similarity relationships between documents. By doing so, AR enhances the likelihood of including relevant documents that may have been omitted in the initial retrieval, thereby improving the overall recall of the system. By constructing the corpus graph offline, overheads of applying AR (as compared to traditional reranking) are minimal, since lookups can be conducted in a constant and minimal time.[1]

Previous studies have demonstrated that AR consistently enhances recall across a variety of retrieval and reranking models [7,11,12], underscoring its potential as a robust enhancement to traditional ranking pipelines. However, existing work in AR made the Probably Ranking Principle (PRP [30]) assumption that a relevance score is a function solely of the query and document. LLM reranking models use signals from an entire ranked list to determine relevance, breaking the PRP assumption. Therefore, in this work, we build on the foundations of AR to explore their applicability to listwise Large Language Model (LLM) rerankers.

The primary challenge in integrating AR with LLM rerankers lies in the fundamental difference between pointwise and listwise ranking approaches. So far, AR has been designed for pointwise ranking methods that score each document independently based on its relevance to the query like BERT, or MonoT5 [6,23]. In contrast, LLM rerankers excel in listwise settings, where the ranking function considers the relative order of documents as a whole rather than assigning individual scores. This discrepancy necessitates the adaptation of AR to accommodate the listwise nature of LLM rerankers.

[1] In contrast, traditional PRF methods, such as RM3 [1], rewrite and re-execute the query. This is both a costly process and one that can lead to query drift.

In this paper, we make several key contributions to address this challenge:

- We propose SlideGar to modify AR for compatibility with ranking functions that output only a ranked order of documents, rather than discrete relevance scores.
- We conduct extensive experiments across a diverse array of LLM rankers and substrates to evaluate the effectiveness and efficiency of AR in enhancing LLM-based reranking performance. Our experimental results show that SlideGar can improve the nDCG@10 by up to 13.23% and recall by 28.02% over the state-of-the-art listwise rankers.
- To further understand the computational overheads, we perform ablation studies and find that SlideGar adds negligible latency cost (only 0.02% of rankers latency) in the re-ranking pipelines.

Our findings reveal that integrating AR with LLM rankers improves recall and nDCG scores across most corpus graphs and LLM configurations. These results affirm the viability of AR as a valuable enhancement to LLM-based reranking systems, offering a pathway to more effective document ranking, even in settings where the first stage is limited (e.g., due to legacy systems or due to the cost of deploying a semantic first stage.). We release our code at https://github.com/Mandeep-Rathee/llmgar.

2 Related Work

2.1 Retrieval and Ranking

Cascading approaches have a long history in information access systems [18]. In retrieval systems, cascading often manifests itself as a retrieve-then-rerank pipeline. Sparse retrieval algorithms (such as BlockMax-WAND [4]) and dense retrieval algorithms (such as HNSW [16]) can produce an initial set of (potentially approximate) top-k search results in sublinear time with respect to the size of the corpus, allowing them to scale to massive corpora. These initial retrieval results are limited by the representation bottleneck (e.g., the terms matched in a bag-of-words representation), and therefore often benefit from an additional iteration of refinement before being presented to a user—often through a cross-encoding neural model, such as one based on BERT [23] or an LLM [34]. By considering the content of the query and document jointly, they can often produce a higher-quality relevance ranking. These models are not capable of scaling to large corpora, given that they operate in linear (or superlinear) time (usually with high constant factors). The central limitation of cascading approaches is the bounded recall problem; documents filtered out in previous stages have no chance of being seen by the reranker. In the context of search systems, this has spawned substantial research into stronger first-stage retrieval systems, such as those that leverage dense [8] or learned sparse vectors [21]. Despite considerable progress in this space, these methods still struggle with certain queries [36] due to their representation bottlenecks. Furthermore, such systems often depend on an entirely new core engine, which can be operationally challenging when existing infrastructure has been built around a lexical-based engine.

2.2 Adaptive Retrieval

To overcome the recall limitations of cascading approaches, adaptive reranking [12] techniques have been proposed. These methods utilize an index structure based on document-document similarities and leverage relevance estimations obtained during the reranking process to retrieve and score documents that were not captured in the initial retrieval stage. Drawing inspiration from the Cluster Hypothesis [5], adaptive reranking posits that documents in close proximity to those with high relevance scores are also likely to be relevant; by pulling in these similar documents when reranking, recall limitations of the first stage can be mitigated. By using a document-to-document similarity graph ("corpus graph", akin to HNSW [16]), this process can be accomplished with minimal overheads.

Several strategies for adaptive reranking have been proposed. For example, GAR [12] and QUAM [29] employ an alternating strategy that scores batches of documents from both the initial retrieval pool and those sourced from the corpus graph. Beyond cross-encoders, adaptive reranking techniques have also been successfully applied to bi-encoders [7,11] and ensemble models [37]. For instance, LADR [7] uses an efficient lexical model to identify promising "seed" documents, which are then used to explore the corpus graph further. AR is inherently designed for pointwise ranking methods. In this work we extend AR to listwise settings, where the ranking function considers the relative order of documents as a whole rather than assigning individual scores.

2.3 Pseudo-Relevance Feedback (PRF)

Another line of work, pseudo-relevance feedback (PRF), tries to improve recall by providing feedback from the top-k initial retrieved documents. Examples of methods based on this approach include Rocchio [32], KL expansion [38], relevance modeling [19], LCE [20], and RM3 expansion [1]. These methods result in the *query drift* problem. Recent works [14,15] have proposed Generative Relevance Feedback (GRF). The goal is to enhance the original query by using an LLM. The expanded query can incorporate the original terms along with useful terms generated from various subtasks, such as keywords, entities, chain-of-thought reasoning, facts, news articles, documents, and essays. However, such methods of expanding LLM-based queries are costly, since the number of output tokens is high.

2.4 LLM Rerankers

Recent works, such as [33,34], have proposed **RankGPT**, zero-shot LLMs for document ranking task. The zero-shot LLM-based ranker shows significant improvements in ranking tasks over the various benchmark datasets. The LLM-based ranker is instructed by providing a prompt and takes the query and the text of the documents as input (based on the size of the context) and ranks these documents. Recently, Large Language Models (LLMs) have been shown to be highly effective in the reranking phase. **RankLLaMA** [9] leverages the LLaMA architecture to provide nuanced relevance estimations, thereby enhancing the precision of reranked document lists. Recent works proposed open-source fine-tuned listwise LLM rankers, like **RankZephyr** [27] and **RankVicuna** [26]. RankVicuna is built upon Vicuna[2] model and distilled using RankGPT-3.5 [34] as a teacher for some queries from the MSMARCO passage dataset. RankZephyr is an improved version of RankVicuna and further fine-tuned using RankGPT-4 [34] only for a smaller set of queries. The initial retrieval is done using BM25 for these rankers. Another notable model, **LiT5** [35], is based on the T5 architecture and is optimized for listwise ranking tasks. Given that these models exhibit a limited context length, various approaches have been proposed to aggregate sub-lists from an initial ranker into a final ranked list [25,28].

Despite the advancements introduced by these LLM-based rerankers, integrating adaptive retrieval (AR) techniques with listwise ranking outputs remains a unexplored. Our work addresses these gaps by proposing an adaptive retrieval mechanism specifically designed for LLM rerankers, thereby enhancing both recall and ranking effectiveness.

3 SlideGar: Sliding Window Based Adaptive Retrieval

3.1 Preliminaries

Recall from Sect. 2.2 that Graph-based Adaptive Retrieval (Gar) and Query Affinity Modelling based Adaptive Retrieval (Quam) augment the set of documents for reranking by selecting the neighbors of top-ranked documents by leveraging a corpus graph (constructed offline) to find neighbors in constant time. To formalize and set the stage for our proposed method, let R_0 be the initial retrieval pool, b is the batch size, c is the ranking budget, and G is the corpus graph. Gar and Quam start with taking the top b (batch size for pointwise ranker) documents from the initial retrieval R_0 and ranking by a pointwise ranker like MonoT5 [23]. Next, considering the top-ranked documents, the corpus graph G is explored to select b neighboring documents. Subsequent iterations alternate between the initial retrieval and the graph neighborhood until the reranking budget c is met.

3.2 The SlideGar Algorithm

For a given query q, the standard listwise reranking pipeline takes the retrieval results as input. Since the LLM rankers are limited by the context size[3], an effective sliding window strategy is proposed [34], based on mainly two hyper-parameters, window size w and step size b. The LLM ranker takes w documents (in back-to-first order) and reranks them, and then slides the window by step (of size b). The sliding window strategy continues until all documents in the list have been processed and reranked.

[2] https://huggingface.co/lmsys/vicuna-7b-v1.5.

[3] Although recent advances in LLMs enable very large context sizes [2], we continue to use the strategies proposed in prior work so as to remain comparable.

Algorithm 1 SLIDEGAR: Sliding Window based Graph Adaptive Retrieval

Input: Initial ranking R_0, window size w, step size b, budget c, corpus graph G
Output: Reranked pool R_1

$R_1 \leftarrow \emptyset$ ▷ Reranking results
$P \leftarrow R_0$ ▷ Reranking pool
$F \leftarrow \emptyset$ ▷ Graph frontier
$L \leftarrow$ SELECT(top w from R_0) ▷ Prepare first window
do
  $B \leftarrow$ LISTWISERANKER(L, subject to w and b) ▷ e.g., RankZephyr
  $R_0 \leftarrow R_0 \setminus B$ ▷ Discard batch from initial ranking
  $L_1 \leftarrow B[1:b]$ ▷ Keep top b docs for next window
  $R_1 \leftarrow R_1 \cup B[b+1:|B|]$ ▷ Add remaining to results
  $F \leftarrow$ NEIGHBOURS$(B, G) \setminus (R_1 \cup L_1)$ ▷ Update frontier
  $P \leftarrow \begin{cases} R_0 & \text{if } P = F \\ F & \text{if } P = R_0 \end{cases}$ ▷ Alternate between initial ranking and frontier
  $L_2 \leftarrow$ SELECT(top b from P)
  $L \leftarrow L_1 \cup L_2$ ▷ Prepare window for next round
while $|R_1| < c - b$
$R_1 \leftarrow R_1 \cup L_1$

Since GAR and QUAM are based on the pointwise ranker, they are not directly applicable to the listwise rankers. In our work, we propose the SLIDEGAR (Sliding-window-based Graph Adaptive Retrieval) algorithm to use adaptive retrieval for listwise rankers. Algorithm 1 presents the workings of SLIDEGAR, also visualized in Fig. 1. For query q, let R_0 and R_1 be the initial and final pool of ranked documents. The initial pool R_0 is the ranking given the retriever, e.g., BM25 and R_1 is initialized empty. Let w be the sliding window size and b be the step size. Similar to GAR, we consider P as the reranking pool of documents (initialized with R_0, $P \leftarrow R_0$), which will be dynamically updated. We consider graph frontier F, which will contain graph neighborhood documents. Contrary to the standard sliding window listwise rankers, SLIDEGAR starts in left-to-right order and takes top w documents (denoted by L) from the R_0 and ranks them using the listwise ranker (e.g., RankZephyr). Let B be the ranked list (without scores). These w documents are removed from R_0 since they are ranked. Next, let L_1 be the list of top b documents from B and taken to the next window. We add the remaining documents to R_1. Since the ranker does not provide any score, we use the reciprocal of the rank as a pseudo-score to the documents in B.

The graph frontier F is now updated with the neighbors (which are not in R_1 and L_1) of documents from B. The neighbors are prioritized by the scores of the source documents in B. For the next windows, we follow the conventions proposed by [12] and alternate between the R_0 and the graph frontier F. We take the top b documents from F (denoted by L_2) and prepare the window for the next ranking round ($L \leftarrow L_1 \cup L_2$). SLIDEGAR is generalizable for any choice of w and b, but for simplicity, we assume that $b = w/2$. We keep alternating between the initial ranking R_0 and graph frontier F until R_1 has $c - b$ documents and then finally the documents from L_1 are appended on the top of R_1. Similar to

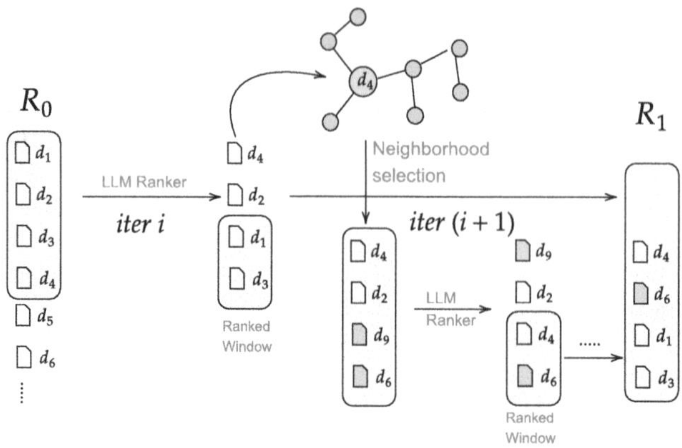

Fig. 1. The SlideGar algorithm visualized. LLM Ranker ranks the list of documents (window) and then SlideGar leverages LLM Ranker feedback and looks for the neighbors of d_4 and d_2, and carries both documents and their neighbors in the next window. The neighborhood documents are highlighted in green. The remaining documents, d_1 and d_3, are added to R_1.

standard listwise LLM ranker, SlideGar needs $\lceil \frac{c-w}{b} \rceil + 1$ LLM calls to process c documents with window size w and step size b.

SlideGar has some similarities to existing methods. Compared to Gar [12], it does not assume scores are independent, making it suitable for listwise reranking methods. When compared with TDPart [25], it does not perform partitioning and pulls in documents missing from the initial results.

4 Experimental Setup

Our experiments answer the following research questions about SlideGar:

- **RQ1:** What is the impact of adaptive retrieval in the ranking performance of LLM rerankers?
- **RQ2:** How sensitive are the LLM based rerankers to the graph depth (k)?
- **RQ3:** What are the additional computational overheads by SlideGar in comparison to standard LLM based reranking?

4.1 Dataset and Evaluation

In our experiments, we use the MSMARCO [22] passage corpus and evaluate performance on TREC Deep Learning 2019 (DL19) and 2020 (DL20) query sets. The DL19 and DL20 datasets have 43 and 54 queries respectively. To further study the generalization abilities of our method, we use MSMARCO-passage-v2 corpus and test on TREC Deep Learning 2021 (DL21) and 2022 (DL22). Since this corpus contains duplicate documents (same text assigned to different document IDs), we de-duplicate the whole corpus and create BM25 [31] and

TCT [8] indexes. We also de-duplicate the qrels from the test sets DL21 and DL22. We release de-duplicate corpus indexes and qrels with our code. The DL21 and DL22 datasets have 53 and 76 queries respectively.

We use nDCG@10 and Recall@c to evaluate different ranking approaches, where c is the reranking budget. We use the PyTerrier [13] framework for evaluating runs and follow the standard ranking pipeline notation from PyTerrier, for example, **BM25»RankZephyr** represents the ranking pipeline where retrieval is done using BM25 and ranking by RankZephyr model. We re-use the BM25 [31] and TCT [8] based document corpus graphs from GAR [12], consisting of the 16 nearest neighbors of each document in the corpus. For MSMARCO-passage-v2, we build both BM25 and TCT-based corpus graphs with 16 neighbors. We release the corpus graphs with our code.

4.2 Retrieval and Ranking Models

Retrievers: We perform experiments with both sparse and dense retrieval methods. For sparse retrieval, we use BM25 [31] using a PISA [17] index. For dense retrieval, we use TCT [8] and retrieve (exhaustively) top $c \in [50, 100]$ documents using the TCT-ColBERT-HNP [8] model from *huggingface*[4].

Rankers: We use different listwise LLM rankers, both zero-shot and fine-tuned. We use zero-shot rankers from **RankGPT** [34] mainly based on `gpt-3.5-turbo` and `gpt-4o` models from OpenAI [24]. We use Azure API[5] for both `gpt-3.5-turbo` (gpt-35-turbo in Azure) and `gpt-4o` models. We also use fine-tuned listwise LLM rankers, mainly **RankZephyr** [27] and **RankVicuna** [26]. For all these listwise rankers, we use the implementation from the open-source Python library, called Rerankers [3] which incorporates the RankLLM[6] (a better variant of RankGPT). These models use the sliding window strategy over a list of documents and generate ranking (also called permutation). It is important to note that these rankers do not provide scores. We use default settings where the sliding window size is 20 and the step size is 10. We use the context size of 4096 for all rankers.

Baselines: We compare our approach to the baseline (non-adaptive) LLM reranking systems. Furthermore, we compare with a variation of SLIDEGAR that uses RM3 [1] (denoted by $\text{SLIDEGAR}_{\text{RM3}}$) as its source of additional relevant documents, rather than the corpus graph. Specifically, after reranking the current window by the LLM, we expand the query using top-b documents and for the subsequent window, retrieve the remaining b documents using the expanded query from the index (mainly BM25). Note that the LLM always uses the original query for reranking; the RM3-expanded query is only used to retrieve additional results. For $\text{SLIDEGAR}_{\text{RM3}}$, we set both *fb_docs* and *fb_terms* to 10, and the *original_query_weight* to 0.6.

---
4 https://huggingface.co/castorini/tct_colbert-v2-hnp-msmarco.
5 https://learn.microsoft.com/en-us/azure/ai-services/openai/overview.
6 https://github.com/castorini/rank_llm/tree/main.

Other Hyperparameters: We denote the reranking budget by c, window size w, step size by b, and graph depth (number of neighbors in the corpus graph) by k. We choose the reranking budget c of 50 and 100. We set w=20, b=10 and the depth of the graph k=16. For further ablation study, we vary graph depth k in [2,16] (multiples of 2).

5 Results and Analysis

We now discuss our results and provide a detailed analysis. We denote our sliding window and graph-based retrieval method by SlideGar and the corresponding graph in the subscript, for example, $\text{SlideGar}_{\text{BM25}}$ represents the SlideGar with a BM25-based graph. We incorporate SlideGar with all ranking methods.

5.1 Effectiveness

To answer **RQ1**, we propose the SlideGar algorithm and assess the effectiveness over different reranking budgets, initial retrievers, and rerankers. We report the performance of different reranking approaches in Table 1. We observe that SlideGar shows a significant improvement over the standard ranking. Across all rankers, the improved recall remains in the same range for the given reranking budget. However, since we use the window level ranking to look for the neighbors, we find that the better ranker provides slightly better recall. Hence, RankGPT-4o and RankZephyr show slightly better recall improvements than RankGPT-3.5 and RankVicuna.

The most significant improvements are at the ranking budget of 50 and when the sparse retriever is used. On DL19, Recall@50 improves between 22% and 28% across rankers, where BM25»RankGPT-4o with $\text{SlideGar}_{\text{TCT}}$ shows the greatest improvements. Additionally, the fine-tuned RankZephyr improves nDCG@10 the most for the BM25»RankZephyr (from 0.670 to 0.747) with $\text{SlideGar}_{\text{TCT}}$. Recall@100 improves the highest (up to 20%) for RankGPT-4o and the lowest (up to 14%) for RankVicuna. We observe similar trends for the DL20 dataset. Our adaptive RM3 baseline, $\text{SlideGar}_{\text{RM3}}$, also improves recall (significant in 13 of 16 comparisons), but is not able to successfully re-rank these documents to yield an improved nDCG@10. This observation suggests that the corpus graph exploration approach is preferable to that of traditional PRF variants in terms of both efficiency and effectiveness. We also find that $\text{SlideGar}_{\text{TCT}}$ shows better performance in comparison to $\text{SlideGar}_{\text{BM25}}$ and $\text{SlideGar}_{\text{RM3}}$, since the dense retriever-based corpus graph is capable of providing complementary signals to the lexically retrieved documents.

When the dense retriever, TCT, is used for the initial ranking, we continue to see improvements, though the relative differences are lower due to the stronger initial result set. Both RankGPT-4o and RankZephyr improve recall@50 between 9%-10% and RankVicuna and RankGPT-3.5 between 6%-7% on the DL19. Surprisingly, the nDCG@10 improved from 0.792 to 0.816 for the RankGPT-4o and from 0.771 to 0.800 for RankZephyr with $\text{SlideGar}_{\text{TCT}}$, achieving the highest

Table 1. The performance comparison of different LLM-based rankers on TREC DL19 and DL20. Significant improvements (paired t-test, $p < 0.05$, using Bonferroni correction) with the reranking baseline are marked with * in the superscript. Boldface indicates the strongest result within each group.

	DL19				DL20			
	$c = 50$		$c = 100$		$c = 50$		$c = 100$	
pipeline	nDCG	R@c	nDCG	R@c	nDCG	R@c	nDCG	R@c
BM25»RankGPT-4o	0.704	0.389	0.745	0.497	0.688	0.465	0.712	0.569
w/ SlideGAR$_{RM3}$	0.712	*0.441	0.729	*0.538	*0.747	*0.530	*0.760	*0.660
w/ SlideGAR$_{BM25}$	0.741	*0.445	0.742	*0.543	0.718	0.499	*0.750	*0.615
w/ SlideGAR$_{TCT}$	**0.762**	***0.498**	**0.771**	***0.601**	***0.779**	***0.578**	***0.771**	***0.678**
BM25»RankGPT-3.5	0.623	0.389	0.642	0.497	0.585	0.465	0.615	0.569
w/ SlideGAR$_{RM3}$	0.639	*0.422	0.651	*0.549	0.596	*0.512	0.599	*0.600
w/ SlideGAR$_{BM25}$	0.651	*0.434	0.655	*0.549	0.604	0.495	0.603	0.597
w/ SlideGAR$_{TCT}$	**0.660**	***0.480**	**0.659**	***0.589**	***0.652**	***0.560**	**0.646**	***0.671**
BM25»RankZephyr	0.670	0.389	0.714	0.497	0.684	0.465	0.708	0.569
w/ SlideGAR$_{RM3}$	0.711	*0.438	0.740	*0.553	0.699	*0.508	0.720	*0.608
w/ SlideGAR$_{BM25}$	0.705	*0.430	0.715	*0.546	0.703	0.498	0.727	*0.612
w/ SlideGAR$_{TCT}$	***0.747**	***0.481**	**0.747**	***0.590**	***0.772**	***0.574**	***0.779**	***0.681**
BM25»RankVicuna	0.633	0.389	**0.663**	0.497	0.646	0.465	0.663	0.569
w/ SlideGAR$_{RM3}$	0.645	0.418	0.658	0.528	0.653	*0.515	0.663	0.593
w/ SlideGAR$_{BM25}$	0.650	*0.427	0.650	*0.543	0.639	0.485	0.654	0.596
w/ SlideGAR$_{TCT}$	**0.655**	***0.478**	0.661	***0.571**	**0.661**	***0.570**	**0.677**	***0.672**
TCT»RankGPT-4o	0.792	0.506	0.775	0.610	0.779	**0.652**	**0.787**	0.713
w/ SlideGAR$_{BM25}$	0.809	0.541	**0.793**	0.645	**0.783**	0.638	0.771	0.723
w/ SlideGAR$_{TCT}$	**0.816**	***0.561**	0.771	***0.666**	0.777	0.643	0.755	***0.733**
TCT»RankGPT-3.5	0.755	0.506	0.753	0.610	**0.736**	**0.652**	**0.738**	0.713
w/ SlideGAR$_{BM25}$	**0.765**	0.525	0.747	0.641	0.721	0.641	0.702	***0.737**
w/ SlideGAR$_{TCT}$	0.758	***0.539**	**0.765**	***0.647**	0.697	0.645	0.680	*0.729
TCT»RankZephyr	0.771	0.506	0.763	0.610	0.772	**0.652**	**0.772**	0.713
w/ SlideGAR$_{BM25}$	0.787	0.534	0.788	**0.655**	0.768	0.638	0.766	**0.734**
w/ SlideGAR$_{TCT}$	**0.800**	***0.552**	**0.789**	*0.653	**0.775**	0.645	0.770	*0.728
TCT»RankVicuna	0.746	0.506	0.746	0.610	0.729	**0.652**	0.719	0.713
w/ SlideGAR$_{BM25}$	0.756	0.520	0.741	0.636	**0.736**	0.641	**0.736**	0.732
w/ SlideGAR$_{TCT}$	**0.763**	***0.542**	**0.754**	***0.647**	0.733	0.649	0.718	***0.730**

reported effectiveness that we are aware of on this benchmark. This demonstrates the ranking capabilities of the LLM-based rerankers and the adaptive retrieval helps in finding missing relevant documents. We see statistically significant improvements in Recall@100 for all rankers with SlideGAR$_{\text{TCT}}$. However, the Recall@50 does not improve for the DL20 dataset. We suspect that the reason for this observation is twofold. First, the absolute value of the recall in DL20 with TCT is higher than that in DL19, giving less potential room for improvement. This problem is also exacerbated by a lower average number of relevant documents per query in DL20 than DL19 (30.9 vs 58.1). Second, the LLM rerankers may be up against their maximum possible effectiveness on the collection; some relevant documents may be incorrectly ranked poorly by these models, thereby never allowing them to be used as a feedback signal to find additional ones. This observation is challenging to validate, however, because it would require exhaustively scoring the MSMARCO corpus to see where the remaining relevant documents are ranked—which would be very expensive.

Overall, these results indicate that adaptive retrieval techniques, particularly SlideGAR$_{\text{TCT}}$, substantially enhance the performance of listwise LLM rankers.

5.2 Generalization on Different Corpus

In Table 2, we report the results on the MSMARCO-passage-v2 corpus using DL21 and DL22 as test sets. We use RankGPT-3.5 and RankZephyr as rankers. Unsurprisingly, SlideGAR also shows significant improvements on the new corpus. Specifically, on DL21, SlideGAR$_{\text{TCT}}$ improves Recall@50 from 0.248 to 0.356 (43% improvement) with BM25»RankGPT-3.5 and from 0.248 to 0.365 (47% improvement) with BM25»RankZephyr. Similar improvements are observed when the budget is set to 100. Our method also shows robust performance on the DL22 dataset, where the initial retrieval effectiveness is considerably lower (Recall@50 is 0.119 and Recall@100 is 0.171). In particular, SlideGAR$_{\text{TCT}}$ improves Recall@50 by 28% with BM25»RankGPT-3.5 and by 47% with BM25»RankZephyr. Similar trends are observed when the budget increases to 100. Additionally, recall improvements are evident when the dense (TCT) retriever is used at the initial stage. Overall, we observe better ranking and recall improvements when the fine-tuned listwise ranker, RankZephyr, is utilized.

5.3 Effect of Graph Depth

A key hyperparameter of SlideGAR is the graph depth k, which we now investigate to answer **RQ2**. The adaptive retrieval is mainly based on the number of neighbors, k, in the corpus graph. In this section, we study the behavior of different ranking approaches when graph depth k varies. We do not consider RankGPT-4o for this ablation study due to its high cost. Towards this, we vary the graph depth $k \in [2, 16]$ (by multiples of 2) and fix the reranking budget c to 50. We report the performance across different metrics in Fig. 2. We observe

Table 2. The performance comparison of different LLM-based rankers on TREC DL21 and DL22. Significant improvements (paired t-test, $p < 0.05$, using Bonferroni correction) with the reranking baseline are marked with * in the superscript. Boldface indicates the strongest result within each group.

	DL21				DL22			
	$c = 50$		$c = 100$		$c = 50$		$c = 100$	
pipeline	nDCG	R@c	nDCG	R@c	nDCG	R@c	nDCG	R@c
BM25»RankGPT-3.5	0.579	0.248	**0.591**	0.337	0.381	0.119	**0.415**	0.171
w/ SlideGar$_{RM3}$	0.572	*0.273	0.579	0.361	0.379	0.123	0.395	0.180
w/ SlideGar$_{BM25}$	0.580	*0.328	0.574	*0.414	0.368	0.128	0.377	0.184
w/ SlideGar$_{TCT}$	**0.594**	***0.356**	0.586	***0.458**	**0.386**	***0.153**	0.396	***0.225**
BM25»RankZephyr	0.638	0.248	0.663	0.337	0.439	0.119	0.486	0.171
w/ SlideGar$_{RM3}$	0.652	*0.282	0.678	*0.371	0.458	0.132	0.492	*0.192
w/ SlideGar$_{BM25}$	0.671	*0.325	0.685	*0.429	0.418	0.131	0.459	*0.192
w/ SlideGar$_{TCT}$	***0.692**	***0.365**	**0.699**	***0.481**	***0.490**	***0.175**	**0.518**	***0.266**
TCT»RankGPT-3.5	**0.695**	0.406	**0.695**	0.527	**0.585**	0.274	**0.589**	0.356
w/ SlideGar$_{BM25}$	0.692	0.427	0.693	*0.565	0.556	0.250	0.544	0.357
w/ SlideGar$_{TCT}$	0.677	***0.439**	0.658	***0.584**	0.552	**0.284**	0.532	**0.376**
TCT»RankZephyr	0.733	0.406	0.734	0.527	0.652	0.274	**0.663**	0.356
w/ SlideGar$_{BM25}$	**0.759**	0.424	**0.760**	*0.569	0.657	0.256	0.662	0.362
w/ SlideGar$_{TCT}$	0.745	***0.445**	0.753	***0.576**	**0.661**	***0.297**	0.656	***0.388**

that the recall@50 improves for SlideGar as we go deeper into the graph neighborhood. Though, the BM25 variant, SlideGar$_{BM25}$ seems to be robust to the number of neighbors. A similar trend can be seen in the nDCG metrics. On the other hand, the SlideGar$_{TCT}$ is able to find further relevant documents as we go deeper in the graph. In particular, recall@50 improves for both RankVicuna and RankGPT-3.5, and a similar effect can be seen in nDCG@50 even though the nDCG@10 does not vary. Unsurprisingly, RankZephyr shows good improvements as k starts from 2 to 10 and then becomes stable. Since, in the alternate iteration, SlideGar selects b (step size, which we set to 10) documents from the neighborhood, SlideGar might end up taking all b documents from a single source document (the top-ranked document) from the current window.

5.4 Computational Overhead

We re-iterate that SlideGar alternate between initial retrieved documents and neighborhood documents after each window is processed and has the same number of reranking operations (number of LLM calls depending upon window size

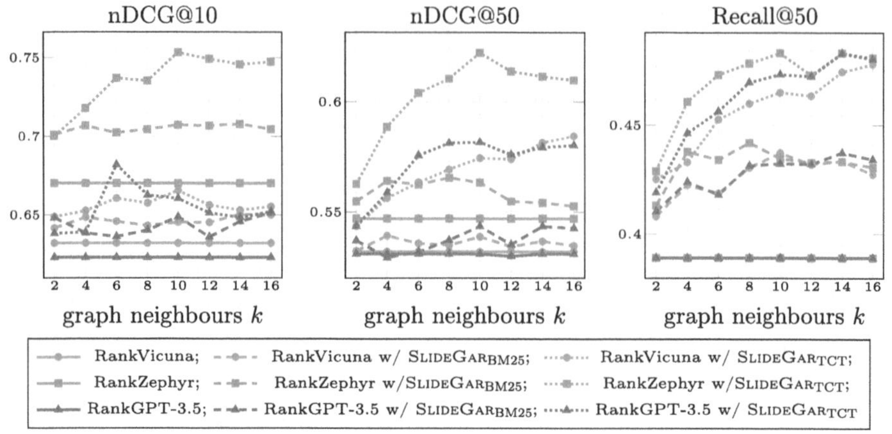

Fig. 2. Effect of SlideGar on different ranking pipelines on **TREC DL19** dataset when the number of graph neighbors k varies and window size is 20 and step size 10.

w and step size b) as standard listwise ranking. The cost of ranking all windows dominates the total cost of reranking the pipeline (including the cost of preparing windows). To answer **RQ3**, we are interested to see what the additional latency cost our adaptive sliding window approach, SlideGar, adds to the total LLM reranking latency. Toward this, we conduct an experiment on the DL19 dataset with the reranking budget c of 50 and 100. We use BM25 as initial retrieval and $\text{SlideGar}_{\text{BM25}}$ variant of our adaptive method. In Table 3, we report the mean latency (in ms) per query for reranking and SlideGar component during reranking. For this experiment, we only use RankZephyr and RankVicuna rankers and do not consider the API-based rankers. We report the mean and variance over 5 consecutive runs. We conduct our experiment on the NVIDIA V100 GPU with 32G RAM. We observe that RankVicuna and RankZephyr take around 13797 ms and 14884 ms, respectively, to process 50 documents per query. On the other hand, the SlideGar component adds only around 2.6 ms which is just 0.02% of the latency cost of LLM ranker. Similarly, when reranking 100 documents, SlideGar adds negligible latency. Overall, our adaptive retrieval approach adds a minimal latency cost while providing better recall.

Table 3. Mean Latency (ms/query) of LLM Rerankers at $c = 50$ and $c = 100$.

	RankVicuna		RankZephyr	
budget	Ranking	SlideGar	Ranking	SlideGar
50	13797 ± 2.57	2.63 ± 0.01	14884 ± 1.90	2.67 ± 0.01
100	30923 ± 0.89	6.73 ± 0.01	33474 ± 7.04	6.97 ± 0.01

6 Conclusion

We augment existing adaptive ranking algorithms to work with listwise LLM reranking models. We find that our proposed method, SlideGar, is able to successfully overcome the *bounded recall problem* from first-stage retrievers by successfully leveraging feedback signals from an LLM. Also, the computational overhead of applying SlideGar is minimal compared to a typical LLM reranking pipeline. In our opinion this work enables the broader adoption of LLM reranking, such as in cases where the first stage is unsuccessful or systems are limited by legacy first-stage (lexical) keyword-based retrieval systems.

References

1. Abdul-Jaleel, N., et al.: Umass at trec 2004: Novelty and hard. Computer Science Department Faculty Publication Series, p. 189 (2004)
2. Chen, S., Wong, S., Chen, L., Tian, Y.: Extending context window of large language models via positional interpolation. arXiv preprint arXiv:2306.15595 (2023)
3. Clavié, B.: rerankers: A lightweight python library to unify ranking methods (2024). https://arxiv.org/abs/2408.17344
4. Ding, S., Suel, T.: Faster top-k document retrieval using block-max indexes. In: Ma, W., Nie, J., Baeza-Yates, R., Chua, T., Croft, W.B. (eds.) Proceeding of the 34th International ACM SIGIR Conference on Research and Development in Information Retrieval, SIGIR 2011, Beijing, China, July 25-29, 2011, pp. 993–1002. ACM (2011). https://doi.org/10.1145/2009916.2010048, https://doi.org/10.1145/2009916.2010048
5. Jardine, N., van Rijsbergen, C.J.: The use of hierarchic clustering in information retrieval. Inf. Storage Retr. **7**(5), 217–240 (1971)
6. Khattab, O., Zaharia, M.: Colbert: Efficient and effective passage search via contextualized late interaction over bert. In: Proceedings of the 43rd International ACM SIGIR Conference on Research and Development in Information Retrieval, pp. 39–48 (2020)
7. Kulkarni, H., MacAvaney, S., Goharian, N., Frieder, O.: Lexically-accelerated dense retrieval. In: Chen, H., Duh, W.E., Huang, H., Kato, M.P., Mothe, J., Poblete, B. (eds.) Proceedings of the 46th International ACM SIGIR Conference on Research and Development in Information Retrieval, SIGIR 2023, Taipei, Taiwan, July 23-27, 2023, pp. 152–162. ACM (2023). https://doi.org/10.1145/3539618.3591715, https://doi.org/10.1145/3539618.3591715

8. Lin, S., Yang, J., Lin, J.: In-batch negatives for knowledge distillation with tightly-coupled teachers for dense retrieval. In: Rogers, A., et al., (eds.) Proceedings of the 6th Workshop on Representation Learning for NLP, RepL4NLP@ACL-IJCNLP 2021, Online, August 6, 2021, pp. 163–173. Association for Computational Linguistics (2021). https://doi.org/10.18653/V1/2021.REPL4NLP-1.17
9. Ma, X., Wang, L., Yang, N., Wei, F., Lin, J.: Fine-tuning llama for multi-stage text retrieval. In: Proceedings of the 47th International ACM SIGIR Conference on Research and Development in Information Retrieval, pp. 2421–2425 (2024)
10. Ma, X., Zhang, X., Pradeep, R., Lin, J.: Zero-shot listwise document reranking with a large language model. arXiv preprint arXiv:2305.02156 (2023)
11. MacAvaney, S., Tonellotto, N.: A reproducibility study of PLAID. In: Yang, G.H., Wang, H., Han, S., Hauff, C., Zuccon, G., Zhang, Y. (eds.) Proceedings of the 47th International ACM SIGIR Conference on Research and Development in Information Retrieval, SIGIR 2024, Washington DC, USA, July 14-18, 2024, pp. 1411–1419. ACM (2024). https://doi.org/10.1145/3626772.3657856, https://doi.org/10.1145/3626772.3657856
12. MacAvaney, S., Tonellotto, N., Macdonald, C.: Adaptive re-ranking with a corpus graph. In: Hasan, M.A., Xiong, L. (eds.) Proceedings of the 31st ACM International Conference on Information & Knowledge Management, Atlanta, GA, USA, October 17-21, 2022, pp. 1491–1500. ACM (2022). https://doi.org/10.1145/3511808.3557231
13. Macdonald, C., Tonellotto, N., MacAvaney, S., Ounis, I.: Pyterrier: Declarative experimentation in python from BM25 to dense retrieval. In: Demartini, G., Zuccon, G., Culpepper, J.S., Huang, Z., Tong, H. (eds.) CIKM '21: The 30th ACM International Conference on Information and Knowledge Management, Virtual Event, Queensland, Australia, November 1 - 5, 2021, pp. 4526–4533. ACM (2021). https://doi.org/10.1145/3459637.3482013
14. Mackie, I., Chatterjee, S., Dalton, J.: Generative and pseudo-relevant feedback for sparse, dense and learned sparse retrieval. arXiv preprint arXiv:2305.07477 (2023)
15. Mackie, I., Chatterjee, S., Dalton, J.: Generative relevance feedback with large language models. In: Proceedings of the 46th International ACM SIGIR Conference on Research and Development in Information Retrieval, pp. 2026–2031 (2023)
16. Malkov, Y.A., Yashunin, D.A.: Efficient and robust approximate nearest neighbor search using hierarchical navigable small world graphs. IEEE Trans. Pattern Anal. Mach. Intell. **42**(4), 824–836 (2020)
17. Mallia, A., Siedlaczek, M., Mackenzie, J.M., Suel, T.: PISA: performant indexes and search for academia. In: Clancy, R., Ferro, N., Hauff, C., Lin, J., Sakai, T., Wu, Z.Z. (eds.) Proceedings of the Open-Source IR Replicability Challenge co-located with 42nd International ACM SIGIR Conference on Research and Development in Information Retrieval, OSIRRC@SIGIR 2019, Paris, France, July 25, 2019. CEUR Workshop Proceedings, vol. 2409, pp. 50–56. CEUR-WS.org (2019). https://ceur-ws.org/Vol-2409/docker08.pdf
18. Matveeva, I., Burges, C., Burkard, T., Laucius, A., Wong, L.: High accuracy retrieval with multiple nested ranker. In: Efthimiadis, E.N., Dumais, S.T., Hawking, D., Järvelin, K. (eds.) SIGIR 2006: Proceedings of the 29th Annual International ACM SIGIR Conference on Research and Development in Information Retrieval, Seattle, Washington, USA, August 6-11, 2006, pp. 437–444. ACM (2006). https://doi.org/10.1145/1148170.1148246
19. Metzler, D., Croft, W.B.: A Markov random field model for term dependencies. In: Proceedings of the 28th Annual International ACM SIGIR Conference on Research and Development in Information Retrieval, pp. 472–479 (2005)

20. Metzler, D., Croft, W.B.: Latent concept expansion using markov random fields. In: Proceedings of the 30th Annual International ACM SIGIR Conference on Research and Development In Information Retrieval, pp. 311–318 (2007)
21. Nguyen, T., MacAvaney, S., Yates, A.: A unified framework for learned sparse retrieval. In: Kamps, J., et al., (eds.) Advances in Information Retrieval - 45th European Conference on Information Retrieval, ECIR 2023, Dublin, Ireland, April 2-6, 2023, Proceedings, Part III. Lecture Notes in Computer Science, vol. 13982, pp. 101–116. Springer (2023). https://doi.org/10.1007/978-3-031-28241-6_7
22. Nguyen, T., et al.: MS MARCO: A human generated machine reading comprehension dataset. In: Besold, T.R., Bordes, A., d'Avila Garcez, A.S., Wayne, G. (eds.) Proceedings of the Workshop on Cognitive Computation: Integrating neural and symbolic approaches 2016 co-located with the 30th Annual Conference on Neural Information Processing Systems (NIPS 2016), Barcelona, Spain, December 9, 2016. CEUR Workshop Proceedings, vol. 1773. CEUR-WS.org (2016). https://ceur-ws.org/Vol-1773/CoCoNIPS_2016_paper9.pdf
23. Nogueira, R.F., Jiang, Z., Pradeep, R., Lin, J.: Document ranking with a pre-trained sequence-to-sequence model. In: Cohn, T., He, Y., Liu, Y. (eds.) Findings of the Association for Computational Linguistics: EMNLP 2020, Online Event, 16-20 November 2020. Findings of ACL, vol. EMNLP 2020, pp. 708–718. Association for Computational Linguistics (2020). https://doi.org/10.18653/V1/2020.FINDINGS-EMNLP.63
24. OpenAI: Openai api. https://openai.com (2024)
25. Parry, A., MacAvaney, S., Ganguly, D.: Top-down partitioning for efficient list-wise ranking. arXiv preprint arXiv:2405.14589 (2024)
26. Pradeep, R., Sharifymoghaddam, S., Lin, J.: Rankvicuna: Zero-shot listwise document reranking with open-source large language models. arXiv preprint arXiv:2309.15088 (2023)
27. Pradeep, R., Sharifymoghaddam, S., Lin, J.: Rankzephyr: Effective and robust zero-shot listwise reranking is a breeze! arXiv preprint arXiv:2312.02724 (2023)
28. Qin, Z., et al.: Large language models are effective text rankers with pairwise ranking prompting. arXiv preprint arXiv:2306.17563 (2023)
29. Rathee, M., MacAvaney, S., Anand, A.: Quam: Adaptive retrieval through query affinity modelling. arXiv preprint arXiv:2410.20286 (2024)
30. Robertson, S.E.: The probability ranking principle in IR. J. Document. **33**(4), 294–304 (1977)
31. Robertson, S.E., Zaragoza, H.: The probabilistic relevance framework: BM25 and beyond. Found. Trends Inf. Retr. **3**(4), 333–389 (2009)
32. Rocchio, J.: Relevance feedback in information retrieval. The SMART Retrieval System-Experiments in Automatic Document Processing/Prentice Hall (1971)
33. Sun, S., Zhuang, S., Wang, S., Zuccon, G.: An investigation of prompt variations for zero-shot llm-based rankers. arXiv preprint arXiv:2406.14117 (2024)
34. Sun, W., et al.: Is chatgpt good at search? investigating large language models as re-ranking agents. arXiv preprint arXiv:2304.09542 (2023)
35. Tamber, M.S., Pradeep, R., Lin, J.: Scaling down, litting up: Efficient zero-shot listwise reranking with seq2seq encoder-decoder models. arXiv:2312.16098 (2023)

36. Weller, O., et al.: Followir: Evaluating and teaching information retrieval models to follow instructions. arXiv preprint arXiv:2403.15246 (2024)
37. Yang, Y., Carlson, P., He, S., Qiao, Y., Yang, T.: Cluster-based partial dense retrieval fused with sparse text retrieval. In: Yang, G.H., Wang, H., Han, S., Hauff, C., Zuccon, G., Zhang, Y. (eds.) Proceedings of the 47th International ACM SIGIR Conference on Research and Development in Information Retrieval, SIGIR 2024, Washington DC, USA, July 14-18, 2024, pp. 2327–2331. ACM (2024). https://doi.org/10.1145/3626772.3657972
38. Zhai, C., Lafferty, J.: Model-based feedback in the language modeling approach to information retrieval. In: Proceedings of the Tenth International Conference on Information and Knowledge Management, pp. 403–410 (2001)

Lost but Not Only in the Middle
Positional Bias in Retrieval Augmented Generation

Jan Hutter(✉), David Rau, Maarten Marx, and Jaap Kamps(✉)

University of Amsterdam, Amsterdam, The Netherlands
{j.w.j.hutter,d.m.rau,maartenmarx,j.kamps}@uva.nl

Abstract. Large language models (LLMs) are known to exhibit positional bias, the tendency of models to perform differently based on where relevant information appears within the input context. Understanding this bias is important in a retrieval augmented generation (RAG) setting, as it impacts how retrieved passages are taken into account by the model. We systematically investigate positional bias in a RAG setting, by evaluating four LLMs using three different types of distractor documents, assessing their ability to extract relevant information from the input context. Our findings reveal significant positional bias depending on the type of context documents used and the total amount of documents in the context. Furthermore, the results show that positional bias in state-of-the-art LLMs is not limited to information located in the middle of the input context. By analyzing the models' attention, we identified patterns between the model's accuracy in responding to questions and the correct attribution of attention to relevant information in the context. Our code is available at https://github.com/Janhutter/LBNOITM.

1 Introduction

In this research, we will focus on retrieval augmented generation (RAG), an implementation where the performance of using a large context as input is particularly important. RAG is used to provide context to a language model with respect to a query that has been prompted and yields significant improvement over the base LLM [6]. The context consists of several documents retrieved from a collection, such as excerpts from Wikipedia. However, inputting the context documents into the prompt directly can result in reduced accuracy, as seen in [7]. Their research shows that placing one document which is relevant to the question, in the middle of a context with several 'distractor', or irrelevant, documents decreases accuracy on a question-answering task. It shows that the positioning of information inside the input is indicative of the performance of an LLM on knowledge-intensive tasks. This could be due to the mismatch between the task LLMs are trained on and the retrieval-augmented setup that is used during generation [10]. Parametric, or non-retrieval-augmented, LLMs are trained on datasets where the most informative tokens are the most recent ones, resulting in decreased accuracy when important information is placed in the middle of the input.

C. Hauff et al. (Eds.): ECIR 2025, LNCS 15572, pp. 247–261, 2025.
https://doi.org/10.1007/978-3-031-88708-6_16

This paper investigates *how does positional bias impact the performance of retrieval-augmented large language models?* We conduct a systematic analysis of the occurrence of positional bias in the prototypical RAG setting of knowledge-intensive, open-domain question-answering tasks. Rather than a SotA study, we conduct a scientific experiment, where we manipulate the context input in a systematic way (our independent variables) and measure the effect on performance (our dependent variable). Specifically, we examine the influence of the position of the relevant document with respect to multiple distractor documents. We will refer to this influence of position of a relevant document within the context as positional bias. We systematically vary the position of the single relevant document among different contextual settings. These settings exist of three different types of distractor documents, and two context lengths of 5 and 10 documents. Specifically, these shorter contexts simulate a more reasonable scenario and allow for extended and complementary results of Liu et al. [7]. By exhausting all combinations of the positions of the relevant document, context types, and context lengths, this approach enables a controlled examination of positional bias on RAG performance. Finally, after examining the positional bias within each contextual setting, the inspection of the attention values of the used models will provide insights into the mechanisms underlying the manifestation of positional bias.

2 Related Research

The observation that modern transformers can naturally cope with many context documents for answer generation in open domain QA tasks was central to the development of RAG [2,3]. As Peysakhovich and Lerer [10] state: "Since the performance of the RAG system is bounded by the recall of the retriever and thus the number of documents returned, larger context windows can naturally improve the performance of these systems." This highlights the importance of utilizing the context completely and increasing its size to improve performance in a RAG setting. Peysakhovich and Lerer [10] show that utilizing attention values of a prompted model to change the position of documents in the context can considerably improve performance, and reduce positional bias. By performing one or more steps of decoding and sorting the documents by the attention values that the model attributes to them, ensures that the documents are sorted by the inherent importance assigned by the model.

Our work is inspired by Liu et al. [7], who found that performance can degrade significantly when changing the position of relevant information in the long context input. They observed a *Lost in the Middle* effect with the best performance when relevant information occurs at the beginning or end of the input context. Similarly, Tang et al. [13] in *Found in the Middle* demonstrate positional bias in passage ranking tasks. Using five different LLMs, they show that their approach, *permutation self-consistency*, consistently improves document reranking and reduces the impact of positional bias. Documents that initially receive a lower score due to their position in the context are more accurately identified and reranked with this method. Our paper contributes to this with an exhaustive analysis over different positions of the relevant information, context lengths, and distractor types, in the context of the task and models of a typical RAG setting [11].

Instead of changing only external aspects of the model, there is research on changing properties inside the architecture of LLMs to decrease positional bias. Yu et al. [16] conclude that "attention weights manifest position bias within transformers. Additionally, experiments demonstrate that, besides position embeddings, the causal mask also contributes to position bias." By modifying attention values, the positional bias is reduced.

3 Method

To evaluate how the position of relevant information within a context affects the model's performance in a question-answering setting, we systematically vary the position of the oracle document, the document that holds the information to answer the question, in the context of different types of distractor documents. Additionally, by varying the length of the context in which this oracle document is located, we assess the model's ability to identify and interpret important information from varying lengths of text sections. Lastly, we will examine the internal attention values received during the generation of answers by the tested models to gain insights into positional bias. By analyzing the patterns in these attention values, we aim to uncover the underlying mechanisms that contribute to positional bias in LLMs.

In our approach to multi-context RAG, the inputs to the generative language model are as follows: (i) a question, (ii) a context consisting of n amount of documents where exactly one contains information relevant to the question, and n-*1* distractor documents, and (iii) an instruction to help answer the question. The models should then, according to the instruction (iii), answer the question (i) while using relevant information from the context (ii). Building on these aspects, we construct a setup closely aligned with the one proposed by [7], to highlight positional bias in RAG. Specifically, we show this by controlling a single factor: the position of the document that contains the relevant information, which we will call the oracle document.

4 Experimental Setup

Datasets. To investigate the implications of positional bias in RAG, we make use of the KILT-NQ and KILT-HotpotQA datasets [9]. In our implementation, both datasets leverage the same collection for retrieval in the RAG setting. The collection consists of the publicly available KILT collection [9], where the documents are split into nonoverlapping chunks of 100 words, each prefixed with its title, which is similar to that found in Rau et al. [11]. The structure of the answers in the dataset differs from each other, increasing the generalizability. The NQ and HotpotQA datasets contain answers labels which are quite concise, consisting of on average 3 and 2 words, respectively. NQ allows for more flexibility in answering, having on average two answer labels per question. Multiple answers are not allowed by HotpotQA. Another difference in answer labels of the datasets is the type of answer, where the NQ dataset has extractive answers and the HotpotQA dataset has short abstractive answers.

Context Types. By using different configurations of context type and position at which the oracle is injected, we can simulate a variety of scenarios. We have chosen three different context types, or distractor documents, in this research. First, *random distractor* consists of inserting the oracle document in a context comprised of uniformly sampled documents from the total collection, simulating a scenario where distractor documents are not relevant to the question. This represents the expected upper bound, as it assumes the easiest scenario for distinguishing relevant information from distractions. Secondly, *hard distractor* includes inserting the oracle document among relevant documents that do not contain the answer, simulating a realistic scenario where documents are closely related to the question but do not always provide the answer. These 'relevant but incorrect' documents are gathered using a BM25 retriever with a top-k of 1,000. To ensure that the retrieved documents did not contain the answer to the question for which they are retrieved, documents that contained the answer were removed from the ranking. Lastly, *relevant context* consists of inserting the oracle document among relevant documents retrieved by BM25, simulating a scenario where documents are related to the question and are able to contain the answer to the question. Resulting in a context where more than one document contains the answer. In related research like [7], context lengths of up to 30 documents are used. Their shortest context length is 10 documents, which we will use for comparison. However, increasing this context length given to different models from 5 to 10 or even 15 documents, no significant improvement in match and recall score was observed. Utilizing more documents in the context did show unwanted behavior: not all the models used in our research are able to tokenize the complete context because it exceeds the maximum length that these models support. For this reason, only context lengths of 5 and 10 documents are used in our experiments. Furthermore, increasing the context length can reduce effectiveness and efficiency, introducing the need for compression techniques [12].

Baselines. To consistently compare performance of the three 'lost-in-the-middle' test scenarios to an upper and lower bound, we make use of 'closed book' and 'oracle' settings. Closed book denotes a setting where the model is not provided with any additional information alongside the question; it relies solely on internally parameterized knowledge to generate an answer. This is the expected lower bound of our experiments.The oracle baseline represents a scenario where the LLM receives one or two documents per question, the oracle documents, that explicitly contain the answer, setting an expected upper bound for performance. These baselines help compare to scenarios when minimal positional bias occurs.

Prompting. After the selection of documents, an instruction is prepended to every question-context pair to inform the model about the method of generation.

For all the three aforementioned tests, this instruction is: "*Instruction: You are a helpful assistant. Your task is to extract relevant information from provided documents and to answer questions as short as possible.*" For the closed book baseline, the instruction was: "*Instruction: You are a helpful assistant. Answer the questions as short as possible.*" Using the combined context, four different decoder-only transformers are used to generate responses, which were then evaluated according to their similarity to the annotated answers, also called answer labels.

Metrics. Match Score is a variant of standard Exact Match [e.g., 8]. As models without fine-tuned decoders can generate verbose answers, the Match Score indicates whether the label is contained (as an exact match) in the generated answer. Because the Match Score is only accurate with short answer labels that consist of only a couple of words, recall is also used. Formally, in Match Score (Eq. 1) and Recall (Eq. 2) the amount of answer labels are denoted by M, the total amount of questions by N, the normalized answer labels split into a set of words by A_{ij}, and the normalized response of an LLM split into a set of words by R_i.

$$\text{Match Score} = \frac{1}{N}\sum_{i=1}^{N} \max_{j\in\{1,\ldots,M\}} \left(\frac{|A_{ij} \cap R_i|}{|R_i|}\right) \tag{1}$$

$$\text{Recall} = \frac{1}{N}\sum_{i=1}^{N} \max_{j\in\{1,\ldots,M\}} \left(\frac{|A_{ij} \cap R_i|}{|A_{ij}|}\right) \tag{2}$$

Utilizing these two metrics to assess model performance in a question-answering context will provide a robust approximation of their effectiveness and allow us to use them as indicators to evaluate their positional bias.

Models. To make sure the effects of positional bias are still present in modern decoder-based LLMs, we employ Mixtral-8x7B-Instruct [4], Llama-7B-chat-hf [14], and Solar-10.7-Instruct [5]. Figure 1 shows the performance of TinyLlama-1.1B-Chat [17] on the random distractor test, illustrating the extent to which positional bias can impact a model's performance in a RAG setting. Seeing the performance of TinyLlama in Fig. 1, reveals significant positional bias in a smaller model, highlighting the need to explore this phenomenon in greater depth with the three larger models. The recall scores are omitted from Fig. 1, as the results demonstrated similar trends and results to those observed with the match score.

5 Experimental Results

Resilience to Changes in the Positioning of Essential Information. Figures 2 and 3 compare the per model performance on the three tasks. In the random distractor test utilizing a context length of 5 or 10 documents, the three larger models—Llama2, Solar, and Mixtral—show no significant positional bias. This shows that these models are able to discern important information in

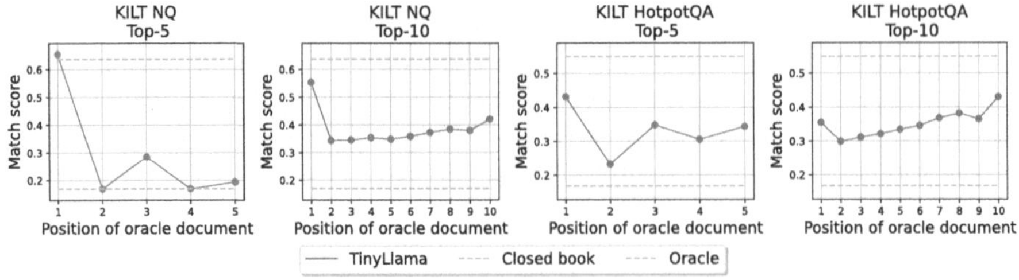

Fig. 1. The effect of changing the position of the oracle document in a context of random distractor documents, our 'random distractor' test scenario, when generating with TinyLlama on the NQ (left) and HotpotQA (right) datasets.

the input quite consistently. However, this consistency does not extend to the hard distractor and relevant context tasks, where the models show significant positional bias. We observe that these models show significantly decreased performance in both settings compared to the oracle setting. TinyLlama shows significant positional bias in all three test scenarios, showing large decrease in performance when the oracle document is not in the first position for all settings, except for the 10 document context on KILT-HotpotQA. In this setting, TinyLlama shows significantly increased performance when the oracle document is at the last position instead of the first.

A commonality between both context lengths is the increased performance of the Mixtral model on the random distractor task built from KILT-NQ compared to oracle performance. This same phenomenon was shown by TinyLlama in the 5 document random distractor task. These findings seem counterintuitive because adding irrelevant documents to the context seems to improve the performance of the Mixtral and TinyLlama models on the question-answering task.

The Figures show that all four models suffer from significant positional bias in both the hard distractor and relevant context tasks. The Llama2 and Solar models exhibit decreased performance when the relevant information appears at the end of the input (primacy bias), while Mixtral shows the opposite pattern, performing worse when the relevant information is at the beginning (recency bias). Specifically, Mixtral's performance improves by 6.81% when the oracle document is at the last position instead of the first, in the hard distractor task with five documents on KILT-NQ. For the random distractor task under the same conditions, this fluctuation is more modest, at approximately 1.95%. For Llama2 the performance decreases by 12.45% in the 5 document hard distractor task on KILT-NQ, and decreases by only 1.16% in the random distractor task under the same conditions. The performance change in of the Solar model is less significant, decreasing 2.72% in the hard distractor task and 1.34% in the random distractor task.

Comparing the hard distractor and relevant context tests and both datasets side to side shows that the relevant context test has an average which is slightly higher for every model than that found on the hard distractor test. Similarly, we see slightly less positional bias on the HotpotQA relevant context test than the HotpotQA hard distractor test. Furthermore, we see that the models show less positional bias on the HotpotQA than on the NQ dataset. In addition, using a larger context with 10 documents instead of 5 results in less significant changes in performance when changing the position of the oracle document in the context.

Figure 4 plots the results for the three larger models on the hard distractor task. The most notable result from this setting is the reversed 'lost in the middle' effect of Llama2 on the 5 document context constructed from HotpotQA. We see increased performance at positions 2, 3 and 4 compared to 1 and 5. This trend can also be found in the 10 document context, but having a smaller decrease in performance at the last positions as well. Similarly, Tinyllama showed increase performance in the middle (position 3) of the 5 document random distractor task when comparing to positions 1, 3, and 4.

Not Only 'Lost in the Middle'. It is interesting to compare our findings for a typical RAG setting to the findings of [7]. There are some differences in the experimental setup. First, our results cover shorter context lengths at more frequent intervals. [7] use context lengths from 10 to up to 40 documents, evaluating every five positions. Our results cover every position, with up to 10 documents. When inspecting the types of distractor documents, [7] use documents which match our description of hard distractor, where documents were retrieved which were 'most relevant to the query and do not contain any of the NaturalQuestions-annotated answers'.

For the typical RAG setting, our results show behavior which contradicts their title 'Lost in the Middle'. The performance of the Mixtral model on the hard distractor and relevant context tests, when comparing to the other models, shows us that positional bias not only occurs in the middle of an LLM input context. Similarly, a positional bias of the Llama2 model on the HotpotQA hard distractor test was observed, which was the inverse of the 'Lost in the Middle' effect. Hence, the positional bias pattern depends on the used model, and not all LLMs exhibit the U-shaped performance over the position of relevant information in the input context.

6 Attention Analysis

Using the attention values of the models while generating, we can reason about what tokens in the input are important for a certain generated answer. Furthermore, the models' attention values allow for a comparison between the model's internal mechanisms and the external occurrence of positional bias. The attention values that are visualized are calculated during generation and are of the shape:

(attention steps, layers, batch size, number of heads, sequence length)

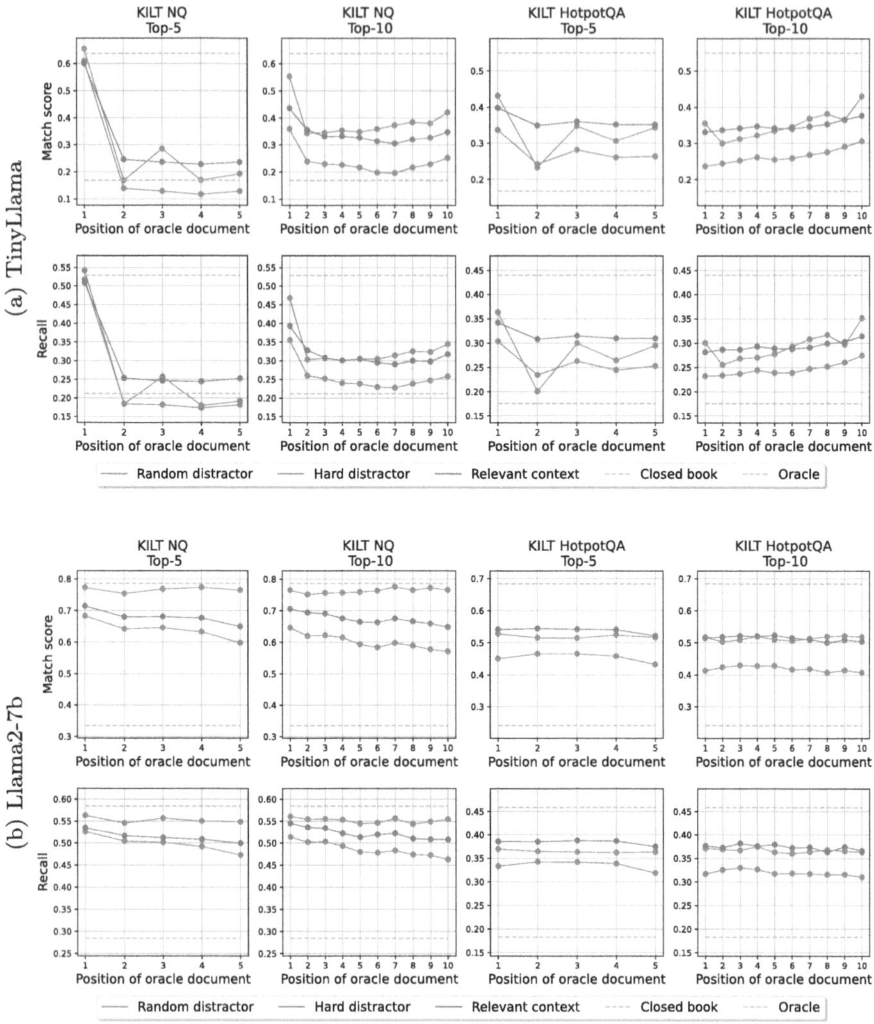

Fig. 2. Performance of TinyLlama and Llama2-7b models on the three different test scenarios constructed from KILT-NQ and KILT-HotpotQA.

We averaged over the attention steps for all the generated tokens, then took the attention for the middle and last attention layers. These values are then averaged over the number of attention heads. Consequently, using the amount of tokens for the instruction, documents, and question, their respective attention can be calculated. This allows us to see what the model calculated attention values are for every single document in the input context. The first layer is being left out in our results, as preliminary results showed that these layers only attend to the most recent document for every context type and model. This is similar to earlier findings: 'The attention maps in the first two layers (layers 0

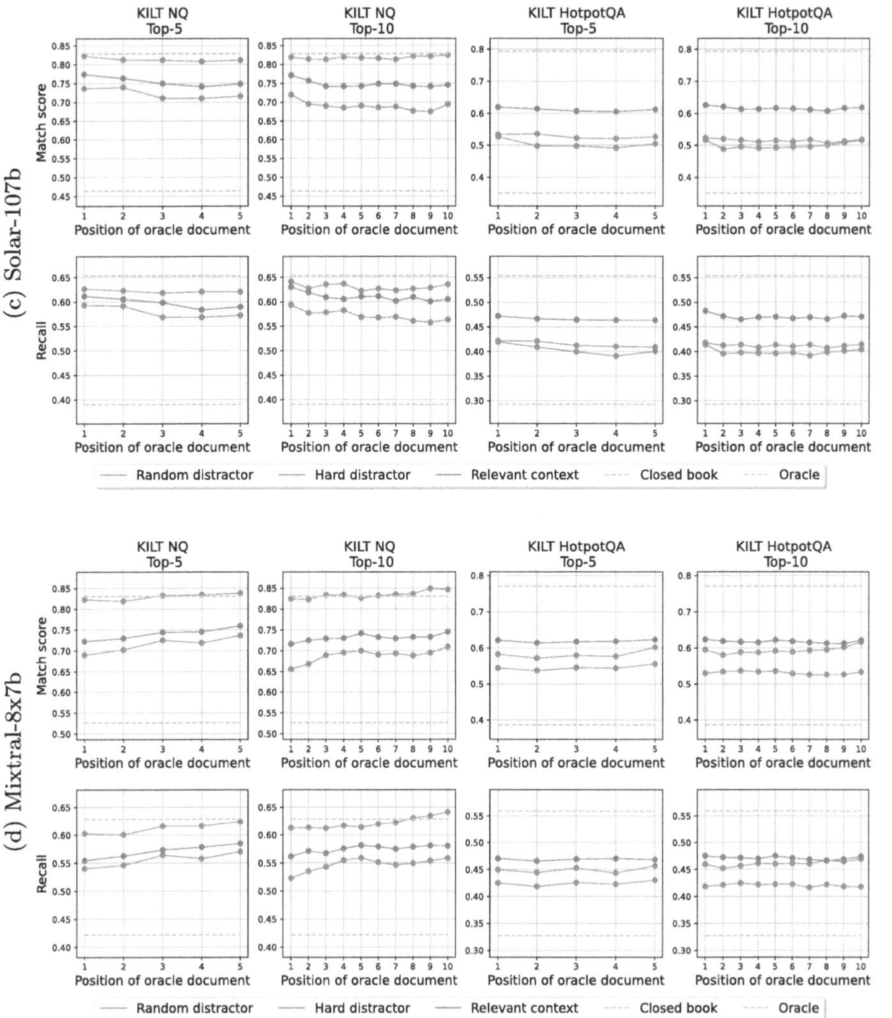

Fig. 3. Performance of Solar-107b and Mixtral-8x7b models on the three different test scenarios constructed from KILT-NQ and KILT-HotpotQA.

and 1) exhibit the "local" pattern, with recent tokens receiving more attention' [15]. We mainly focus on the middle layer, which shows more interpretable results similar to the results by [10]. They found that 'the differential attention to true vs. distractor documents happens mostly in the middle layers.'

This is similar to the visualizations made in [10], but where they only use the attention for the first generated token and averaged over the heads and layers. To distinguish the layers, all generated tokens are included to increase the amount of data points. Taking inspiration from their approach for mechanistic

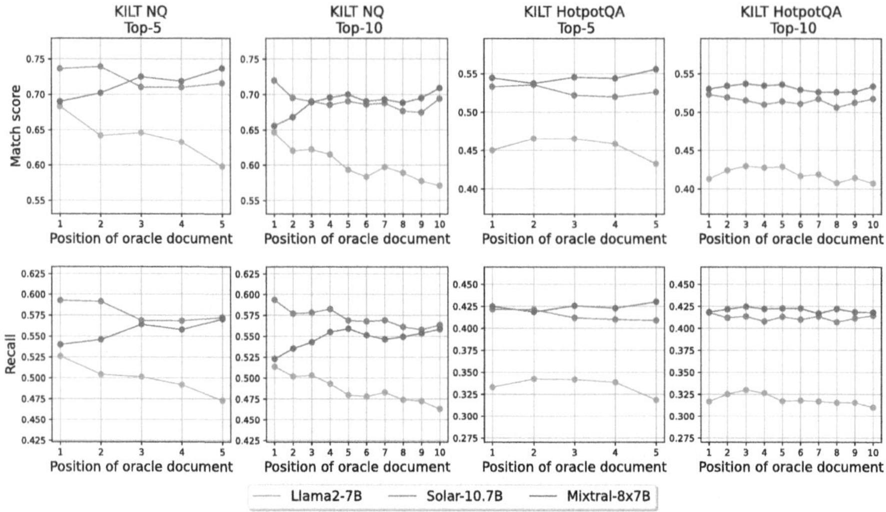

Fig. 4. The effect of changing the position of the oracle document in the context of hard distractor documents on the tested models for the NQ (left) and HotpotQA (right) datasets. We can observe the reversed 'lost-in-the-middle' effect of Llama2-7B on the KILT-HotpotQA dataset.

interpretability, the specific calculation listed above is utilized to look into TinyLlama and Llama2. Their attention values are discussed after generating 600 examples for every test scenario constructed from KILT-NQ and KILT-HotpotQA. We also differentiate our work by visualizing the model's attention over the quality of the answer it is generating. This allows us to explore potential patterns between response quality and positional bias. Analyzing these patterns allows us to investigate how incorrect attention attribution might affect the correctness of the model's response.

Using the generated examples and the models' attention values, we will first look at the positions of documents which received the highest attention on average across the position of the oracle document in the context. Secondly, we will take 50 uniformly sampled questions and answers according to their correctness, where correctness is attributed by the match score. These visualizations will help look into the effect of the attribution of attention toward the correctness of the output of LLMs. The rationale for selecting 600 examples for analysis, instead of the whole dataset, was to reduce computational requirements. Furthermore, 600 samples would result in at least 50 correct and incorrect samples per test scenario for further analysis, where 50 samples would be enough to find regularities in the output attention values.

Attention Across the Answer Quality. On the random distractor test where TinyLlama showed strong positional bias, its attention values show a similarly large pattern between the quality of the answer and the fidelity of the attribution of attention towards the position of the oracle document, as seen in Fig. 5 (top).

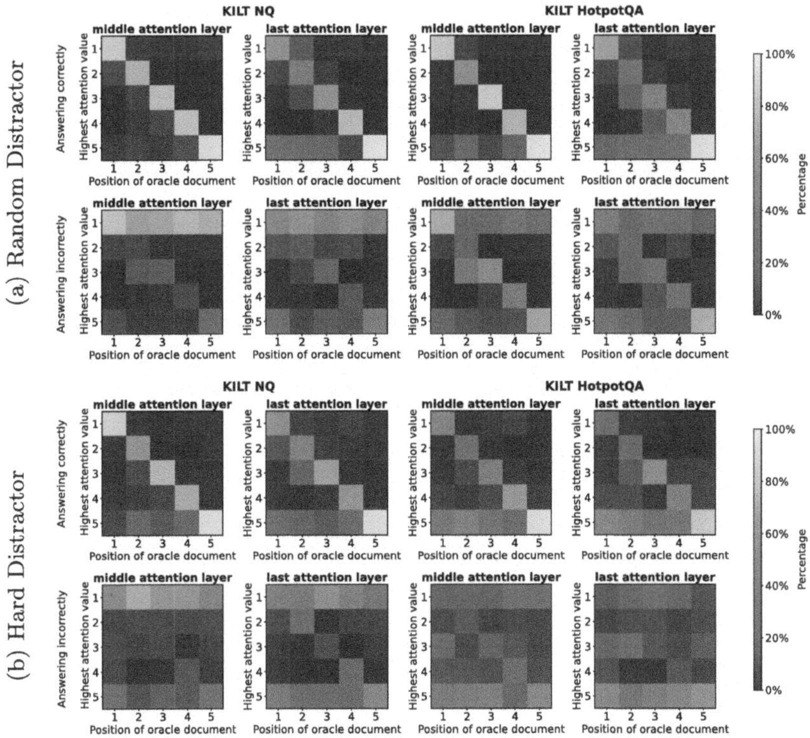

Fig. 5. The attention values of Tinyllama on the random distractor (top) and hard distractor (bottom) test. The variance in the position of the oracle document is plotted on the x-axis, and the position of the document with the most attention on average is plotted on the y-axis.

This pattern is stronger on KILT-NQ than KILT-HotpotQA. In the former, we see that the attention values of the middle layer follow the position of the oracle document closely when answering correctly. This pattern is displayed by the clear diagonal in the plot. When the answer to the question is incorrect, the model attributes most of its attention to the first document.

TinyLlama shows similar changes in attention values in the hard distractor task, as in the random distractor task as found in Fig. 5 (bottom). These results are very similar to those of the relevant context test, which have been omitted for brevity. However, TinyLlama shows higher attention values at the last position regardless of the position of the oracle document on the hard distractor task than the random distractor task. This finding contrasts with the findings of positional bias found in TinyLlama, where the model only showed primacy bias without significant changes over the different test scenarios.

Llama2 displays exceptionally consistent attention in the middle layer when changing the position of the oracle document in the random distractor test, found in Fig. 6 (top). This plot shows distinct colors for the five document positions,

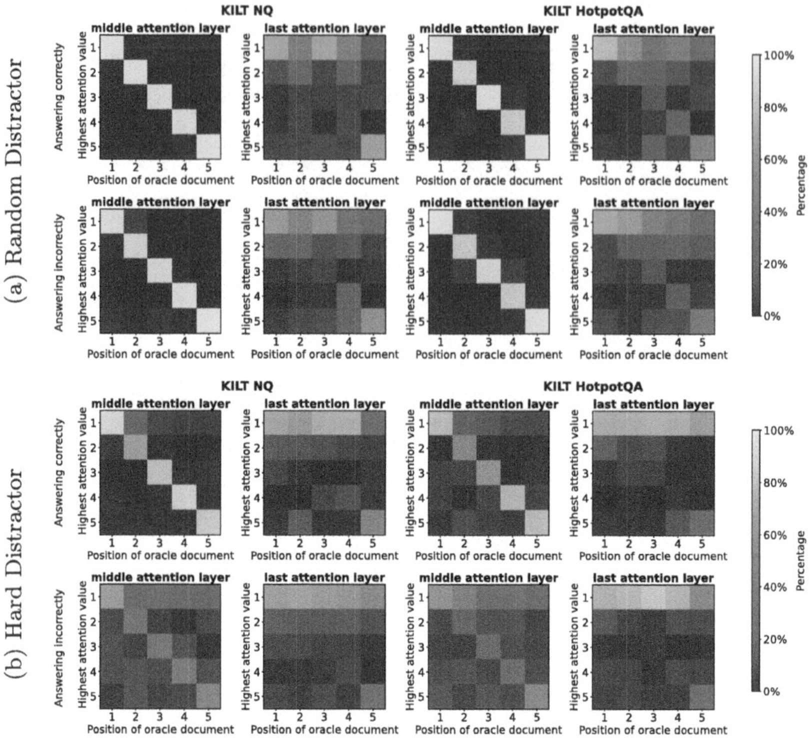

Fig. 6. The attention values of Llama2 on the random distractor (top) and hard distractor (bottom) test. The variance in the position of the oracle document is plotted on the x-axis, and the position of the document with the most attention on average is plotted on the y-axis.

as the oracle document receives the highest attention when varying its location. These results align with the minimal occurrence of positional in this setting as found in Fig. 2, meaning that in this setting and context length, there is a pattern between the amount of positional bias and the correct attribution of attention. Llama2 shows no pattern between the quality of the answer and the correctness of the attribution of attention. This would be explained by the insignificant amount of positional bias that this model shows on this test. Figure 6 (bottom) illustrates Llama2's attention during the hard distractor task, which closely aligns with its attention in the relevant context task; the latter has been omitted for brevity. In both the hard distractor and relevant context tests, Llama2 demonstrates an attribution of attention that is less closely aligned with the position of the oracle document compared to the random distractor task.

7 Discussion and Conclusions

The goal of this research was to investigate positional bias in state-of-the-art LLMs, and expand upon the findings by [7] using RAG on two question answering datasets, both split into three identical tests.

We found that the smaller used model, TinyLlama, showed significant positional bias regardless of the type of distractor documents. The larger models showed positional bias to a lesser extent and could quite confidently distinguish the oracle document from the random distractors. The models showed quite significant positional bias on the hard distractor and relevant context tests. Both Llama2-7b-chat-hf and Mixtral-8x7b-instruct showed positional bias in the input context locations, unlike the findings of [7]. Rather than only *Lost in the Middle*, we observed varying patterns better described as *Lost but Not Only in the Middle*. More generally, understanding the exact positional bias of particular models in a RAG setting is important when designing a RAG framework for real-life applications.

Our findings are supported by the analysis of the attention values in TinyLlama and Llama2, revealing a significant pattern between the correctness in the placement of attention and the quality of the answer. This finding would be explained by the models incorrectly identifying relevant information in the context using its attention mechanism, which then causes decreased performance, or positional bias. However, we need to be careful with describing this as the actual cause of positional bias, with other research doubting the truthfulness of using attention as an explanation [1].

Concluding, our results both quantify the impact and importance of positional bias in RAG, but also show that further research on the phenomenon of positional bias in LLMs is needed. In future research, we are extending the experiments to further models and other RAG settings (including the TREC RAG track datasets). We are also extending the analysis of attention and understanding of how this impacts the models internally, as our controlled experiments relating attention to performance can also help understand the types of explanations we can derive from such attention analysis [1, e.g.,].

Acknowledgement. Our research was greatly facilitated by the RAG library Bergen [11] providing an adaptable framework of state-of-the-art transformer models in combination with a wide range of retrievers, re-rankers, and datasets.

David Rau and Jaap Kamps are partly funded by the Netherlands Organization for Scientific Research (NWO CI # CISC.CC.016). Jaap Kamps is further supported by (NWO NWA # 1518.22.105), the University of Amsterdam (AI4FinTech program), and ICAI (AI for Open Government Lab). Views expressed in this paper are not necessarily shared or endorsed by those funding the research.

References

1. Bibal, A., et al.: Is attention explanation? An introduction to the debate. In: Muresan, S., Nakov, P., Villavicencio, A. (eds.) Proceedings of the 60th Annual Meeting of the Association for Computational Linguistics (Volume 1: Long Papers), Dublin, Ireland, pp. 3889–3900. Association for Computational Linguistics (2022). https://doi.org/10.18653/v1/2022.acl-long.269. https://aclanthology.org/2022.acl-long.269/
2. Dehghani, M., Azarbonyad, H., Kamps, J., de Rijke, M.: Learning to transform, combine, and reason in open-domain question answering. In: Culpepper, J.S., Moffat, A., Bennett, P.N., Lerman, K. (eds.) Proceedings of the Twelfth ACM International Conference on Web Search and Data Mining, WSDM 2019, Melbourne, VIC, Australia, 11–15 February 2019, pp. 681–689. ACM (2019). https://doi.org/10.1145/3289600.3291012
3. Izacard, G., Grave, E.: Leveraging passage retrieval with generative models for open domain question answering. In: Merlo, P., Tiedemann, J., Tsarfaty, R. (eds.) Proceedings of the 16th Conference of the European Chapter of the Association for Computational Linguistics: Main Volume, pp. 874–880. Association for Computational Linguistics, Online (2021). https://doi.org/10.18653/v1/2021.eacl-main.74. https://aclanthology.org/2021.eacl-main.74/
4. Jiang, A.Q., et al.: Mixtral of experts. CoRR abs/2401.04088 (2024). https://doi.org/10.48550/arxiv.2401.04088
5. Kim, S., et al.: SOLAR 10.7B: scaling large language models with simple yet effective depth up-scaling. In: Yang, Y., Davani, A., Sil, A., Kumar, A. (eds.) Proceedings of the 2024 Conference of the North American Chapter of the Association for Computational Linguistics: Human Language Technologies (Volume 6: Industry Track), Mexico City, Mexico, pp. 23–35. Association for Computational Linguistics (2024). https://doi.org/10.18653/v1/2024.naacl-industry.3. https://aclanthology.org/2024.naacl-industry.3/
6. Lewis, P., et al.: Retrieval-augmented generation for knowledge-intensive NLP tasks. In: Larochelle, H., Ranzato, M., Hadsell, R., Balcan, M., Lin, H. (eds.) Advances in Neural Information Processing Systems, vol. 33, pp. 9459–9474. Curran Associates, Inc. (2020). https://proceedings.neurips.cc/paper_files/paper/2020/file/6b493230205f780e1bc26945df7481e5-Paper.pdf
7. Liu, N.F., et al.: Lost in the middle: how language models use long contexts. Trans. Assoc. Comput. Linguist. **12**, 157–173 (2024). https://doi.org/10.1162/tacl_a_00638. https://aclanthology.org/2024.tacl-1.9/
8. Mallen, A., Asai, A., Zhong, V., Das, R., Khashabi, D., Hajishirzi, H.: When not to trust language models: investigating effectiveness of parametric and non-parametric memories. In: Rogers, A., Boyd-Graber, J., Okazaki, N. (eds.) Proceedings of the 61st Annual Meeting of the Association for Computational Linguistics (Volume 1: Long Papers), Toronto, Canada, pp. 9802–9822. Association for Computational Linguistics (2023). https://doi.org/10.18653/v1/2023.acl-long.546. https://aclanthology.org/2023.acl-long.546/
9. Petroni, F., et al.: KILT: a benchmark for knowledge intensive language tasks. In: Toutanova, K., et al. (eds.) Proceedings of the 2021 Conference of the North American Chapter of the Association for Computational Linguistics: Human Language Technologies, pp. 2523–2544. Association for Computational Linguistics, Online (2021). https://doi.org/10.18653/v1/2021.naacl-main.200. https://aclanthology.org/2021.naacl-main.200/

10. Peysakhovich, A., Lerer, A.: Attention sorting combats recency bias in long context language models. CoRR abs/2310.01427 (2023). https://doi.org/10.48550/arxiv.2310.01427
11. Rau, D., et al.: BERGEN: A Benchmarking Library for Retrieval-Augmented Generation (2024). https://github.com/naver/bergen. gitHub repository
12. Rau, D., Wang, S., Déjean, H., Clinchant, S., Kamps, J.: Context embeddings for efficient answer generation in retrieval-augmented generation. In: Proceedings of WSDM 2025. ACM (2025)
13. Tang, R., Zhang, C., Ma, X., Lin, J., Ture, F.: Found in the middle: permutation self-consistency improves listwise ranking in large language models. In: Duh, K., Gomez, H., Bethard, S. (eds.) Proceedings of the 2024 Conference of the North American Chapter of the Association for Computational Linguistics: Human Language Technologies (Volume 1: Long Papers), Mexico City, Mexico, pp. 2327–2340. Association for Computational Linguistics (2024). https://doi.org/10.18653/v1/2024.naacl-long.129. https://aclanthology.org/2024.naacl-long.129/
14. Touvron, H., et al.: Llama 2: open foundation and fine-tuned chat models. CoRR abs/2307.09288 (2023). https://doi.org/10.48550/arxiv.2307.09288
15. Xiao, G., Tian, Y., Chen, B., Han, S., Lewis, M.: Efficient streaming language models with attention sinks. In: The Twelfth International Conference on Learning Representations, ICLR 2024, Vienna, Austria, 7–11 May 2024. OpenReview.net (2024). https://openreview.net/forum?id=NG7sS51zVF
16. Yu, Y., et al.: Mitigate position bias in large language models via scaling a single dimension. CoRR abs/2406.02536 (2024). https://doi.org/10.48550/arxiv.2406.02536
17. Zhang, P., Zeng, G., Wang, T., Lu, W.: Tinyllama: an open-source small language model. CoRR abs/2401.02385 (2024). https://doi.org/10.48550/arxiv.2401.02385

Biased PromptORE: Enhancing Relation Extraction in Gendered Languages and Complex Texts The Case of Spanish Documents from the XVIth Century

Michel Boeglin[1], David Kahn[2,3], Héctor López Hidalgo[4], Josiane Mothe[4,5], Diego Ortiz[4], and David Panzoli[2,4](✉)

[1] IRIEC – Univ. Paul-Valéry, Montpellier, France
michel.boeglin@univ-valery.fr
[2] INU Jean-Francois Champollion, Albi, France
{david.kahn,david.panzoli}@univ-jfc.fr
[3] FRAMESPA, UMR5136 CNRS, Albi, France
[4] IRIT, UMR5505 CNRS, Toulouse, France
{hector.hidalgo,josiane.mothe,diego.ortiz,david.panzoli}@irit.fr
[5] INSPE, Univ. de Toulouse, Toulouse, France

Abstract. Semantic relations between entities such as places, people, and concepts are a widely used method for representing knowledge. Current approaches to relation extraction typically rely on Large Language Models (LLMs) including BERT. PromptORE (Prompt-based Open Relation Extraction) was developed to enhance relation extraction using LLMs for general-purpose documents. However, it is less effective when applied to historical texts, particularly in languages other than English. In this paper, we present an adaptation of PromptORE for specialized documents, specifically digital transcripts of trials from the Spanish Inquisition. Our approach fine-tunes transformer models with their pre-training objective on the data for inference, a process we call "biasing." This addresses challenges such as intricate entity arrangements and gender-related issues in Spanish texts. We address these challenges through prompt engineering. Our method is evaluated using encoder-based models and validated through expert assessments. In addition, we use a binary classification to assess performance. Our results demonstrate significant accuracy improvements, with Biased PromptORE models outperforming baseline PromptORE models by up to 50%.

Keywords: Information systems · Information extraction · Large language models · Prompt engineering · Historical documents

1 Introduction

Semantic relations between entities, such as places, people, and concepts, are commonly used to represent knowledge. In natural language processing (NLP),

C. Hauff et al. (Eds.): ECIR 2025, LNCS 15572, pp. 262–275, 2025.
https://doi.org/10.1007/978-3-031-88708-6_17

the task of extracting these relationships is known as relation extraction, which involves identifying and classifying relationships between named entities in texts. Automated relation extraction is typically implemented as a pipeline, where entities are first recognized, then their relationships are identified [18]. Alternatively, relation extraction can be implemented as a joint extraction process, where entities and relations are identified simultaneously [12]. While joint extraction is more complex, it is considered more effective, as it better captures dependencies between entities [24]. Recent relation extraction methods are based on Large Language Models (LLMs) including BERT-models [14], which have consistently achieved state-of-the-art performance in various NLP tasks [20]. However, most of these methods are optimized for contemporary texts [26]. When applied to domain-specific documents, they encounter unique challenges related to specialized language and complex syntax.

Documents from expert domains often contain specialized language, terminology, or idioms that may not be captured by modern LLMs or general-purpose dictionaries. As a result, modern NLP tools may struggle to accurately extract entities from these types of documents. This issue has been observed in domains such as legal [3] and medical texts [13], where specific entity recognition systems have been developed. For instance, the medical field has created tailored named entity recognizers like BioBERT [13] to handle domain-specific challenges.

In fields like History, however, documents can span various periods and sources, making the development of a universal named entity recognizer even more challenging. A recognizer designed for one period may not generalize well across different periods. For example, the NewsEye project [17] focuses specifically on entity extraction from 19th-century newspapers, highlighting the difficulty of creating tools that can handle diverse historical texts.

Documents in specialized fields may also exhibit unique syntactic structures.

The sentence below is extracted from *El proceso de Pedro de Cazalla* [1], an inquisitorial trial from the 16th century. This text, written in Old Castilian, reflects the lexical and syntactic features typical of Spanish from that era, as shown in nits translation in Appendix A.

"Y que después de esto, que podría haber un año que el dicho Padilla fue a Pedrosa a conocer al dicho Pedro de Cazalla, y que una tarde el dicho Pedro de Cazalla le llevó al dicho Padilla a casa de esta confesante, y esta confesante llamó a Catalina Román, para que también le conociese, y que allí no se hablaron en ninguna cosa de estas opiniones, por estar su madre de esta confesante delante; y que luego el mismo día, porque él se quiso despedir e irse, esta confesante y la dicha Catalina Román se salieron a las eras; tornó a decir que los que se salieron a las eras eran los dichos Pedro de Cazalla y Padilla, y que al tiempo que Catalina Román salieron con ellos, hasta la puerta de la dicha casa, y allí se despidieron con regocijo, dándose a entender, los unos a los otros, que estaban en esto del purgatorio; y que después de esto otra vez fue el dicho Padilla a Pedrosa, *y fue a visitar al dicho Pedro de Cazalla, el cual, delante de esta confesante y de la dicha Catalina Román, riñó muy de veras al dicho Padilla, diciéndole que hablaba muy rotamente de estas cosas con todas las gentes."*

Since Nebrija's first grammar of Castilian was published in 1492 until the establishment of the Real Academia Española (RAE) in 1713, the Spanish language underwent significant changes. These changes were syntactic, lexical, and prosodic. During this period, a system of phoneme accentuation and standardized punctuation emerged. Between 1726 and 1815, the RAE reformed Spanish orthographic rules, establishing most of the modern rules [19]. Back then, punctuation rules were different from today's standards and were designed mainly for reading aloud. In contemporary texts, punctuation serves different purposes, such as rhythm, flow, and disambiguation of meaning. Nowadays texts favor short sentences. The unique sentence above contains over fifteen entities (underlined in the text), many of which are nested within others.

Isolated entities present another challenge to address. For example, in the following sentence of the Cazalla trial, *"Pasó ante mí; Sebastián de Landeta, Notario."* (He/she testified before me; Sebastián de Landeta, Notary.), the notary named Sebastián de Landeta confirms that the testimony was indeed given by the witness. The isolated entity here has a relationship with the document itself.

All of this makes automatic relation extraction very challenging, even for the most advanced techniques. Liu et al. [15] stated that relation extraction methods are typically designed for triplets consisting of two entities and their relationship within one sentence, usually short.

In this paper, we present a step-by-step solution to deal with complex sentences. First we present in Sect. 2 the state-of-the-art methods upon which we base our original approach. Section 3 introduces our research methodology and our hypothesis of work. Then, a description of our method follows in Sect. 4. Finally, we present and discuss our findings in Sect. 5 and give our conclusions in Sect. 6.

2 Related Work

Since 2019, language models such as BERT, RoBERTa, and SpanBERT have significantly advanced relation extraction (RE) tasks by leveraging rich semantic representations learned from large-scale corpora. These models, built on transformer-based architectures and enhanced with auxiliary embeddings (e.g., position and segment embeddings), excel at capturing complex language patterns and dependencies. Among MLMs, BERT has become a cornerstone for RE, primary due to its masked language modeling objective, which enables the prediction of masked tokens in context. For example, Soares et al.[21] introduced entity markers and a novel training strategy, "matching the blanks," to significantly improve few-shot RE performance while reducing the dependency on extensive labeled datasets. SpanBERT[11] extended BERT by incorporating span-level masking and boundary representations, which proved effective for span-based selection tasks. LUKE [25] further enhanced the capabilities of BERT by introducing an entity-aware self-attention mechanism, explicitly modeling interactions between entities and their surrounding context.

Recently, prompt engineering has emerged as a powerful technique for adapting pre-trained language models to specific RE tasks. By crafting targeted natural language instructions, prompt engineering guide the model toward the desired task, reducing the need for fine-tuning. Recent work in this area [15] has shown the versatility of MLM-based objectives, such as those used by BERT-like models. It employs token corruption with a [MASK] token; the model then reduces the cross-entropy loss function by predicting the masked tokens, reframing RE as a prompt engineering problem. Some techniques have expanded on this by incorporating special continuous tokens to refine and improve the prompts [22], while others use token positions and tagging strategies to enhance relation extraction [4]. However, to our knowledge, PromptORE [6] is the only approach that does not require model fine-tuning to perform relation extraction.

PromptORE operates on a dataset $\mathcal{D}$ where each sentence contains two entities. Relationships are encoded using a prompt structure $\mathcal{P}$. The hidden state of the [MASK] token in the final BERT encoder layer ($\mathcal{P}$, [MASK]) is compared with other [MASK] hidden states among all sentences using cosine similarity to cluster the relationships. This technique capitalizes on prompt design, a critical element in relation extraction with transformer models. Notably, PromptORE applies BERT's pre-trained MLM objective without requiring fine-tuning [6].

Considering information extraction from historical documents, most studies focus on texts from the 19th and 20th centuries [17]. This includes studies using GPT models. This focus is largely due to the availability of more consistent and abundant source material [7]. However, to the best of our knowledge, no research has yet addressed the complexities of documents from earlier periods, such as the Modern Age.

Despite their success, current relation extraction techniques often fail to address challenges such as isolated entities, long-range dependency relationships between entities, and specific semantic nuances that are particularly prevalent in texts predating the 19th century. These shortcomings are critical when dealing with documents from specialized fields such as law, history, or science, where conventional models may struggle with the linguistic and structural peculiarities.

Building on the PromptORE framework and considering the unique challenges of historical documents, we hypothesize that additional MLM pretraining, focused on a specific semantic domain, can enhance the effectiveness of prompt engineering. In the following sections, we introduce an original approach, the *bias-prompt-extract technique*, which incorporates a "biasing" process to fine-tune the LLM with a MLM objective for specialized historical documents.

3 Methodology

Spanish historical documents present new challenges that existing relation extraction techniques have yet to solve. We propose a novel approach based on prompt engineering to extract relations from such texts.

We hypothesize that biasing language models toward the target knowledge domain will improve prompt-based relation extraction by addressing difficult entity placements in the text and accommodating gender-specific language aspects of Spanish. Here we biased the model by fine-tuning it with an MLM objective. Additionally, we hypothesize that isolated entities, often overlooked, can be effectively extracted through a tailored prompt, which we refer as "anaphoric prompting."

3.1 Overview

Previous work that used PromptORE on English and general-domain documents demonstrated that encoder-based LLMs like BERT, do not require fine-tuning for relation extraction [6]. However, for specialized domains, such as historical Spanish documents, the general linguistic knowledge captured during the model's pretraining may not be sufficient to effectively capture complex entity relations or deal with domain-specific language structures because of the scarcity of such documents in the training.

We hypothesize that biasing a pre-trained model towards a specific semantic field enhances the model's contextual understanding, enabling it to process domain-specific texts more effectively. By exposing the model to texts from the target domain during additional pre-training (biasing), we align its internal representations with the specific syntactic and semantic features of these documents. This tailored understanding can improve the effectiveness of prompts, especially in dealing with entity positioning challenges and complex grammatical rules, such as those related to gendered languages like Spanish. Additionally, biasing the model helps mitigate the occurrence of uninformative predictions, such as irrelevant punctuation symbols, by guiding the model towards more contextually appropriate predictions.

Thus, biasing the model towards the language and domain of the target texts not only makes it more attuned to the particularities of the historical corpus but also provides a foundation to handle challenges like long-range dependencies and isolated entities which are not typically encountered in modern or general domain texts. It can also help in gender distinction.

3.2 PromptORE

In PromptORE each sentence of the dataset $\mathcal{D}$ contains exactly two entities. The relationships between entity-pairs are encoded with a prompt structure $\mathcal{P}$. The [MASK] token's hidden state from the last computation of the BERT encoder ($\mathcal{P}$, [MASK]) is compared with all other [MASK] hidden states across sentences using cosine similarity to cluster the relationships.

The technique leverages the prompting design that is crucial for relation extraction in transformer models. Initially developed for English, it uses BERT's MLM pretraining objective without fine-tuning.

3.3 Specific Adaptations for Spanish

The original PromptORE design is not directly applicable to Spanish. Specifically, it struggles with gendered language, as it was created for English. In our preliminary tests on Spanish historical documents, pre-trained Spanish transformers predominantly predicted punctuation marks instead of meaningful relations. For instance, the most likely tokens were symbols like ";" and "," or conjunctions like "y" (and) and "o" (or). While grammatically correct, these outputs were not useful for relation extraction. Using BETO [2], a BERT model trained for Spanish, the model produced similar results. With Spanish RoBERTa Base [8], outputs were slightly different but still uninformative. These observations indicate the need to redesign prompts for Spanish-language documents.

The authors of PromptORE evaluated their approach with the FewRel dataset [10], which contains entity-relationship triplets from various domains. However, this dataset consists of general content and does not reflect the complexity of historical documents.

3.4 Addressed Challenges

Encoder models like BERT [5] and RoBERTa [16], are pretrained using a MLM objective to learn the general representation of a language, including its semantics and grammar [23]. These models have proven effective for relation extraction tasks. PromptORE, selected as our baseline, works with sentences containing two entities and assumes that the sentence conveys a relationship between them. It processes with a structured prompt (see Sect. 3.2), and the [MASK] hidden states are compared using cosine similarity, with relationships clustered via an unsupervised classifier like K-means.

PromptORE was evaluated on FewRel [6], but historical Spanish documents do not follow the same sentence structure (see Appendix A). To enhance relation extraction for such texts, we propose biasing BERT and RoBERTa with the MLM objective using a corpus of texts from a specific historical period and domain. Our proposal addresses the following challenges:

- Gendered language in Spanish, which the original English-based technique does not account for.
- Sentences with more than two entities, or sentences containing only one entity.
- Isolated entities, where the relationship exists between the entity and the document itself, a feature of legal documents from the Modern Age (see Sect. 1).

4 Biased-PromptORE Technique

To address these issues, our Biased PromptORE method consists of four phases (See Fig. 1):

- Composition Phase: We prepare the data by building a binomial dataset for benchmarking and selecting texts useful for the biasing phase.
- Biasing Phase: We apply the MLM objective to BETO and RoBERTa models in Spanish, creating a biased model using training selected data.
- Prompting Phase: Our newly designed prompts improve on the original PromptORE for handling Spanish documents from the Modern Age.

– Extraction Phase: We extract the relations on the test part of the binomial dataset created in the composition phase using on the one hand the Biased PromptORE and on the other hand the baseline PromptORE and compare the results.

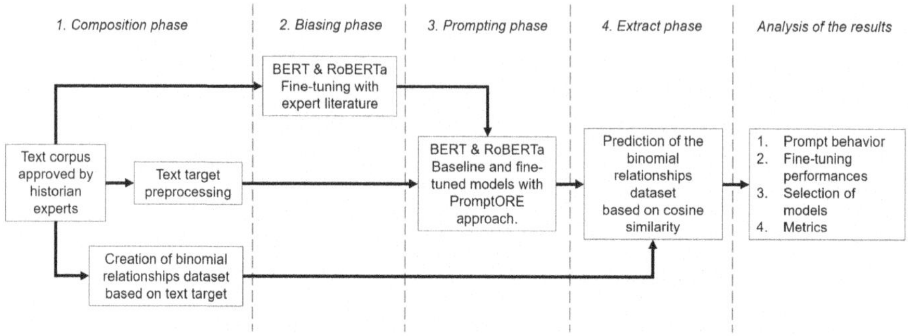

Fig. 1. An overview of our bias-prompt-extract technique for complex historical documents. The biasing phase is inserted between the composition phase (selection of documents from the period) and the prompting phase (testing PromptORE baselines, biased BERT and RoBERTa models).

4.1 Composition Phase

To bias the LLM, we created a corpus centered on the same semantic domain (Inquisitorial Spanish religious history of the XVIth century) with the help of historians. It consists of 5 long (12,661 words on average) historical documents, ranging from legal records (trials of heresy by the Spanish Inquisition) to religious literature (Catholic doctrine books). We divided the corpus between the target text (test) and expert books (training). The target text is the one on which we want to extract relations while the expert books are used to fine-tune BERT and RoBERTa models.

Target Text. The target text is an inquisitorial trial popularly known as "El Proceso de Pedro de Cazalla" [1]. The historians chose this document because it is quite rich with named entities and complex relationships, including the issues we mentioned in the methodology section. The document was first normalized, by removing word contractions and modernizing terms. In addition, named entities were extracted by the historian team. We also use this document in the second dataset to show that knowing the text beforehand improves relation extraction.

Relation extraction from this document faces the challenges we explained in Sect. 3.4) which are common in Spanish texts from the Modern Age.

We found that 75% of the sentences contain fewer than 4 entities, with most having 1, 2, or 3 entities, while a few outliers have up to 130 entities, making parsing difficult.

Expert Books. The expert books consist in a collection of different manuals and expert articles; they are used here to bias BERT and RoBERTa models. The expert book set consists of the following:

- The Bible: We used a popular reformist bible of the XVIth century called "La Biblia del Oso". Historians recommended this tome for MLM fine-tuning. The tokens from this document follow almost a normal distribution and it is the longest, therefore, quite appropriated for the biasing phase.
- Ecclesiastic and Inquisitorial concept dictionaries: These dictionaries contain expert definitions and semantic related fields that will be used to bias the models.
- The trial of Cazalla and the Catechism of Pineda: There are XVIth century legal and religious texts, respectively. Both share the same vocabulary and semantic fields. The trial of Cazalla is used both as a target text and as part of the expert books. We hypothesize the structure and grammar of this text will help in the extraction task.

For the relation extraction task, we constructed a proprietary dataset. The dataset includes annotated sentences with named entities and relations focusing on the types of relations relevant to our domain, such as judicial and religious relations. Historian domain experts annotated these sentences to evaluate accuracy. Additionally, we pre-processed expert-authored books for the biasing phase, which helped further refine the model's ability to handle domain-specific linguistic features (Table 1).

Table 1. Distribution of words by sentence in the Expert Books dataset.

Title	Mean	Std	Median	Total
Bible book	23.97	20.04	22	31,095
Ecclesiastical Dictionary	19.83	14.29	16.00	17,543
Catechism of Pineda	20.04	14.06	16	892
Process of Cazalla	77.13	105	48	1115

4.2 Biasing Phase

We used RoBERTa and BETO encoder models that are already pretrained for Spanish. We selected three models: BETO [2], RoBERTa Base and RoBERTa Large [8]. BETO is a fully trained BERT model in Spanish developed by the University of Computation of Santiago de Chile. RoBERTa Base and Large in Spanish were trained within the MarIA project [8]. We biased the models with the expert books dataset. The training setting is as follows: we use a learning rate of 5e−5 for BETO and RoBERTa Base and 5e−6 for RoBERTa Large; all the models were trained for 5 epochs, the losses were 0.0484 for BETO, 0.0627 for RoBERTa Base, and 0.9899 for RoBERTa Large. In this, we follow what has been done in previous studies [2,5,8,16].

4.3 Prompting Phase

The original PromptORE technique aims to design the simplest prompt as possible. For a given dataset $\mathcal{D}$ composed of S sentences as stated in Eq. 1, given a sentence s with 2 entities (e_1, e_2) and one relation (r) as stated in Eq. 2, the simplest prompt defined in PromptORE is given in Eq. 3. Asking BERT or RoBERTa encoders to predict a grammatically correct word, the answer would be a verb in English [6].

$$\mathcal{D} = S = [s_1, s_2, \ldots, s_n] \tag{1}$$

$$s = (e_1, e_2, r) \tag{2}$$

$$\mathcal{P} = (s, e_1, [MASK], e_2) \tag{3}$$

There is no specific prompt designed for Spanish using PromptORE technique [9]. We created five prompts (Eqs. 4–8) specifically designed for Spanish. Equations 4–6 corresponds to the translation of the original PromptORE with non-fixed gendered prompts for Spanish, while Eqs. 7–8 are fixed gendered prompts.

$$\mathcal{P}_0 = (s, e_1, [MASK], e_2, [SEP]) \tag{4}$$

$$\mathcal{P}_1 = (s, \text{“la relación entre”}, e_1, \text{“y”}, e_2, \text{“es una relación de”}, [MASK], [SEP]) \tag{5}$$

$$\mathcal{P}_2 = (s, \text{“la relación entre”}, e_1, \text{“y”}, e_2, \text{“es”}, [MASK], [SEP]) \tag{6}$$

$$\mathcal{P}_3 = (s, \text{“la relación entre”}, e_1, \text{“y”}, e_2, \text{“es la”}, [MASK], [SEP]) \tag{7}$$

$$\mathcal{P}_4 = (s, \text{“la relación entre”}, e_1, \text{“y”}, e_2, \text{“es el”}, [MASK], [SEP]) \tag{8}$$

Several sentences contain isolated entities; based on a manual analysis by historians, the corresponding relations were all between entities and the text itself (for example: “Paso ante mi, Sebastián Landeta.” *Step before me, Sebastian Landeta.*). We need a specific type of prompting design to extract such relationships. We designed an *anaphoric prompt* to extract them (See Eq. 9). It includes a “phantom” entity, which refers contextually to the sentence itself.

$$\begin{aligned}\mathcal{P}_{Anaphoric} = (s, &\text{“la relación entre”}, e_1, \\ &\text{“y la frase anterior es una relación de”}, \\ &[MASK], [SEP])\end{aligned} \tag{9}$$

Finally, to deal with three entities within a sentence, we choose a combinatory strategy without repetition. This allows us to implement data augmentation for the corpus, multiplying by 3 the number of sentences with 3 entities.

4.4 Relation Extraction Phase

PromptORE technique uses the MLM objective of BETO and RoBERTa Base to predict the most semantically plausible word for the [MASK] token. The final hidden state of the [MASK] token-which is the ultimate representation derived from the model's embedding process-is extracted and analyzed using cosine similarity. The analysis is performed by an unsupervised classifier, such as K-means clustering [6].

In our experiments, before clustering the hidden states the historians checked the outputs of the models to select the most semantically accurate word. We took the ten most probable words predicted by the models and asked the historians to select the 2 models (one for BETO, the other for RoBERTa) to be tested with the binomial classification benchmark. The baseline model had the natural tendency to generalize the context due to their pretrain objective. On the other hand, the biased models were more specific in the selection of words. Finally, the large version of RoBERTa performed poorly. One of the reasons could be that the model has too many parameters considering the relatively low amount of data used to bias it.

5 Results

5.1 Manual Analysis of the Results of Prompting

We test every prompt $\mathcal{P}_n$ with both BETO and RoBERTa. $\mathcal{P}_1$ to $\mathcal{P}_4$ work for 2–3 entities and $\mathcal{P}_{Anaphoric}$ captures isolated entities. Then, we clustered the selected model outputs with K-means clustering as depicted in PromptORE. We set K=2 to test our approach on the binomial dataset.

By analyzing manually the results, we could observe the following:

- $\mathcal{P}_1$ was the most successful of the prompts. It handles genders yielding a rich, accurate semantic selection of words. BETO gives also the most consistent results with $\mathcal{P}_1$.
- $\mathcal{P}_2$ is quite successful when focusing on the semantic field. Nevertheless, $\mathcal{P}_2$ output has a greater degree of randomness than $\mathcal{P}_1$.
- $\mathcal{P}_3$ and $\mathcal{P}_4$ are successful regarding gender aspects. Some words produced by the model are better than $\mathcal{P}_1$ and $\mathcal{P}_2$ because they specify a feminine or masculine relationship.
- The Anaphoric prompt is appropriate for this type of legal text; it correctly identifies the targeted relationships. This result is consistent across the four tests performed. The "phantom entity" allows the biased models to catch the semantic relationship and cluster it accurately.

To summarize the manual analysis we conduced, $\mathcal{P}_1$ is a general and stable option for the biased models, $\mathcal{P}_3$ and $\mathcal{P}_4$ is useful in contexts where the relationships between entities can be defined as grammatically feminine or masculine in Spanish.

Table 2. Results with the binary dataset from our Target Text. Best results are in bold font.

Model	Accuracy	Precision	Recall	F1 Score
BERT (BETO) baseline	0.16	0.06	0.03	0.04
BERT (BETO) biased	**0.66**	**0.65**	**0.58**	**0.77**
RoBERTa Base baseline	0.30	0.33	0.36	0.34
RoBERTa Base biased	0.46	0.47	0.44	0.45

5.2 Evaluation of Relation Extraction

Table 2 reports the comparison between the biased PromptORE models we developed and the PromptORE considered as baselines. As we can see, baseline BETO and both, RoBERTa Base and Large baselines, performed poorly. The outputs predicted for the [MASK] token in the classification were too general, giving low accuracy, precision, recall, and F-1 scores.

Biased PromptORE models are significantly more effective across all metrics. Biased BETO consistently outperforms the other models. Also, RoBERTa Base is improved when biased, whatever the measure. Biased PromptORE RoBERTa Base improves by an average of 20% all its scores compared to the baseline. The best model, Biased PromptORE BETO, improves accuracy up to 50%, resulting in an overall F1 score of 77%. This model was also selected by the historians as the model to use for relation extraction because of its performance and accuracy. This model, combined with $\mathcal{P}_1$ and $\mathcal{P}_a naphoric$ prompts made much more effective relation extraction within the target text.

We found that using combinatory rules with no repetition for the sentences that contain 3 entities also works quite well, as the model can accurately predict different types of relationships among entities within the same sentence. Nevertheless this approach will yield incorrect results for more than 3 entities within the same sentence. Having more than 3 entities may yield uninformative or non-accurate results because the further apart the entities are, the less likely they have a meaningful relationship within the text.

For biased RoBERTa Large, its poor performance may be due to the tradeoff between the relatively small amount of data used and the size of the model itself, which may be too large to properly being biased as mentioned in [9].

When analyzing the confusion matrices for both the PromptORE baselines and biased PromptORE models, the results demonstrate significant improvements for both models. This indicates that biasing and fine-tuning in a downstream task are particularly beneficial for complex historical documents. For the baselines, many relationships are misclassified. The general Spanish language knowledge of these models renders them overly general and unsuitable for the specific task. However, when biased, both BETO and RoBERTa outperform the baselines by a significant margin, providing richer and more accurate predictions for the [MASK] token.

There is still room for improvement. BETO tends to classify everything as a legal relationship, showing a preference for this type of relationship. The reason behind this could be that some sentences are so contextually specific that, to a non-expert, they appear to describe legal relationships. However, knowing that these models have not undergone further fine-tuning for relation extraction suggests additional possibilities, such as further fine-tuning the models to better process the dataset.

6 Conclusions and Future Work

As we presented throughout this paper, our biased PromptORE modification improves the relation extraction for documents from expert fields, specifically historical ones, and our design of prompt outperforms the simple from PromptORE for Spanish language. This paper also developed a methodology that could be followed for other specific domains and other languages.

Our goal now is to analyze specific mistakes made by the LLM and try to find patterns in those mistakes. We are also developing a multiclass dataset benchmark, also developed by historical documents and hope the results will be enhanced by this approach. The problem of dealing with relation extraction beyond the triplet method still remains an important field of research that other authors tried to solve. Another perspective to this work is to evaluate on other data collections to understand better the strengths and weaknesses of Biased PromptORE. We envision to use SciERC and CoNLL04 datasets for this purpose.

A Translation of a Paragraph from *El proceso de Pedro de Cazalla* [1]

The following text is a tentative English translation of the paragraph from *El proceso de Pedro de Cazalla* [1]:

> *"And that after this, which could have been a year ago, the foresaid Padilla went to Pedrosa to meet the foresaid Pedro de Cazalla, and that one afternoon the foresaid Pedro de Cazalla took the foresaid Padilla to the house of this witness, and this witness called Catalina Román so that she could also meet him, and that they did not discuss any of these matters there, as the witness's mother was present; and that later the same day, because he wanted to say goodbye and leave, this witness and the foresaid Catalina Román went out to the fields; it was said again that those who went out to the fields were the foresaid Pedro de Cazalla and Padilla, and that at the time when Catalina Román left with them, up to the door of the foresaid house, and there they said goodbye with joy, understanding each other, that they were in purgatory about this matter; and that after this, the foresaid Padilla went again to Pedrosa and visited*

the foresaid Pedro de Cazalla, who, in front of this witness and the foresaid Catalina Román, severely reprimanded the foresaid Padilla, telling him that he spoke very carelessly about these things with everyone."

References

1. Proceso de fe de pedro de cazalla. Archivo Histórico Nacional (España), Sección inquisición, legajo 1864, expediente 2
2. Cañete, J., Chaperon, G., Fuentes, R., Ho, J.-H., Kang, H., Pérez, J.: Spanish pre-trained BERT model and evaluation data. arXiv preprint arXiv:2308.02976 (2023)
3. Chalkidis, I., Fergadiotis, M., Malakasiotis, P., Aletras, N., Androutsopoulos, I.: Legal-BERT: the muppets straight out of law school. arXiv preprint arXiv:2010.02559 (2020)
4. Chen, X., et al.: KnowPrompt: knowledge-aware prompt-tuning with synergistic optimization for relation extraction. In: Proceedings of the ACM Web Conference 2022, pp. 2778–2788 (2022)
5. Devlin, J., Chang, M.-W., Lee, K., Toutanova, K.: BERT: pre-training of deep bidirectional transformers for language understanding. arXiv preprint arXiv:1810.04805 (2018)
6. Genest, P.-Y., Portier, P.-E., Egyed-Zsigmond, E., Goix, L.-W.: Promptore - a novel approach towards fully unsupervised relation extraction. In: Proceedings of the 31st ACM International Conference on Information and Knowledge Management, pp. 561–571 (2022)
7. González-Gallardo, C.-E., Boros, E., Girdhar, N., Hamdi, A., Moreno, J.G., Doucet, A.: Yes but.. can ChatGPT identify entities in historical documents? arXiv preprint arXiv:2303.17322 (2023)
8. Gutiérrez-Fandiño, A., et al.: Maria: Spanish language models. arXiv preprint arXiv:2107.07253 (2021)
9. Géron, A.: Aprende machine learning con scikit-learn, keras y tensorflow. Anaya, España (2020)
10. Han, X., et al.: Fewrel: a large-scale supervised few-shot relation classification dataset with state-of-the-art evaluation. arXiv preprint arXiv:1810.10147 (2018)
11. Joshi, M., Chen, D., Liu, Y., Weld, D.S., Zettlemoyer, L., Levy, O.: SpanBERT: improving pre-training by representing and predicting spans. Trans. Assoc. Comput. Linguist. **8**, 64–77 (2020)
12. Kumar, S.: A survey of deep learning methods for relation extraction. arXiv preprint arXiv:1705.03645 (2017)
13. Lee, J., et al.: BioBERT: a pre-trained biomedical language representation model for biomedical text mining. Bioinformatics **36**(4),1234–1240 (2020)
14. Li, J., Sun, A., Han, J., Li, C.: A survey on deep learning for named entity recognition. IEEE Trans. Knowl. Data Eng. **34**(1), 50–70 (2020)
15. Liu, P., Yuan, W., Jinlan, F., Jiang, Z., Hayashi, H., Neubig, G.: Pre-train, prompt, and predict: a systematic survey of prompting methods in natural language processing. ACM Comput. Surv. **55**(9), 1–35 (2023)
16. Liu, Y., et al.: RoBERTa: a robustly optimized BERT pretraining approach. arXiv preprint arXiv:1907.11692 (2019)
17. Oberbichler, S., et al.: Integrated interdisciplinary workflows for research on historical newspapers: Perspectives from humanities scholars, computer scientists, and librarians. J. Am. Soc. Inf. Sci. **73**(2), 225–239 (2022)

18. Pawar, S., Palshikar, G.K.: Bhattacharyya, P.: Relation extraction: a survey. arXiv preprint arXiv:1712.05191 (2017)
19. Petrucci, A.: Alfabetismo, escritura, sociedad. Barcelona, Gedisa (1999)
20. Rothman, D., Gulli, A.: Transformers for Natural Language Processing: Build, Train, and Fine-Tune Deep Neural Network Architectures for NLP with Python, PyTorch, TensorFlow, BERT, and GPT-3. Packt Publishing Ltd., Birmingham (2022)
21. Soares, L.B., FitzGerald, N., Ling, J., Kwiatkowski, T.: Matching the blanks: distributional similarity for relation learning (2019). arXiv preprint arXiv:1906.03158. Accessed 20 Dec 2024
22. Son, J., Kim, J., Lim, J., Lim, H.: Grasp: guiding model with relational semantics using prompt. arXiv preprint arXiv:2208.12494 (2022)
23. Vaswani, A., et al.: Attention is all you need. In: Advances in Neural Information Processing Systems, vol. 30 (2017)
24. Wang, H., Qin, K., Zakari, R.Y., Lu, G., Yin, J.: Deep neural network-based relation extraction: an overview. Neural Comput. Appl. 1–21 (2022)
25. Yamada, I., Asai, A., Shindo, H., Takeda, H., Matsumoto, Y.: Luke: deep contextualized entity representations with entity-aware self-attention. arXiv preprint arXiv:2010.01057 (2020)
26. Zhao, X., et al.: A comprehensive survey on deep learning for relation extraction: recent advances and new frontiers. arXiv preprint arXiv:2306.02051 (2023)

LSTM-Based Selective Dense Text Retrieval Guided by Sparse Lexical Retrieval

Yingrui Yang[1(✉)], Parker Carlson[2], Yifan Qiao[3], Wentai Xie[2], Shanxiu He[2], and Tao Yang[2]

[1] Coursera Inc., Mountain View, USA
yingruiyang@ucsb.edu
[2] University of California at Santa Barbara, Santa Barbara, USA
{parker_carlson,wentaixie,shanxiuhe,tyang}@ucsb.edu
[3] Apple Inc., Cupertino, USA
yifan_qiao@apple.com

Abstract. This paper studies fast fusion of dense retrieval and sparse lexical retrieval, and proposes a cluster-based selective dense retrieval method called CluSD guided by sparse lexical retrieval. CluSD takes a lightweight cluster-based approach and exploits the overlap of sparse retrieval results and embedding clusters in a two-stage selection process with an LSTM model to quickly identify relevant clusters while incurring limited extra memory space overhead. CluSD triggers partial dense retrieval and performs cluster-based block disk I/O if needed. This paper evaluates CluSD and compares it with several baselines for searching in-memory and on-disk MS MARCO and BEIR datasets.

Keywords: Partial dense retrieval · selective fusion with sparse retrieval · efficient document search

1 Introduction and Related Work

Dense and sparse retrievers are two main categories of retrieval techniques for text document search. As shown in previous literature [21,24,25], combining sparse and dense retrieval scores with linear interpolation can boost search relevance. It is important to optimize for efficiency when combining the above two retrieval approaches on a low-cost CPU-only platform. This is desirable in a large-scale search system which employs a multi-stage search architecture, and runs partitioned first-stage retrieval in parallel on a massive number of inexpensive CPU-only machines. It is also critical for search on personal devices such as phones with limited computing/memory resources or battery use constraints.

Dense retrieval can be accelerated with approximate nearest neighbor (ANN) search using partial IVF cluster search [18,23] or proximity-graph-based navigation (e.g. HNSW) [29]. However, there is a significant relevance tradeoff for

C. Hauff et al. (Eds.): ECIR 2025, LNCS 15572, pp. 276–292, 2025.
https://doi.org/10.1007/978-3-031-88708-6_18

a reduced search cost in these efforts. For example, RetroMAE [41], a state-of-the-art dense retriever, is slow on CPU without compression, the use of 5% IVF cluster search reduces CPU time substantially, but there is a 7.5% drop in MRR@10. The use of OPQ quantization [13], implemented in FAISS [18], further reduces memory space to 1.2GB but causes an extra 4.6% MRR@10 drop. Notice that the idea of cluster-based retrieval was explored for traditional information retrieval and selective search [1,4,15].

LADR [20], as a follow-up study of GAR [28], has investigated partial dense retrieval by using the results of a sparse retriever as a seed to select embeddings based on a document-to-document proximity graph. This strategy follows the previous graph-based ANN approaches [29,36]. Using sparse retrieval results as a seed allows quick narrowing of the search scope at a low CPU cost. While LADR and HNSW have demonstrated their high efficiency to search within a time budget, the use of a document-wise similarity graph adds a significant online space requirement on a low-cost computing platform. For example, the proximity graph can take an extra 4.3GB of memory space [20] for 8.8M MS MARCO passages.

Another limitation of LADR and HNSW is the assumption of in-memory access to proximity graphs and dense vectors. For large datasets or higher-dimension embedding vectors, some or all of the embeddings and proximity graphs will have to be stored on disk, especially when such applications desire uncompressed embeddings for better relevance. For example, the embedding dimension of the recent RepLLaMA dense retriever [27] is 4096, based on LLaMA-2 [38], which leads to 145GB storage space for MS MARCO passages. It is possible that more advanced dense models may be developed to take advantages of large language models in the future with a higher embedding dimensionality (e.g. up to 12,288 [2]) for better relevance. For large on-disk search, random access of dense embeddings and/or a document-level proximity graph can incur substantial fine-grained I/O overhead. DiskANN [17] and SPANN [5] are two on-disk ANN search algorithms for general data applications which do not leverage sparse retrieval. They have not been studied in the context of text document retrieval where the importance of semantic text matching presents a unique challenge and requires new design considerations.

One recently published contemporary work, CDFS [42], addresses the above problem. One weakness of CDFS is that it uses probabilistic thresholding to select dense clusters based on a strong assumption that the order statistics of query document ranking is independently and identically distributed. This assumption is only true when the query document similarity scores for relevant and irrelevant documents follow the same distribution, which is rare because they are usually distributed differently.

This paper proposes a lightweight approach called CluSD (Cluster-based Selective Dense retrieval) guided by sparse retrieval results. Unlike CDFS, CluSD does not follow any statistical distribution assumption. CluSD limits query-document similarity computations through a two-stage cluster selection algorithm via an LSTM model, which exploits inter-cluster distances and the overlapping degree between top sparse retrieval results and dense clusters. When dense clusters are not available in memory, CluSD selects and loads a limited number of clusters with efficient disk block I/O. Thus, it can handle a large dataset with high-dimension embeddings that cannot fit in memory, with less I/O overhead compared to a graph-based ANN approach.

2 Selective Dense Cluster Retrieval

Problem Definition. Given query q for searching a collection of D text documents: $\{d_i\}_{i=1}^{D}$, retrieval obtains the top k relevant documents from this collection based on the similarity between query q and document d_i using their representation vectors. With a lexical representation, each representation vector of a document or a query is sparse. Sparse retrieval uses the dot product of their lexical representation vectors as the similarity function and implements its search efficiently using an inverted index: $L(q) \cdot L(d_i)$ where $L(.)$ is a lexical representation which is a vector of weighted term tokens for a document or a query. For BM25-based lexical representation [34], $L(.)$ treats each document or a query as a bag of terms. Term weights are scaled based on the frequency of terms within a document and across all documents in the collection. For a learned sparse representation [8,11,25,30], a document or a query is encoded using a trained neural ranking model, which produces modified representations of document or query containing both original and expanded terms. The encoding functions for a query and document can be different. Without the loss of generality, we assume they are the same in this paper. With a single-vector dense representation, each representation vector of a document or a query is a dense vector. Dense retrieval computes the following rank score: $R(q) \cdot R(d_i)$ where $R(.)$ is a dense representation vector of a fixed size [19].

Following the work of [12,21,24], linear interpolation is used to ensemble dense and sparse retrieval scores. The fused rank score for document d_i in responding query q is: $\alpha L(q) \cdot L(d_i) + (1-\alpha) R(q) \cdot R(d_i)$ where α is a coefficient in a range between 0 and 1. Our goal is to maintain a relevance competitive to full dense retrieval, while minimizing the time and memory overhead of cluster-based dense retrieval, guided by sparse retrieval results.

Design Considerations. We aim to identify the opportunity to skip a substantial portion of dense retrieval by taking the IVF clustering approach [18] that groups similar documents. This helps CluSD to recognize similar relevant documents during inference while avoiding the need of a document-level proximity graph. The number of clusters needs to be as large as possible (meaning a small size per cluster) to avoid unnecessary I/O and computing cost. That imposes a significant challenge for accurate and fast selection of dense clusters. To address this our work proposes a two-stage LSTM-based selection algorithm that exploits the overlap of sparse retrieval results with the potential dense clusters.

2.1 Steps of Online Inference in CluSD

Online inference of CluSD has the following steps: **Step 1**: Conduct sparse retrieval to obtain top-k results. **Step 2**: Conduct a two-stage selection process. Stage I selects the top-n clusters from the given N dense embedding clusters, discussed in detail in Sect. 2.2. Stage II applies an LSTM model to the n top clusters to choose a limited number of dense clusters to evaluate, discussed in Sect. 2.3. **Step 3**: Expand the top-k sparse retrieval results to include the documents vectors of these dense clusters and fuse their sparse and dense scores with linear interpolation.

Time and Space Cost for CluSD. The embedding space is $O(D)$ for a corpus with D vectors. For MS MARCO, this cost is around 27GB without compression. Quantization can reduce the total size to about 0.5GB to 1GB, depending on codebook parameters. Then this cost becomes a fraction of sparse retrieval index cost. Excluding embeddings, the extra space overhead for CluSD is to maintain the similarity among centroids of N clusters. We only maintain the top m cluster neighbors for each cluster in terms of their centroid similarity. For our evaluation, $m = 128$ and this reduces the extra space overhead as $O(N)$, which is negligible compared to the embedding space with $N << D$. For example, with $N = 8192$, the inter-cluster graph for MS MARCO passages takes about 5MB with quantization. Excluding sparse retrieval cost, the time cost of CluSD is dominated by the cost of cluster selection and dense similarity computation of selected dense clusters. The former is about $O(n)$ while the latter is proportional to the number of clusters selected, which is capped by $O(\frac{D}{N}n)$. In our evaluation of MS MARCO and BEIR, n is chosen to be 32, and the actual number of clusters selected and visited is about 22.3. Thus, the overall latency time overhead introduced by CluSD is fairly small.

2.2 Stage I of Step 2: Selection of Top n Candidate Clusters

Given N clusters grouped using dense document embeddings, we devise a strategy to conduct a preliminary selection that prunes low-scoring clusters that contain only less relevant documents. A two-stage approach is motivated by the fact that it is too time consuming for an LSTM model, used in the second stage, to examine all clusters. Thus, we choose the top-n candidate clusters first.

To prioritize clusters, one idea is to adopt their query-centroid distance used in IVF cluster selection. We find this does not work well for CluSD. For example, if we rank clusters based on the query-centroid distance for MS MARCO passages, around 10% of top-10 dense results reside in clusters ranked beyond the 175th cluster, which means we would have to select 175 clusters in order to recover 90% of top-10 documents. Visiting all of these clusters is still too slow and CluSD only visits about 22.3 clusters on average for MS MARCO Dev set.

Our method for the preliminary selection at this stage is to rely on the degree of overlap between dense clusters and the top sparse retrieval results. Specifically, we divide k top sparse retrieval results into v top sparse result bins denoted $B_1, \cdots, B_v$. We assign each embedding cluster C_i with a priority vector $(P(C_i, B_1), \cdots, P(C_i, B_v))$ where $P(C_i, B_j) = |C_i \cap B_j|$, which is the number of documents in top sparse bin B_j included in cluster C_i. A larger $P(C_i, B_j)$ value indicates a higher priority for C_i with respect to sparse result position bin B_j. In our evaluation with retrieval depth $k = 1000$ for the MS MARCO and BEIR datasets, we use $v = 6$ result bins, divided into the top-10 results, top 11–25, top 26–50, top 51–100, top 101–200, top 201–500, and top 501 to k documents. The main reason to divide sparse results into bins is to reduce the time complexity of LSTM for cluster selection with a shorter sequence. The reason to have a non-uniform bin size is that the importance of rank positions derived by sparse retrieval is nonlinear. For example, the top 1 to 50 results are much more important than the results ranked from 50 to 1,000, requiring finer grained partitioning. Then, we perform multikey sorting based on cluster priority vectors $(P(C_i, B_1), \cdots, P(C_i, B_v))$. The clusters are sorted first based on $(P(C_i, B_1)$, then $(P(C_i, B_2)$ if there's a tie. If there is a tie between two clusters, then we rank them by their centroid distance to the query. The top-n ranked candidate clusters will be the input for the LSTM model described below.

2.3 Stage II of Step 2: Selection with LSTM

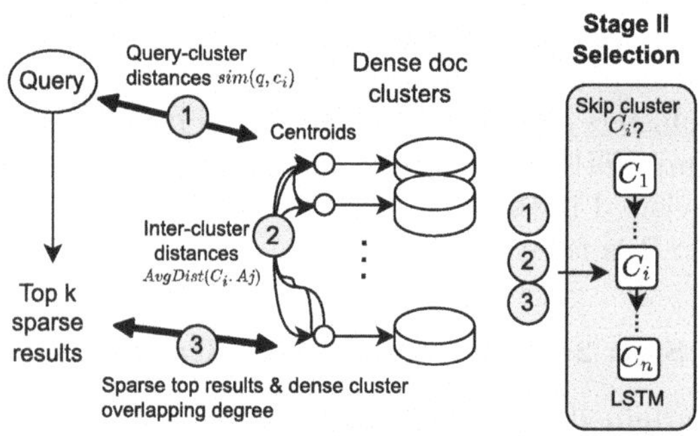

Fig. 1. Illustration of CluSD and its features.

To select which dense clusters we visit, we adopt a simple LSTM model [16] which takes a sequence of clusters as input along with their features. We choose an LSTM because it can capture cluster selection order as a sequence. Selected clusters can influence what will be selected next based on their semantic distance and other features. This allows CluSD to select clusters accurately to minimize I/O and computing cost. The time complexity of the LSTM model is limited by $O(n)$, where the n top clusters are given from Stage I.

The LSTM sequentially visits each cluster, denoted C_1 to C_n. For each cluster C_i, the LSTM predicts a score $f(C_i)$ such that $0 \leq f(C_i) \leq 1$. If $f(C_i) \geq \Theta$, where Θ is a set prediction threshold, then cluster C_i should be visited. Θ controls the tradeoff between efficiency and relevance; in practice, Θ can be tuned based on the overall latency requirement. In our experiments, the default setting of CluSD is to use a threshold of 0.02, yielding 22.3 clusters selected on average. Due to the page limit, we have omitted a study of the impact of varying this threshold on the number of clusters selected.

As depicted in Fig. 1, the feature input vector for the current cluster C_i for the above LSTM computation to produce $f(C_i)$ is composed of the following three groups of features:

- **Query-cluster similarity:** $sim(q, c_i)$. The similarity distance of this query q with the centroid c_i of cluster C_i.
- **Inter-cluster similarity:** $AvgDist(C_i, Aj)$ for $1 \leq j \leq u$. Given the n sorted clusters derived from Stage I, we uniformly partition these n consecutive clusters into u consecutive cluster bins: $\{A_1, A_2, \cdots, A_u\}$. Then we define

  $$AvgDist(C_i, Aj) = \frac{1}{|A_j|} \sum_{c_l \in A_j} sim(c_i, c_l)$$

  where c_l is the centroid representing cluster C_l in bin A_j. The above formula computes the mean distance between centroid c_i and the centroid of each cluster within Bin A_j.
  The purpose of this feature group is to capture the distances of a cluster to previously examined clusters and unexamined clusters following the LSTM flow. When the mean distance between cluster c_i and clusters in a previously-examined bin is close, if many clusters in that bin have been selected, cluster c_i may have a good chance to be selected. In our evaluation with MS MARCO and BEIR datasets, we found that $u = 6$ is appropriate. As discussed earlier in Sect. 2.1, to reduce the extra space cost, we only maintain top-m cluster neighbors of each cluster where m is not large (128 in our evaluation).
- **Cluster overlap:** $P(C_i, B_j)$ and $Q(C_i, B_j)$ for $1 \leq j \leq v$.
  This group of features captures the position-weighted and score-weighted overlap degree of this embedding cluster C_i with the top sparse results. Stage I in Sect. 2.2 divided the top-k sparse retrieval results into v position bins $B_1, \cdots B_v$ and defined $P(C_i, B_j)$ as the count-based overlap measure. Now, we define a score-weighted overlap degree $Q(C_i, B_j)$ as the average sparse rank score of documents that are in both C_i and sparse position bin B_j. Namely,

  $$Q(C_i, B_j) = \frac{\sum_{d \in C_i \cap B_j} SparseRankScore(d)}{|C_i \cap B_j|}.$$

  Clusters with high Q scores for the top bins contain many top sparse results and may be selected.

Training of LSTM. We assume that the training data contains a set of queries and relevance-labeled documents, that allows the computation of a ranking metric such as MRR or NDCG. For example, from MS MARCO training set we randomly sample 5000 training instances, where each training instance contains a query and one or more relevant documents. To train the LSTM for cluster selection, for each training instance, we mark the positive and negative examples for top cluster selection. If a cluster contains one of top-10 dense retrieval results, we mark this cluster as positive otherwise negative. This approximately indicates that such an embedding cluster should be visited for a fusion.

3 Evaluations

Datasets and Measures. Our evaluation uses the MS MARCO dataset with 8.8 million passages [3,7] for training and in-domain search, and the 13 publicly available BEIR datasets [37] for zero-shot retrieval. The size of BEIR data sets ranges from 3,633 to 5.4M documents. To report the mean latency and 99 percentile latency, we test queries multiple times using a single thread on a 2.25GHz AMD EPYC 7742 CPU server with PCIe SSD.

Models and Parameters. For sparse retrieval, we use SPLADE [9,10,22] with hybrid thresholding [32] and BM25-guided traversal [31,33]: 1) SPLADE-HT1 with index space 3.9GB and 31.2 ms retrieval latency. We also test on three other sparse models: 2) uniCOIL [11,25] with 1.2GB index and 35.5ms latency; 3) LexMAE [35] with 3.7GB index and 180ms latency; 4) BM25-T5 that applies BM25 with DocT5Query document expansion [6]. BM25-T5 has a 1.2GB index and 9.2ms latency. For dense retrieval, we adopt the recent dense models SimLM [39], RetroMAE [41], and RepLLaMA [27]. We use K-means in the FAISS library [18] to derive dense embedding clusters. RetroMAE-2 [26] is not used because its checkpoint is not released.

Our evaluation implementation uses C++ and Python. To train the CluSD model, we sample 5000 queries from the MS MARCO training set and the hidden dimension for LSTM is 32. The models are trained for 150 epochs. For sparse and dense model interpolation, we use min-max normalization to rescale the top results per query. For BM25-T5, the interpolation weights are 0.05 and 0.95 for sparse and dense scores, respectively. For other sparse retrieval models, they are 0.5 and 0.5. The results are marked with tag † when statistically significant drop is observed compared to the baseline marked with ▲ at 95% confidence level.

Time Budget. We mainly use a time budget of about 50ms including the sparse retrieval time for the average latency, and the use of this budget allows us to choose the algorithm parameters accordingly. We choose this budget because in a search system running as an interactive web service, a query with multi-stage processing needs to be completed within a few hundred milliseconds. A first-stage retriever operation for a data partition needs to be completed within several tens of milliseconds without query caching.

Table 1. Cluster-based in-memory search with and without space constraints

		MSMARCO Dev		DL19	DL20	BEIR	Latency
	% D	MRR@10	R@1K	NDCG@10	NDCG@10	NDCG@10	ms
Uncompressed flat setting. Embedding space 27.2GB							
D=RetroMAE	100	0.416†	0.988	0.720	0.703	0.482	1674.1
▲ S + D	100	0.425	0.988	0.740	0731	0.520	1705.3
S + CDFS	0.45	0.424	0.987	–	–	0.517	46.0
S + CluSD	0.3	0.426	0.987	0.744	0.734	0.518	44.4
OPQ $m = 128$. **Embedding space 1.1GB**							
▲ S + D-OPQ	100	0.416	0.988	0.737	0.732	0.515	600.1
S+D-IVF	10	0.404†	0.987	0.713	0.722	0.513	126.4
S+D-IVF	5	0.394†	0.987	0.687	0.706	0.507	80.0
S+D-IVF	2	0.374†	0.986	0.656	0.700	0.499	52.5
S+CDFS	0.45	0.415	0.986	0.740	0.730	–	43.3
S + CluSD	0.3	0.417	0.986	0.742	0.735	0.514	42.6

Cluster-Based Retrieval with In-Memory Data. Table 1 compares CluSD with a few baselines for cluster-based selective retrieval in searching MS MARCO and BEIR datasets with and without compression. The dense model is RetroMAE and sparse model is SPLADE-HT1. IVFOPQ uses a proportion of top dense clusters sorted by the query-centroid distance. OPQ quantization from FAISS is configured with the number of codebooks as $m = 128$ or 64. We report performance of IVF clustering using the top 10%, 5%, or 2% clusters respectively. The notation "S+D" in Table 1 (and later in other tables) reports the fused performance of a sparse model with linear interpolation. Marker "%D" means the approximate percentage of document embeddings evaluated based on the number of clusters selected. The last column marked with "Latency" is the mean single-query time for MS MARCO Dev set. CluSD uses setting $N = 8192$ and $n = 32$, which results in selection of 22.3 clusters on average per query and allows CluSD to meet the latency requirement. The results in Table 1 shows that under the same time budget, CluSD outperforms partial dense retrieval with IVF search in relevance. Additionally, CluSD performs slightly better than CDFS in terms of both relevance and latency because CluSD's LSTM effectively selects less clusters to search than CDFS.

CluSD vs. Graph Navigation. Table 2 compares CluSD with selective dense retrieval based on proximity graph navigation including HNSW and LADR under time budget around 50ms. HNSW's expansion factor parameter (ef) is set as 1024. For LADR, we use its default setting with seed = 200, number of neighbors = 128 and use an exploration depth of 50 to meet the latency budget. The takeaway from this table is that relevance of CluSD is better than HNSW while CluSD has similar or slightly better relevance without incurring significant extra space cost, compared to LADR. When the time budget is around 25ms, the takeaway is similar.

Table 2. CluSD vs. graph navigation methods

	MSMARCO Dev		**DL19**	**DL20**	**Latency**	**Space**
	MRR@10	R@1K	NDCG	NDCG	Total(ms)	GB
Time Budget = 50 ms, S=SPLADE-HT1						
D=SimLM	$0.411^{\dagger}$	$0.985^{\dagger}$	0.714	0.697	1674.1	27.2
S	$0.396^{\dagger}$	$0.980^{\dagger}$	0.732	0.721	31.2	3.9
▲ S + D	0.424	0.989	0.740	0.726	1705.0	31.1
HNSW	$0.409^{\dagger}$	$0.978^{\dagger}$	0.669	0.695	54.4	40.3
S + HNSW	0.420	0.987	0.718	0.723	59.6	40.3
S + LADR	0.422	$0.984^{\dagger}$	0.743	0.728	43.6	35.4
S + CluSD	0.426	0.987	0.744	0.724	46.3	31.1

Table 3. Zero-shot performance in average NDCG@10 on BEIR datasets.

	BM25	SimLM	RetroMAE	SPLADE-HT1	SPLADE-HT1+SimLM ▲ flat	rrk	**CluSD flat**	**CluSD m=128**	**CluSD m=64**	SPLADE-HT1+RetroMAE ▲ flat	rrk	**CDFS flat**	**CluSD flat**	**CluSD m=128**	**CluSD m=64**
Avg.	0.440	0.429	0.482	0.500	0.518	0.496	0.516	0.514	0.511	0.520	0.483	0.517	0.518	0.514	0.506

Detailed Zero-Shot Performance with BEIR. Table 3 lists the performance of CluSD with SimLM and RetroMAE after SPLADE-HT1 retrieval in searching each of 13 BEIR datasets. The LSTM model of CluSD is trained with the MS MARCO training set, and it is directly applied to each BEIR dataset to select clusters without further tuning. As a baseline, column "▲ flat" is to fuse the uncompressed full dense retrieval with SPLADE-HT1 sparse retrieval. Column marked "rrk" is to rerank top 1,000 sparse retrieval results interpolated with their dense scores. Columns with "CluSD-flat", "CluSD-m = 128" and "CluSD-m = 64" are CluSD results using uncompressed or compressed embeddings with m = 128 or m = 64. This table also includes a column marked with "CDFS-flat" which is the selective fusion outcome with CDFS using uncompressed RetroMAE embeddings.

Table 3 shows that the relevance of CluSD with selective fusion of sparse and dense results is higher than each individual retriever, and CluSD with limited dense retrieval performs closely to oracle "▲ flat", and delivers good NDCG@10 under two compression settings. CluSD is competitive to CDFS for the fusion of SPLADE and RetroMAE.

CluSD vs. Baselines with In-Memory or On-Disk Data. Table 4 examines the performance of CluSD fused with SPLADE when MS MARCO passage embeddings cannot fit in memory and are stored on an SSD disk. We use $N = 65,000$ for CluSD, and the extra space for quantized inter-cluster distances takes about 40MB. This setting here is larger than the $N = 8192$ setting in Table 1 and we choose that because a smaller cluster size gives a flexibility in reducing the disk I/O size when the total number of clusters selected by

CluSD is controlled. We also compare other selective dense retrieval baselines using reranking or proximity graph navigation. The mean response time (MRT) and 99th percentile (P99) include the total sparse and dense retrieval time for methods noted "S+*", otherwise only the dense retrieval time. The MRT cost breakdown for I/O and CPU time is also listed. Times reported are in milliseconds. "%D" is the approximate percentage of document embeddings evaluated based on the number of clusters fetched from the disk.

Reranking simply fetches top-k embeddings from the disk for fusion. LADR is designed for in-memory search [20] and we run it by assuming memory can sufficiently host the proximity graph while letting LADR access embeddings from the disk. Excluding the I/O cost portion of LADR in this table, the in-memory search cost of LADR is marked in the "CPU" column. We test two configurations of LADR, its default with 128 neighbors, 200 seed documents, and a search depth of 50, as well as a faster configuration, selected such that its CPU time is similar to CluSD. LADR_{fast} uses 128 neighbors, 50 seed documents, and a depth of 50. DiskANN [17] assumes the graph and original embeddings are on disk while the memory hosts compressed embeddings for quick guidance. SPANN [5] searches disk data in a cluster-based manner based on query-centroid distances. Both DiskANN and SPANN are designed for on-disk search from scratch without fusing with others, and thus we simply fuse their outcome with SPLADE results. HNSW [29] is not included because its in-memory relevance is similar as DiskANN while it is not designed to be on disk.

Table 4. Search when uncompressed embeddings are on disk for SPLADE + SimLM

		Relevance		Latency (ms)		Breakdown	
S=SPLADE-HT1	%D	MRR@10	R@1K	MRT	P99	I/O	CPU
S+Rerank	0.01	0.421	$0.980^{\dagger}$	158.7	217.3	125.2	33.5
S+LADR_{fast}	0.03	0.419	$0.980^{\dagger}$	346.1	1013	1003	41.5
S+$\text{LADR}_{default}$	0.10	0.422	0.984	674.9	2849	2788	61.3
DiskANN	–	$0.398^{\dagger}$	$0.970^{\dagger}$	276.5	321	272	5.2
S+ DiskANN	–	$0.417^{\dagger}$	0.984	307.7	352.2	272	36.4
SPANN	–	$0.396^{\dagger}$	$0.965^{\dagger}$	123.6	159.6	–	–
S+SPANN	–	0.420	0.989	154.8	190.8	–	–
▲ S+CluSD	0.05	0.425	0.986	55.6	94.4	17.5	38.1

CluSD significantly outperforms all other methods by leveraging block I/O. CluSD is 2.2x faster than the next fastest system (SPANN), while having a noticeably higher relevance. DiskANN performs similarly to SPANN in terms of relevance, but is 4.97x slower than CluSD. Reranking has a lower MRR@10 and recall@1k while being 2.85x slower. Both variants of LADR have a lower MRR@10 than CluSD, but are 6.2x and 12.1x slower on average for LADR_{fast}

and $\text{LADR}_{default}$ respectively. We find that each I/O operation has about a 0.15ms queueing and other software overhead on our tested PCIe SSD. Thus, more fine-grained operations in reranking yields more overhead. Proximity graph methods such as LADR rely on large amounts of fine-grained I/O operations and suffer significantly in terms of latency when embeddings are stored on disk. In comparison, CluSD takes advantage of block I/O operations because it fetches documents by clusters and thus CluSD issues fewer I/O requests to search more documents in less time.

CluSD with LLaMA-2 Based Dense Retrieval. Table 5 demonstrates that CluSD effectively supports selective RepLLaMA dense retrieval [27] to take advantage of LLMs on CPUs. The "Latency" column includes sparse retrieval CPU time when "S+" is marked in the method name. S=SPLADE-HT1. RepLLaMA uses 145GB embedding space with 4K dimensionality. All rows except the last row conduct in-memory search assuming data fits into memory. Row 1 shows it takes 7.9 s CPU time to conduct in-memory full RepLLaMA retrieval. Row 2 shows performance of RepLLaMA using OPQ quantization but without IVF selective search. Row 3 shows performance of RepLLaMA using OPQ and IVF selective search. Use of these optimization techniques leads to a significant relevance degradation. As a reference, Row 4 lists the fusion of SPLADE and full dense retrieval. CluSD visits a small number of clusters from the 60,000 dense RepLLaMA clusters. CluSD with SPLADE-HT1 takes about 39ms for in-memory search shown in Row 6, and 60ms with on-disk search shown in Row 8. RepLLaMA inference requires five Nvidia 32GB V100 GPUs to host data and take 37ms GPU time. Thus, CluSD can deliver 1 to 2 orders of magnitude of improvement in either total latency or infrastructure cost. Rows 5 and 7 show the performance of CDFS, which is similar to CluSD for in-memory search, but 23.3% slower than CluSD for on-disk search while achieving a similar relevance.

Table 5. Selective fusion with dense RepLLaMA and sparse SPLADE

Method	MRR@10	R@1K	Latency	Space
1. RepLLaMA	$0.412^{\dagger}$	0.990	7.9 s	145 GB
2. RepLLaMA-OPQ	$0.384^{\dagger}$	0.990	666 ms	2.4 GB
3. RepLLaMA-IVFOPQ	$0.365^{\dagger}$	$0.915^{\dagger}$	97 ms	6.1 GB
4. S + RepLLaMA	0.426	0.994	7.9 s	149 GB
5. S+CDFS	0.425	0.987	41 ms	149 GB
6. S+CluSD	0.424	0.986	39 ms	149 GB
7. S+CDFS (on-disk)	0.425	0.987	74 ms	149 GB
8. ▲ S+CluSD (on-disk)	0.424	0.986	60 ms	149 GB

For the BEIR datasets, RepLLaMA achieves an average NDCG of 0.561. While significantly faster than RepLLaMA, CluSD achieves an average NDCG of 0.541 while CDFS reaches 0.554. However, CluSD's LSTM model for RepLLaMA was used in a zero-shot manner after being trained using SimLM on MS MARCO passage. In the future we expect that further improvement in CluSD is possible if its LSTM is trained using RepLLaMA.

CluSD with Other Sparse Models and Quantization Methods. Table 6 examines CluSD's fusion of SimLM with other three sparse models. Last column lists in-memory dense retrieval portion of latency. For each sparse model, CluSD boosts overall relevance. As CluSD's cluster selection relies on the overlap feature of top sparse results with dense clusters, the guidance from the fastest BM25 results is less accurate than other sparse models. On the other hand, with a guidance from the slowest but stronger LexMAE retrieval, CluSD yields the highest MRR@10, comparable to full search.

Table 6. CluSD with LexMAE, uniCOIL & BM25 sparse models

Sparse model	**LexMAE**		**uniCOIL**		**BM25T5**		**Dense**
	MRR	R@1K	MRR	R@1K	MRR	R@1K	**latency**
Uncompressed SimLM embeddings with 27.2 GB space							
S	0.425	0.988	$0.347^{\dagger}$	$0.957^{\dagger}$	$0.259^{\dagger}$	$0.935^{\dagger}$	–
▲ S + D	0.430	0.990	0.413	0.985	0.415	0.985	1674 ms
S + Rerank	0.432	0.988	0.415	$0.957^{\dagger}$	0.409	$0.935^{\dagger}$	21.9 ms
S + CluSD	0.431	0.989	0.414	0.980	0.410	0.980	12.1 ms
OPQ m=128. SimLM embedding space 1.1 GB							
S + D-IVFOPQ	$0.383^{\dagger}$	0.988	$0.354^{\dagger}$	$0.961^{\dagger}$	$0.369^{\dagger}$	$0.975^{\dagger}$	21.9 ms
S + CluSD	0.429	0.989	0.404	0.980	$0.402^{\dagger}$	$0.975^{\dagger}$	9.8 ms

CluSD with Advanced Quantization Models. Table 7 shows the use of two other recent quantization methods DistillVQ [40] and JPQ [43] with in-memory CluSD for MS MARCO Passage Dev set. For JPQ, we use its released model and compression checkpoints and follow its default setting with $m = 96$. For DistillVQ, we use its released code to train compression on the RetroMAE model using its default value $m = 128$. While switching to a different quantization affects compression and relevance, this result shows that CluSD is still effective and outperforms other baselines.

Design Options for CluSD. Table 8 investigates the impact of the following design options for cluster selection in CluSD with the MS MARCO Dev set. For

Table 7. Use of CluSD with DistillVQ and JPQ quantization

Quantization	DistillVQ			JPQ		
S=SPLADE-HT1	MRR@10	R@1K	Latency	MRR@10	R@1K	Latency
IVF (2%)	0.365†	0.899†	20.8 ms	0.326†	0.917†	21.3 ms
S+ IVF (2%)	0.392†	0.987	52.0 ms	0.379†	0.985	52.5 ms
▲ S+CluSD	0.417	0.987	40.9 ms	0.392	0.985	42.4 ms

Stage I selection, we explore two options: 1) SortByDist: Choose top clusters sorted by the distance of their centroids to the query embedding. 2) SortByOverlap: Choose top clusters sorted by $P(C_i, B_j)$ features discussed in Sect. 2. Namely, use the overlap degree of clusters with top result bins of sparse retrieval. For Stage II selection, we explore three options: 1) XGBoost: Use the boosting tree XGBoost model to detect if a cluster should be visited. The features used are the same ones discussed in Sect. 2. This is a pointwise approach, as we view query-cluster pairs as independent samples, and the decision on one cluster does not affect other clusters for the same query. 2) RNN: Use a vanilla RNN model. The feature and sequence setup is the same as our LSTM model. 3) LSTM: the model we used in CluSD. We also assess the impact of removing a feature group used in CluSD: 1) w/o inter-cluster dist: removing the inter-cluster distance feature group; 2) w/o S-C overlap: removing the overlap-degree feature of top sparse results with each cluster.

Table 8. Design options for CluSD. ▲ is the default choice.

Avg #clusters targeted	3		5		Time
	MRR@10	R@1K	MRR@10	R@1K	ms
Stage II is removed. Stage I options only					
SortByDist	0.297	0.655	0.331	0.740	0
▲ SortByOverlap	0.384	0.867	0.401	0.917	0.2
Stage I=SortByDist; Stage II model options					
XGBoost	0.398	0.888	0.404	0.929	192
RNN	0.404	0.908	0.406	0.938	2.8
LSTM	0.406	0.923	0.406	0.943	2.8
Stage II=SortByOverlap; Stage II LSTM feature group options					
w/o inter-cluster dist	0.402	0.915	0.406	0.941	–
w/o S-C overlap	0.289	0.638	0.325	0.730	–
▲ Default	0.408	0.943	0.410	0.953	2.8

The last row marked "Default" is the final version used by CluSD with Sort-ByOverlap and LSTM. To have a fair comparison, this table assumes on average either 3 or 5 clusters are targeted by CluSD. For XGBoost, RNN or LSTM, we set their prediction threshold so that the average number of clusters is close to the targeted number. The last column of this table is the time spent to select clusters. Table 8 shows that the RNN or LSTM based prediction achieves higher MRR@10 and recall compared to XGBoost. Feature exploration indicates that the overlap degree is important for cluster selection while the inter-cluster distance feature group is also useful.

Impact of Cluster Partitioning Size. CluSD pre-partitions the embeddings into N clusters. Figure 2 shows the impact when N is 4096 and 8192 for MS MARCO Dev. Stage I of CluSD uses $n = 128$ in this figure. A solid line refers to the use of uncompressed dense embeddings while a dash-dot line refers to OPQ quantization with m = 128, and a dotted line refers to OPQ quantization with m = 64. The x-axis is the average number of clusters selected by CluSD. We vary the selection threshold parameter Θ in CluSD to achieve the different average number of clusters selected. With both N values, MRR@10 exceeds 0.420 when more than 10 clusters are selected. As more clusters are selected, the recall ratio increases. Different N values impact latency. For example, when 10 clusters are selected on average, CluSD scans through 0.1% and 0.2% of embeddings, corresponding to N = 4096 and 8192, respectively. When $N = 8192$, its latency is reduced because CluSD time complexity is proportional to the size per cluster and the number of clusters selected.

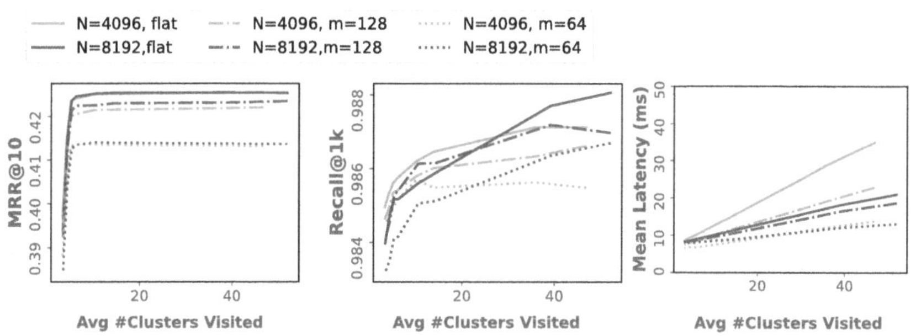

Fig. 2. CluSD relevance and latency vs. the average number of clusters selected

4 Concluding Remarks

CluSD is a lightweight cluster-based partial dense retrieval with fast CPU response time and competitive relevance. Compared to CDFS [42], CluSD does not make any strong statistical assumptions about the sparse ranking result distribution. The evaluation results show that CluSD can effectively select top dense clusters with a performance reasonably competitive to CDFS in relevance and latency even CluSD incurs small LSTM computing overhead. More studies and tuning could be conducted in the future in evaluating CluSD and CDFS.

The evaluation shows CluSD significantly outperforms cluster-based partial dense retrieval with IVF in relevance. Under the same time budget, CluSD with negligible extra space overhead delivers better or similar relevance than HNSW and LADR that rely upon a document-level proximity graph. Further, CluSD is up-to 12.1x faster than them when embeddings do not fit in memory. When compared to on-disk ANN methods DiskANN and SPANN [14,17], CluSD is 2.2x and 4.97x faster respectively, while achieving better relevance by leveraging sparse retrieval. Compared to graph navigation approaches, CluSD does not need a document-level proximity graph, and it conducts faster block I/O when fetching clustered document embeddings from disk.

Acknowledgments. We thank anonymous referees for their valuable comments. This work is supported in part by U.S. NSF IIS-2225942 and has used the computing resource of the ACCESS program supported by NSF. Any opinions, findings, conclusions or recommendations expressed in this material are those of the authors and do not necessarily reflect the views of the U.S. NSF.

References

1. Altingovde, I.S., Demir, E., Can, F., Ulusoy, O.: Incremental cluster-based retrieval using compressed cluster-skipping inverted files. ACM Trans. Inf. Syst. **26**(3) (2008). https://doi.org/10.1145/1361684.1361688
2. Brown, T., et al.: Language models are few-shot learners. In: Larochelle, H., Ranzato, M., Hadsell, R., Balcan, M., Lin, H. (eds.) Advances in Neural Information Processing Systems. vol. 33, pp. 1877–1901. Curran Associates, Inc. (2020). https://proceedings.neurips.cc/paper_files/paper/2020/file/1457c0d6bfcb4967418bfb8ac142f64a-Paper.pdf
3. Campos, D.F., et al.: MS MARCO: A human generated machine reading comprehension dataset. NIPS (2016)
4. Can, F., Altingövde, I.S., Demir, E.: Efficiency and effectiveness of query processing in cluster-based retrieval. Inf. Syst. **29**, 697–717 (2004)
5. Chen, Q., et al.: SPANN: Highly-efficient billion-scale approximate nearest neighborhood search. In: Beygelzimer, A., Dauphin, Y., Liang, P., Vaughan, J.W. (eds.) Advances in Neural Information Processing Systems (2021). https://openreview.net/forum?id=-1rrzmJCp4
6. Cheriton, D.R.: From doc2query to docTTTTTquery (2019). https://api.semanticscholar.org/CorpusID:208612557
7. Craswell, N., Mitra, B., Yilmaz, E., Campos, D.F., Voorhees, E.M.: Overview of the TREC 2020 deep learning track. ArXiv **abs/2102.07662** (2020)
8. Dai, Z., Callan, J.: Context-aware term weighting for first stage passage retrieval. SIGIR (2020)

9. Formal, T., Lassance, C., Piwowarski, B., Clinchant, S.: From distillation to hard negative sampling: Making sparse neural IR models more effective. Proceedings of the 45th International ACM SIGIR Conference on Research and Development in Information Retrieval (2022)
10. Formal, T., Piwowarski, B., Clinchant, S.: SPLADE: Sparse lexical and expansion model for first stage ranking. SIGIR (2021)
11. Gao, L., Dai, Z., Callan, J.: COIL: revisit exact lexical match in information retrieval with contextualized inverted list. NAACL (2021)
12. Gao, L., Dai, Z., Fan, Z., Callan, J.: Complementing lexical retrieval with semantic residual embedding. ECIR abs/2004.13969 (2021)
13. Ge, T., He, K., Ke, Q., Sun, J.: Optimized product quantization for approximate nearest neighbor search. In: Proceedings of the IEEE Conference on Computer Vision and Pattern Recognition (CVPR) (2013)
14. Gollapudi, S., et al.: Filtered-diskANN: graph algorithms for approximate nearest neighbor search with filters. In: Proceedings of the ACM Web Conference 2023, pp. 3406–3416. WWW '23, ACM, New York, NY, USA (2023)
15. Hafizoglu, F., Kucukoglu, E.C., Altingovde, I.S.: On the efficiency of selective search. In: Jose, J.M., et al. (eds.) ECIR 2017. LNCS, vol. 10193, pp. 705–712. Springer, Cham (2017). https://doi.org/10.1007/978-3-319-56608-5_69
16. Hochreiter, S., Schmidhuber, J.: Long short-term memory. Neural Comput. **9**(8), 1735–1780 (1997)
17. Jayaram Subramanya, S., Devvrit, F., Simhadri, H.V., Krishnawamy, R., Kadekodi, R.: DiskANN: fast accurate billion-point nearest neighbor search on a single node. In: Wallach, H., Larochelle, H., Beygelzimer, A., d'Alché-Buc, F., Fox, E., Garnett, R. (eds.) Advances in Neural Information Processing Systems. vol. 32. Curran Associates, Inc. (2019). https://proceedings.neurips.cc/paper_files/paper/2019/file/09853c7fb1d3f8ee67a61b6bf4a7f8e6-Paper.pdf
18. Johnson, J., Douze, M., Jégou, H.: Billion-scale similarity search with GPUs. IEEE Trans. Big Data **7**(3), 535–547 (2019)
19. Karpukhin, V., et al.: Dense passage retrieval for open-domain question answering. EMNLP'2020 ArXiv abs/2010.08191 (2020)
20. Kulkarni, H., MacAvaney, S., Goharian, N., Frieder, O.: Lexically-accelerated dense retrieval. In: Proceedings of the 46th International ACM SIGIR Conference on Research and Development in Information Retrieval, pp. 152–162. SIGIR '23, Association for Computing Machinery, New York, NY, USA (2023)
21. Kuzi, S., Zhang, M., Li, C., Bendersky, M., Najork, M.: Leveraging semantic and lexical matching to improve the recall of document retrieval systems: A hybrid approach. arXiv preprint arXiv:2010.01195 (2020)
22. Lassance, C., Clinchant, S.: An efficiency study for splade models. SIGIR (2022)
23. Lewis, P., Oğuz, B., Xiong, W., Petroni, F., tau Yih, W., Riedel, S.: Boosted dense retriever. Proceedings of NAACL, pp. 3102–3117 (2022)
24. Li, H., et al.: To interpolate or not to interpolate: PRF, dense and sparse retrievers. SIGIR (2022)
25. Lin, J.J., Ma, X.: A few brief notes on Deepimpact, COIL, and a conceptual framework for information retrieval techniques. ArXiv abs/2106.14807 (2021)
26. Liu, Z., Xiao, S., Shao, Y., Cao, Z.: RetroMAE-2: duplex masked auto-encoder for pre-training retrieval-oriented language models. In: Rogers, A., Boyd-Graber, J., Okazaki, N. (eds.) Proceedings of the 61st Annual Meeting of the Association for Computational Linguistics (Volume 1: Long Papers), pp. 2635–2648. Association for Computational Linguistics, Toronto, Canada (2023)

27. Ma, X., Wang, L., Yang, N., Wei, F., Lin, J.: Fine-tuning llama for multi-stage text retrieval (2023)
28. MacAvaney, S., Tonellotto, N., Macdonald, C.: Adaptive re-ranking with a corpus graph. In: Proceedings of the 31st ACM International Conference on Information & Knowledge Management, pp. 1491–1500. CIKM '22, Association for Computing Machinery, New York, NY, USA (2022)
29. Malkov, Y.A., Yashunin, D.A.: Efficient and robust approximate nearest neighbor search using hierarchical navigable small world graphs. IEEE Trans. Pattern Anal. Mach. Intell. **42**(4), 824-836 (2020)
30. Mallia, A., Khattab, O., Tonellotto, N., Suel, T.: Learning passage impacts for inverted indexes. SIGIR (2021)
31. Mallia, A., Mackenzie, J., Suel, T., Tonellotto, N.: Faster learned sparse retrieval with guided traversal. In: Proceedings of the 45th International ACM SIGIR Conference on Research and Development in Information Retrieval, pp. 1901–1905 (2022)
32. Qiao, Y., Yang, Y., He, S., Yang, T.: Representation sparsification with hybrid thresholding for fast SPLADE-based document retrieval. ACM SIGIR'23 (2023)
33. Qiao, Y., Yang, Y., Lin, H., Yang, T.: Optimizing guided traversal for fast learned sparse retrieval. In: Proceedings of the ACM Web Conference 2023, pp. 3375–3385 (2023)
34. Robertson, S.E., Zaragoza, H.: The probabilistic relevance framework: BM25 and beyond. Found. Trends Inf. Retr. **3**, 333–389 (2009)
35. Shen, T., et al.: LexMAE: Lexicon-bottlenecked pretraining for large-scale retrieval. In: The Eleventh International Conference on Learning Representations (2023). https://openreview.net/forum?id=PfpEtB3-csK
36. Tan, S., Xu, Z., Zhao, W., Fei, H., Zhou, Z., Li, P.: Norm adjusted proximity graph for fast inner product retrieval. In: Proceedings of the 27th ACM SIGKDD Conference on Knowledge Discovery & Data Mining, pp. 1552–1560. KDD '21, Association for Computing Machinery, New York, NY, USA (2021)
37. Thakur, N., Reimers, N., Rücklé, A., Srivastava, A., Gurevych, I.: BEIR: a heterogeneous benchmark for zero-shot evaluation of information retrieval models. In: NeurIPS (2021)
38. Touvron, H., et al.: Llama 2: Open foundation and fine-tuned chat models (2023)
39. Wang, L., et al.: SimLM: Pre-training with representation bottleneck for dense passage retrieval. ACL (2023)
40. Xiao, S., et al.: Distill-VQ: learning retrieval oriented vector quantization by distilling knowledge from dense embeddings. SIGIR (2022)
41. Xiao, S., Liu, Z., Shao, Y., Cao, Z.: RetroMAE: pre-training retrieval-oriented transformers via masked auto-encoder. EMNLP (2022)
42. Yang, Y., Carlson, P., He, S., Qiao, Y., Yang, T.: Cluster-based partial dense retrieval fused with sparse text retrieval. In: Proceedings of the 47th International ACM SIGIR Conference on Research and Development in Information Retrieval, pp. 2327–2331. SIGIR '24, ACM, New York, NY, USA (2024)
43. Zhan, J., Mao, J., Liu, Y., Guo, J., Zhang, M., Ma, S.: Jointly optimizing query encoder and product quantization to improve retrieval performance. CIKM (2021)

Context Example Selection for LLM Generated Relevance Assessments

Jack McKechnie(✉), Graham McDonald, and Craig Macdonald

University of Glasgow, Glasgow, UK
j.mckechnie.1@research.gla.ac.uk,
{graham.mcdonald,craig.macdonald}@glasgow.ac.uk

Abstract. Evaluating information retrieval systems typically requires large test collections with human-annotated relevance judgements. However, labelling large collections for relevance is an expensive and time-consuming process. Recently, previous work has explored the possibility of using large language models (LLMs) to generate relevance labels. Typically, such works deploy zero-shot prompting or in-context learning (ICL) strategies. However, ICL requires careful selection of representative examples to demonstrate the judging task to the LLM. In this work, we investigate strategies for selecting informative examples to include as context to help an LLM to generate discriminative relevance assessments. Our experiments using the TREC DL 2019 and 2020 test collections show that including informative examples as context can lead to notable improvements in the quality of generated relevance assessments. In particular, we found that our SAME QUERY RANDOM DOCUMENT example selection strategy resulted in +37% κ agreement between LLM-generated and human-labelled relevance assessments, and +17% τ agreement in the subsequent ranking of systems by the generated qrels, compared to a zero-shot baseline strategy. Such improvements in LLM-generated qrels can help to reduce the amount of human relevance assessments that are required to create test collections and thus facilitate quicker and easier information retrieval research.

Keywords: Evaluation · LLM Evaluation · In-Context Learning

1 Introduction

Cranfield-style [11] Information Retrieval (IR) test collections contain a corpus of documents, a set of queries, and a set of query-document relevance assessments, also known as qrels. Qrels are a set of mappings between pairs of queries and documents with a human-labelled relevance assessment. However, judging every document in the corpus for every query is prohibitively expensive [30]. This is especially true given that recent test collections that are used to train and test neural IR models [28,41,53] contain millions of documents [5,19]. Therefore, when creating test collections, top-k pooling using the results from a diverse set of retrieval systems [30] is performed. This is to amass a 'good enough' subset

C. Hauff et al. (Eds.): ECIR 2025, LNCS 15572, pp. 293–309, 2025.
https://doi.org/10.1007/978-3-031-88708-6_19

of the corpus to have manually judged for relevance. Human relevance assessors must be paid for their work and budgets for creating test collections are limited, leading to tradeoffs when using top-k pooling. Either a large number of documents can be judged for a small number of queries, or vice versa. Each approach has its advantages and disadvantages [3,6,10,33,59]. However, the limiting factor when creating a test collection is almost always the cost of human relevance assessment [9].

On the other hand, automatic relevance assessment has gained attention in the IR evaluation literature. This is due to recent advances in the training and architectures [15,57] of Large Language Models (LLMs) that have yielded models that are capable of a great variety of text-based tasks [7,16,40]. This has prompted efforts to investigate if LLMs can be used to judge the relevance of documents to a query and thus remove or reduce the need for human relevance assessors [1,20,54–56]. The LLM-generated qrels from such previous studies are found to agree well with human relevance assessments. Either zero-shot prompting is used, where the relevance assessment task is described but no context examples are given, or k-shot prompting is used, where a small number of demonstration examples are given to the model. This k-shot prompting leverages in-context learning (ICL) to show the language model how to perform the given task using only a small number of example demonstrations [8]. ICL literature has shown that certain examples provide more utility in learning the downstream task than others and that selecting examples in a principled manner can result in improved task performance [23,32,50]. However, to the best of our knowledge, no previous study on using LLMs for relevance assessment has thoroughly investigated *which* context examples should be selected and *how* they should be selected to produce the highest quality qrels.

Therefore, in this paper, we perform a study of context example selection methods for generating relevance assessments with LLMs. We evaluate seven methods for selecting k-shot context examples and thoroughly investigate the impact that each of these selection methods has on the quality of the generated qrels. High-quality LLM qrels should be able to rank retrieval systems by their performance under a given metric and should be able to correctly identify significant differences between retrieval systems. The results of such an evaluation should be as similar as possible to the results that would be obtained when evaluating the systems with human-labelled qrels. That is to say that LLM qrels should be able to rank retrieval systems from best to worst and be able to distinguish pairs of systems as effectively as human-assessed qrels. We use the Llama3 [17] model to generate qrels for the TREC Deep Learning 2019 [13] and 2020 [14] test collections using a variety of different example selection methods. We provide analysis of the quality of LLM-generated qrels using measures of agreement with human assessors, measures of the correlation of the ranking of systems when evaluating with human or LLM qrels, and a measure of the *discriminative power* [20] of the qrels to distinguish statistically significant differences between systems. Using the results of this analysis, we make recommendations for IR practitioners on how to select the context examples that are the most likely to be useful for developing high-quality LLM-generated qrels.

We find that selecting examples that have previously been judged, by humans, for the query that the LLM is being asked to make a relevance judgement for is the most important factor when selecting context examples. Additionally, we see that providing context examples that are relevant to the query is more important than providing examples of all relevance labels. The contributions of this work are as follows:

- We provide the first in-depth analysis of different context selection strategies for generating qrels via LLM.
- Based on this analysis, we make recommendations on the best context selection strategies to use for generating qrels via an LLM.

2 Related Work

This section introduces previous work on IR evaluation, the use of Large Language Models in IR evaluation, and In-Context Learning.

Automatic Evaluation in IR. The offline evaluation of information retrieval systems has a long and rich history [9,24,25,27,30]. Traditionally, a variety of retrieval systems are run over a query set and top-k pooling is performed to obtain a set of documents to be judged by human reviewers [30]. However, collecting human-generated relevance assessments is both time-consuming and expensive [30]. Thus, efforts have been made to develop evaluation protocols that circumvent the need for human-labelled relevance assessments [26,39,49,51,60,61] by using statistical properties of the pools collected by a diverse set of retrieval systems. Much of the work done in this area is carried out under the assumption that documents retrieved by many systems are likely to be relevant. Soboroff et al. [49] investigated selecting documents randomly drawn from the pools and assigning relevance labels based on the percentage of documents judged as relevant for that query. Soboroff et al. [49] observed that the ranking of systems by their performance under mean average precision remains largely unchanged, except for the highest-performing systems. Other work has developed similar approaches that create pseudo-relevance labels without the need for human reviewers based on the statistical properties of the top-k pool [39,51,60,61]. In this work, we deploy an approach that is similar to Soboroff et al. [49] as a baseline, denoted as POPULARITY-BIASED, to represent this family of non-LLM automatic judgement approaches [39,49,51,60,61]. Such automatic evaluation methods generate relevance assessments that use frequency of retrieval as a proxy for relevance [4]. In contrast, we aim to generate relevance assessments according to the semantic relevance of query-document pairs.

LLMs for IR Evaluation. Large Language Models, such as GPT [2], BERT [15], and Llama [17] are Transformer-based [57] models that are pretrained on vast amounts of textual data. Such models can achieve impressive performance on a diverse set of tasks. LLMs have been applied to IR tasks in a variety of ways; for example, T5 [43] has been fine-tuned to produce ad-hoc search models [41,53,62]. The impact on the final evaluation of systems when *augmenting* qrels with LLMs has been recently investigated. MacAvaney and Soldaini [35] proposed a collection of models that use one known relevant document to estimate the gain value of an unlabelled document for use in precision-based metrics. Upadhyay et al. [55] and Abbasiantaeb et al. [1] both simulated situations where unjudged documents are believed to be relevant and use LLMs to 'fill in the holes' in the qrels. Using such LLM augmented qrels is found to be an effective way of mitigating problems caused by unjudged documents. However, whilst context examples are used to generate LLM augmented qrels, the impact of those context examples and how they should be chosen has not previously been investigated. Therefore, in this work, we study *strategies* for selecting context examples when generating qrels.

Thomas et al. [54] used an in-house version of GPT-4 [2] to generate qrels for the TREC Robust dataset [58]. Their approach focused on the development of a high-quality zero-shot prompt by breaking down different aspects of relevance assessment and performing a leave-one-out analysis to develop their prompt. They found that their approach was useful in a commercial setting as labels generated by their method more accurately reflected searchers' intents than third-party crowd-sourced labels. The prompt developed by Thomas et al. [54] is used as the base of our prompt in this study. Upadhyay et al. [56] presented an open-source version of the work done by Thomas et al. [54] using commercially available models. Results show that the proposed prompting technique transfers to the commercially available models and thus use this as part of the evaluation for the TREC 2024 RAG track [42]. The studies carried out by Thomas et al. [54] and Upadhyay et al. [56] concentrated on the development of a high-quality prompt. In this work, we rather focus on studying what methods to use to select the context examples used when generating qrels via LLMs.

The use of LLMs for evaluation in IR remains a hotly debated topic. Faggioli et al. [20] presented a number of different perspectives on the topic. Three frameworks under which LLMs can be used for the evaluation of IR systems are presented, which range from entirely human evaluation to entirely LLM evaluation. We focus on the entirely LLM-based evaluation framework in this paper. Faggioli et al. [20] presented initial results using zero-shot and few-shot examples. Prompting with few-shot examples leverages in-context learning by using a small number of demonstrative examples to help show an LLM how to effectively perform the downstream task of relevance assessment. The decision of which examples are used in such learning processes greatly influences overall task performance [23,32,50]. Faggioli et al.. [20] use examples from the same query as the one that is being judged. We evaluate a similar context example selection strategy in this study, denoted as SAME QUERY, LABEL PREVALENCE. Using agreement measures, rank correlation coefficients, and a metric for the 'discriminative power' of the produced qrels, Faggioli et al. [20] showed that the use of LLMs for automatic IR evaluation is feasible. On the other hand, Soboroff [48]

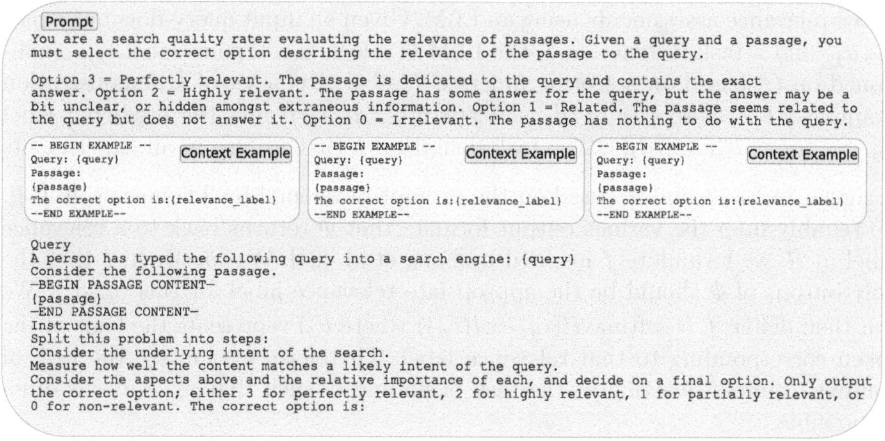

Fig. 1. The prompt used in all experiments. Values in curly braces are replaced with their corresponding content.

argued that LLMs should not be used for creating relevance assessments as the model used becomes the upper bound evaluated against, rather than human performance. Soboroff [48] advocated for an entirely human evaluation framework (in the nomenclature of [20]), whereas others in the community have been more open to using LLMs for relevance assessment [1,54,55], with several workshops recently being organised on the topic [44–46]. It is for this reason that a systematic study of context example selection studies is needed to provide insights on which context examples should be used and where should they come from in order to generate the highest-quality qrels as possible.

3 Context Example Selection

In the following, we detail the process of using an LLM to generate relevance assessments, and introduce the context sample selection strategies that we test.

3.1 Generation of Qrels by LLMs

Generating qrels using an LLM involves prompting the model with instructions on how to perform the task i.e. relevance assessment, a set of k demonstration examples, and a candidate query-document pair to be judged. This combination is input to the LLM, which then outputs a response. The response is then mapped back to the set of possible answers; binary or graded relevance assessments. Let R be the set of n possible relevance labels, e.g. $R = \{0, 1, 2, 3\}$ when using DL19 and DL20. Next, we define A as a set of query-relevance assessments. Each element $a \in A$ is a three-tuple consisting of $\langle q_i, d_j, r_{ij} \rangle$ where q_i is a query, d_j is a document, and $r_{ij} \in R$ is the relevance assessment given to document d_j for query q_i. Using this notation, we define more formally how we generate query-relevance assessments using an LLM. Given an input query-document pair $\langle q_i, d_j \rangle$ and a task instruction I, the LLM Φ selects a relevance label $\hat{r_{ij}}$, conditioned on C, based on a scoring function f. $C \subseteq A$ is a set of k demonstration examples. Therefore $P(r|\langle q_i, d_j \rangle) \triangleq f(\Phi(r, C, \langle q_i, d_j \rangle))$ and the predicted label $\hat{r_{ij}} = \underset{r \in R}{\text{argmax}}\, P(r|\langle q_i, d_j \rangle)$. The task definition I that we use in all experiments is shown in Fig. 1 and is based on the prompt developed by Thomas et al. [54]. To reliably map the various output formats that Φ returns back to a relevance label in R, we formulate f inspired by Fang et al. [21]. We specify in I that the only output of Φ should be the appropriate relevance label for this $\langle q_i, d_j \rangle$. We can then define $f \triangleq \text{softmax}(l(r_0) \ldots l(r_n))$ where $l(\cdot)$ represents the logit of the token corresponding to that relevance label. Therefore, we can see the task of context example selection as the task of finding C such that $\hat{r_{ij}} = r_{ij}$ as often as possible.

3.2 Context Example Selection Strategies

Assuming that $\langle q_1, d_2 \rangle$ is the query-document pair that is to be judged by Φ, we outline our nine context example selection strategies below:

ZERO-SHOT: This is the simplest of all of the strategies for generating relevance assessments with LLMs. In the ZERO-SHOT strategy $C = \emptyset$ i.e., no context examples are given to Φ, only the task instruction I and the query document pair $\langle q_1, d_2 \rangle$. This strategy can be thought of as a baseline to be improved upon via the selection of C and we expect that it will be the lowest-performing LLM qrel generation strategy as no context examples are given.

POPULARITY-BIASED: This is essentially the strategy used by Soboroff et al. [49]. For each query q_i, we take the number of times that each document d_j is retrieved across all systems submitted to TREC Deep Learning 2019/2020 for evaluation and call this the popularity score. We then calculate the percentage of documents that have a non-zero relevance assessment for q_i in the official qrels, call this p. The relevance label of the top $p\%$ of documents with the highest popularity score is set to 1 and to 0 for all other documents. This strategy is also a baseline, as there is no LLM used for relevance assessment. Thus, we expect this strategy to perform worse than all of the LLM assessment strategies as there is no concept of semantic relevance, only the popularity of being retrieved.

RANDOM QUERY RANDOM DOCUMENT: In this strategy, we populate C with $\langle q_i, d_j \rangle$ that are randomly chosen. There is no control over how such pairs are selected, other than we ensure that we do not include $\langle q_1, d_2 \rangle$ in C. This strategy is used to test if examples from a diverse set of queries allow the model to better learn how to judge relevance via ICL. One might expect that a diverse set of examples is helpful to show the model what is and what is not relevant to a query.

SAME QUERY RANDOM DOCUMENT: Here, we randomly sample from A_{q_1}, where $A_{q_1} = \{\langle q_i, d_j, r_{ij} \rangle \in A | i \neq 1, j \neq 2\}$ to populate C. This strategy is used to investigate if the context examples from the same query as the query to be judged provide the model with a more specific knowledge of what is relevant or not to that specific query, we would expect that this strategy will allow for the model to leverage such specific knowledge for quality relevance assessments.

SAME QUERY, LABEL PREVALENCE: Random samples from A_{q_1} are also taken in this strategy, but the sampling probability is weighted by the prevalence of each relevance label in A_{q_1}. For example, if more documents are judged as highly relevant for q_1 than are judged as partially relevant, then highly relevant documents are more likely to be included in C than partially relevant documents. It is therefore expected that using examples that are likely to be of the same relevance label as d_1 will lead to the model better learning how d_1 should be judged. We use this to draw conclusions on whether providing context examples that are more likely to have the same relevance label as the query-document pair to be judged leads to improved judgments.

ALL RELEVANT: We select elements from A_{q_1} where $r_{ij} > 0$ to populate C. Using this strategy we test if context examples should describe to the model what *is* relevant, rather than what is *not* relevant. One might expect that by showing the model only relevant examples the model will gain an understanding of what relevance to this specific query looks like.

ALL NON-RELEVANT: Elements from A_{q_1} where $r_{ij} > 0$ are selected to populate C. Converse to the ALL RELEVANT selection strategy, we test here if context examples should describe to the model what *is not* relevant, rather than what is *is* relevant and we expect that this may lead to the model learning how to judge relevance for the given query by elimination.

We investigate the context example selection strategies through the experiments that are introduced in Sect. 4.

4 Experimental Setup

Our experiments investigate the context example selection strategies through the following research questions:

- **RQ0** What is the effect of removing qrels to be used as a set of possible context examples?
- **RQ1** What source should context examples be selected from to generate qrels?
- **RQ2** What strategy should be used to select context examples to use when generating qrels?

Test Collections: To perform our experiments we use two query sets based on the MS MARCO passage retrieval collection [5], which contains ~8.8M short textual passages. We use TREC Deep Learning 2019 [13] and 2020 [14] in all of our experiments as these contain much deeper relevance assessments than the original qrels released with MS MARCO. Additionally, the relevance labels in this dataset are graded, rather than the binary labels of the original qrels. These datasets allow us to test how well each context example selection strategy can handle more fine-grained relevance assessments. For brevity, we henceforth refer to these datasets as DL19 and DL20 respectively. Specifically, we use the `msmarco-passage/trec-dl-(2019/2020)/judged` versions, available through the ir-datasets package [36]. We hold out 10% of the qrels for each dataset as a training set to use as k-shot examples. To ensure that our held out training set maintains the same distribution of relevance labels across each query, we employ a stratified sampling strategy. This method ensures that the proportion of each documents judged with each relevance label for each query reflects that of the original dataset. The effect of removing this 10% of qrels on the ranking of systems is investigated in Sect. 5. To test our generated qrels, we use runs submitted to the corresponding TREC track. This results in 32 runs for DL19 and 59 runs for DL20. Note that we remove one run that was submitted to the 2019 Deep Learning track (UNH_exDL_bm25) as this run is a clear outlier, achieving an nDCG@10 score ~ 0.35 lower than the next lowest scoring run.

Baselines: The ZERO-SHOT and POPULARITY-BIASED strategies are the baseline relevance assessment generation strategies strategies. These strategies involve no context example selection and therefore we compare the strategies outlined in Sect. 3.2 with these baselines.

Model Details: We use the Llama 3 [17][1] model in all of our experiments, accessed through the Transformers library[2]. We use LLama 3 as it is a modern and high-performing model that is open source, leading to more reproducible results. We take Llama 3 to be a representative example of the LLMs used for relevance assessment in the literature. We believe that the risk of data contamination from MS MARCO is low and not of concern, as per previous literature [38]. The prompt used in all experiments remains the same, with the exception of the additional of k context examples. The prompt used is shown in Fig. 1.

Evaluation Metrics: To evaluate our experiments we compare qrels generated by our approaches to the original, human-labelled, qrels released by TREC. Specifically, we compare the agreement between both sets of qrels via Cohen's Kappa [12]. We use both a graded relevance assessment setting where all relevance labels in the datasets are considered and a binary setting where all relevance labels >0 are set to 1. We include the binary setting as we expect that determining relevance on a binary scale is easier for the LLM than on a graded scale. We also compare qrels via how well they rank systems. We use both Kendall's τ [31] and Spearman's ρ [47] rank correlation coefficients to examine

[1] https://huggingface.co/meta-llama/Meta-Llama-3-8B-Instruct.
[2] https://huggingface.co/docs/transformers/index.

how the ranking of systems compares when using LLM-labelled and human-labelled qrels. We rank systems by their performance under nDCG@10 [29] using ir_measures [34]. Our final evaluation metric is used to measure the 'discriminative power' of a set of qrels. We use the 'sensitivity' metric proposed in Faggioli et al. [20], designed to measure the number of significant differences between pairs of systems found by a set of qrels. The sensitivity metric is formulated as $\frac{\#\,of\,distinguished\,pairs}{total\,pairs}$ where a distinguished pair is a pair of systems that are significantly different with $p < 0.05$. We report this metric as Δ sensitivity, i.e. the difference in sensitivity when using LLM-generated qrels and official, human-labelled, qrels. Negative values indicate that LLM-generated qrels find less pairs with significant differences than human-labelled qrels and vice versa. For significance testing with agreement measures, we use a paired t-test and when significance testing with correlation measures we use the Fisher Z-Transform [22] to calculate p-values. In both cases, we use $p < 0.05$ and Bonferroni correction [18] for multiple tests.

We run the RANDOM QUERY RANDOM DOCUMENT experiments three times and report the mean and variance as one would expect that this approach would result in the highest variance. We observe low variance and thus omit repetition of other experiments due to the computational cost of repeating experiments.

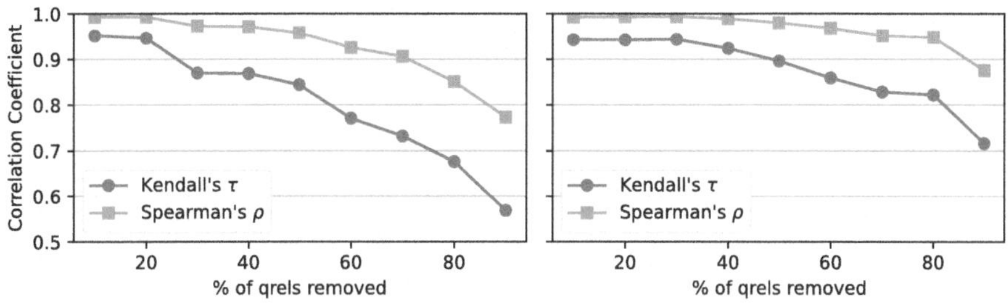

Fig. 2. Effect of removing different percentages of qrels. Left: DL19 Right: DL20.

5 Results

In this section, we address each of the research questions outlined in Sect. 4 and draw further analyses on the effectiveness of each of the context example selection strategies tested.

Firstly, we address **RQ0**, in order to verify the impact of withholding a small training set to use as context examples. To this end, we remove an increasing percentage of the overall qrels and measure how well correlated the rankings of systems are when using 100% of the qrels vs. using the qrels with the subset removed. We rank each system by nDCG@10. Figure 2 shows the obtained correlations (both Spearman's ρ and Kendall's τ), for both DL19 (left) and DL20 (right). On the x-axis, we show the percentage of the qrel labels removed, while the y-axis is the correlation of the ranking of systems. From the figures, we can see that removing even up to 40% of the qrels using our stratified sampling approach still results in strong correlation of ranking of systems. This observation is true for both DL19 and DL20, though correlations are generally higher when using DL20. In our experiments testing context example selection strategies, only 10% of the qrels are removed to be used as source of examples. The change in the ranking of systems is minimal in this setting and we therefore believe it a valid experimental setup to hold out 10% of the qrels as context examples.

Observation 1

TREC DL19 and DL20 are robust when removing up to 40% of qrels, showing minimal change the ranking of systems.

Table 1. Experimental results on DL19. Bold values indicate the highest-performing approach in a given metric. Values marked with * and † have significant differences when compared to the ZERO-SHOT approach and POPULARITY-BIASED approaches respectively (t-test, $p < 0.05$, Bonferonni correction).

Strategy	Shots	Graded κ	Binary κ	τ	ρ	ΔSens.
ZERO-SHOT	0	0.1225	0.3356	0.8022†	0.9327†	−0.3742
POPULARITY-BIASED	–	0.1740	0.3253	−0.1315*	−0.2079*	−0.6279
ALL RELEVANT	1	0.2258*	0.4725	0.9140*	0.9867*†	−0.0581
	2	0.2548*	0.4951*	0.9140†	0.9855*†	−0.0279
	3	0.2778*	**0.4995***	0.9269†	0.9899*†	**0.0495**
ALL NON-RELEVANT	1	0.1781*†	0.3876*	0.8194†	0.9464†	−0.0946
	2	0.1910*	0.4136*†	0.9140†	0.9778†	−0.1763
	3	0.2134*	0.3992*†	0.9269†	0.9887*†	−0.0538
RANDOM QUERY RANDOM DOCUMENT	1	0.1920*	0.3612	0.8839†	0.9786†	−0.1806
	2	0.1938*	0.3678*	0.9369†	0.9902*†	−0.2072
	3	0.2053*	0.3839*	0.9093†	0.9800†	−0.2473
SAME QUERY RANDOM DOCUMENT	1	0.2327*	0.4047*	0.9128†	0.9813†	−0.0365
	2	0.2464*	0.4391*	0.9140†	0.9843*†	−0.0516
	3	**0.2823***	0.4605*	**0.9398†**	**0.9911*†**	−0.0043
SAME QUERY, LABEL PREVALENCE	1	0.2241*	0.3981*	0.8839†	0.9774†	−0.0172
	2	0.2390*	0.4251*	0.8968†	0.9802†	−0.0774
	3	0.2815*	0.4662*	0.9140†	0.9859*†	−0.0215

We now move on to determine the best strategy for selecting context examples; more specifically, if context examples should come from the same query as the query-document pair that is to be judged or from another, different, query. **RQ1** can be answered by comparing the RANDOM QUERY RANDOM DOCUMENT strategy with our other LLM qrel generation strategies as the source used in RANDOM QUERY RANDOM DOCUMENT contains different queries, whereas the source for all other context example selection strategies is the same query as is to be judged. Tables 1 & 2 contain results for the DL19 and DL20 query sets respectively. Each row in the table represents a context example selection strategy and each row for the same strategy indicates a varying number of examples used, indicated by the 'shots' column. We report (i) agreement with official qrels (binary and graded) in terms of Cohen's κ, (ii) correlations, in terms of Spearman's ρ and Kendall's τ of the system rankings obtained under nDCG@10 using the obtained qrels compared with the official qrels, and (iii) the sensitivity, i.e. the difference in the percentage of pairs of systems that are distinguished compared to the official qrels. For instance, the ZERO-SHOT labeller, such as that used in [20], obtained low agreement (graded κ of only 0.1225) with the official qrels, but was slightly better at distinguishing relevant from not-relevant (binary κ 0.3356); despite these low agreements, the correlation of system rankings was strong (Spearman's $\rho > 0.8$). In terms of sensitivity, the ZERO-SHOT approach identifies ~37% less distinguished pairs of systems than the official qrels, indicating that they have less discriminative power rendering significance testing between systems less reliable. Results for the RANDOM QUERY RANDOM DOCUMENT strategy are the average of three repetitions. On inspection of the variances across the repetitions, we observe up to 0.0025 on DL19 and up to 0.0033 on DL20. One would expect that the RANDOM QUERY RANDOM DOCUMENT approach has the highest variance amongst all the approaches that we test as there is randomness in both the query and the document being selected. Given that we observe such little variance for this approach, we do not repeat our other experiments, due to their computational cost.

On inspection of the results in the tables, we observe that, for both DL19 and DL20, selecting context examples from the same query as the query that is being judged always performs better than using context examples from a random query. This is true in terms of agreement, ranking of systems and of discriminative power. For example, using the SAME QUERY RANDOM DOCUMENT results in an 11.4% increase in Kendall's τ over the RANDOM QUERY RANDOM DOCUMENT sample selection approach when using one example and is the highest performing strategy overall on DL19 when using three examples. We posit that this is because documents that are judged from the same query are likely to be similar to each other as they are drawn from the same pool of documents, created by numerous retrieval systems. Therefore, the model can use similar examples to better judge relevance for a given query. However, the inclusion of even one context example, from any source, results in improvements over the ZERO-SHOT approach (22% increase in Graded κ on DL20).

Observation 2

Including context examples from any query is better than zero-shot.

Examples that are potentially quite unrelated, either lexically or semantically, to the query-document pair to be judged are useful as the task of relevance assessment is demonstrated to the model. This observation concurs with previous ICL studies; using examples, of any kind, is generally better than using none [52]. In general, we observe that the performance increases as more examples are used.

For example, ~6% more pairs of significant differences between systems can be identified when using 3 vs. 1 context examples when using the SAME QUERY, LABEL PREVALENCE approach. This indicates that the best context selection strategy is one where more examples are used rather than less. We can therefore conclude that, of the context example sources that we test here, the best source is from the same query as the one that is to be judged. Additionally, selecting examples from any source is better than not selecting any at all. To concretely answer **RQ1**, context examples should be selected from the same query as the query that is to be judged.

Observation 3

Selecting context examples that have judged w.r.t. the query in the query-document pair that is being judged leads to better LLM-generated qrels.

Table 2. Experimental results on DL20. Notation as per Table 1.

Strategy	Shots	Graded κ	Binary κ	τ	ρ	ΔSens.
ZERO-SHOT	0	0.1407	0.3110	0.7884†	0.9297†	−0.3284
POPULARITY-BIASED	–	0.2253	0.3310	0.2683*	0.3549*	−0.4155
ALL RELEVANT	1	0.2525*	0.4722	**0.8980**†	**0.9840***†	−0.0818
	2	0.2759*	0.5049	0.8828†	0.9752†	−0.0695
	3	**0.2926***	**0.5105**	0.8866†	0.9790†	**−0.0794**
ALL NON-RELEVANT	1	0.1740*	0.3612*†	0.8504†	0.9627†	−0.1747
	2	0.1882*	0.3830*†	0.8641†	0.9698†	−0.1718
	3	0.2012*	0.3968*†	0.8512†	0.9611†	−0.1484
RANDOM QUERY RANDOM DOCUMENT	1	0.1716	0.3332	0.8348†	0.9589†	−0.1794
	2	0.1627*	0.3275*	0.8274†	0.9500†	−0.2156
	3	0.1683*	0.3356*	0.8374†	0.9597†	−0.1905
SAME QUERY RANDOM DOCUMENT	1	0.2158*	0.3919*	0.8796†	0.9758†	−0.1022
	2	0.2225*	0.4070*	0.8551†	0.9645†	−0.1385
	3	0.2476*	0.4310*	0.8424†	0.9521†	−0.1648
SAME QUERY, LABEL PREVALENCE	1	0.2047*	0.3749	0.8173†	0.9489†	−0.1402
	2	0.2215*	0.4037*	0.8714†	0.9677†	−0.1560
	3	0.2393*	0.4223*	0.8784†	0.9761†	−0.0806

Next, we address **RQ2** by evaluating what is the best strategy for selecting context examples given a specific source to select from. We do so by comparing the ALL RELEVANT, ALL NON-RELEVANT, SAME QUERY RANDOM DOCUMENT, and SAME QUERY, LABEL PREVALENCE context example selection strategies. All of these strategies select context documents that have been labelled for the same query that is to be judged and the selection strategy is the only difference. We can see that for DL19, in general, the SAME QUERY RANDOM DOCUMENT approach performs the best, although it under-performs in terms of discriminative power in comparison to the ALL RELEVANT strategy. The ALL RELEVANT selection strategy is by far the best at judging if a document is relevant or not, but cannot make

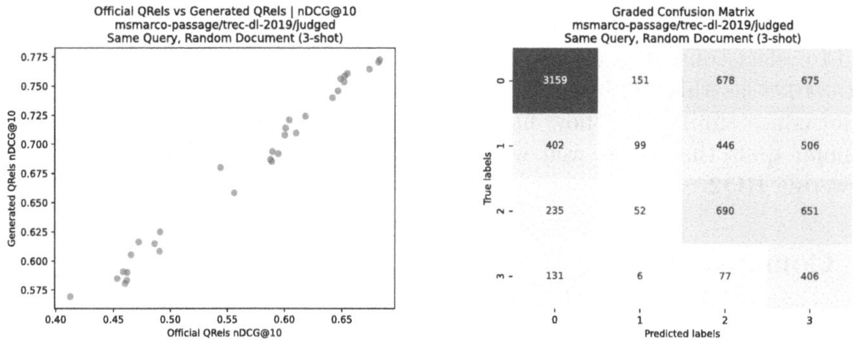

(a) System Rankings Comparison (b) Confusion Matrix

Fig. 3. Comparisons between human and LLM-generated qrels on DL19. LLM judgements are from the SAME QUERY RANDOM DOCUMENT strategy.

the more subtle graded decision well. This explains why ALL RELEVANT is better at distinguishing significant differences (distinguishing ~5% *more* pairs than the human-labelled qrels), per-query metric scores are relatively different between systems, but more granular relevance assessments are required to appropriately rank systems and therefore SAME QUERY RANDOM DOCUMENT has higher correlation coefficients than ALL RELEVANT. We can visually see the effectiveness of the SAME QUERY RANDOM DOCUMENT selection strategy in Fig. 3. The left-hand plot shows nDCG@10 values for when systems submitted to TREC are evaluated by the human vs LLM-generated qrels. Performance using the human-labelled qrels is given on the x-axis and performance using the LLM-generated qrels is given on the y-axis. Each point is a retrieval system. Figure 3(a) shows visually that the SAME QUERY RANDOM DOCUMENT strategy ranks systems well, as witnessed by the tight distribution of points from bottom left to top right of the plot.

Observation 4

The SAME QUERY RANDOM DOCUMENT strategy produces qrels that can rank systems by their retrieval performance comparably to human-labelled qrels.

Additionally, Fig. 3(b) shows that the SAME QUERY RANDOM DOCUMENT strategy rarely makes the mistake of judging a highly relevant document as partially relevant, assigning labels of 1 or 2 to query-document pairs with a label of 3, 6 and 77 times respectively. This strategy produces high-quality, granular qrels on DL19. However, for DL20, the highest performing context selection strategy is always ALL RELEVANT, with SAME QUERY RANDOM DOCUMENT being the second best. We attribute the differences here to how different the query sets are; for example, DL20 has many more documents judged as relevant than DL19 in general [13,14,37]. We can therefore conclude that the strategy that should be used to select context examples when generating qrels is SAME QUERY RANDOM DOCUMENT as this produces qrels that are able to rank systems in terms of performance similarly to how humans do (via high τ) and are able to create granular qrels that agree well with the relevance labels assigned by humans, answering **RQ2**.

6 Conclusions

This work has provided a comprehensive study of context example selection strategies for use when creating relevance judgements with large language models. We surveyed existing literature and identified seven context example selection strategies which we compared and contrasted against each other, and against automatic IR evaluation methods from the literature. Through careful and thorough comparisons, we found that context examples should be selected from the same query as the query to be judged and that many context examples should be used. However, many open questions remain about using LLMs for relevance assessment. Future work in this direction may include the evaluation of the impact of using the same LLM for retrieval and evaluation, the effect of new systems being introduced to the pools, and other new ways of selecting high-quality context examples.

Disclosure of Interests. The authors have no competing interests to declare that are relevant to the content of this article.

References

1. Abbasiantaeb, Z., Meng, C., Azzopardi, L., Aliannejadi, M.: Can we use large language models to fill relevance judgment holes? arXiv preprint arXiv:2405.05600 (2024)
2. Achiam, J., et al.: GPT-4 technical report. arXiv preprint arXiv:2303.08774 (2023)
3. Arabzadeh, N., Vtyurina, A., Yan, X., Clarke, C.L.: Shallow pooling for sparse labels. Inf. Retrieval J. **25**(4) (2022)
4. Aslam, J.A., Savell, R.: On the effectiveness of evaluating retrieval systems in the absence of relevance judgments. In: Proceedings of SIGIR (2003)
5. Bajaj, P., et al.: MS MARCO: A human generated machine reading comprehension dataset. arXiv preprint arXiv:1611.09268 (2016)
6. Boytsov, L., Belova, A., Westfall, P.: Deciding on an adjustment for multiplicity in IR experiments. In: Proceedings of SIGIR (2013)
7. Brants, T., Popat, A., Xu, P., Och, F.J., Dean, J.: Large language models in machine translation. In: Proceedings of EMNLP-CoNLL (2007)
8. Brown, T.B., et al.: Language models are few-shot learners. In: Proceedings of NeurIPS (2020)
9. Buckley, C., Dimmick, D., Soboroff, I., Voorhees, E.: Bias and the limits of pooling for large collections. Inf. Retrieval **10** (2007)
10. Carterette, B., Gabrilovich, E., Josifovski, V., Metzler, D.: Measuring the reusability of test collections. In: Proceedings of WSDM (2010)

11. Cleverdon, C.: The Cranfield tests on index language devices. In: ASLIB. vol. 19 (1967)
12. Cohen, J.: A coefficient of agreement for nominal scales. EPM **20**(1) (1960)
13. Craswell, N., Mitra, B., Yilmaz, E., Campos, D., Voorhees, E.M.: Overview of the TREC 2019 deep learning track. arXiv preprint arXiv:2003.07820 (2020)
14. Craswell, N., Mitra, B., Yilmaz, E., Campos, D., Voorhees, E.M.: Overview of the TREC 2019 deep learning track. arXiv preprint arXiv:2003.07820 (2020)
15. Devlin, J.: BERT: Pre-training of deep bidirectional transformers for language understanding. arXiv preprint arXiv:1810.04805 (2018)
16. Dong, L., et al.: Unified language model pre-training for natural language understanding and generation. In: Proceedings of NeurIPS (2019)
17. Dubey, A., et al.: The Llama 3 herd of models. arXiv preprint arXiv:2407.21783 (2024)
18. Dunn, O.J.: Multiple comparisons among means. JASA **56**(293) (1961)
19. Ekstrand, M.D., McDonald, G., Raj, A., Johnson, I.: Overview of the TREC 2022 fair ranking track. arXiv preprint arXiv:2302.05558 (2023)
20. Faggioli, G., et al.: Perspectives on large language models for relevance judgment. In: Proceedings of SIGIR (2023)
21. Fang, J., Meng, Z., Macdonald, C.: Trace the evidence: Constructing knowledge-grounded reasoning chains for retrieval-augmented generation. arXiv preprint arXiv:2406.11460 (2024)
22. Fisher, R.A.: Frequency distribution of the values of the correlation coefficient in samples from an indefinitely large population. Biometrika **10**(4) (1915)
23. Gonen, H., Iyer, S., Blevins, T., Smith, N.A., Zettlemoyer, L.: Demystifying prompts in language models via perplexity estimation. arXiv preprint arXiv:2212.04037 (2022)
24. Harman, D.: Overview of the first TREC conference. In: Proceedings of SIGIR (1993)
25. Harman, D.: Overview of the third text retrieval conference (TREC-3). DIANE Publishing (1995)
26. Hauff, C., Hiemstra, D., Azzopardi, L., De Jong, F.: A case for automatic system evaluation. In: Proceedings of ECIR (2010)
27. Hawking, D., Craswell, N., Thistlewaite, P., Harman, D.: Results and challenges in web search evaluation. Comput. Netw. **31**(11-16) (1999)
28. Hofstätter, S., Lin, S.C., Yang, J.H., Lin, J., Hanbury, A.: Efficiently teaching an effective dense retriever with balanced topic aware sampling. In: Proceedings of SIGIR (2021)
29. Järvelin, K., Kekäläinen, J.: Cumulated gain-based evaluation of IR techniques. TOIS **20**(4) (2002)
30. Jones, S.: Report on the need for and provision of an "ideal" information retrieval test collection (1975)
31. Kendall, M.G.: Rank correlation methods (1948)
32. Liu, J., Shen, D., Zhang, Y., Dolan, W.B., Carin, L., Chen, W.: What makes good in-context examples for GPT-3? In: Proceedings of DeeLIO@ACL (2022)
33. Losada, D.E., Parapar, J., Barreiro, A.: Cost-effective construction of information retrieval test collections. In: Proceedings of CERI (2018)
34. MacAvaney, S., Macdonald, C., Ounis, I.: Streamlining evaluation with IR-measures. In: Proceedings of ECIR (2022)
35. MacAvaney, S., Soldaini, L.: One-shot labeling for automatic relevance estimation. In: Proceedings of SIGIR (2023)

36. MacAvaney, S., Yates, A., Feldman, S., Downey, D., Cohan, A., Goharian, N.: Simplified data wrangling with IR_datasets. In: Proceedings of SIGIR (2021)
37. Mackie, I., Dalton, J., Yates, A.: How deep is your learning: The DL-HARD annotated deep learning dataset. In: Proceedings of SIGIR (2021)
38. Meng, C., Arabzadeh, N., Askari, A., Aliannejadi, M., de Rijke, M.: Query performance prediction using relevance judgments generated by large language models. arXiv preprint arXiv:2404.01012 (2024)
39. Nuray, R., Can, F.: Automatic ranking of information retrieval systems using data fusion. Inf. Process. Manage. **42**(3) (2006)
40. Pilault, J., Li, R., Subramanian, S., Pal, C.: On extractive and abstractive neural document summarization with transformer language models. In: Proceedings of EMNLP (2020)
41. Pradeep, R., Nogueira, R., Lin, J.: The expando-mono-duo design pattern for text ranking with pretrained sequence-to-sequence models. arXiv preprint arXiv:2101.05667 (2021)
42. Pradeep, R., et al.: Ragnarök: A reusable RAG framework and baselines for TREC 2024 retrieval-augmented generation track. arXiv preprint arXiv:2406.16828 (2024)
43. Raffel, C., et al.: Exploring the limits of transfer learning with a unified text-to-text transformer. JMLR **21**(140) (2020)
44. Rahmani, H.A., et al.: LLM4Eval: large language model for evaluation in IR. In: Proceedings of SIGIR (2024)
45. Rahmani, H.A., et al.: Report on the 1st workshop on large language model for evaluation in information retrieval (LLM4Eval 2024) at SIGIR 2024. arXiv preprint arXiv:2408.05388 (2024)
46. Rahmani, H.A., et al.: LLMJudge: LLMs for relevance judgments. arXiv preprint arXiv:2408.08896 (2024)
47. Schweizer, B., Wolff, E.F.: On nonparametric measures of dependence for random variables. Ann. Stat. **9**(4) (1981)
48. Soboroff, I.: Don't use LLMs to make relevance judgments. arXiv preprint arXiv:2409.15133 (2024)
49. Soboroff, I., Nicholas, C., Cahan, P.: Ranking retrieval systems without relevance judgments. In: Proceedings of SIGIR (2001)
50. Sorensen, T., et al.: An information-theoretic approach to prompt engineering without ground truth labels. arXiv preprint arXiv:2203.11364 (2022)
51. Spoerri, A.: Using the structure of overlap between search results to rank retrieval systems without relevance judgments. Inf. Process. Manag. **43**(4) (2007)
52. Su, Y., Tai, Y., Ji, Y., Li, J., Yan, B., Zhang, M.: Demonstration augmentation for zero-shot in-context learning. arXiv preprint arXiv:2406.01224 (2024)
53. Tamber, M.S., Pradeep, R., Lin, J.: Scaling down, LiTting up: Efficient zero-shot listwise reranking with seq2seq encoder-decoder models. arXiv preprint arXiv:2312.16098 (2023)
54. Thomas, P., Spielman, S., Craswell, N., Mitra, B.: Large language models can accurately predict searcher preferences. In: Proceedings of SIGIR (2024)
55. Upadhyay, S., Kamalloo, E., Lin, J.: LLMs can patch up missing relevance judgments in evaluation. arXiv preprint arXiv:2405.04727 (2024)
56. Upadhyay, S., Pradeep, R., Thakur, N., Craswell, N., Lin, J.: UMBRELA: UMbrela is the (open-source reproduction of the) Bing RELevance assessor. arXiv preprint arXiv:2406.06519 (2024)
57. Vaswani, A.: Attention is all you need. In: Proceedings of NeurIPS (2017)
58. Voorhees, E.M.: The TREC robust retrieval track. In: ACM SIGIR Forum. vol. 39 (2005)

59. Voorhees, E.M., Soboroff, I., Lin, J.: Can old TREC collections reliably evaluate modern neural retrieval models? arXiv preprint arXiv:2201.11086 (2022)
60. Wu, S., Crestani, F.: Data fusion with estimated weights. In: Proceedings of CIKM (2002)
61. Wu, S., Crestani, F.: Methods for ranking information retrieval systems without relevance judgments. In: Proceedings of ACM/SIGAPP (2003)
62. Zhuang, H., et al.: RankT5: fine-tuning t5 for text ranking with ranking losses. In: Proceedings of SIGIR (2023)

Enhancing FEVER-Style Claim Fact-Checking Against Wikipedia: A Diagnostic Taxonomy and a Generative Framework

Anton Chernyavskiy[1(✉)], Dmitry Ilvovsky[1], and Preslav Nakov[2]

[1] HSE University, Moscow, Russia
{acherniavskii,dilvovsky}@hse.ru
[2] Mohamed bin Zayed University of Artificial Intelligence, Abu Dhabi, United Arab Emirates
preslav.nakov@mbzuai.ac.ae

Abstract. Fact-checking is a crucial yet challenging task that continues to gain importance. In an effort to address this issue, the FEVER large-scale dataset was developed to facilitate evidence-based fact-checking using Wikipedia as a reference. Despite numerous proposed approaches and evaluations on this dataset, a comprehensive understanding of the errors made by these approaches is still lacking. Here, we aim to bridge this gap. We introduce a diagnostic taxonomy and a generative framework to enhance FEVER-style fact-checking. We establish a taxonomy of errors and we construct a diagnostic dataset that enables the analysis of the errors made by state-of-the-art models as well as their distribution within the FEVER dataset. Additionally, we provide a set of prompts to generate examples within this taxonomy. Our experiments demonstrate promising results through the utilization of these generated examples for fine-tuning.

Keywords: Fact-checking · Diagnostic Taxonomy · Diagnostic Dataset · Synthetic Data · Natural Language Inference · Wikipedia

1 Introduction

The task of fact-checking is challenging and complex. There is a growing demand for evidence-based fact-checking, as it provides users understandable and interpretable results. However, manual verification by experts is not possible due to large volumes, and thus automatic methods are being actively developed [14,15,19]. In order to facilitate the training of models, the FEVER dataset was proposed, which is currently the largest fact-checking dataset against textual data from Wikipedia [20]. Various models were investigated and evaluated on this dataset during the competitions [30].

As in general in machine learning, the models for fact-checking are compared by some target metric. In the case of FEVER, the FEVER-score is used, which

C. Hauff et al. (Eds.): ECIR 2025, LNCS 15572, pp. 310–325, 2025.
https://doi.org/10.1007/978-3-031-88708-6_20

aggregates the score for correct evidence retrieval and classification accuracy. This can cause problems, since the metric is sensitive to shifts and biases in the dataset. For example, the dataset may be biased towards specific classes, leading to a lack of generalizability, especially if there are consistent labeling errors. While state-of-the-art models may demonstrate high classification performance on such validation sets, these models may still have substantial limitations when applied to real-world tasks. To address this, a subsequent iteration of the FEVER competition was organized, where participants were asked to come up with adversarial attacks to identify weaknesses in the current systems [21]. Similar ideas are being actively explored in other NLP tasks. For instance, adversarial attacks and diagnostic datasets for NLI [22], hate speech [23], and question answering [24] have been proposed. Nevertheless, adversarial attacks alone do not provide a comprehensive understanding of model limitations, and as of yet, there has been no proposal for a diagnostic dataset for FEVER (evidence-based fact checking tasks). Our contributions can be summarized as follows:

- We introduce a diagnostic taxonomy for FEVER-style tasks.
- We create a diagnostic dataset using the proposed taxonomy to test models weaknesses[1].
- We propose a generative framework, which can be used to improve the quality of the final systems. To this end, we develop algorithms and a set of prompts that are used to generate a synthetic training set based on the taxonomy.
- We offer a detailed analysis of the FEVER corpus based on the taxonomy and we present diagnostic results for current models. We demonstrate that the generative framework can improve models in different aspects.

2 Related Work

The fact-checking problem has been studied within various task formulations and competitions. One of the popular directions is explainable fact-checking, where evidence is either knowledge graphs [4], previously checked claims [3] or text databases (Wikipedia) [30]. FEVER [20] is a large-scale dataset for fact-checking against Wikipedia. Several iterations of the FEVER competitions were organised and different approaches were proposed. The best of them have a multi-stage structure: Document Retrieval, Sentence Retrieval, and Natural Language Inference (NLI) [6,7]. In this paper, we focus on the NLI part only. For this, previous work used Random Forest [25], ESIM [26], LSTM [5], and BERT [27–29]. The best results were achieved using Transformers.

In machine learning, models are evaluated and compared using evaluation measures such as accuracy and F1 score. However, these measures do not indicate the weaknesses of the models and could potentially lead to inaccurate general comparison due to dataset biases. To address this, different types of adversarial attacks were proposed. They have the target to change the examples of the original dataset, so that the model starts making incorrect predictions for them.

[1] We release our dataset and code at https://github.com/aschern/FEVERDiagnostics.

Wallace et al. [24] proposed universal adversarial triggers, utilizing the gradient-guided search for SNLI and question-answering tasks. Morris et al. [8] presented a framework for adversarial attacks on BERT for GLUE tasks. Nie et al. [22] developed a benchmark for NLI (AdversarialNLI), where examples were constructed using a human-and-model-in-the-loop procedure. Wang et al. [10] constructed a multi-task benchmark, AdvGLUE, which includes 14 types of textual adversarial attacks, including word-level, sentence-level, and human-level.

There are also a range of studies of adversarial attacks for the FEVER task. Thorne et al. [31] proposed lexically-informed WordNet attacks and model-targeting attacks with paraphrases for FEVER task. As part of the FEVER2.0 competition, a "breakers" phase was introduced, requiring participants to produce adversarial examples that would trigger classification errors in existing systems [21]. This phase revealed that models were vulnerable to temporal (such as replacing specific dates with intervals) and multi-hop evidence attacks. Niewinski et al. [32] proposed a system based on GPT-2 to generate adversarial examples. Kim and Allan [33] used arithmetic operations and template-based transition reasoning to create examples. However, it remains challenging to ensure the high quality of adversarial attacks and to use them to provide a comprehensive understanding of the model drawbacks. Another promising direction is the development of diagnostic datasets.

A diagnostic dataset comprises the general set of tests that can help to exactly determine the weaknesses of a model. To analyze the linguistic capabilities and the world knowledge of models, expertly constructed diagnostic datasets were developed as part of the GLUE and SuperGLUE benchmarks for the MultiNLI [34] and RTE [9], respectively. Joshi et al. [35] presented a taxonomic hierarchy for NLI task, along with the TaxiNLI dataset, which consists of examples labeled with the taxonomy categories such as linguistic, logical, and knowledge-based types. Ribeiro et al. [36] created a task-agnostic NLP checklist comprising eleven types of tests, including: temporal aspects, negations, robustness, coreference, and suggested a scheme for generating such examples using a template-based approach. Similarly, Röttger et al. [23] implemented a template-based approach to generate examples for evaluating 29 model functionalities in the hate speech detection task. The creation of diagnostic datasets presents challenges due to the limited manual expert annotation, which makes it difficult to create large datasets. Therefore, many datasets are created or extended using templates and rules, which may result in limited variability and unconfirmed correctness.

The alternative to template-based generation is the use of Large Language Models (LLMs), such as GPT-3, which demonstrated strong results across various text generation tasks [2]. It can be also useful for generating synthetic training datasets, which is beneficial in zero-shot or few-shot learning scenarios in the absence of labeled training data, e.g., due to the complexity of collection (low-resource languages) or the need for extensive human annotation [11,13,37]. However, these studies indicate that models trained solely on this generated synthetic data typically do not yield quality comparable to that of such trained using real human-annotated data.

Table 1. Selection of tests from our diagnostic dataset. Each test consists of `Claim` (C), `Evidence` (E) concatenated with `Title` using the separator # and `Label` (REFUTES or SUPPORTS). There are 10 classes in the taxonomy.

Class	Example	Label
Names	C: Bring It On stars Eliza Dunst.	
	E: Bring It On # The film stars Kirsten Dunst, Eliza Dushku, and Gabrielle Union.	REF.
Numbers	C: Cannes Film Festival lasted 12 days.	
	E: 2002 Cannes Film Festival # It started on 15 May and ran until 26 May.	SUP.
Negations	C: Dark matter may not be real.	
	E: Dark matter # However, no dark matter particle has been conclusively identified.	SUP.
Geography	C: West Virginia borders Pennsylvania and Ohio to the north.	
	E: West Virginia # It is bordered by Ohio to the northwest, Pennsylvania to the north.	REF.
Relations	C: James Franco is the brother of Tom Franco.	
	E: Tom Franco # Tom Franco is the brother of actors James Franco and Dave Franco.	SUP.
Time	C: Ian Brennan was born in the 1980s.	
	E: Ian Brennan # Ian Brennan was born in April 23, 1978.	REF.
Quantifiers	C: Edgar Wright is only a songwriter.	
	E: Edgar Wright # He is an English director, screenwriter, producer, and actor.	REF.
Overload	C: Adam DeVine portrayed Haley Dunphy on a critically acclaimed sitcom.	
	E: Adam DeVine # He has portrayed Andy Bailey, the boyfriend of Haley Dunphy ...	REF.
Syn/Ant	C: Cadet Kelly was an original work.	
	E: Cadet Kelly # It is a 2002 Disney Channel Original Movie starring Hilary Duff.	SUP.
Meaning	C: Cara Delevingne was a child.	
	E: Cara Delevingne # Delevingne won the "Model of the Year" award ...	SUP.

No diagnostic dataset has yet been proposed for FEVER-style tasks. There is also a lack of research on the errors made by existing models and the specific characteristics of the training dataset. We aim to bridge this gap. Our diagnostic taxonomy is generally applicable, and can be utilized in various evidence-based fact-checking problems, such as FAVIQ [38], SciFact [39]. We also introduce a generative framework that can improve the predictions of the current models.

3 Diagnostic Dataset

3.1 Taxonomy

We propose a diagnostic taxonomy designed to evaluate models for FEVER-style tasks, which consists of ten distinct categories: *Synonym/Antonym, Names, Negations, Quantifiers, Relations, Numbers, Time, Meaning, Geography, Overload.* These categories cover tests in lexical, logical, and factual knowledge domains and are based on the taxonomies outlined in previous studies [35,36]. It also incorporates classes of adversarial attacks proposed in FEVER2.0, such as geography-based, arithmetic-based and temporal-based attacks [21]. Thus, we get categories that are widely represented in FEVER-style datasets, and their definitions allow for achieving a high level of accuracy in their annotation.

Table 1 presents examples for each class. In the FEVER-style format, each example is structured by concatenating the entities `Claim` and `Evidence` and `Title` via a separator, and is labeled as either SUPPORTS or REFUTES. We exclude the NOT ENOUGH INFO examples from our analysis due to unique specificity and complexity of category annotation.

Synonym/Antonym. The class contains examples that require a semantic understanding of words for the classification. So, `Claim` may be constructed from `Evidence` by substituting some words with their synonyms or antonyms. To increase the classification complexity, we paraphrase `Claim` in some cases.

Names. This class includes examples that test the model's ability to compare the acting entities mentioned in the `Claim` and in the `Evidence`. Here, only part of the name or of the title can differ. Moreover, the [Evidence] may contain multiple entities with their shuffled parts.

Negations. These examples assess the ability of the model to accurately interpret negations. Note that the presence of a negation does not automatically result in a REFUTES label. The negation may be doubled or used in combination with an antonym.

Quantifiers. Examples of this class test the impact of "quantifiers" on model predictions. For example, these can include words such as "some", "any", "few", as well as "only", and "no one".

Relations. This class contains examples where the order of the entities in `Claim` and `Evidence` in a particular relation is changed. So, involved entities can alter their roles, where the action remains the same. It is forbidden to change the form of this action, that is, the active form of the verb to a passive one. SUPPORTS label remains only in the case of symmetrical relations. For example, such relations are "played with" and "is a relative of".

Numbers/Arithmetic. The class is based on numbers and includes several subcategories that focus on different aspects of numerical reasoning. The first one evaluates the model's capability to compare numbers taking into account modifiers, such as "more" or "less". The second subcategory involves counting entities in `Evidence` to check `Claim`. For example, by counting "trophies" from the list of trophies in `Evidence`, `Claim` "won two trophies" can be supported or refuted. The third subcategory tests the model's ability to perform arithmetic operations on numbers in `Evidence`. Note that not all numbers in `Evidence` may be relevant; the model should select only those that are necessary. For example, `Evidence` "released five studio albums, two compilation albums, and one live DVD" contains only two numbers related to albums. Moreover, across all categories, the numbers may appear in numerical, textual, or mixed formats.

Time. This class tests the model's ability to analyze the relationship between time statements mentioned in Claim and Evidence. These statements can be defined concretely like "between 1990 and 1995", "before 1900" (which is similar to the Numbers class); or they can be expressed more abstractly like "in the 20th century", "in the 1990s" and finally "100 years ago". In the last case, we consider the date of the statement to be the moment of its verification.

Meaning. The class includes examples where the model is required to understand the semantics of the words and underlying concepts and perform logical inferences based on this understanding. So, this class consists of several subcategories based on concept understanding, object properties and event relations [16–18]. It is much more challenging than Synonym/Antonym as it involves a deeper level of logical analysis. For example, using `Evidence` stating that "people got married in 1992" (event relation), the claim "met first time before 1992" can be confirmed, and the claim "met first time in 1892" can be refuted (concept understanding). In the second case, an additional understanding of the duration of life is required.

Geography. Examples of this class test the model's knowledge and understanding of geography by comparing location or neighborhood data in `Claim` and `Evidence`. For example, `Evidence` can provide explicit information that "country X is adjacent to country Y from the North" or can use implicitly a well-known fact like "China does not border Europe". Additionally, this class tests that the model may understand variations in spelling such as "U.S. vs. United States", and the ability to make a logical conclusion. For example, `Evidence` "the event was held in London" confirms the `Claim` "the event was held in Great Britain".

Overload. This class of examples is characterized by a large amount of superfluous `Evidence`, containing a lot of irrelevant information for verification, e.g., long enumerations. In addition, this irrelevant information may have a large overlap in words with the verifiable `Claim`, especially when it describes another entity (a person, an event, etc.).

3.2 Golden Set

Each example in the golden set in our diagnostic dataset is constructed using the examples from the development FEVER set. To this end, we manually augment examples according to the requirements of the diagnostic taxonomy. We construct 20 examples for each class with a total of 200 examples. Thus, unlike the original FEVER test dataset, our dataset is balanced across the taxonomy classes by definition. This set is used for testing purposes only.

4 Generative Framework

Creating examples manually for each category is a challenging task, as for most of them it is not enough to change one word to a synonym/antonym. In this section, we present faster, automated approaches for generating diagnostic examples based on the proposed taxonomy. These examples can be used both for validation and for improvement the quality of predictions of individual classes, as the increased amount of training data can enhance these predictions. We generate examples of taxonomy classes based on word augmentations (names, time, negations) using manual rules, since for such classes this process is easily formalized. For the remaining classes, with the exception of *Overload*, we utilize GPT-3.5 and class-specific prompts, which comprise the task description and a selection of the hardest examples from the golden set. We do not generate the Overload class due to its specifics.

4.1 Generation Using Rules

Negations. To generate examples of this class, we take the current `Evidence` (paired with the corresponding `Title`), search for a verb in it and then add the negation to create the `Claim`. Additionally, with some probability, we replace the following adjective (which is detected by the POS tag) with its antonym using WordNet synsets [40]. We preserve the grammar, bringing the new word to the form of the original one.

Time. To construct examples of this class, first we select a set of pieces of evidence that contains information about time. To this end, we apply a pre-trained named entity recognition (NER) model and we search for entities of the "Date" class. Among them, we select single-word entities (years) and we perform random augmentation of their representation. This involves either changing the year format to the 1990s, or introducing an interval (using *before, after, since, between*) with a random shift of the year. The label is then transformed accordingly.

Names. We generate examples of this class by first selecting a list of all named entities using a pre-trained NER model and grouping them into two categories: (*i*) "Nationalities or religious or political groups", "Any named language"; "Names", "Companies, agencies, institutions, etc.", (*ii*) "Events, Titles of books, songs, etc., Objects, vehicles, foods, etc.". For substitutions of entities from the first group, we keep only entities consisting of one word. As a replacement, we choose a random entity with the same label. For the second group, we either replace the entire entity with another entity from the same group, or just replace part of the entity with the corresponding part of another entity from the pre-filtered set of entities of the same length.

4.2 GPT-Based Generation

We use the GPT-3.5 language model [2] (namely, text-davinci-003) to generate full FEVER-style examples, that is `Title`, `Claim`, `Evidence`, `Label` are generated at the same time. To this end, we compose class-specific prompts. Prompts include a general description of the task as well as a sub-sample of examples of the taxonomy class from the golden dataset, along with a description of this class.

This description highlights the class specifics and imposes restrictions. These can be both general restrictions, e.g., that the length of `Claim` and `Evidence` must be at least seven words, or class-specific restrictions. We generate sets by groups of five examples at a time. In total, we generated around 400 examples per each taxonomy class.

We highlight the important features of the prompts for some classes[2]: (*i*) we generate examples of the *Numbers* class with three different prompts in a balanced ratio: arithmetic, comparing, counting; (*ii*) to improve the quality of generation and enhance the meaningfulness of the task and examples, we add the `Relation` entity to the Relations class; (*iii*) we also add the `Word` entity to generate instances of the Meaning class. By the meaning of this `Word`, the `Claim` is checked.

We attempted to additionally generate the Chain-of-Thought (CoT) [1] to provide an instruction for choosing the labels. However, this did not improve the quality of the generated dataset. Consequently, there was no improvement of performance metrics on the golden set for the models trained using this data.

There are some general limitations and problems that we encountered when creating the dataset:

- The *input size* (prompt length) is limited. If there are a lot of examples in the prompt, the model forgets the context and the task description.
- The limited *output size* is also an issue because high variability could not be guaranteed. Examples are generated in small batches which have pre-existing variability, but there is nothing about the relationships between batches.
- It is difficult to maintain the *balance of labels* in the dataset. The possible option is to generate examples with a separate label. Although the requirement to generate only one label can be specified at the end of the prompt, according to our preliminary analysis, this yielded more errors.
- Despite explicitly stating some *requirements* at the end of the prompt, they were often not fulfilled. For instance, when generating a Relations class, the restriction that "the active form of the verb should not be changed to the passive form" was often not fulfilled, resulting in many examples with the wrong SUPPORTS label.

Although many GPT models may suffer from hallucinations, hallucinations in such examples are not a problem since the purpose of the task is to determine the relationship between `Claim` and `Evidence`. For example, for the category *Time*, `Evidence` "He was born in 1978" may contradict the `Claim` "He was born in the 1980s". This means that the model must learn the REFUTES relation regardless of the *correctness* of the year "1978".

5 Experiments and Results

In this section, we begin by analyzing the FEVER dataset using the proposed diagnostic taxonomy. We then identify errors in the model trained on the original

[2] Full prompts can be found at https://github.com/aschern/FEVERDiagnostics.

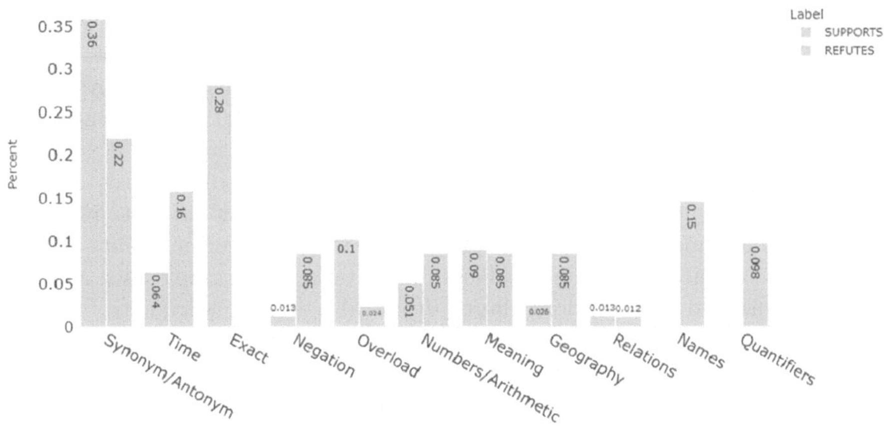

Fig. 1. Distribution of taxonomy classes for SUPPORTS/REFUTES examples in FEVER dataset. Exact is a special subcategory of Synonym/Antonym, meaning an exact matching of `Claim` and `Evidence`.

FEVER dataset. Following this, we move to a generative framework. First, to assess the usefulness of the generated examples, we demonstrate that validation on these examples yields metrics comparable to those obtained on a manually created diagnostic dataset. Next, we use these generated examples for model training and demonstrate that the observed errors across different taxonomy classes are not caused by inherent limitations of the model's architecture but rather by shortcomings in the training dataset—specifically, the insufficient representation of certain taxonomy classes. Finally, we demonstrate that adding such examples to the training process not only addresses these class-specific issues but also improves the overall performance of the final model on the original FEVER test set.

5.1 FEVER Dataset

We analyzed the FEVER dataset using our proposed taxonomy. To this end, we made a multi-label markup of 100 randomly sampled examples from the SUPPORTS and REFUTES classes (such an amount of examples is sufficient to check for potential imbalance in the taxonomy classes). Examples that do not belong to any taxonomy class were labeled to the Other class. These includes examples in which `Evidence` is combined from several sentences from different sources.

Figure 1 shows a histogram of frequency of taxonomy classes for examples labeled as SUPPORTS. It demonstrates that the majority of them are from Synonym/Antonym and Exact categories. Exact is a separate category, which we selected from Synonym/Antonym class. It means that `Evidence` and `Claim` are just a copies without valuable modifications, so it is the easiest category and has almost zero training value. Other categories have only 36% in sum. This imbalance can cause problems in the model trained on these two easiest classes will learn almost no complex dependencies at all. Notably, Park et al. [38] have demonstrated that training on the FAVIQ dataset yields superior metrics when compared to training on the FEVER for real-world problems. The proposed taxonomy almost completely covers the FEVER dataset, with the share of the Other class being less than 2%.

The distribution for REFUTES examples is more balanced than for SUPPORTS class. Synonym/Antonym again is the largest class. Time and Names are the second and third classes by popularity. Other classes have almost the same share except Overload and Relations. Relations is the least presented class and nearly absent both in SUPPORTS and REFUTES examples.

5.2 Experimental Setup

According to prior research and our preliminary study, the near-state-of-the-art encoder Transformer-based model for the FEVER task, specifically the RoBERTa [21,41], was selected for this study. Specifically, it was demonstrated [42] that in the NLI mode for single-sentence evidence (not covering multi-hop scenarios, which are beyond the scope of this study), RoBERTa-large achieves state-of-the-art results, outperforming even larger models such as T5 and DeBERTa.

We used two variants of RoBERTa model. The "base" version was used to analyze the quality of generated examples starting from the standard pre-trained model. The "large" version was used in other experiments. It was pre-trained on the FEVER dataset using all three classes: SUPPORTS, REFUTES, NOT ENOUGH INFO. To address class imbalance, we augmented the dataset by duplicating examples that had the REFUTES label. The model was fine-tuned for 3 epochs with early stopping, the learning rate of 2e-5 and the batch size of 6 (these hyperparameters were optimized based on performance metrics from a validation set, which comprised 10% of the total dataset). We report the best score achieved across 3 runs.

5.3 Error Analysis

Table 2 shows the metrics for the RoBERTa-large: accuracy and F1 score. In terms of accuracy, the model performs best on Synonym/Antonym examples, primarily because most of these examples have the SUPPORTS label. The errors in the Quantifiers category are also relatively low (accuracy of 0.85). The hardest class in our taxonomy in terms of accuracy is Meaning. Regarding the F1 score, which accounts for class imbalance, the most difficult classes are Overload, Negations (with a majority of errors in REFUTES examples) and Geography, while Time and Quantifiers are the easiest.

Similar results were obtained using generated examples (refer to Table 3). We did not generate examples for Synonym/Antonym and Quantifiers due to their already high performance. High accuracy but low F1 score for Relations class demonstrates problems with its generation where most of the examples are

Table 2. Metrics for RoBERTa-large on the diagnostic dataset.

Class	Size	Accuracy	F1
Names	20	0.55	0.38
Numbers	22	0.59	0.38
Negations	20	0.60	0.42
Geography	21	0.57	0.41
Relations	22	0.50	0.33
Time	20	0.60	0.60
Quantifiers	20	0.85	0.55
Overload	20	0.50	0.32
Synonym/Antonym	20	0.90	0.54
Meaning	20	0.45	0.39

Table 3. Results for RoBERTa-large on generated set.

Class	Size	Accuracy	F1
Names	1079	0.65	0.50
Numbers	624	0.59	0.38
Negations	1044	0.64	0.29
Geography	408	0.73	0.51
Relations	445	0.82	0.31
Time	2395	0.38	0.31
Meaning	339	0.65	0.49

Table 4. Metrics for RoBERTa-base fine-tuned using the generated dataset without using the FEVER data. The number of training epochs, batch size, and learning rate are specified in the *Params* column.

Class	Accuracy	F1	Params
Names	0.50	0.23	8, 8, 2e−5
Numbers	0.64	0.64	8, 8, 2e−5
Negations	0.90	0.90	8, 8, 2e−5
Geography	0.62	0.42	5, 8, 2e−5
Relations	0.59	0.56	8, 8, 2e−5
Time	0.70	0.70	8, 8, 2e−5
Meaning	0.65	0.63	8, 8, 2e−5

Table 5. Results for RoBERTa-large pre-trained on FEVER and fine-tuned on our generated dataset. The comparison with the original model is also shown with underscript.

Class	Accuracy	F1
Names	$0.55_{0.00}$	$0.38_{0.00}$
Numbers	$0.68_{0.09}$	$0.62_{\uparrow 0.24}$
Negations	$0.75_{\uparrow 0.15}$	$0.75_{\uparrow 0.32}$
Geography	$0.76_{\uparrow 0.19}$	$0.52_{\uparrow 0.11}$
Relations	$0.50_{0.00}$	$0.33_{0.00}$
Time	$0.75_{\uparrow 0.15}$	$0.75_{\uparrow 0.15}$
Meaning	$0.65_{\uparrow 0.20}$	$0.65_{\uparrow 0.26}$

labeled as SUPPORTS. Moreover, the model achieved higher score for Meaning, which indicates that GPT-3.5 was unable to construct complex examples.

We can conclude that the model trained on the FEVER dataset learns the word matching that may involve paraphrasing or the use of synonyms and antonyms. It has no tendency to make mistakes on examples from some specific class within our taxonomy (based on words or values and logic), showing a relatively uniform distribution of errors across categories.

5.4 Training with Generated Data

Separate Classes. We used the RoBERTa-base model to evaluate the quality of the generated data. In each experiment, the model was trained and evaluated exclusively on examples from a specific class. For each class, we selected optimal parameters using the validation subset of our dataset. These parameters, along with the experimental results, are shown in Table 4. Our experiments confirm that the generated examples are highly useful. The obtained results are even higher than for RoBERTa-large fine-tuned on the original FEVER dataset (Table 2). This improvement can be attributed to the focused nature of our experiments, which concentrated on separate classes. Among the categories, the most challenging ones for the model to learn were Geography (GPT-based generation) and Names (rule-based generation).

Joint Training. In this experiment, we fine-tuned RoBERTa-large, which had been pre-trained on the original FEVER dataset, using all the generated data. The performance results on the diagnostic dataset are presented in Table 5. For some classes, we see relatively low scores compared to retraining the source RoBERTa-base for specific classes. This is likely due to class imbalance, as we chose a dataset size sufficient enough to learn a specific class.

Finally, we fine-tuned RoBERTa-large model on the generated data using the original FEVER validation subset for early stopping and evaluated it on the development FEVER set. Since the training exclusively involved the classes SUPPORTS and REFUTES (there was no NOT ENOUGH INFO), we limited our evaluation to these two categories. For this purpose, we presumed the existence of an oracle capable of distinguishing them from NOT ENOUGH INFO. We observed an improvement in accuracy from 0.87 to 0.91.

6 Discussion and Limitations

We proposed a taxonomy that can highlight the weaknesses of models for solving FEVER-style fact-checking tasks. Our analysis revealed that the FEVER dataset mainly consists of simple examples that do not present significant training complexity. This leads to poor performance of the trained models. Through experiments, we demonstrated that we can improve the quality of such models for the individual classes of the taxonomy by incorporating the additional training data. Furthermore, training on all classes simultaneously improved performance on the original development set, though it occasionally led to declines in performance for certain classes.

To address this issue, we propose to classify each example into one or more relevant categories from our taxonomy. It would allow to handle each example in a specific manner, when appropriate. Furthermore, it allows us to annotate full training set for FEVER-style tasks with our taxonomy classes and to generate new training examples for specific classes.

As the taxonomy class cannot be labeled automatically by the set of manual rules, we tried to train RoBERTa-base using the generated data (here, we experimented with the "base" version, as the dataset size is small). It took `Claim`, `Evidence`, `Title`, `Label` as an input and the taxonomy class as output. The variability of the generated examples was not substantial. This resulted in the model easily learning the features of the generated data, achieving an accuracy of 0.92 on the validation set. However, it achieved 0.53 on the diagnostic dataset, with the highest error for the Meaning class. Moreover, the model was almost unable to generalize to the original FEVER data, which had been labeled manually and achieved an accuracy of only 0.34. Thus, a more variable dataset is needed: this can be achieved by removing all rule-based examples, as they have the same "template structure", and replacing them with GPT-3.5-based examples which also requires improving the variability of the generated examples and minimizing generation errors, such as failure to meet specified requirements.

7 Conclusion and Future Work

We presented a diagnostic taxonomy and a generative framework to enhance FEVER-style fact-checking. We established a taxonomy of errors and constructed a diagnostic dataset that enables the analysis of the errors made by state-of-the-art models and their distribution within the FEVER dataset (we demonstrated that it is critically imbalanced). We further provided a set of prompts to generate examples within this taxonomy. Our experiments demonstrated promising results when using these generated examples for fine-tuning models, leading to improved quality in specific taxonomy classes and an overall performance enhancement of more than 4%.

In future work, we plan to experiment with stronger generation models to improve the quality of the generated data. Another direction is to conduct a similar analysis for these decoder models as final classifiers when fine-tuning or in few-shot/zero-shot setup.

Acknowledgments. The article was prepared within the framework of the HSE University Basic Research Program.

This research was supported in part through computational resources of HPC facilities at HSE University.

Limitations. Our dataset is relatively small, but it is typical for diagnostic datasets. Note also that this is a test dataset: since we have many classes, its total size is not that small. There is a need for a more comprehensive taxonomy and also for a larger number of examples. Moreover, we focus on English only; there is a need for such research for other languages.

Ethics and Broader Impact. Our dataset can be misused to unfairly moderate content. Thus, we ask researchers and potential users to exercise caution. The intended use is to help improve research systems by means of providing a way to analyze their limitations.

References

1. Chen, J., Chen, L., Huang, H., Zhou, T.: When do you need chain-of-thought prompting for ChatGPT?. CoRR, abs/2304.03262 (2023). https://doi.org/10.48550/arXiv.2304.03262
2. Brown, T., et al.: Language models are few-shot learners. In: Advances in Neural Information Processing Systems, vol. 33, pp. 1877–1901 (2020)
3. Nakov, P., et al.: The CLEF-2021 CheckThat! Lab on detecting check-worthy claims, previously fact-checked claims, and fake news. In: Hiemstra, D., Moens, M.-F., Mothe, J., Perego, R., Potthast, M., Sebastiani, F. (eds.) ECIR 2021. LNCS, vol. 12657, pp. 639–649. Springer, Cham (2021). https://doi.org/10.1007/978-3-030-72240-1_75
4. Gad-Elrab, M., Stepanova, D., Urbani, J., Weikum, G.: Tracy: tracing facts over knowledge graphs and text. IN: Proceedings of the World Wide Web Conference, pp. 3516–3520 (2019)
5. Nie, Y., Chen, H., Bansal, M.: Combining fact extraction and verification with neural semantic matching networks. In: Proceedings Of The AAAI Conference On Artificial Intelligence, pp. 6859–6866 (2019)
6. Bekoulis, G., Papagiannopoulou, C., Deligiannis, N.: A review on fact extraction and verification. ACM Comput. Surv. (CSUR) **55**, 1–35 (2020)
7. Guo, Z., Schlichtkrull, M., Vlachos, A.: A survey on automated fact-checking. Trans. Assoc. Comput. Linguist. **10**, 178–206 (2022). https://doi.org/10.1162/tacl%5C_a%5C_00454
8. Morris, J., et al. TextAttack: a framework for adversarial attacks, data augmentation, and adversarial training in NLP. In: Proceedings of the 2020 Conference on Empirical Methods in Natural Language Processing: System Demonstrations, pp. 119–126 (2020)
9. Wang, A., et al.: SuperGLUE: a stickier benchmark for general-purpose language understanding systems. In: Advances in Neural Information Processing Systems, vol. 32 (2019)
10. Wang, B., et al.: Adversarial GLUE: a multi-task benchmark for robustness evaluation of language models. In: Advances in Neural Information Processing Systems (2021)
11. Meyer, S., Elsweiler, D., Ludwig, B., Fernandez-Pichel, M., Losada, D.: Do we still need human assessors? Prompt-based GPT-3 user simulation in conversational AI. In: Proceedings of the 4th Conference on Conversational User Interfaces (2022). https://doi.org/10.1145/3543829.3544529
12. Wang, S., Liu, Y., Xu, Y., Zhu, C., Zeng, M.: Want to reduce labeling cost? GPT-3 can help. arXiv abs/2108.13487 (2021)
13. Bonifacio, L., Abonizio, H., Fadaee, M., Nogueira, R.: InPars: unsupervised dataset generation for information retrieval. In: Proceedings of the 45th International ACM SIGIR Conference on Research and Development in Information Retrieval, pp. 2387–2392 (2022). https://doi.org/10.1145/3477495.3531863
14. Nakov, P., et al.: Automated fact-checking for assisting human fact-checkers. In: Proceedings of the 30th International Joint Conference on Artificial Intelligence, pp. 4551–4558 (2021)
15. Gu, Z., Fan, J., Tang, N., Nakov, P., Zhao, X. , Du, X.: PASTA: table-operations aware fact verification via sentence-table cloze pre-training. In: Proceedings of the 2022 Conference on Empirical Methods In Natural Language Processing (2022)

16. Liu, H., Singh, P.: ConceptNet - a practical commonsense reasoning tool-kit. BT Technol. J. **22**, 211–226 (2004)
17. Speer, R., Chin, J., Havasi, C.: ConceptNet 5.5: an open multilingual graph of general knowledge. In: AAAI, pp. 4444–4451 (2017)
18. Jackendoff, R. Semantic Structures. MIT Press, Cambridge (1990)
19. Guo, Z., Schlichtkrull, M., Vlachos, A.: A survey on automated fact-checking. Trans. Assoc. Comput. Linguist. **10**, 178–206 (2022). https://aclanthology.org/2022.tacl-1.11
20. Thorne, J., Vlachos, A., Cocarascu, O., Christodoulopoulos, C., Mittal, A.: The Fact Extraction and VERification (FEVER) shared task. In: Proceedings of the First Workshop on Fact Extraction and VERification (FEVER), pp. 1–9 (2018). https://aclanthology.org/W18-5501
21. Thorne, J., Vlachos, A., Cocarascu, O., Christodoulopoulos, C., Mittal, A.: The FEVER2.0 shared task. In: Proceedings of the Second Workshop on Fact Extraction And VERification (FEVER). pp. 1–6 (2019). https://aclanthology.org/D19-6601
22. Nie, Y., Williams, A., Dinan, E., Bansal, M., Weston, J., Kiela, D.: Adversarial NLI: a new benchmark for natural language understanding. In: Proceedings of the 58th Annual Meeting of the Association For Computational Linguistics, pp. 4885–4901 (2020). https://aclanthology.org/2020.acl-main.441
23. Röttger, P., Vidgen, B., Nguyen, D., Waseem, Z., Margetts, H., Pierrehumbert, J.: HateCheck: functional tests for hate speech detection models. In: Proceedings of the 59th Annual Meeting of the Association For Computational Linguistics and the 11th International Joint Conference on Natural Language Processing (Volume 1: Long Papers), pp. 41–58 (2021). https://aclanthology.org/2021.acl-long.4
24. Wallace, E., Feng, S., Kandpal, N., Gardner, M., Singh, S.: Universal adversarial triggers for attacking and analyzing NLP. In: Proceedings of the 2019 Conference on Empirical Methods in Natural Language Processing and the 9th International Joint Conference on Natural Language Processing (EMNLP-IJCNLP), pp. 2153–2162 (2019). https://aclanthology.org/D19-1221
25. Reddy, A., Rocha, G., Esteves, D.: DeFactoNLP: fact verification using entity recognition, TFIDF vector comparison and decomposable attention. In: Proceedings of the First Workshop on Fact Extraction and VERification (FEVER), pp. 132–137 (2018). https://aclanthology.org/W18-5522
26. Hidey, C., Diab, M.: Team SWEEPer: joint sentence extraction and fact checking with pointer networks. In: Proceedings of the First Workshop on fact Extraction and VERification (FEVER), pp. 150–155 (2018). https://aclanthology.org/W18-5525
27. Stammbach, D., Neumann, G.: Team DOMLIN: exploiting evidence enhancement for the FEVER shared task. In: Proceedings of the Second Workshop on fact Extraction and VERification (FEVER), pp. 105–109 (2019). https://aclanthology.org/D19-6616
28. Chernyavskiy, A., Ilvovsky, D.: Extract and aggregate: a novel domain-independent approach to factual data verification. In: Proceedings of the Second Workshop on Fact Extraction and VERification (FEVER), pp. 69–78 (2019). https://aclanthology.org/D19-6612
29. Subramanian, S., Lee, K.: Hierarchical evidence set modeling for automated fact extraction and verification. In: Proceedings of the 2020 Conference on Empirical Methods in Natural Language Processing (EMNLP), pp. 7798–7809 (2020). https://aclanthology.org/2020.emnlp-main.627

30. Thorne, J., Vlachos, A., Christodoulopoulos, C., Mittal, A.: FEVER: a large-scale dataset for fact extraction and verification. In: Proceedings of the 2018 Conference of the North American Chapter of the Association for Computational Linguistics: Human Language Technologies, Volume 1 (Long Papers), pp. 809–819 (2018). https://aclanthology.org/N18-1074
31. Thorne, J., Vlachos, A., Christodoulopoulos, C., Mittal, A.: Evaluating adversarial attacks against multiple fact verification systems. In: Proceedings of the 2019 Conference on Empirical Methods in Natural Language Processing and the 9th International Joint Conference on Natural Language Processing (EMNLP-IJCNLP), pp. 2944–2953 (2019). https://aclanthology.org/D19-1292
32. Niewinski, P., Pszona, M., Janicka, M.: GEM: generative enhanced model for adversarial attacks. In: Proceedings of the Second Workshop on Fact Extraction and VERification (FEVER), pp. 20–26 (2019). https://aclanthology.org/D19-6604
33. Kim, Y., Allan, J.: FEVER breaker's run of team NbAuzDrLqg. In: Proceedings of the Second Workshop on Fact Extraction and VERification (FEVER), pp. 99–104 (2019). https://aclanthology.org/D19-6615
34. Wang, A., Singh, A., Michael, J., Hill, F., Levy, O., Bowman, S.: GLUE: a multi-task benchmark and analysis platform for natural language understanding. In: Proceedings of the 2018 EMNLP Workshop BlackboxNLP: Analyzing and Interpreting Neural Networks for NLP, pp. 353–355 (2018). https://aclanthology.org/W18-5446
35. Joshi, P., Aditya, S., Sathe, A., Choudhury, M.: TaxiNLI: taking a ride up the NLU hill. In: Proceedings of the 24th Conference on Computational Natural Language Learning, pp. 41–55 (2020). https://aclanthology.org/2020.conll-1.4
36. Ribeiro, M., Wu, T., Guestrin, C., Singh, S.: Beyond accuracy: behavioral testing of NLP models with checklist. In: Proceedings of the 58th Annual Meeting of the Association for Computational Linguistics, pp. 4902–4912 (2020). https://aclanthology.org/2020.acl-main.442
37. Wang, Y., et al.: PromDA: prompt-based data augmentation for low-resource NLU tasks. In: Proceedings of the 60th Annual Meeting of the Association for Computational Linguistics (Volume 1: Long Papers), pp. 4242–4255 (2022). https://aclanthology.org/2022.acl-long.292
38. Park, J., Min, S., Kang, J., Zettlemoyer, L., Hajishirzi, H.: FaVIQ: FAct verification from information-seeking questions. In: Proceedings of the 60th Annual Meeting of the Association for Computational Linguistics (Volume 1: Long Papers), pp. 5154–5166 (2022). https://aclanthology.org/2022.acl-long.354
39. Wadden, D., et al.: Fact or fiction: verifying scientific claims. In: Proceedings of the 2020 Conference on Empirical Methods in Natural Language Processing (EMNLP), pp. 7534–7550 (2020). https://aclanthology.org/2020.emnlp-main.609
40. Miller, G.: WordNet: a lexical database for English. In: Human Language Technology: Proceedings of a Workshop Held at Plainsboro, New Jersey, 8–11 March 1994 (1994). https://aclanthology.org/H94-1111
41. Aly, R., et al.: The Fact Extraction and VERification Over Unstructured and Structured information (FEVEROUS) shared task. In: Proceedings of the Fourth Workshop on Fact Extraction and VERification (FEVER), pp. 1–13 (2021)
42. DeHaven, M., Scott, S.: BEVERS: a general, simple, and performant framework for automatic fact verification. In: Proceedings of the Sixth Fact Extraction and VERification Workshop (FEVER), pp. 58–65. Association for Computational Linguistics. https://doi.org/10.18653/v1/2023.fever-1.6

Evaluating Auto-complete Ranking for Diversity and Relevance

Sonali Singh(✉), Sachin Farfade, and Prakash Mandayam Comar

Amazon, New York, USA
{ssonl,sfarfade,prakasc}@amazon.com

Abstract. Traditional Query Auto-completion (QAC) systems optimise for query relevance based on past user interactions. This approach excels at surfacing frequently searched queries, but ensuring a diverse range of suggestions and incorporating new products or trends often requires post-processing heuristics. This limitation stems from relying on user search logs, which may not fully capture the evolving product landscape. This paper presents a comparison of traditional state-of-the-art (SOTA) methods with Large Language Models (LLMs) for query auto-completion. The LLMs, with their ability to understand language and product information, can generate a wider range of relevant and diverse suggestions, encompassing the entire product catalog and potentially including entirely new queries. Here, we study the trade-off between latency, relevance and comprehensiveness in the QAC systems. Our experiments on real world data show that the LLM based QAC system offers a significant 38% boost in diversity with no significant change in relevance metrics. However, their high compute and memory demands make them less suitable for real-time applications. We introduce a heuristic approach that integrates the strengths of existing methods with the power of LLMs. This combined approach has been shown to yield a measurable 0.13% increase in sales revenue in live A/B testing.

Keywords: query auto-completion · large language models · relevance · diversity · latency

1 Introduction

The QAC (Query Auto-completion) system [9] acts as an essential entry point for e-commerce sites, assisting millions of users in quickly finalizing their search queries while also uncovering new and trending product-related keywords they hadn't encountered before. These QAC systems typically employ ranking models trained to rank popular queries, matching the given prefix using Learning to Rank (LTR) [10] algorithms. The process begins by identifying candidate queries matching the prefix, followed by ranking the top candidates to maximize the query acceptance rate.

Traditional QAC systems leverage Learning-to-Rank (LTR) algorithms trained with various loss functions:- pointwise, pairwise and listwise [10,32] to

C. Hauff et al. (Eds.): ECIR 2025, LNCS 15572, pp. 326–340, 2025.
https://doi.org/10.1007/978-3-031-88708-6_21

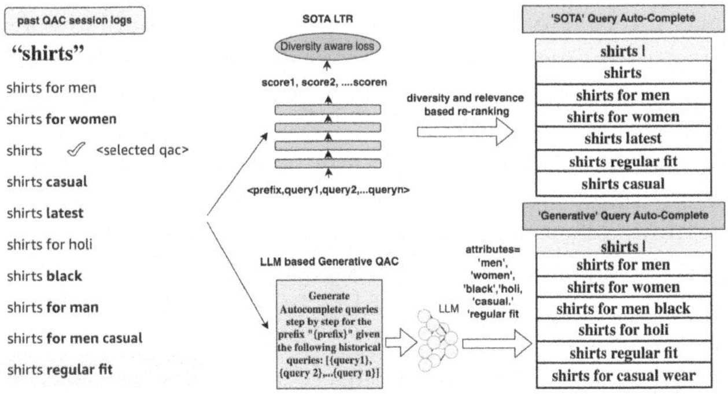

Fig. 1. Schematic illustrating enhancing QAC experience balancing relevance and diversity using SOTA LTR losses vs Generative QAC by prompting LLMs.

optimize relevance based on user search logs. While LTR excels at ranking frequently searched queries, it struggles to capture the full product range. New additions like "vacation dresses" might not surface prominently unless explicitly searched. Large Language Models (LLMs) offer a compelling alternative. By understanding both language and product information, LLMs have the ability to generate a broader array of relevant and varied suggestions for every prefix, covering the entire product catalog. This eliminates the need for separate diversity algorithms like Maximal Marginal Relevance [11] or Determinantal Point Processes [17], which are often employed with LTR to improve suggestion variety. Furthermore, LLMs possess the capability to handle novel query possibilities more effectively by generating new queries through the employment of effective prompting techniques. This shift from LTR to LLM-based approaches holds promise for a more comprehensive and dynamic QAC system, potentially leading to a more satisfying user experience. Large Language Models (LLMs) offer a significant improvement over traditional QAC systems that rely on user search logs. Trained on massive amounts of text data, LLMs excel at generating well-formed queries, unlike the potentially misspelled or incomplete queries often found in search logs. Beyond synthetic query generation, LLMs can also be leveraged to rank these suggestions. Efficacy of LLM as zero-shot rankers is demonstrated for both document ranking and recommendation tasks [14] through the utilization of pairwise-prompting techniques [22]. However, the document and item ranking contexts are typically broader than e-commerce QAC systems. The effectiveness of LLMs as rankers with minimal input, such as just a prefix or a few characters, remains unexplored. To address these concerns, we propose a comparative analysis of LLM performance against state-of-the-art (SOTA) neural network (NN) based ranking algorithms as illustrated in Fig. 1. This analysis will help us understand the strengths and limitations of LLMs in the specific context of e-commerce QAC ranking.

We leverage internal e-commerce search logs and publicly available AOL search logs [19] to conduct a series of experiments aimed at addressing several critical questions within the domain of query auto-completion (QAC) ranking:

- **CQ1**: Can Language Models based ranking methods outperform traditional prefix to query ranking strategies in e-commerce QAC?
- **CQ2**: Can LLMs effectively tackle diversity issues through prompting, or is a diversity-aware loss indispensable in ranking tasks?
- **CQ3**: Can the generative nature of LLM be successfully adopted in real-time auto-complete application?
- **CQ4**: Does LLM based understanding of prefixes and search queries bring value to QAC system?

Our observations reveal that, although LLMs demonstrate exceptional performance in diversifying QAC, they only marginally surpass the relevance benchmark. Moreover, it fails to meet the latency requirements of large e-commerce store. Hence, we propose an approach to leverage the inherent generative capabilities of the LLM methodology to create high-quality, diverse candidate queries which are then merged into the organic candidate list and subsequently ranked by the SOTA NN ranking model. We empirically demonstrate that this approach leads to a significant reduction in redundant and low-quality queries in QAC which lead to 0.13% revenue improvement in online A/B test.

2 Related Work

Query Auto-completion (QAC) systems encompass two approaches: retrieving and ranking historical customer queries that match the user's prefix, or employing a generative approach to generate a list of completions. Traditionally, retrieval and ranking relied on heuristics, with Most Popular Completion algorithms being a prevalent choice [3]. Subsequently, learning-based methods emerged, often incorporating time-sensitive signals and user feedback to enhance the ranking process [8,15]. Generative heuristic approaches, on the other hand, employed techniques such as templates and n-grams [4,26] for QAC. Later advancements embraced Generative Neural Language Models (LMs) [16,28] to synthesize QAC suggestions. Cai et al. [9] provide a comprehensive survey of these traditional QAC approaches.

Our research explores, compares, and contrasts traditional ranking approaches for QAC with Large Language Models (LLMs) that can be strategically employed to enhance the quality of suggestions in QAC within the e-commerce domain without compromising query clicks and downstream sales. Initially, we assess the performance of traditional ranking approaches by employing well-established learning-to-rank pairwise and listwise objective functions [7,10,12] to optimize for relevance. We also benchmark on state-of-the-art (SOTA) diversity-enhancing techniques in the traditional ranking setting by introducing additional objective functions within pairwise loss [24] or by directly optimizing diversity aware listwise losses [31]. [24] proposes improving QAC quality through multi-objective ranking, boosting navigational queries via pairwise ranking. Additionally, we investigate the application of diversification-aware listwise techniques in QAC, drawing inspiration from diversifying web search results. We adapt this through a listwise Learning to Rank (LTR) framework [10] to score queries, incorporating a query interaction layer inspired by the Document Interaction Network (DIN) [20] to generate higher-order features for queries in the list. Benchmarks and investigations into effective neural architectures and loss functions for Learning to Rank are presented in [23]. Building upon these insights, we utilize the optimization metric as a loss function, as suggested by [6]. Furthermore, we explore the integration of LLMs as rankers within QAC systems. The utilization of LLMs as rankers has been examined in the context of document retrieval ranking and recommendation tasks. Initial efforts focused on fine-tuning pre-trained language models for supervised ranking tasks with ranking loss [18,33]. With the recent notable performance of LLMs in Natural Language Processing (NLP) tasks, there has been growing interest in assessing their zero-shot performance in retrieval and ranking capabilities. [25] investigates the use of ChatGPT as a re-ranking agent. Additionally, [14] formalizes the recommendation task as a conditional ranking task, considering sequential historical interactions as conditions. The items retrieved by candidate generation models are treated as candidates and further ranked by constructing natural language prompts that incorporate historical interactions and candidates. [22] introduces the Pairwise Prompting technique (PRP), which reduces the complexity of employing LLMs as a ranking agent and demonstrates competitive performance using open-sourced LLMs. Furthermore, [34] compares all existing LLM-based zero-shot ranking approaches and introduces the novel Setwise approach which improves the effectiveness of the listwise prompting approach.

3 QAC Problem Formulation

QAC can be approached as either a Learning-to-Rank (LTR) task or a generative task. We delineate both formulations as follows:

QAC as a LTR Problem: QAC can be conceptualized as an LTR task where, for a given prefix p, the initial step involves retrieving a set of top-matched candidates L. Subsequently, the objective is to learn a scoring mechanism s_k for each completion $k \in L$ with the highest score assigned to the most relevant query. These scores are acquired through training on appropriate ranking losses, optimizing for relevance and other pertinent business criteria. The determination of completion relevance typically hinges on user acceptance, specifically the likelihood of a user accepting or clicking on one of the queries in the QAC.

QAC as a Generative Problem: In this formulation, given a prefix p and a context denoted by C, where C could represent a list of historical queries L, a set of instructions, or other relevant information, the objective is to generate a ranked list of the top K suggestions that align with the provided prefix.

4 SOTA Rankers with Ranking Loss

Pairwise and listwise ranking methodologies represent widely embraced strategies within the framework of LTR methods. We conduct experiments applying these methodologies to QAC with the following details:

Pairwise Ranking: Pairwise ranking losses, such as RankNet, LambdaRank, and LambdaMART [7], are widely adopted for pairwise ranking tasks. In our experiments, we choose to optimize a pairwise cross-entropy loss based on LambdaRank [7], as it is easily adaptable to different scoring architectures and business needs, with the goal of enhancing customer behavior.

Listwise Ranking: While pairwise loss remains a popular choice in designing ranking systems due to its simpler architectures and fast inference, it lacks the context of other queries in the list, which can be a limitation. Listwise losses, such as Softmax loss, ListMLE, ListNET, and approxNDCG [5,27,30], tend to mitigate these issues.

Handling Diversity: In the context of e-commerce QAC, we define diversity as the presence of varied product types, categories, and attributes (navigational terms) within the queries. Conversely, for generic web search QAC, diversity often implies the inclusion of varied topics or subtopics. We leverage distinct objective functions that optimize for relevance and diversity, employing pairwise and listwise loss mechanisms. For the pairwise loss, we integrate the diversification strategy proposed by [24], which introduces a query quality (QQ) objective.

Recent works on diversification directly optimize diversification metrics [31] to introduce diversity in listwise loss. We investigate optimizing an approximate version of the diversity metric αNDCG through a listwise approach.

5 LLMs as Rankers

LLMs have demonstrated effective zero-shot ranking performance in document ranking benchmarks [14,25]. However, their performance in QAC ranking tasks remains unexplored. E-commerce QAC are constructed to be concise, often lacking grammatical correctness. For instance, a completion such as "men's running shoes" may lack clarity, whereas a more meaningful sentence could be "I want to buy running shoes for men." LLMs may not perform optimally on such proprietary search query datasets of companies. We explore leveraging LLMs as rankers by employing both generative and ranking loss functions, aiming to mitigate these challenges and enhance their performance in QAC ranking tasks.

5.1 Generative Loss

LLMs are trained to maximize log-likelihood using cross-entropy loss for next-word prediction tasks. A straight forward approach to leverage LLMs for ranking is to formulate the ranking task as a generative sequence-to-sequence problem.

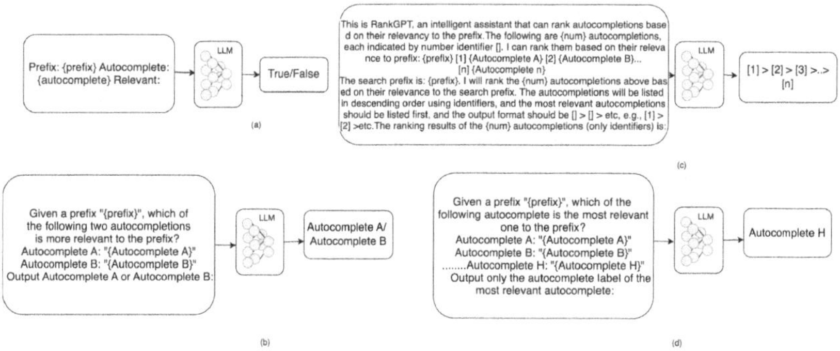

Fig. 2. Prompting techniques for ranking QAC using sequence-to-sequence LLMs: (a) Pointwise prompting for prefix-candidate relevance; (b) Pairwise prompting to generate label for more relevant candidate; (c) Listwise prompting to re-rank and generate sorted indices; (d) Setwise prompting to output most relevant candidate label from a set.

Various prompting styles can be used for fine-tuning sequence-to-sequence LLM based ranking for QAC as explained below (Fig. 2).

Pointwise Prompting: In this approach, we employ pointwise training of sequences similar to [18], where for a given prefix p and completion c, the large language model (LLM) is prompted to generate either "true" or "false" as output, indicating whether the completion c is relevant to the query prefix p. These generated tokens are further converted to relevance scores by applying the softmax function over the probabilities of generating the "true" or "false" token, as indicated by the following equation:

$$\text{score}(c|p) = \frac{e^{P_\theta(\text{"true"}|p,c)}}{e^{P_\theta(\text{"true"}|p,c)} + e^{P_\theta(\text{"false"}|p,c)}} \tag{1}$$

Here, $P_\theta(\text{"true"}|p,c)$ and $P_\theta(\text{"false"}|p,c)$ represent the probabilities of the LLM generating the "true" and "false" tokens, respectively, given the prefix p and completion c, parameterized by the LLM parameters θ.

Pairwise Prompting: The pairwise prompting [22] technique has emerged as one of the most effective methods for document ranking. In this approach, the model learns to rank the more relevant candidate given a pair of candidates (c_i, c_j) and a prefix p. The pairwise P_θ^{PRP} prompting involves prompting the large language model (LLM) twice with an ordering and a swapped ordering of the candidate pairs to obtain rankings (r_i, r_j). If the output remains consistent, scores are assigned based on the following label:

$$score(r_i > r_j|p, c_i, c_j) = \frac{1}{2}\left[P_\theta^{PRP}(p, c_i, c_j) + 1 - P_\theta^{PRP}(p, c_j, c_i)\right] \tag{2}$$

where $P_\theta^{PRP}(p, c_i, c_j)$ is an indicator function that takes the value 1 if the LLM outputs the label for c_i as more relevant, and 0 otherwise. Sorting techniques such as heap sort and bubble sort having $\mathcal{O}(n \log n)$ complexity are used at the time of inference to rank the entire list of items.

Listwise Prompting: The listwise prompting [21] approach ranks candidates by considering the entire context of the list. For short QAC queries, LLMs can be fine-tuned to generate a ranked list of suggestions. Instead of generating complete sequences, an alternative approach generates labels for sequences in ranked order to avoid queries outside the candidate list. Given a list of candidates with labels [1], [2], [3]...[n], the listwise prompting given to the LLM $P_\theta^{LIST}(p, c_1, c_2, ...c_n)$ generates a permutation according to the following equation:

$$P_\theta^{LIST}(p, c_1, c_2, ..., c_n) = [1] > [2] > [3] > [4].. > [n] \tag{3}$$

To accommodate lists that exceed the context window size, the list is partitioned into chunks, and the most relevant candidate is inserted into each chunk. During inference, a sliding window strategy [34] is employed.

Setwise Prompting: The setwise prompting technique [34] was introduced to achieve the power of efficiency of pairwise prompting and retain the advantage of the listwise technique that provides a larger context. As the name indicates, setwise prompting involves comparing a set of items at a time, which sits in between the pairwise and listwise approaches. This suitably eliminates the issue of the list of items being unable to fit in the context window and also the added complexity of pairwise inference.

5.2 Ranking Loss

LLMs can be leveraged as feature generators with traditional ranking loss. Here, we plug in encoder of LLM as the feature generator network and add the traditional score generator network to generate the score per candidate. This combination of network is then trained on listwise loss based on the scores predicted by the network. Note that this method tends to be even more computationally complex as each candidate is scored individually unlike pairwise or listwise prompting techniques. Another variant to this approach is, instead of using additional network on top of LLM to generate the score, directly compute the probability of sequence (candidate) generation using teacher forcing [29] at the time of decoding and use these probability as scores in the ranking loss.

6 Diversification in QAC Using LLMs

LLMs with their world knowledge are expected to understand the issue of diversity if prompted with suitable instructions. This translates the difficulty to comprehend the task of diversity from equations to simple english prompts illustrated in Fig. 3.

Chain of Thought [COT] Prompting: A logical thought process to generate relevant and diverse QAC for a prefix is to rank the most popular autocomplete with a popular product type followed by the next popular product type

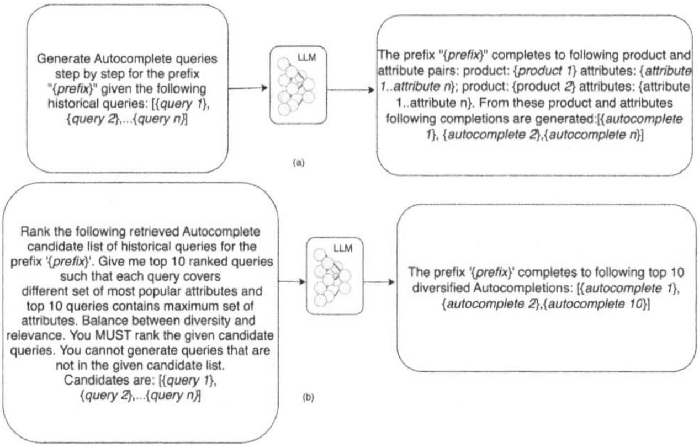

Fig. 3. Diversity-aware prompts for fine-tuning LLMs. (a) Chain of thought prompting to generate diverse attributes and auto-completions given prefix and corresponding candidates. (b) Prompt based instruction fine-tuning where the LLM is instructed to re-rank candidates balancing relevance and diversity.

or attribute. Hence, we design a prompt with instruction to order the candidates along-with mentioning attributes in the candidate indicating the navigational component of the candidate.

Instruction Fine-Tuning [Instruct-FT]: In this approach, we design a prompt instructing the model to rank the candidates while balancing relevance and diversity. We employ a heuristic algorithm to generate training samples, where we first sort the items by their popularity, as labeled from logs, and then sequentially include the next candidate only if it adds either a new attribute or product type. This carefully curated training data is then utilized for fine-tuning the model.

Diversity Aware Loss: Instead of using generative loss, we use diversity aware loss for training and use LLM as a feature generator with a traditional ranking network as score generator. Jointly, we fine-tune the complete network on both relevance and diversity.

7 Evaluation and Results

7.1 Experiment Setup

Datasets

E-commerce Dataset: Anonymized logs from an e-commerce store are utilized for generating training samples. A week's worth of customer logged data is used to create a mapping of prefixes to candidates by compiling the top 100 clicked candidates for each prefix. Each row of the evaluation sample contains a prefix, selected keyword, 100 candidates, past searches, device type, and other relevant details. Prefixes with fewer than 10 candidates are excluded from the evaluation. Additionally, prefixes without a click relevance label (i.e., sessions where no auto-complete candidate was selected) from the logs are omitted. The data is randomly down-sampled to obtain 80k prefixes for training and 30k prefixes for testing, selected from different days within a month. Features used for training all models with ranking losses are calculated with approach mentioned in [24]. For training the LLMs, numerical features are discarded, and the prefix and candidates are used as context, which is fine-tuned with the selected candidate as the label.

AOL Search Logs: We utilize publicly available query topic analysis data [1], derived from AOL search logs [19]. This dataset contains the top 1000 user search queries with a primary and secondary topic, along with associated scores with a total of 318,023 queries. In the absence of prefix for a query, we randomly select a prefix for each query, following a strategy similar to [16]. Each searched query is treated as the clicked query, and a list of the top 100 clicked queries starting with the same prefix from these logs is added as candidates. 90k prefixes are randomly sampled for training, and 30k prefixes are set aside for testing. We create a query relevance matrix, containing 30 topics, by combining click-based relevance and aggregating the top 29 most frequently occurring topics per prefix-candidate list, used for both loss computation and evaluation. We utilize a 384-dimensional embedding from the Sentence Transformer model "multi-qa-MiniLM-L6-cos-v1" [2], designed for semantic search, as a feature for each candidate. Our feature set also includes aggregated click information, cosine similarity of the query embedding with the two previous queries within a 300-second timeframe, the ratio of prefix to query length, a binary feature indicating the presence of recent past two searches, and the time difference between the ongoing search and the two prior searches.

Model Architecture: We employ DeepPLTR [32], a RankNet-based Siamese NN model, for pairwise loss and moDPLTR [24] for diversity-aware pairwise loss. ListLTR, trained on softmax loss, has a neural ranker architecture with interaction layers [23]. DiALTR, the diversity-aware version for listwise ranking, retains ListLTR's architecture with a change in loss. For supervised LLM fine-tuning with ranking loss (referred as ListFlanT5-XLLTR), we use FLAN-T5-XL's [29] encoder as the embedding generator, retaining ListLTR's scoring architecture. We also train ListFlanT5-XLRankGen model with a combination of ranking (softmax) and generative loss, where ranking loss scores are computed by candidate generation probability and has architecture same as FLAN-T5-XL. For all prompt-based fine-tuning, we use FLAN-T5-XL with LoRA [13] for consistency.

Evaluation Metric: We utilize MRR (Mean Reciprocal Rank) and NDCG (Normalized Discounted Cumulative Gain) to evaluate click-based relevance, and αNDCG to assess diversity. MRR and NDCG are computed on the top 10 candidates per prefix, using the clicked candidate as the true label. αNDCG is computed with the top 29 most frequent topics as true labels along-with click labels obtained from logs for each prefix and candidate list. For each evaluation row, the topics are generated by mining attributes from e-commerce QAC candidates using an internal attribute tagger tool, and binary relevance labels are assigned if the topic/attribute is present within the candidate.

7.2 Results

Table 1. Relevance metrics comparison of traditional and LLM-based ranking networks relative to DeepPLTR baseline. Relative lifts on e-commerce data; absolute numbers on public AOL dataset.

Model	E-commerce Dataset		AOL Dataset			
	MRR	NDCG@10	MRR	NDCG@10	Latency	Parameters
Ranking loss						
DeepPLTR	-	-	0.384	0.438	2 ms	50k
ListLTR	+9.31%	+5.28%	0.417	0.469	12 ms	600k
ListFlanT5-XLLTR	+10.68%	+5.69%	0.423	0.494	3 s	3B
Ranking loss+Generative loss						
ListFlanT5-XLRankGen	−5.32%	−3.14%	0.341	0.406	12 s	3B
Generative Loss						
FlanT5-XL-pointwise	−10.71%	−5.79%	0.303	0.379	2.1 s	3B
FlanT5-XL-pairwise	−2.13%	−1.32%	0.371	0.432	22.34 s	3B
FlanT5-XL-listwise	+3.25%	+1.79%	0.403	0.447	15.89 s	3B
FlanT5-XL-setwise	+11.78%	+6.43%	0.539	0.519	10.13 s	3B

CQ1: Ranking Performance: Table 1 reports ranking metric results for fine-tuned models on small NN networks and Flan-T5-XL network. We observe that ListLTR's performance is significantly better than DeepPLTR on MRR and NDCG metrics. This can be attributed to the interaction network and listwise loss, which better captures the interaction among queries mapped to a prefix and understands the graded relevance across all queries in the list. In contrast, the pairwise loss compares two queries at a time during training, and at inference, the score for each query in the list is computed independently. Both DeepPLTR and ListLTR models have inference latencies in the order of a few milliseconds, enabling real-time inference for production QAC models. Their high ranking performance is also attributed to high-quality customer behavior features collected and refreshed daily. The ListFlanT5-XLLTR model shows a marginal improvement over the ListLTR network on relevance. This model uses the Flan-T5-XL network's encoder for obtaining query embeddings and ListLTR as the score generator. The addition of embeddings from Flan-T5-XL provides a limited lift, as customer behavior (CB) features alone seem rich enough to bring maximum improvement in relevance. However, this network's efficiency is much poorer due to slow inference, making it unsuitable for deployment in a commercial real-time QAC setup.

As a next step, we try a combination of ranking and generative loss with the ListFlanT5-XLRankGen model, which uses the probability of generating a candidate as the score and employs listwise softmax loss with these scores in addition to generative loss for prefix-to-query generation. Note that in this sequence-to-sequence setup, each prefix learns to generate a single query, and hence there is no interaction between queries at the time of scoring. The reported results are far poorer than the baseline, mainly because the queries are scored individually, and the knowledge imported from LLMs fails to outperform the CB features.

With Flan-T5-XL as the LLM ranking network and different prompting techniques to mimic the different LTR learning setups, we observe that pointwise and pairwise prompting techniques do not improve relevance over baselines. Moreover, with pairwise prompting, inference latency is huge as multiple calls are needed for sorting all the items in the list. Interestingly, listwise and setwise prompting techniques improve relevance over the baseline. However, FlanT5-XL-listwise relevance lift is unable to beat the lift with traditional ListLTR models. The FlanT5-XL-setwise model appears to be the best in terms of the lift in MRR and NDCG@10. The setwise prompting technique uses only the labels of the candidate queries at the time of training and inference, which simplifies the sequence-to-sequence task of reproducing the complete list of re-ranked candidates. Moreover, similar to listwise, it compares and contrasts a set of candidates together, making it efficient to capture listwise context. However, the inference latency for the setwise technique is in the order of a few seconds (far better than pairwise and listwise prompting techniques) but very poor compared to the baseline.

Table 2. Ranking and diversity performance of traditional and LLM-based ranking networks relative to DeepPLTR baseline. Relative lifts on e-commerce data; absolute numbers on public AOL dataset.

Model	E-commerce Dataset			AOL Dataset		
	MRR	NDCG@10	αNDCG@10	MRR	NDCG@10	αNDCG@10
Ranking loss						
DeepPLTR	-	-	-	0.384	0.438	0.578
moDPLTR	+2.48%	+1.03%	+2.24%	0.381	0.435	0.585
DiALTR	+9.23%	+5.16%	+22.83%	0.409	0.463	0.681
ListFlanT5-XLLTR	+10.47%	+5.33%	+28.91%	0.413	0.467	0.595
Generative Loss						
FlanT5-XL-setwise [COT]	+1.32%	+0.54%	+38.5%	0.386	0.459	0.713
FlanT5-XL-setwise [Instruct-FT]	+9.86%	+5.32%	+29.33%	0.406	0.455	0.692

CQ2: Diversity Performance: Table 2 compares the performance of traditional models trained on diversity-aware ranking losses with Flan-T5-XL fine-tuned with prompts for diverse QAC ranking. While moDPLTR results in a marginal uplift in relevance and diversity metrics over the baseline, DiALTR significantly improves both metrics compared to the baseline. DiALTR uses approximate αNDCG as a loss function with listwise context during training, leading to better performance compared to the pairwise multi-objective fine-tuning in the case of moDPLTR. ListFlanT5-XLLTR, with Flan-T5-XL as the embedding generator and approximate αNDCG as the loss function, gives performance very similar to DiALTR, with no significant jump in the diversity metric, consistent with the trends observed in the relevance benchmarks reported in Table 2. FlanT5-XL-setwise [COT] results in a significant improvement in diversity metrics as it induces chain-of-thought prompting to first output a sequence of unique attributes and product phrases, followed by the construction of fluent queries. However, as this prompting forces the generation of sequences instead of labels, it leads to many new queries that may not be part of historical queries, resulting in a drop in ranking metrics. FlanT5-XL-setwise [Instruct-FT], on the other hand, instructs the model to re-rank the queries such that both relevance and diversity are balanced, with only re-ranked labels as output. We observe a marginal improvement in relevance and diversity metrics for this model compared to DiALTR. Although generative fine-tuning helps improve the diversity metric significantly, the improvement in the relevance metric is not enough to compensate for the latency and cost of inference.

CQ3: Latency vs Performance Discussion: Setwise prompting with Flan-T5-XL demonstrates competitive performance compared to traditional listwise models on relevance and diversity metrics. However, the relevance lift is insignificant to replace traditional models due to huge latency. Distilling the LLM-based models, to improve latency while maintaining good relevance and diversity performance could lead to further relevance deterioration. Therefore, we leverage the diversity improvement of FlanT5-XL-setwise [COT] in an offline setting for QAC candidate generation, retaining the online performance of the traditional ranking model.

CQ4: Online A/B Test: Utilizing the FlanT5-XL-setwise [COT] model, we generated diverse sequences incorporating attributes and categories for 100k prefixes. These candidate sequences were augmented with the existing pool of popular candidates and re-ranked for QAC using SOTA methods. The modified candidate generation setup underwent a month-long live A/B test on a large e-commerce store's search page, resulting in a 0.13% revenue lift and a 15% quality improvement in diversity and fluency compared to the baseline.

7.3 Conclusion

This study compares traditional models with Large Language Models (LLMs) on various parameters like model architecture, loss functions, and diversity objectives. Evaluating their generative capability and ranking potential through prompt/fine-tuning methods, we found LLMs superior in enhancing QAC diversity and relevance. However, they failed to meet real-time QAC latency requirements. To address this, we propose a practical approach combining language model capabilities with SOTA method efficiency, boosting revenue in a live A/B test. The study highlights LLMs' potential in QAC, paving the way for e-commerce search to anticipate user needs and deliver a comprehensive shopping experience.

References

1. AOL query log analysis (2017). https://github.com/wasiahmad/aol_query_log_analysis/
2. Sentence transformers (2022). https://huggingface.co/sentence-transformers/multi-qa-MiniLM-L6-cos-v1
3. Bar-Yossef, Z., Kraus, N.: Context-sensitive query auto-completion. In: Proceedings of the 20th International Conference on World Wide Web (WWW), pp. 107–116 (2011)
4. Bhatia, S., Majumdar, D., Mitra, P.: Query suggestions in the absence of query logs. In: Proceedings of the 34th International ACM SIGIR Conference on Research and Development in Information Retrieval, pp. 795–804 (2011)
5. Bruch, S., Wang, X., Bendersky, M., Najork, M.: An analysis of the softmax cross entropy loss for learning-to-rank with binary relevance. In: Proceedings of the 2019 ACM SIGIR International Conference on Theory of Information Retrieval, pp. 75–78 (2019)
6. Bruch, S., Zoghi, M., Bendersky, M., Najork, M.: Revisiting approximate metric optimization in the age of deep neural networks. In: Proceedings of the 42nd International ACM SIGIR Conference on Research and Development in Information Retrieval, pp. 1241–1244 (2019)
7. Burges, C.J.: From ranknet to lambdarank to lambdamart: an overview. Learning **11**(23–581), 81 (2010)
8. Cai, F., de Rijke, M.: Learning from homologous queries and semantically related terms for query auto completion. Inf. Process. Manag. **52**(4), 628–643 (2016)
9. Cai, F., de Rijke, M.: A survey of query auto completion in information retrieval. Found. Trends® Inf. Retrieval **10**(4), 273–363 (2016)
10. Cao, Z., Qin, T., Liu, T.Y., Tsai, M.F., Li, H.: Learning to rank: from pairwise approach to listwise approach. In: Proceedings of the 24th International Conference on Machine Learning, pp. 129–136 (2007)
11. Carbonell, J., Goldstein, J.: The use of MMR, diversity-based reranking for reordering documents and producing summaries, pp. 335–336 (1998)
12. Chen, W., Liu, T.Y., Lan, Y., Ma, Z.M., Li, H.: Ranking measures and loss functions in learning to rank. In: Advances in Neural Information Processing Systems, vol. 22 (2009)
13. Ding, N., et al.: Parameter-efficient fine-tuning of large-scale pre-trained language models. Nat. Mach. Intell. **5**(3), 220–235 (2023)
14. Hou, Y., et al.: Large language models are zero-shot rankers for recommender systems. In: Goharian, N., et al. (eds.) ECIR 2024, Part II. LNCS, vol. 14609, pp. 364–381. Springer, Cham (2024). https://doi.org/10.1007/978-3-031-56060-6_24

15. Jiang, J.Y., Ke, Y.Y., Chien, P.Y., Cheng, P.J.: Learning user reformulation behavior for query auto-completion. In: Proceedings of the 37th International ACM SIGIR Conference on Research and Development in Information Retrieval, pp. 445–454 (2014)
16. Kim, G.: Subword language model for query auto-completion. In: Proceedings of the 2019 Conference on Empirical Methods in Natural Language Processing and the 9th International Joint Conference on Natural Language Processing (EMNLP-IJCNLP), pp. 5022–5032. Association for Computational Linguistics (2019)
17. Kulesza, A., Taskar, B.: Determinantal Point Processes for Machine Learning. Now Publishers Inc., Hanover (2012)
18. Nogueira, R., Jiang, Z., Lin, J.: Document ranking with a pretrained sequence-to-sequence model. In: Findings of the Association for Computational Linguistics: EMNLP 2020 (2020)
19. Pass, G., Chowdhury, A., Torgeson, C.: A picture of search. In: Proceedings of the 1st International Conference on Scalable Information Systems, pp. 1–es (2006)
20. Pasumarthi, R.K., Zhuang, H., Wang, X., Bendersky, M., Najork, M.: Permutation equivariant document interaction network for neural learning to rank. In: Proceedings of the 2020 ACM SIGIR on International Conference on Theory of Information Retrieval, pp. 145–148 (2020)
21. Pradeep, R., Sharifymoghaddam, S., Lin, J.: Rankvicuna: zero-shot listwise document reranking with open-source large language models. arXiv preprint arXiv:2309.15088 (2023)
22. Qin, Z., et al.: Large language models are effective text rankers with pairwise ranking prompting. In: Findings of the Association for Computational Linguistics: NAACL 2024 (2024)
23. Qin, Z., et al.: Are neural rankers still outperformed by gradient boosted decision trees? (2021)
24. Singh, S., Farfade, S., Comar, P.M.: Multi-objective ranking to boost navigational suggestions in ecommerce autocomplete. In: Companion Proceedings of the ACM Web Conference 2023. WWW '23 Companion, pp. 469–474 (2023)
25. Sun, W., Yan, L., Ma, X., Ren, P., Yin, D., Ren, Z.: Is ChatGPT good at search? Investigating large language models as re-ranking agent. In: Proceedings of the 2023 Conference on Empirical Methods in Natural Language Processing (2023)
26. Szpektor, I., Gionis, A., Maarek, Y.: Improving recommendation for long-tail queries via templates. In: Proceedings of the 20th International Conference on World Wide Web, pp. 47–56 (2011)
27. Valizadegan, H., Jin, R., Zhang, R., Mao, J.: Learning to rank by optimizing NDCG measure. In: Advances in Neural Information Processing Systems, vol. 22 (2009)
28. Wang, S., Guo, W., Gao, H., Long, B.: Efficient neural query auto completion. In: Proceedings of the 29th ACM International Conference on Information and Knowledge Management, pp. 2797–2804 (2020)
29. Wei, J., et al.: Finetuned language models are zero-shot learners. In: International Conference on Learning Representations (2022)
30. Xia, F., Liu, T.Y., Wang, J., Zhang, W., Li, H.: Listwise approach to learning to rank: theory and algorithm. In: Proceedings of the 25th International Conference on Machine Learning, pp. 1192–1199 (2008)
31. Yan, L., Qin, Z., Pasumarthi, R.K., Wang, X., Bendersky, M.: Diversification-aware learning to rank using distributed representation. In: Proceedings of the Web Conference 2021, pp. 127–136 (2021)
32. Yuan, K., Kuang, D.: Deep pairwise learning to rank for search autocomplete. arXiv preprint arXiv:2108.04976 (2021)

33. Zhuang, H., et al.: Rankt5: fine-tuning t5 for text ranking with ranking losses. In: Proceedings of the 46th International ACM SIGIR Conference on Research and Development in Information Retrieval, pp. 2308–2313 (2023)
34. Zhuang, S., Zhuang, H., Koopman, B., Zuccon, G.: A setwise approach for effective and highly efficient zero-shot ranking with large language models. In: Proceedings of the 47th International ACM SIGIR Conference on Research and Development in Information Retrieval, pp. 38–47 (2024)

Semantically Proportioned nDCG for Explaining ColBERT's Learning Process

Ariane Mueller(✉) and Craig Macdonald

University of Glasgow, Glasgow, Scotland, UK
{ariane.mueller,craig.macdonald}@glasgow.ac.uk

Abstract. In recent years, dense retrieval approaches have brought significant improvements in information retrieval effectiveness. However, these approaches are significantly more complex than traditional retrieval techniques, decreasing their transparency and explainability. ColBERT, a popular dense retrieval approach, uses a late interaction mechanism (MaxSim) to match the embeddings of tokens in queries and documents. Compared to single-representation dense retrieval approaches, this late interaction offers transparency, in that it allows to observe token-level MaxSim matches between the query and document, and even measure the extent that tokens were matched in exact "lexical" matches, or "semantic" matches. However, previous measures for semantic matching do not take into account how semantic matching influences the model's effectiveness, i.e. whether it is beneficial or harmful. We propose S-nDCG, which weights the relevance gains in nDCG by the semantic or lexical proportion of matching. In combination with nDCG, this allows to quantify the extent to which a (ColBERT-based) model's effectiveness can be attributed to semantic or lexical matching, respectively. Through experiments on TREC Deep Learning track queries, we examine how ColBERT's effectiveness is obtained over training iterations, and find that generally ColBERT learns to rely more heavily on semantic matching throughout training. Surprisingly though, there is a significant decrease of semantic matching during the first 5k training iterations, indicating that even though BERT's pre-trained token embeddings provide a base for the semantic matching of several different token types, they likely lead to a lot of inaccurate semantic matches at the beginning of training that are addressed by fine-tuning for retrieval.

1 Introduction

In recent years, dense retrieval approaches have achieved significant gains in retrieval effectiveness compared to traditional approaches [11,16,27], such as BM25 [20]. One of their main advantages is their ability to match query and document tokens based on their meaning ("semantic matching") as opposed to traditional approaches, which solely rely on spelling to match terms ("lexical matching"). Semantic matching addresses the potential vocabulary mismatch

C. Hauff et al. (Eds.): ECIR 2025, LNCS 15572, pp. 341–356, 2025.
https://doi.org/10.1007/978-3-031-88708-6_22

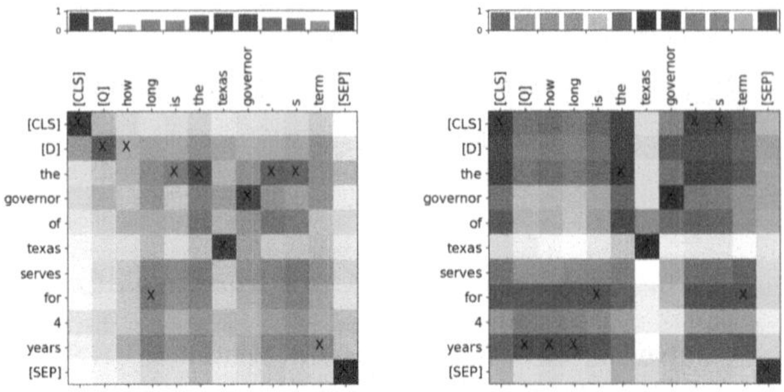

Fig. 1. An illustration of ColBERT's matching behaviour at the beginning (left) and end (right) of training. The matrix on the bottom visualises the matching process. The x-axis represents the query tokens, while the y-axis represents the document tokens. The most similar document token for each query token, and thus the one it is matched with, is marked with an 'X'. The bar chart above the matrix shows the MaxSim contributions of the different matches to the document's final score. At the beginning of training, i.e. solely relying on BERT's pre-training, the model is able to match both "texas" and "governor" lexically. However, the tokens "how" and "long" in the question are not matched with appropriate tokens in the answer, i.e. "4" or "years". Fine-tuning gives ColBERT the ability to match such tokens with a more complex semantic relationship, as both "how" and "long" are matched with "years" at the end of training.

between the query and relevant documents (e.g. between 'car' and 'vehicle') and helps to resolve polysemous words (e.g. 'bow'), improving retrieval effectiveness [26].

Many approaches for dense retrieval use a single embedding to represent each query and passage. Similarity measurements are then typically obtained using cosine similarity. Moreover single representation makes it more difficult to explain *why* some passages have been scored highly [7]. Instead, ColBERT [11] offers an alternative dense retrieval mechanism, in that it creates *token-level* embeddings for each token in a query or document. These are then matched together using a late interaction mechanism called MaxSim. ColBERT is attractive because it typically exhibits good zero-shot effectiveness [21,22,25], and passage scores can be explained by observing which contextualised query token embeddings matched with which contextualised document token embeddings [14].

We argue that the effectiveness of ColBERT's contextualised token-level embeddings is likely inherited from the underlying language model (e.g. BERT is trained using masked language modelling). Such a language model is trained so that identical and similar tokens appearing in similar contexts should have similar embeddings. Indeed, it has been observed that much of the scores from ColBERT are obtained from exact matches - the same token appearing in both query and document [6,17,26]. However, ColBERT models have the ability to learn to perform various types of semantic (i.e. non-lexical) matches. For instance, query tokens (such as 'how', 'long') may be matched with document tokens (such as

'4', 'years') [26] as shown in Fig. 1. In this paper, we aim to determine when such characteristics emerge from ColBERT's training, and what their impact is upon effectiveness. We build upon existing work that has quantified the presence of different types of semantic matching in ColBERT [26], but going further, integrate this quantification within the context of effectiveness, specifically nDCG.

In short this paper contributes: (1) novel S-nDCG and L-nDCG (Semantic/Lexical normalised Discounted Cumulative Gain) measures for assessing the impact of lexical and semantic matching on nDCG effectiveness measurements for ColBERT-based models; (2) a study training and evaluating ColBERT wrt. the introduced measures. S-nDCG and L-nDCG *supplement* nDCG to quantify semantic/lexical matching contributions to overall effectiveness.

In experiments conducted using ColBERT on the TREC DL 2019&2020 datasets, we find that generally, the proportion of the model's effectiveness that can be attributed to semantic matching increases throughout training, indicating that ColBERT's fine-tuning does improve its semantic matching ability compared to BERT. However, we also observe a notable drop in the semantic matching proportion in the very early stages of training, meaning that the pre-trained BERT embeddings may lead to inaccurate semantic matching, which the model 'unlearns' in this phase.

The remainder of this paper is structured as follows: Sect. 2 briefly introduces ColBERT and gives an overview of previous analyses of the model. In Sect. 3, we introduce our novel measure S-nDCG as well as the semantic matching proportion (SMP) metric it is based on. Section 4 introduces our research questions and experimental setup; the results of which we describe in Sect. 5. The final section summarises and concludes this work.

2 ColBERT

We first provide an understanding of ColBERT as a base for our evaluation and analysis of the model in Sects. 4 to 5. We also give an overview over previous analyses of ColBERT and related models in Sect. 2.2.

2.1 Model Architecture

ColBERT follows a bi-encoder [8] architecture and encodes the document and query separately but using the same language model (typically BERT [5]). However, unlike single representation models (e.g. DPR [10], ANCE [27], etc.), it does not aggregate the obtained token embeddings into single vectors for documents and queries, but instead retains one embedding per input token. Thus, ColBERT can be described as a mapping function $f : V^n \rightarrow \mathbb{R}^{n \times e}$ from a query q, consisting of the terms $\{t_1, ..., t_n\}$ from the vocabulary V, represented as $|q|$ tokens, to embeddings $\{E_{q_1}, ..., E_{q_{|q|}}\}$ of dimension e, or from a document consisting of $|d|$ tokens to a set of embeddings $\{E_{d_1}, ..., E_{d_{|d|}}\}$. The original embeddings retain the large dimension of the underlying language model (e.g. 768 for `bert-base`). To make the model more memory-efficient, ColBERT applies a linear layer to compress the created embeddings to size 128 (denoted as 'Lin'). These compressed embeddings are then normalised to facilitate the computation of their cosine similarity as a dot-product. The query embeddings E_q are thus produced by the following pipeline:

$$E_q = Normalise(Lin(BERT(\text{`}[CLS][Q]q_0q_1...q_l[SEP]\#\#...\#\text{'}))) \quad (1)$$

Here, [Q] is a special token marking the sequence as a query and "#" denotes [MASK] tokens used for padding and query augmentation.

The document embeddings E_d are produced by a slightly different pipeline:

$$E_d = Normalise(Lin(BERT(\text{`}[CLS][D]d_0d_1...d_n[SEP]\text{'}))) \quad (2)$$

There are no '#' characters added to the end of the input as documents are not augmented with additional [MASK] tokens like queries.

To score candidate documents, ColBERT applies a late interaction mechanism using the MaxSim operator. In short, the score for a document d wrt. a query q, $S_{q,d}$, is the sum of the maximum similarity of each query token embedding with any token embedding in the document. The similarity is computed as the dot-product between the query and document token embeddings[1]. Hence, the score of a document is defined as:

$$S_{q,d} = \sum_{i=1,...,|q|} max_{j=1,...,|d|} E_{q_i} \cdot E_{d_j}^T \quad (3)$$

where E_{q_i} and E_{d_j} are the embeddings of the i-th query token and the j-th document token for a query of length $|q|$ and a document of length $|d|$, respectively [11]. Overall, the multiple representations and the late interaction mechanism allow the query and document embeddings to interact, which gives ColBERT an effectiveness comparable to many cross-encoders [11].

2.2 Related Work

Due to the superior performance of BERT-based retrieval models [11,16], there are numerous works investigating the origin of their effectiveness in more detail.

Several works investigated BERT-based retrieval wrt. the model's input. Möller et al. [15] assessed the gradients of a RoBERTa-based bi-encoder to determine which input tokens had the largest influence on the document's final score. Other works focussed on certain textual properties of the input, such as term frequency and term proximity [2,19,23]. MacAvaney et al. [12] used probing to identify certain 'preferences' of ColBERT for textual attributes like fluency, succinctness and factuality, whereas Rau and Kamps [18] tested the sensitivity of

[1] As the embeddings are normalised to an L2 norm of 1, the dot-product is equal to the cosine similarity of the two vectors. Khattab and Zaharia [11] evaluate both squared L2 distance and cosine similarity as similarity measures. However, as we focus on re-ranking in this project, we use cosine similarity as per [11].

a BERT-based cross-encoder in regards to word order, in order to quantify to what degree the model relies on context modelling.

Several works have analysed (Col-)BERT's behaviour wrt. lexical and semantic matching. Qiao et al. [17] investigated the matching behaviour of a BERT-based cross-encoder wrt. non-stopword terms and found that lexically matched terms have the biggest influence on the final score. Formal et al. [6] came to a consistent conclusion for ColBERT as they measured the degree to which the model applies lexical matching. A similar problem is addressed by Jiang et al. [9], who measured to what degree BERT-based re-rankers rely on exact matching and how they improve over BM25. In contrast to this, Wallat et al. [24] focussed on semantic matching and investigated its contribution to the effectiveness of a BERT-based bi-encoder via probing.

Most relevant to this work, Wang et al. [26] measured both lexical and semantic matching proportions (as further explained in Sect. 3.1), and to what degree ColBERT relies on either type of matching. The authors reported a decrease in effectiveness of 25% when forcing ColBERT to rely solely on lexical matching. However, the achieved nDCG decreases by more than 80% when the model relies only on semantic matching (of text tokens) [26]. The authors also conducted a more detailed analysis of the extent of semantic matching utilised by ColBERT with the Semantic Matching Proportion (SMP) measure. The analysis revealed that the model matches 37–38% of text tokens semantically. However, the large majority of these tokens are low-IDF and stopword tokens, while token types such as question, subword, medium-IDF and numeric tokens are subject to semantic matching less frequently, or in the case of high-IDF tokens barely at all.

In summary, there are several works investigating the origin of ColBERT's or related models' effectiveness, particularly wrt. their input [2,12,18] and matching behaviour [6,26]. However, all of the mentioned works focus on the fully trained model. Wang et al. [26] do show the effectiveness curve of ColBERT during training; however, they only conduct their analysis of its origin on the final, trained model. This leaves unclear how ColBERT's behaviour develops throughout training, i.e. how its (semantic) matching behaviour changes as the model is fine-tuned on retrieval data and how much is inherited from BERT's pre-training. Additionally, previous measures do not directly relate ColBERT's matching behaviour to its effectiveness and thus do not distinguish between beneficial and harmful matches. Therefore, it remains unclear to what degree the effectiveness of the model can be attributed to lexical or semantic matching.

3 Quantifying Semantic Matching

We aim to produce insights into how ColBERT builds and improves upon BERT's pre-training. For this, we build upon Wang et al.'s [26] work (further described in Sect. 3.1) to quantify the extent to which ColBERT's effectiveness can be attributed to semantic or exact matching throughout training. To gain further insights into how the model's semantic matching behaviour and effectiveness emerge, we also measure the attribution for certain token types throughout training, which allows us to analyse ColBERT's learning behaviour, as further elaborated in Sect. 3.2.

3.1 Semantic Match Proportion

To study how semantic matching impacts the effectiveness of ColBERT, we need to quantify the proportion of semantic matching utilised by the model. Semantic Matching Proportion (SMP), introduced in [25], measures the proportion of all text tokens in a document that are matched by ColBERT's MaxSim to a non-identical query token. The percentage of semantic matching for a document d wrt. a query q can thus be computed as:

$$s(q,d) = \frac{\sum_{i=1,\ldots,|q|} 1[t_{q_i} \neq t_{d_j}] \cdot max_{j=1,\ldots,|d|} E_{q_i}^T E_{d_j}}{\sum_{i=1,\ldots,|q|} max_{j=1,\ldots,|d|} E_{q_i}^T E_{d_j}} \tag{4}$$

Here, t_{q_i} is the token id of the i-th query embedding and t_{d_j} is the token id of the document token that was selected in the MaxSim ($max_{j=1,\ldots,|d|} E_{q_i}^T E_{d_j}$) for the respective query token, with E_{q_i} and E_{d_j} denoting the corresponding embeddings. To obtain a query's SMP the semantic matching percentages for each document in the query's ranking R_k with cutoff k are averaged:

$$SMP(q, R_k) = \frac{1}{k} \sum_{d_r \in R_k} s(q, d_r) \tag{5}$$

As the SMP utilises the MaxSim operator, it measures not only the proportion of semantically matched tokens but is also sensitive to the matches' contribution to the final document score [25]. Thus, matches of more similar embeddings are assigned a higher weight in the final SMP.

However, the measure does not distinguish between accurate and inaccurate matches, i.e. beneficial and harmful semantic matching. Thus, a term is counted as a semantic match even if there is a more accurate lexical match for it; this mismatch would then have a negative impact on model effectiveness while still increasing the SMP. Indeed, in a preliminary study, we quantified the correlation between SMP and nDCG in ColBERT on the TREC 2019 and TREC 2020 query sets. While the two measures generally show a strong positive correlation ($\rho = 0.764$ for TREC 2019 and 0.885 for TREC 2020, ignoring the checkpoint after 0 steps), the model has high SMP but low effectiveness at the beginning of training. This highlights that SMP cannot differentiate between semantic matches that are based on the semantic "understanding" captured by the BERT embeddings and ones that occur due to exact matching not being trained well enough yet.

3.2 S-NDCG

To mitigate the issue highlighted in the previous section and gain an insight into how ColBERT builds and improves upon BERT through semantic matching, we introduce the Semantic normalised Discounted Cumulative Gain (S-nDCG). For a given query q and retrieved documents R_k, the metric is defined as follows:

$$S\text{-}nDCG@k(q, R_k) = \sum_{d_r \in R_k, r \in \{1,\ldots,k\}} \frac{1}{iDCG@k} \cdot \frac{s(q, d_r) \cdot rel(r)}{log_2(r+1)} \tag{6}$$

Here, d_r denotes the document at rank r with $rel(r)$ being its relevance label. Corresponding to the S-nDCG, we also define the counterpart of S-nDCG for lexical matching, L-nDCG, using lexical matching proportion, $l(q, d_r)$, in place of $s(q, d_r)$, where $l(q, d_r) = 1 - s(q, d_r)$. It therefore follows that the following equality holds true:

$$nDCG@k(q, R_k) = L\text{-}nDCG@k(q, R_k) + S\text{-}nDCG@k(q, R_k) \tag{7}$$

The S-nDCG and L-nDCG metrics jointly measure the proportion of semantic/lexical matching applied by a given ColBERT-like model, and its achieved effectiveness. It thus provides an insight into the degree to which the model's effectiveness can be attributed to semantic and lexical matching, respectively. Hence, as the two measures add up to the model's original effectiveness, the S-nDCG/L-nDCG is best interpreted in contrast to the corresponding nDCG. For instance, a S-nDCG of 0 means that no effectiveness can be attributed to semantic matching, i.e. the obtained nDCG was solely achieved with lexical matching. A value of 1 indicates that the model created a perfect ranking based solely on the respective type of matching.[2]

Finally, to gain a further insight into how ColBERT's (semantic) matching behaviour emerges from BERT's pre-training, we further analyse the model's learning process by measuring which token types are learnt at what point during training. For this, we evaluate the model on modified versions of S-nDCG and L-nDCG, which measure the effectiveness that can be attributed to semantic/lexical matching of only a specified token type. For comparability, we consider token types defined by Wang et al. [26], which are listed in Table 1.

4 Experimental Setup

Our experiments address the following research questions:

RQ1: How does semantic matching contribute to ColBERT's effectiveness throughout training?

As shown in [26], ColBERT achieves notably better performance when relying solely on lexical matching than when matching only semantically. Thus, we hypothesis that generally more of the model's effectiveness can be attributed to

[2] We note that S-nDCG (L-nDCG) cannot independently measure the contribution of a particular semantic (lexical) match to nDCG, as if that match was removed, the nature of MaxSim (Eq. (3)) is that another token in the passage would be matched instead for the given query token.

Table 1. The token types we consider for our evaluation for queries (Q) and documents (adapted from [26]). It is to be noted that except for the three IDF classes, none of these groups are mutually exclusive.

	Notation	Token Type	Description
Q	QuesToken	Question tokens	who, what, where, when, why, which, and how
Documents	NumToken	Numeric tokens	Tokens defined as numeric by Python's 'is_numeric()' method of the class String (e.g. integers, floating point numbers)
	StemToken	Stemmed tokens	Tokens in the same form as the matching query token after applying Porter stemming
	Low-IDF	Low IDF tokens	Tokens with IDF below the 25th percentile of the IDF distribution
	Med-IDF	Medium IDF tokens	Tokens with IDF between the 25th and the 75th percentiles of the IDF distribution
	High-IDF	High IDF tokens	Tokens with IDF above the 75th percentile of the IDF distribution

lexical matching than semantic matching. However, we also theorise that ColBERT learns to use semantic matching more frequently as it becomes more accurate and reliable through fine-tuning on retrieval data. Thus, we hypothesis that ColBERT's gain in effectiveness throughout training can mainly be attributed to semantic matching.

RQ2: What tokens does ColBERT learn to match semantically throughout training?

We hypothesis that BERT is able to match tokens lexically as well as simple tokens, such as nouns or very similar tokens (e.g. stems such as 'enter' and 'entered') semantically. However, the model might not match tokens with more complex semantic relationships necessary for high effectiveness (e.g. question words and their corresponding answer such as 'why' and 'because') or match them inaccurately. Hence, ColBERT likely mainly learns to match more (semantically) complex tokens such as questions words or medium-IDF tokens. As this is likely a harder task, we theorise that the matching of these token types is learned later in training, while easier to match tokens (e.g. stems, low-IDF) are learned earlier on, or in the case of tokens that are predominantly matched lexically (e.g. high-IDF tokens, numeric tokens [26]) barely at all.

We evaluate ColBERT as a re-ranker over a pre-computed candidate set wrt. a given query - indeed, this allows experimental results to be obtained without creating a new dense index for every training checkpoint. In accordance to the original paper [11], we apply the model for passage ranking and use a candidate set of the top 1000 passages retrieved by BM25 [20] for each query. The model's performance is measured with nDCG@10, and analysed with L-nDCG@10 and S-nDCG@10, both for all tokens and the different token types shown in Table 1. For our analysis we train the model on the MS MARCO dataset [1] and use the TREC 2019 [4] and TREC 2020 [3] Deep Learning Track datasets for evaluation[3]. Experiments are conducted using the PyTerrier platform [13].

In order to isolate the effect of semantic/lexical matching on ColBERT's performance, we modify the original architecture by removing the linear compression layer (see *Lin*() in Eqs. (1) and (2)) as well as the punctuation filtering step for the document embeddings. We do this as we postulate that the linear layer might affect ColBERT's matching behaviour particularly in the early stages of training as the layer is initialised randomly and thus initially applies a random transformation to the token embeddings, which likely leads to a higher proportion of semantic matching and lower effectiveness. As a consequence the linear layer has to be trained first to mitigate this effect, delaying the training of the actual embeddings and token matching[4]. Our model hence obtains the query and document token embeddings by removing the Linear layers from Eqs. (1) & (2). This allows us to focus on characteristics of ColBERT that are inherited from BERT.

5 Results

The following section first highlights the issue with SMP described in Sect. 3.1 using our experimental results. It then describes and discusses the results of our evaluation of ColBERT wrt. S-nDCG and links them to the model's effectiveness to answer RQ1. In order to produce more detailed insights into the origin of ColBERT's matching behaviour, Sect. 5.2 shows the results of evaluating ColBERT on token type specific S-nDCG and draws conclusions about the extent to which ColBERT's behaviour is inherited from BERT.

[3] All source code for reproducing our experiments can be found at https://github.com/cmacdonald/colbert_sndcg.

[4] A more extensive examination comparing the model without the linear layer to the full model revealed that the linear layer indeed hinders the model in learning semantic matching at the beginning of training, leading to significantly better effectiveness in the model without the linear layer up to 2k training steps on TREC 2020 ($p = 0.002$). Surprisingly, throughout the rest of training, the *full* model relies on semantic matching more, with the final model achieving a significantly higher S-nDCG ($p = 0.017$). However, these differences in matching behaviour do not significantly affect the final model's overall effectiveness.

5.1 RQ1 - Contribution of Semantic Matching to Effectiveness

To understand how ColBERT's semantic matching behaviour develops throughout training and how this affects the model's effectiveness, we measure the proportion of semantic matching utilised by the model throughout training with SMP as shown in Fig. 2.

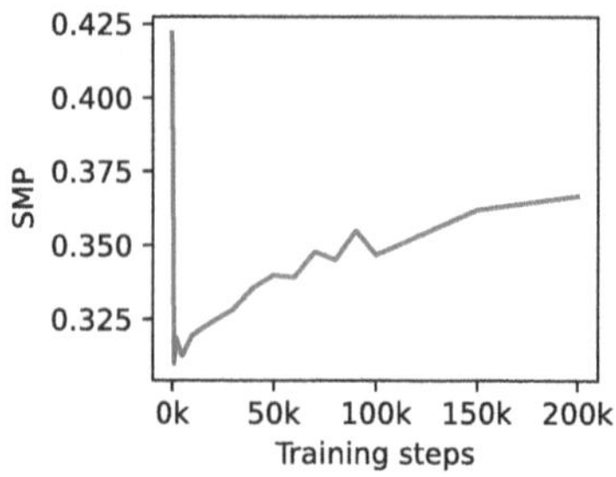

Fig. 2. The semantic matching proportion (SMP) for ColBERT throughout training (TREC 2020).

The figure generally shows an increase in SMP throughout training. However, there is a notable drop at the beginning of fine-tuning, from 0 to approximately 5k iterations, which is surprising as the model's effectiveness *increases* from 0.139 to 0.655 in nDCG during this phase (on TREC 2020). This highlights a shortcoming of SMP, in that it does not distinguish between beneficial and harmful semantic matching and on its own is thus not sufficient to measure the contribution of semantic matching to ColBERT's effectiveness.

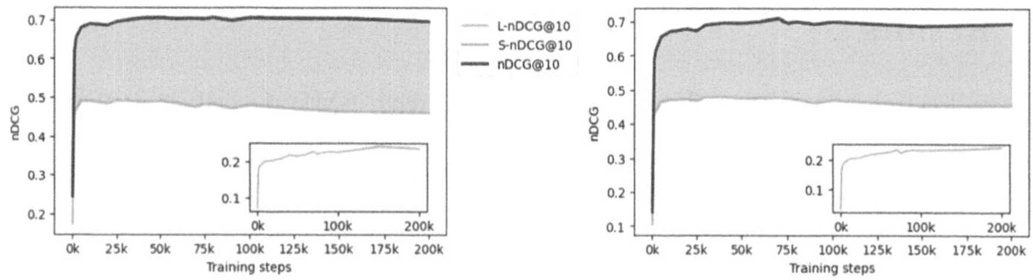

Fig. 3. The S-nDCG@10 and L-nDCG@10 for ColBERT on TREC 2019 (left) and TREC 2020 (right) throughout training. The top line shows the model's achieved effectiveness (nDCG@10), while the bottom line denotes the proportion of effectiveness that can be attributed to lexical matching (L-nDCG). The shaded area in between corresponds to the S-nDCG, i.e. the portion of effectiveness that can be attributed to semantic matching. The subgraphs in the bottom right corners show S-nDCG alone, to better highlight the development of ColBERT's semantic matching ability throughout training.

To measure the degree to which ColBERT's effectiveness can be attributed to semantic or lexical matching, we evaluate the model using the introduced S-nDCG and L-nDCG measures (see Eq. (6)). Figure 3 shows the results of this evaluation throughout training on TREC 2019 (left) and TREC 2020 (right). Similar to the SMP curve in Fig. 2, Fig. 3 shows a general increase of S-nDCG throughout training. However, while there is a drop in SMP at the beginning of training, the S-nDCG *increases* notably between 0 and approx. 5k training steps with a gain of 170% on TREC 2019 and 428% on TREC 2020. Additionally, L-nDCG also shows a notable increase in this phase. As tokens can only be matched either semantically or lexically, these results, in combination with the observed decrease in SMP (i.e. an increase in the proportion of lexical matching), suggest that at the beginning of training, ColBERT learns to rely more on lexical matching. It is postulated that this is likely due to the model 'unlearning' some inaccurate semantic matches. Figure 4 shows matching behaviour for an example document that was the top-ranked document for the query "define bmt medical". In the left figure, we observe that using the raw BERT's embeddings in ColBERT lead to a notable amount of inaccurate semantic matching, which is mitigated in the early stages of fine-tuning (right). This leads to the decrease in SMP but an increase in L-nDCG and S-nDCG as lexical matching becomes more frequently used, and semantic matching, despite being utilised less frequently, becoming more accurate and reliable (thus improving the model's overall effectiveness).

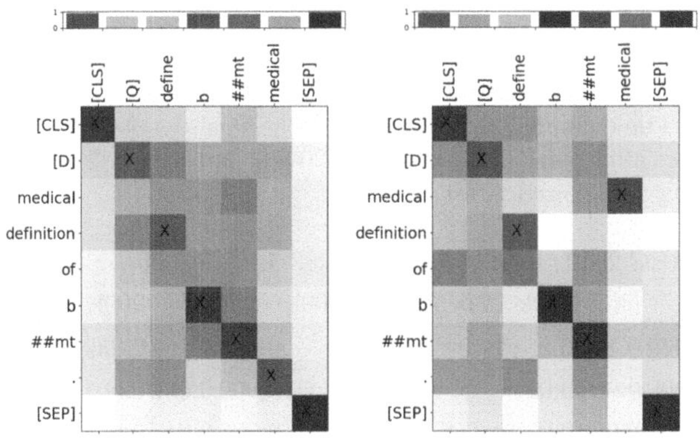

Fig. 4. An example for an inaccurate semantic match ("medical" with ".") being replaced by a lexical match during the early stages of training. The diagram on the left corresponds to ColBERT at the beginning of training (0 steps, i.e. raw BERT), while the right shows the model's matching behaviour after 5k steps.

Throughout the later stages of training the proportion of ColBERT's effectiveness that can be attributed to lexical matching remains approximately stable, while the proportion of semantic matching and the model's effectiveness that can be attributed to this type of matching, increase. For instance, the increase in SMP starting at approx. 10k steps shown in Fig. 2 goes along with the steady increase in S-nDCG and the marginal increase in nDCG in Fig. 3. This confirms our hypothesis that generally, ColBERT learns to utilise semantic matching more frequently and accurately during fine-tuning. It is also to be noted that, as mentioned in Sect. 4, the model still relies on exact matching more heavily than semantic matching with an L-nDCG fluctuating between 0.45 and 0.5 and a maximum S-nDCG of 0.240 on both datasets.

Answer for RQ1. Generally, the proportion of semantic matching utilised by ColBERT and hence also its contribution to the model's effectiveness increase throughout training - a key feature of dense retrieval. Surprisingly though, there is a notable decrease in SMP at the beginning of training, indicating that the model might 'unlearn' some inaccurate semantic matches during this phase. Our results also show that while the S-nDCG generally increases, the majority of ColBERT's effectiveness can still be attributed to lexical matching, which is consistent with the findings of [26].

5.2 RQ2 Semantically Matched Token Types

To gain a more detailed understanding of the origin of ColBERT's matching behaviour and how it emerges from the model's learning process, we evaluate ColBERT using S-nDCG and L-nDCG on only specific token types (listed in Table 1) as described in Sect. 3.2. The results of this evaluation are summarised in Table 2 and visualised for TREC 2020 in Fig. 5.

Table 2. The S-nDCG for ColBERT after 0k, 5k and 200k training iterations for different token types. The columns 'Δ 0–5k' and 'Δ 5k-200k' denote the relative difference between the values after 0 and 5k, and 5k and 200k, respectively. * denotes a significant increase according to a paired t-test.

Token Type	TREC 2019					TREC 2020				
	0k	5k	200k	Δ 0–5k	Δ 5k-200k	0k	5k	200k	Δ 0–5k	Δ 5k-200k
All	0.070	0.190	0.234	1.70*	0.232*	0.036	0.191	0.239	4.28*	0.251*
Numeric	0.000	0.000	0.001	-	1.31	0.000	0.002	0.001	-	−0.229
Stem	0.004	0.022	0.030	4.43*	0.382*	0.005	0.013	0.015	1.77*	0.164
Question	0.008	0.040	0.047	4.04*	0.163*	0.004	0.048	0.059	11.5*	0.229*
Low-IDF	0.069	0.186	0.231	1.68*	0.242*	0.035	0.181	0.227	4.11*	0.251*
Med-IDF	0.001	0.004	0.003	2.37	−0.214	0.001	0.009	0.013	10.8*	0.450
High-IDF	0.000	0.000	0.000	-	-	0.000	0.001	0.000	-	-0.959

The table shows that the relative increase in S-nDCG is statistically significant across *all* token types except the two predominantly lexically matched types, numeric and high-IDF tokens, as well as medium-IDF tokens on TREC 2019. It is also notably higher during the first 5k training iterations than the rest of fine-tuning. This trend is particularly obvious for question and medium-IDF tokens, especially on TREC 2020. On one hand, the increase in token type specific S-nDCG is to be expected based on the rapid increase in general S-nDCG observed in the early stages of training in Fig. 3. On the other hand, the notably larger increase at the beginning of training is surprising for the more semantically complex token types (i.e. medium-IDF and question tokens), as we theorised these would be learned in later stages. A possible explanation for this is that the MaxSim contributions of these token types are generally subject to larger adjustment. Indeed, a qualitative analysis showed that most tokens, even semantically more complex ones (e.g. question tokens), are matched correctly at the beginning of training. However, throughout training, the matches' contributions to the final score increase.

The analysis also revealed that the contributions of lexical and easy semantic matches (c.f. the lexical matches of "b" and "##mt" in Fig. 4) mainly increase in the early stages of fine-tuning (up to 5k iterations). In the later stages of training (100k and 200k steps), the contributions of these easier tokens remain approximately stable, while the contributions of semantically matched (more complex) tokens (e.g. "why" and "because") increase. This indicates that, in contrast to our hypothesis outlined in Sect. 4, BERT's embeddings provide a base to match *all* token types. However, the contributions of these matches to a document's final score are adjusted through fine-tuning which affects the document ranking and improves effectiveness. In particular, the qualitative observations above suggest that ColBERT first learns lexical and easy semantic matching, while adjusting the contributions of more complex semantic matches later in training. In combination with the high delta in S-nDCG observed for question and medium-IDF tokens in Table 2, this suggests that the MaxSim contributions of these more complex token types need to be adjusted more, which could

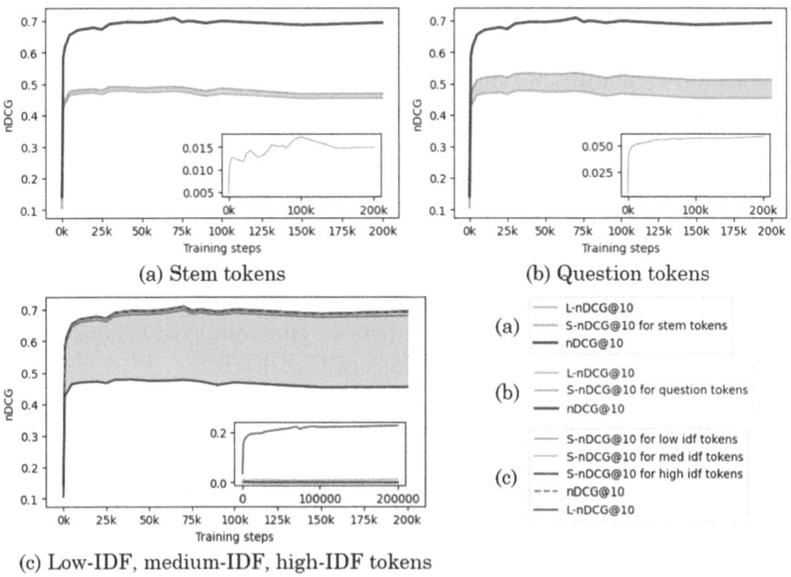

(a) Stem tokens

(b) Question tokens

(c) Low-IDF, medium-IDF, high-IDF tokens

Fig. 5. Visualisations of the L-nDCG@10, S-nDCG@10 and nDCG@10 for the token types with a significant increase on TREC 2020. The shaded areas, as well as the subgraphs in the bottom right corners, show the S-nDCG for the particular token type. The bottom and top lines denote the L-nDCG and nDCG for *all* tokens, respectively.

(partially) account for the gain in S-nDCG throughout training as described in Sect. 5.1 and shown in Fig. 3.

Answer for RQ2. We find that BERT's pre-trained embeddings give ColBERT the (basic) ability to match all token types. However, the contributions of these matches need to be adjusted through fine-tuning. A qualitative analysis reveals that lexical and easy semantic matches are addressed in the early stages of training while more semantically complex matches (i.e. question and medium-IDF tokens) are adjusted in later stages and more heavily, which could largely account for ColBERT's increase in S-nDCG and effectiveness throughout training.

6 Conclusions

In this work, we investigated the (semantic) matching behaviour of ColBERT and how it emerges from BERT's pre-training during fine-tuning. For this, we introduced the S-nDCG and L-nDCG metrics, which quantify to which degree a model's performance can be attributed to semantic and lexical matching, respectively. Upon evaluating ColBERT with these measures, we observe an increase in both measures in the early stages of training. Surprisingly however, this is accompanied by a large *decrease* in the proportion of semantic matching utilised by the model, which indicates that ColBERT may 'unlearn' inaccurate semantic matches during early fine-tuning (learning to rely more on lexical matching at the same time). Generally, we find though that the S-nDCG increases throughout training. In particular, there is an increase in the MaxSim contributions of tokens with complex relationships such as questions words, which is reflected in an increasing attribution of the model's effectiveness to semantic matching throughout training and a general improvement in performance.

While our results here focus on ColBERTv1, we expect that our results would generalise to models trained by ColBERTv2 [21] (which is trained using hard negatives). We have also focussed on a re-ranking setting for empirical efficiency and following previous work [26] - we believe that our conclusions would be highly similar on end-to-end dense settings of ColBERT.

Finally, this work paves the way to applying fixes to ColBERT matching errors. For instance, if an incorrect semantic match is observed, our results suggest that insertion of additional relevance triples earlier in the training process are more likely to address this, but this may depend on the family of semantic match.

Acknowledgements. Ariane Mueller acknowledges partial support from the FUN project (SGA 2024FSTPC2PN30), funded by the OpenWebSearch.eu project (GA 101070014). We would like to express our gratitude to the anonymous reviewers for their feedback, which were helpful for us to improve this paper.

Conflict of Interest. The authors have no competing interests to declare that are relevant to the content of this article.

References

1. Bajaj, P., et al.: MS MARCO: a human generated machine reading comprehension dataset. arXiv preprint arXiv:1611.09268 (2016)
2. Câmara, A., Hauff, C.: Diagnosing BERT with retrieval heuristics. In: Jose, J.M., et al. (eds.) ECIR 2020. LNCS, vol. 12035, pp. 605–618. Springer, Cham (2020). https://doi.org/10.1007/978-3-030-45439-5_40
3. Craswell, N., Mitra, B., Yilmaz, E., Campos, D.: Overview of the TREC 2020 deep learning track. arXiv preprint arXiv:2102.07662 (2021)
4. Craswell, N., Mitra, B., Yilmaz, E., Campos, D., Voorhees, E.M.: Overview of the TREC 2019 deep learning track. arXiv preprint arXiv:2003.07820 (2020)
5. Devlin, J.: BERT: pre-training of deep bidirectional transformers for language understanding. arXiv preprint arXiv:1810.04805 (2018)
6. Formal, T., Piwowarski, B., Clinchant, S.: A white box analysis of ColBERT. In: Hiemstra, D., Moens, M.-F., Mothe, J., Perego, R., Potthast, M., Sebastiani, F. (eds.) ECIR 2021. LNCS, vol. 12657, pp. 257–263. Springer, Cham (2021). https://doi.org/10.1007/978-3-030-72240-1_23
7. Hofstätter, S., Khattab, O., Althammer, S., Sertkan, M., Hanbury, A.: Introducing neural bag of whole-words with colberter: contextualized late interactions using enhanced reduction. In: Proceedings of the 31st ACM International Conference on Information and Knowledge Management, pp. 737–747 (2022)
8. Humeau, S., Shuster, K., Lachaux, M.A., Weston, J.: Poly-encoders: transformer architectures and pre-training strategies for fast and accurate multi-sentence scoring. arXiv preprint arXiv:1905.01969 (2019)
9. Jiang, Z., Tang, R., Xin, J., Lin, J.: How does BERT rerank passages? An attribution analysis with information bottlenecks. In: Proceedings of the Fourth BlackboxNLP Workshop on Analyzing and Interpreting Neural Networks for NLP, pp. 496–509 (2021)
10. Karpukhin, V., et al.: Dense passage retrieval for open-domain question answering. arXiv preprint arXiv:2004.04906 (2020)
11. Khattab, O., Zaharia, M.: ColBERT: efficient and effective passage search via contextualized late interaction over BERT. In: Proceedings of the 43rd International ACM SIGIR Conference on Research and Development in Information Retrieval, pp. 39–48 (2020)
12. MacAvaney, S., Feldman, S., Goharian, N., Downey, D., Cohan, A.: ABNIRML: analyzing the behavior of neural IR models. Trans. Assoc. Comput. Linguist. **10**, 224–239 (2022)
13. Macdonald, C., Tonellotto, N., MacAvaney, S., Ounis, I.: PyTerrier: declarative experimentation in Python from BM25 to dense retrieval. In: Proceedings of the 30th ACM International Conference on Information and Knowledge Management, pp. 4526–4533 (2021)
14. Macdonald, C., Tonellotto, N., Ounis, I.: On single and multiple representations in dense passage retrieval. arXiv preprint arXiv:2108.06279 (2021)
15. Moeller, L., Nikolaev, D., Padó, S.: An attribution method for Siamese encoders. arXiv preprint arXiv:2310.05703 (2023)
16. Nogueira, R., Cho, K.: Passage re-ranking with BERT. arXiv preprint arXiv:1901.04085 (2019)
17. Qiao, Y., Xiong, C., Liu, Z., Liu, Z.: Understanding the behaviors of BERT in ranking. arXiv preprint arXiv:1904.07531 (2019)

18. Rau, D., Kamps, J.: The role of complex NLP in transformers for text ranking. In: Proceedings of the 2022 ACM SIGIR International Conference on Theory of Information Retrieval, pp. 153–160 (2022)
19. Rennings, D., Moraes, F., Hauff, C.: An axiomatic approach to diagnosing neural IR models. In: Azzopardi, L., Stein, B., Fuhr, N., Mayr, P., Hauff, C., Hiemstra, D. (eds.) ECIR 2019. LNCS, vol. 11437, pp. 489–503. Springer, Cham (2019). https://doi.org/10.1007/978-3-030-15712-8_32
20. Robertson, S.E., Walker, S., Jones, S., Hancock-Beaulieu, M.M., Gatford, M., et al.: Okapi at TREC-3. NIST Special Publication Sp **109**, 109 (1995)
21. Santhanam, K., Khattab, O., Saad-Falcon, J., Potts, C., Zaharia, M.: Colbertv2: effective and efficient retrieval via lightweight late interaction. arXiv preprint arXiv:2112.01488 (2021)
22. Thakur, N., Reimers, N., Rücklé, A., Srivastava, A., Gurevych, I.: BEIR: a heterogenous benchmark for zero-shot evaluation of information retrieval models. arXiv preprint arXiv:2104.08663 (2021)
23. Völske, M., et al.: Towards axiomatic explanations for neural ranking models. In: Proceedings of the 2021 ACM SIGIR International Conference on Theory of Information Retrieval, pp. 13–22 (2021)
24. Wallat, J., Beringer, F., Anand, A., Anand, A.: Probing BERT for ranking abilities. In: European Conference on Information Retrieval, pp. 255–273. Springer, Cham (2023)
25. Wang, X., Macdonald, C., Ounis, I.: Improving zero-shot retrieval using dense external expansion. Inf. Process. Manag. **59**(5), 103026 (2022)
26. Wang, X., Macdonald, C., Tonellotto, N., Ounis, I.: Reproducibility, replicability, and insights into dense multi-representation retrieval models: from ColBERT to Col*. In: Proceedings of the 46th International ACM SIGIR Conference on Research and Development in Information Retrieval, pp. 2552–2561 (2023)
27. Xiong, L., et al.: Approximate nearest neighbor negative contrastive learning for dense text retrieval. arXiv preprint arXiv:2007.00808 (2020)

Opt-in Transparent Fairness for Recommender Systems

Bjørnar Vassøy(✉), Benjamin Kille, and Helge Langseth

Norwegian University of Science and Technology, Høgskoleringen 1, 7034 Trondheim, Trøndelag, Norway
bjornar.vassoy@ntnu.no

Abstract. Recommender systems serve large, diverse userbases and are subject to fairness concerns. Seminal work on consumer-side fairness in recommender systems tends to focus on mitigating the influence users' demographic information has on their recommendations. Recent work allows the users to specify which information should be concealed, but these methods are often complex and inadvertently deteriorate the recommendation performance of users who reveal all information. We propose a simple, novel approach that interfaces with trained non-fair models and allows users to opt-in for improved fairness without affecting the recommendations given to others. Unlike previous work, the proposed method includes fairness transparently. Quantitative comparison to the state of the art indicates that the proposed method has a favourable fairness/utility tradeoff.

Keywords: Recommender Systems · Fairness · Transparency

1 Introduction

Biases pose a major challenge for data-driven models. Data-driven models will inadvertently learn potentially undesired correlations. For instance, if one gender dominates the current workforce, a career recommender system may reflect or even reinforce the historical bias. This may hurt models' performances if gender interests have shifted, and users may dislike that their gender affects their recommendations. Fair methods that explicitly hide demographic information optimize "Fairness through unawareness" [5], but information such as gender and age will often correlate with other features, such as users' preferred occupations or hobbies. Further mitigating the influence of demographics can be achieved by optimizing for *neutral representations* [13]. If a representation does not encode demographic information of what it represents, said information does not influence the model. Neutral representation is actively researched in recommender systems and is the *interpretation* of fairness optimized in this work.

Most models optimizing neutral representations extend an underlying model and improve fairness at the cost of accuracy. Some allow users to choose whether

C. Hauff et al. (Eds.): ECIR 2025, LNCS 15572, pp. 357–371, 2025.
https://doi.org/10.1007/978-3-031-88708-6_23

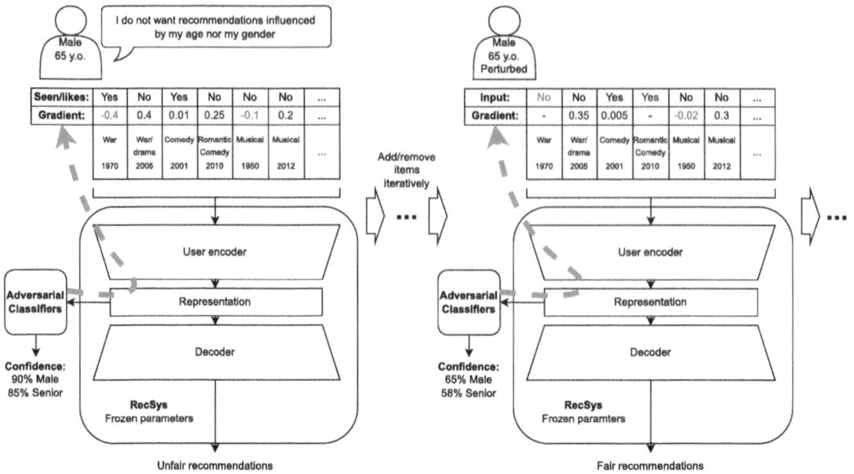

Fig. 1. Overview of how the proposed methods perturb user data to reduce demographic information's effect on recommendations. In each step, we calculate the item gradients given the current (perturbed) input and remove or add a single item. Similar adversarial approaches can be applied to most model types but will not provide transparency for models without a direct link between items as input and representations. Access to differentiable model parameters is required, but they are not adjusted.

they want more fairness, but typically perform worse for users who do not opt-in than standard models. A simple, mostly unexplored strategy circumvents the issue with an auxiliary extension that interfaces with a pre-trained standard model only when users opt-in. Another challenge existing work does not address is transparency in terms of how fairness is enhanced. For instance, can one interpret the changes made to a representation to conceal gender? Our proposed method addresses both challenges and is shown applied to a VAE-based movie recommender system in Fig. 1. By showing users how their input has been altered to conceal demographics, you can give them insight into how the model will perceive them differently. This model insight is further enhanced when showing the users how their recommendations change. We answer these questions:

- How can simple model extensions introduce fairness to a recommender system without negatively affecting users who do not care for fairness?
- How can simple model extensions be transparent about how fairness is introduced and explain the impact of the introduction to individual users?
- How do such model extensions compare to existing work in terms of recommendation quality and fairness?

2 Related Works

This work is mainly inspired by recommender system research focusing on fairness and reducing the influence of demographic information, specifically methods and models that allow users to specify which sensitive attributes to disregard. VAE-based recommender systems are also relevant, along with recommender systems focusing on transparent or explainable fairness.

2.1 Fair Recommender Systems

We focus on the most relevant work on neutral representation [13] as this is the type of fairness optimized by our method. Bose and Hamilton (2019) [3] filter out demographic attributes in link prediction. They extend existing models with filter modules trained by adversarial classifiers to remove individual demographic attributes. The training alternates between updating the adversarial models and updating the underlying model along with the filters. Li et al. (2021) [9] adapt the idea to recommender systems and propose a new way to filter multiple demographic attributes at once. They design a filter for each *combination* of demographic attributes instead of aggregating individually filtered representations. Creager et al. (2019) [4] split a VAE-based classifier's latent space into sensitive and non-sensitive parts. The sensitive part is trained to classify the demographic attributes, and the non-sensitive part is optimized to be independent of the former. The approach inspired Vassøy et al. (2023) [14], who consider recommender systems and explore multiple strategies for ensuring the two parts' independence.

Weinsberg et al. (2012) [15] obfuscate individual user's gender by explicitly extending their training data with items indicative of other genders. This approach inspired Slokom et al. (2021) [12], who also remove items indicative of the user's gender and perturb all users rather than individuals. Both methods perturb the data before training, thus affecting recommendations for all users. The candidate items are identified by training adversarial classifiers to predict user gender given user data.

Zhu et al. (2024) [16] is the only work we found that interfaces with non-fair models without worsening the experience of users who do not hide any attributes. Using information theory, they split user representations into sensitive and non-sensitive parts. The method forms new user embeddings with a network trained to produce good recommendations in the underlying model. If a user disallows the consideration of a demographic attribute, the corresponding part will not be combined into the final user representation.

2.2 Transparently/Explainably Fair Recommender Systems

Ge et al. (2022) [6] explore *explainable fairness* in recommender systems. Their method targets scenarios where quantifiable user and item *features* can be derived from contextual data, like item reviews. It learns how to mitigate popularity bias by changing user and item feature vectors while minimizing the scale of said changes. The optimal changes tell which features affect fairness the most, e.g., more/less of a feature would increase the exposure of an unpopular item or decrease the exposure of a popular item. Examples of possible stakeholders are users and the item providers, e.g., artists in music recommender systems where songs take the role of items. Ludovico et al. [1,2] draw inspiration from counterfactual graph explanation methods to identify graph augmentations that reduce the NDCG disparity among demographic groups. The method is centred on an *augmentation vector* that controls the inclusion of a sampled set of missing user-item pairs/edges in the original graph. This vector is optimized incrementally to decrease the NDCG disparity while minimizing the total augmentation of the graph.

3 Method

The key idea is to train adversarial classifiers to identify how input can be perturbed to obfuscate sensitive attributes in pre-trained models. The approach is adaptable for any gradient-based model and transparent if the input and its perturbations are interpretable. We focus on VAE-based recommender systems and perturb user input by iteratively adding and removing items until the adversarials can no longer identify users' demographics.

3.1 VAE Recommender System

Variational Autoencoders (VAE) are variational Bayesian models first introduced by [8] that have been applied to various tasks. Intuitively, a VAE learns how to compress the data into compact encodings and how to reconstruct the data from such encodings. Different parts of a fitted VAE have been used to solve various tasks: the encoder for producing efficient encodings of complex data like text, the decoder for generating images or text, and the full VAE for removing noise from noisy images or signals.

The classical VAE is a graphical model that relates the latent variable $\boldsymbol{z}$ and the data $\boldsymbol{x}$, and comprises three main parts: an encoder that parameterizes the variational distribution $q(\boldsymbol{z}|\boldsymbol{x})$, a prior distribution over the latent variable $p(\boldsymbol{z})$ and a decoder that parameterizes the likelihood $p(\boldsymbol{x}|\boldsymbol{z})$. The variational distribution $q(\boldsymbol{z}|\boldsymbol{x})$ is introduced to approximate the true, unknown/intractable, posterior $p(\boldsymbol{z}|\boldsymbol{x})$, and the approximation is achieved by minimizing the *distance* between the two distributions. The original work [8] explores minimizing the KL-divergence from $q(\boldsymbol{z}|\boldsymbol{x})$ to $p(\boldsymbol{z}|\boldsymbol{x})$, which is shown to be equivalent to maximizing the Evidence Lower BOund (ELBO), as seen in Eq. 1.

$$\text{ELBO}(q,p) = \mathbb{E}_{q(\boldsymbol{z}|\boldsymbol{x})}\left[\log p(\boldsymbol{x}|\boldsymbol{z})\right] - D_{\text{KL}}\left[q(\boldsymbol{z}|\boldsymbol{x})||p(\boldsymbol{z})\right] \tag{1}$$

Here, $\mathbb{E}_{q(\boldsymbol{z}|\boldsymbol{x})}[\cdot]$ is the expectation over the variational distribution and $D_{\text{KL}}[\cdot]$ is the KL-divergence.

VAEs as recommender systems are comparable to VAEs tasked with denoising input. When presented with a binary vector representing which of the available items a user likes or has interacted with, the model is taught to fill in *missing* entries, i.e., recommendations. Specifically, recommendations are extracted by ranking probabilities, given by the decoder parameterized likelihood $p(\boldsymbol{x}|\boldsymbol{z})$, assigned to items the user has not *liked* or interacted with. The model proposed by [10] explores different choices of likelihood distributions $p(\boldsymbol{x}|\boldsymbol{z})$ as well as strategies for discounting the importance of the second part of the ELBO $D_{\mathrm{KL}}[q(\boldsymbol{z}|\boldsymbol{x})||p(\boldsymbol{z})]$. Our model and the baselines we consider are either extensions of this model or use it as an underlying model.

3.2 Perturb User Input

Using a pre-trained VAE recommender system, we encode the users in the training set and train adversarial models, $C_i(\boldsymbol{z})$, to classify demographic attributes given the encoded user representations, $\boldsymbol{z}$. The adversarial models are multi-layered neural networks to allow gradients to flow from these models through the VAE encoder to the VAE input, $\boldsymbol{x}$. Notably, our method requires access to the underlying model's parameters but does not change them. For any user input, we can then calculate the gradient of the adversarial classification loss, L_{C_i}, with respect to individual items in the user input, $\frac{\partial L_{C_i}}{\partial x_j}$.

If added, items with positive gradients increase the adversarial loss, and removing items with negative gradients will have the same effect. For users who have opted-in to mitigate the influence of their age, the method will continue to add and remove items until meeting one of two criteria: Either the adversarial model responsible for classifying age is *sufficiently unsure* about the user's age, or a predefined budget of item additions and removals has been exhausted. We consider the adversarial sufficiently unsure when the predicted class probability is within an interval centred on the value that would make the adversarial correctly classify the same proportion of young and senior users when used as the binary decision boundary. This interval is designed to counter the effects of imbalanced classes and can be interpreted as "correctly classifying $x\%$ of one age group is equally bad as correctly classifying the same percentage of the other age group". The budget decides how many items can be perturbed, ensuring that the method cannot fully disregard the users' data and preferences to construct perfectly fair but non-personalized user profiles. Whether an item is added or removed is decided by coin-flip when both operations have remaining budgets.

We select items by weighted adversarial loss, and the method only considers the adversarial loss of demographic attributes the user wants to conceal. If gender and age are filtered, the adversarial loss is weighted by adversarial certainty to prioritize the concealment of the most apparent attribute through item selection. After adding or removing an item, the process is repeated by encoding the perturbed user data to get updated adversarial gradients for the next perturbation step. Figure 1 illustrates how the method works.

3.3 Relevance Weighting

Basing the selection of items to be removed or added solely on their adversarial gradients disregards the users' preferences. To personalize the selection of items to *add*, we explore weighting the item gradients by the VAE's output item probabilities for unperturbed users. E.g., an item that the VAE deems relevant may be prioritized over an item with a larger gradient but deemed less relevant. This extension highlights scenarios where the user perturbation adds items that are relevant enough to be recommended. As the VAE is trained to assign high probabilities to the items given as inputs, the assigned probabilities of items added in the perturbation step will not accurately represent the relevance for the user. Instead, we designed a heuristic that replaces the perturbed items' assigned probabilities with the same items' assigned probabilities in the underlying model *if* the items are among the top-k recommended items for individual users. We set $k = 20$, which is twice the number of recommendations for our NDCG evaluation. This way, relevant items added by our method will also be presented as recommendations if their predicted probabilities in the underlying model rank high when combined with the final predicted probabilities of other items. These items could be ranked higher as they are both relevant and improve fairness, but as the optimal relevance improvement is hard to quantify, we rely on the underlying model.

3.4 Double Adversarial

We propose a second extension to address issues with adversarial generalization and combine this with relevance weighting. Another group of adversarials is trained to take over when half of the perturbation budget has been exhausted. This follows the intuition that the first group of adversarials will pick up on the strongest correlations but may ignore weaker ones. An illustrative example could involve two items exclusively liked by one gender, where the users who like the former are a superset of the users who like the latter. A gender classifier that ignores the latter item can still optimally classify the users who like it. Adding input dropout during the adversarials' training mitigated this issue and was applied in all variations of our method.

We can add more levels to the strategy, but each level diminishes returns and increases training complexity significantly as we must perturb users in the training set before each level's adversarial group training.

4 Experiment

4.1 Datasets

The MovieLens 1 million dataset [7] and the LFM-2b[1] [11] datasets were used in our evaluation. We used the two most recent years of data in the LastFM dataset and removed artists listed with fewer than 100 listening events. In both datasets, we adopted the available binary gender labels and split users into binary age groups using 35 as the threshold. More dataset statistics are shown in Table 1.

[1] The dataset is no longer distributed by [11].

Table 1. Dataset statistics. Med I/U is the median number of items per user.

Dataset	#Users	#Items	Med I/U	%Female	%Senior	Included items
MovieLens	6.0k+	3.5k+	57	28.3%	23.6%	Movies rated 4–5/5
LastFM	14.5k+	23.8k+	357	20.9%	8.6%	Artists listened to

Datasplit. Following the evaluation setup of [10], the data is split on a user level. User-level splits are not readily supported by models that rely on learnt user representations, like latent factors in factorization models or user embeddings in deep models. Models that support user-level split can recommend for users not represented in the training data without re-training, and validation on user-level splits can ensure that the model has generalized to unseen users and not overfitted on the observed ones. Generalization is crucial for our methods as we require them to make sense of both new users and perturbed versions of their input.

4.2 Models

All models are VAE-based to support the data split, and the code is public[2].

- **VAErec** [10]: A common VAE-based recommender system that does not attempt to enhance fairness. We apply a multinomial likelihood $p(\boldsymbol{x}|\boldsymbol{z})$ and set β to 0.2 for all experiments. This model is extended or used as an underlying model in all subsequent models.
- **VAEemp** [14]: A fair VAE-based recommender system splitting latent representation into a sensitive and a non-sensitive part exclusively used for recommending. Users cannot specify which attributes will be filtered. The model will always filter out the sensitive attributes it is trained to filter.
- **VAEsm** [9]: Fair recommender system framework that trains separate filters for each combination of the demographic attributes. Users can decide which demographic attributes to filter out to mitigate their influence.
- **VAEafrl***: Baseline inspired by [17], see details below.
- **Perturb-models**: The proposed models. The budget was set to $\min(0.5 * n_{u_i}, 50)$ where n_{u_i} is the number of items seen or liked by user u_i for each operation (addition, removal).
  - **PerturbVan**: The vanilla variation of our method (see Sect. 3.2).
  - **PerturbRel**: Variation of our method that considers relevance when adding items (Sect. 3.3).
  - **Perturb2Adv**: Variation of our method that extends `PerturbRel` by adding a second group of adversarial classifiers (see Sect. 3.4).

[2] https://github.com/BjornarVass/transparent_fair_recsys.

VAEafrl* Description. The framework proposed in [17] applies information theory to split pre-trained user representations into ones that encode individual sensitive attributes and a single non-sensitive representation. A network combines the representations back into user representations that replace the pre-trained ones. The official implementation of this framework targets scenarios where all users are known and must be adapted to generalize to new users for our evaluation. As the sensitive representations were trained to predict the sensitive attributes perfectly, we replaced them with embeddings of the actual labels. We increased the regularization of the adversarial networks used to train the encoder of the non-sensitive representation to improve generalization. Zhu et al. [17] train the network for combining the representations using the underlying model and its recommendation loss. We did extensive testing using BPR-loss, as [17], and also the multinomial loss used in `VAErec`. However, we got noticeably better results when replacing this loss with an MSE loss on the pre-trained and the combined representations. Unlike the model types evaluated in [17], VAEs have a decoder network placed after the representation whose sole input is said representation, i.e., you have to train the first layers of a compound model while keeping the later ones fixed. This is the reverse of the typical fine-tuning scenario, which focuses on the more flexible later layers, and the complexity of this training setup is believed to have caused our poor results. We adopted the MSE-loss in our baseline.

5 Results

Throughout this Section, we will represent demographic skewness by ratios of female/senior users, e.g., the ratio of female users in the dataset or the ratio of senior users who have seen a movie.

5.1 Transparant Fairness Case Study

We provide a case study on a user who uses our methods to conceal their gender and age as an example of how the methods introduce fairness transparently, see Table 2. The user provided a list of 117 movies they enjoyed, and these movies are, on average, liked by senior males, as seen in the top tables. The bottom left table shows how `Perturb2Adv` perturbed the user's data. Despite the individual items indicating otherwise, the adversarial models are fairly confident that the user is a *young* male. This may be because the adversarial finds the specific combination typical of young users or due to the presence of a few items vastly preferred by young users. The method has removed items preferred by male or young users and added items popular among female or senior users. The method converges after reaching the target adversarial intervals in 15 iterations. The bottom right table shows how the top 10 recommendations have changed. Top 1 and 2 recommendations do not move despite being popular among senior male users, while items with similar ratios in lower `VAErec` rankings have fallen up to three positions. We also see that two new items that female users prefer have moved into the top 10 ranking from 11th and 14th, and a movie with a similar profile has moved up one position.

Table 2. Overview of the case study of the transparent fairness offered by our methods. The ratios of female and senior users in the dataset are shown in the top left, and the considered user is summarized by the mean demographic ratios of users who have seen the movies the user has seen in the top right. The bottom left table covers the perturbations made to the user's data to conceal gender and age, while the bottom right table shows how the perturbations affected the user's recommendations.

	G.rat	A.rat		G.rat	A.rat
Userbase	0.283	0.235	Movies seen by user	0.243	0.289

Perturbations						Top 10 Recommendations			
Op	Title	G.rat	A.rat	G.adv	A.adv	Title	G.rat	A.rat	Chng
				0.06	0.12	Bicycle Thief, The (1948)	0.21	0.28	-
Rem	Hands on a Hard Body (1996)	0.27	0.06	0.06	0.15	Seven Samurai (1954)	0.17	0.28	-
Rem	Apocalypse Now (1979)	0.12	0.20	0.07	0.16	Citizen Kane (1941)	0.25	0.24	-
Add	Secrets & Lies (1996)	0.42	0.31	0.08	0.17	Woman in the Dunes (1964)	0.33	0.27	+7
Add	All About Eve (1950)	0.42	0.34	0.09	0.17	To Kill a Mockingbird (1962)	0.34	0.29	+1
Add	A Streetcar Named Desire(1951)	0.34	0.31	0.10	0.17	Seventh Seal, The (1957)	0.19	0.25	-2
Rem	Patton (1970)	0.12	0.31	0.11	0.17	Nights of Cabiria (1957)	0.32	0.32	+7
Add	Buena Vista Social Club (1999)	0.35	0.30	0.12	0.17	Lawrence of Arabia (1962)	0.18	0.35	-3
...	...			...		Casablanca (1942)	0.29	0.30	-1
Rem	Gladiator (2000)	0.19	0.16	0.29	0.21	Taxi Driver (1976)	0.19	0.21	-1

5.2 Comparative Results

We compare NDCG@10 and AUC for all models. AUC is our main fairness metric and is evaluated in the worst-case scenarios where all users have opted to conceal a given combination of demographic attributes, and auxiliary classifier models are trained to identify all demographic attributes. We also report Gender AUC when filtering Age and vice versa since these scores provide insights into the different models.

The `Perturb` models all retain high NDCG scores and see major improvements in AUC compared to `VAErec`. The other models achieved better AUCs but at a greater NDCG cost. Among the `Perturb` models, `PerturbVan` and `PerturbRel` achieve similar AUC, but `PerturbRel` scores better in NDCG, which indicates that NDCG can be improved by considering item relevance without negatively impacting AUC. `Perturb2Adv` has similar or better NDCG than `PerturbVan` while achieving significantly better AUC scores.

`VAEsm` and `VAEafrl*` perform similarly when filtering attributes, with `VAEsm` scoring slightly better in NDCG for single filters and AUC for all filters. `VAEsm` performs worse for users who do not want any filters. `VAEemp` achieved higher NDCG than `VAEsm` and `VAEafrl*` but vary more in AUC. It has a slight edge over the others in Age AUC, especially when only filtering Age, but performs worse for Gender. These differences should be weighed against the fact that you have to train a unique `VAEemp` for each attribute combination your users can choose to conceal instead of a single system for the other options.

Table 3. Comparison of NDCG@10 and **G**ender/**A**ge AUC of all models evaluated on the MovieLens dataset. Each filter column represents a different combination of filters. Models are grouped by functionality: `VAErec` and `VAEemp` have to train specialized models for each filter combination. `VAEsm` allows dynamic filtering, but the training and the integration of the filters affect the performance when no filters are used. The remaining models offer dynamic filters, and the introduced fairness does not affect users who do not opt for it.

Filters	None			Gender			Age			Age and Gender		
Metric	NDCG↑	AUC.G↓	AUC.A↓	NDCG	AUC.G	AUC.A	NDCG	AUC.G	AUC.A	NDCG	AUC.G	AUC.G
VAErec	**0.321**	0.819	0.862									
±	0.009	0.018	0.015									
VAEemp				0.294	0.660	0.852	0.294	0.802	0.619	0.288	0.648	0.625
±				0.008	0.044	0.012	0.009	0.020	0.034	0.010	0.037	0.036
VAEsm	0.292	0.817	0.861	0.281	0.632	0.808	0.281	0.772	0.653	0.272	0.621	0.632
±	0.013	0.020	0.018	0.014	0.026	0.011	0.014	0.029	0.040	0.013	0.032	0.045
VAEafrl*	**0.321**	0.819	0.862	0.277	0.632	0.990	0.276	0.982	0.647	0.272	0.633	0.649
±	0.009	0.018	0.015	0.009	0.036	0.006	0.007	0.007	0.020	0.008	0.038	0.019
Perturb	**0.321**	0.819	0.862	0.305	0.708	0.844	0.299	0.804	0.719	0.278	0.715	0.750
±	0.009	0.018	0.015	0.011	0.021	0.015	0.007	0.021	0.031	0.010	0.030	0.028
PerturbRel	**0.321**	0.819	0.862	0.315	0.716	0.856	0.312	0.812	0.718	0.307	0.712	0.722
±	0.009	0.018	0.015	0.008	0.019	0.013	0.009	0.021	0.022	0.010	0.018	0.023
Perturb2Adv	**0.321**	0.819	0.862	0.303	0.674	0.845	0.304	0.808	0.673	0.295	0.671	0.657
±	0.009	0.018	0.015	0.011	0.022	0.016	0.010	0.018	0.027	0.012	0.026	0.019

The two middle columns show models filtering only one demographic attribute, so only the filtered attribute contributes to the fairness performance. However, the AUC of the non-filtered attribute reveals that the `Perturb` models and `VAEemp` mainly filter the target demographic attribute when compared with `VAErec`, while each filter in `VAEsm` also filters some of the non-target attributes. `VAEafrl*` explicitly adds information on the non-target attribute, which results in almost perfect identifiability.

Table 4. MovieLens NDCG and AUC results when we tune the hyperparameters of `VAEsm` and `VAEafrl*` to match the NDCG achieved by `Perturb2Adv` when filtering both Gender and Age.

Filters	Gender		Age		Age and Gender		
Metric	NDCG↑	AUC.G↓	NDCG↑	AUC.A↓	NDCG↑	AUC.G↓	AUC.A↓
VAEsm $\gamma = 20$	0.301	0.761	0.300	0.795	0.295	0.751	0.770
±	0.012	0.028	0.012	0.019	0.012	0.027	0.026
VAEafrl* $\lambda = 1.2$	0.305	0.675	0.304	0.669	0.295	0.667	0.660
±	0.010	0.024	0.010	0.022	0.012	0.025	0.024
Perturb2Adv	0.303	0.674	0.304	0.673	0.295	0.671	0.657
±	0.011	0.022	0.010	0.027	0.012	0.026	0.019

Table 3 reveals that despite being able to perturb most users to the point where the adversarial classifiers fail to classify them, other approaches achieve better AUC scores. Increasing the allowed budget may make the models successfully perturb some problematic users, but the AUC would still be limited by the static nature of the adversarials. To evaluate whether `Perturb2Adv` achieves good NDCG relative to the AUC improvement, we tuned the hyperparameters of `VAEsm` and `VAEafrl*` to match our NDCG score when filtering both Gender and Age. Table 4 shows that `VAEafrl*` and `Perturb2Adv` are more or less tied for this NDCG value, but `VAEsm` achieves worse AUC scores. We noted that `VAEsm` struggles to improve AUC when targeting higher NDCG ranges and is a better fit in scenarios where fairness is prioritized over recommendation performance.

We performed the same experiments on the LastFM dataset to test the methods' robustness. Despite being more demographically skewed, `VAErec` indicate that the correlation of users within groups is weaker in this dataset. The results in Table 5 show that our methods can drastically improve AUC and stay within two standard deviations of `VAErec`'s NDCG. Interestingly, the `Perturb` models' NDCG scores differ less on this dataset. The other models can achieve better AUC at higher NDCG costs, but tuning `VAEsm` and `VAEafrl*`[3] to match our NDCG when filtering both gender and age again indicates that `Perturb` and `PerturbRel` achieve good tradeoffs.

Table 5. Table that compares NDCG@10 and Gender/Age AUC of all models evaluated on the LastFM dataset.

Filters	None			Gender		Age		Age and Gender		
Metric	NDCG↑	AUC.G↓	AUC.A↓	NDCG	AUC.G	NDCG	AUC.A	NDCG	AUC.G	AUC.A
VAErec	**0.493**	0.756	0.798							
±	0.008	0.012	0.017							
VAEemp				0.444	0.703	0.445	0.772	0.447	0.708	0.763
±				0.009	0.018	0.006	0.017	0.007	0.014	0.013
VAEsm	0.438	0.726	0.772	0.412	0.622	0.426	0.681	0.406	0.623	0.667
±	0.014	0.015	0.024	0.017	0.020	0.016	0.026	0.016	0.014	0.031
VAEafrl*	**0.493**	0.756	0.798	0.431	0.612	0.431	0.662	0.431	0.606	0.653
±	0.008	0.012	0.017	0.006	0.023	0.006	0.042	0.005	0.021	0.039
Perturb	**0.493**	0.756	0.798	0.488	0.677	0.486	0.725	0.481	0.680	0.728
±	0.008	0.012	0.017	0.007	0.016	0.005	0.020	0.006	0.018	0.028
PerturbRel	**0.493**	0.756	0.798	0.487	0.679	0.483	0.742	0.479	0.681	0.744
±	0.008	0.012	0.017	0.006	0.014	0.006	0.015	0.006	0.012	0.013
Perturb2Adv	**0.493**	0.756	0.798	0.485	0.660	0.482	0.714	0.479	0.663	0.721
±	0.008	0.012	0.017	0.007	0.015	0.009	0.027	0.008	0.019	0.022
VAEafrl* $\lambda = 1.0$	**0.493**	0.756	0.798	0.453	0.649	0.453	0.700	0.453	0.643	0.690
±	0.008	0.012	0.017	0.008	0.017	0.007	0.024	0.008	0.019	0.029
VAEsm $\gamma = 250$	0.481	0.776	0.821	0.480	0.700	0.480	0.747	0.478	0.690	0.735
±	0.014	0.015	0.014	0.009	0.021	0.008	0.025	0.009	0.015	0.030

5.3 Ratio Metrics

While all considered models are optimized for neutral representation, which aligns with the AUC metric, it is also interesting to look at how the more neutral representations affect the recommendations. Classifiers that evaluate representation neutrality may pick up a small difference between male and female users, even if the difference does not affect recommendations. So, the performance of methods achieving the same AUC may vary as we compare differences between recommendations to demographic groups.

We propose two intuitive metrics for expressing the differences between the recommendations for two disjoint groups. For a system that is equally likely to

[3] Could not match the `Perturb` models even when completely disregarding fairness, $\lambda = 0$ (NDCG=0.464).

recommend an item to a male user as a female user, each item is expected to be recommended to a ratio of female users equal to the ratio of female users in the userbase. In the *Item Ratio Difference* metric, we quantify the distance from ideal by taking the average absolute difference between item demographic ratios and the userbase demographic ratios. I.e., if an item is recommended to 10 female users and 30 male users, and 10% of users are female, it would contribute $|\frac{10}{10+30} - 0.1| = |0.25 - 0.1| = 0.15$ to the metric. For stability, we only consider items recommended to at least five users, and we consider the top 10 recommendations given to each user.

Table 6. Item Ratio Difference and User Ratio Difference results on the MovieLens dataset when both Gender and Age are filtered in the dynamic models.

	Item Ratio Difference		User Ratio Difference	
	Gender	Age	Gender	Age
VAErec	0.165 ± 0.010	0.146 ± 0.008	0.079 ± 0.006	0.069 ± 0.007
VAEsm	0.107 ± 0.006	0.099 ± 0.005	0.035 ± 0.003	0.033 ± 0.005
VAEafrl*	0.121 ± 0.006	0.108 ± 0.006	0.045 ± 0.005	0.039 ± 0.004
Perturb	0.127 ± 0.012	0.116 ± 0.008	0.046 ± 0.007	0.041 ± 0.005
PerturbRel	0.148 ± 0.011	0.132 ± 0.007	0.068 ± 0.007	0.058 ± 0.005
Perturb2Adv	0.141 ± 0.010	0.123 ± 0.008	0.062 ± 0.006	0.053 ± 0.007
VAEsm $\gamma = 20$	0.157 ± 0.007	0.134 ± 0.008	0.080 ± 0.066	0.066 ± 0.008
VAEafrl*$\lambda = 1.2$	0.137 ± 0.009	0.126 ± 0.010	0.062 ± 0.007	0.054 ± 0.006

We define *user ratio* as the average item demographic ratio of the items a user is recommended and define *User Ratio Difference* as the average absolute difference between the user ratio and the userbase demographic ratio. Table 6 shows the results. Fair models perform better than `VAErec` in both metrics. The ratio metrics mostly align with the AUC performance, with `VAEsm` and `VAEafrl*` achieving the best scores and most other models ordered by their AUC performance. In general, it does not seem as if there is a linear relationship between the AUC improvements and Ratio Difference improvements, as improvements of small AUC scores are reflected in larger Ratio Difference improvements than the improvements seen when improving larger AUC scores equally.

Breaking from the trend, `PerturbVan` performs better than `PerturbRel` and `Perturb2Adv` despite achieving equal or worse AUC scores. The reason for this remains unclear. The key difference between the models is that `PerturbVan` exclusively tries to confuse the adversarial while the others consider user preferences. User preferences derived from the data are correlated with demographic information, so its consideration may influence the rankings. If so, `PerturbVan`'s AUC score must reflect more correlations that do not influence the recommendations. The versions of `VAEsm` and `VAEafrl*` tuned to match the NDCG of `Perturb2Adv` achieve similar or worse Ratio Differences.

A detailed look at how individual Item and User Ratios change between `VAErec` and when using `Perturb2Adv` to filter Gender and Age is shown in Fig. 2 and Fig. 3. The plots show that most items and users with ratios far from the demographic ratios of the userbase see positive improvements, while some close-to-ideal ratios in `VAErec` deteriorate. Said ratios have little or no room for improvement, so minor shifts are often negative and will not move them far from ideal. The discussed trends are especially clear for User Ratios, where much of the 90% confidence interval represents positive improvements, but also in the mean Item Ratio improvements.

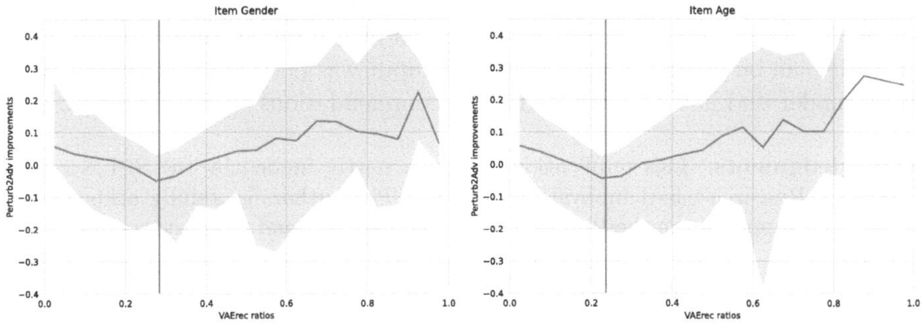

Fig. 2. Plot that shows how *Item Ratios* change between `VAErec` and `Perturb2Adv`. The y-axis shows average improvement towards the demographic composition of the userbase (red line), where *items* have been grouped in 0.05-ratio intervals based on their *Item Ratio* in `VAErec`. The blue band is the 95% confidence interval. (Color figure online)

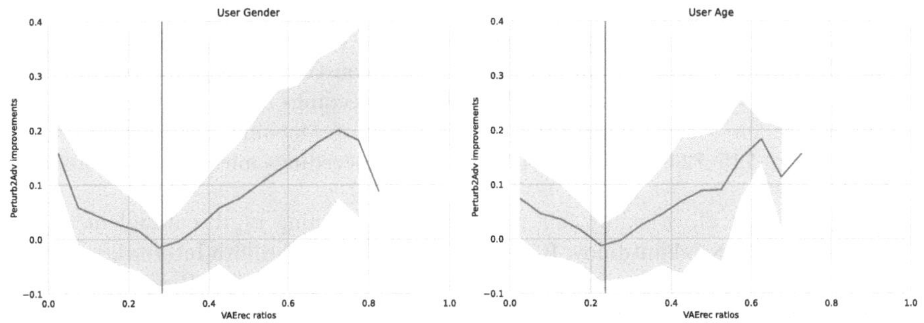

Fig. 3. Plot that shows how *User Ratios* change between `VAErec` and `Perturb2Adv`. The y-axis shows average improvement towards the demographic composition of the userbase (red line), where *users* have been grouped in 0.05-ratio intervals based on their *User Ratio* in `VAErec`. The blue band is the 95% confidence interval. (Color figure online)

6 Conclusion

We propose simple, novel methods that transparently introduce fairness consideration in pre-trained models by perturbing model input to obfuscate user gender and age. Our AUC results show that our methods drastically decrease the demographic information encoded in model representations. While existing complex models can achieve better AUC scores, this always comes at the cost of worse recommendation performance. Our methods achieve similar or better fairness than models tuned to match our recommendation performance. We have further demonstrated that only one existing approach matches ours in not negatively affecting users who do not opt-in for fairness, and our model is alone in being transparent about how fairness has been incorporated. Individual users of our model can be presented with how their input was perturbed to conceal their age or gender and how this affected the recommendations they were given.

Acknowledgments. This publication has been partly funded by the SFI NorwAI, (Centre for Research-based Innovation, 309834). The authors gratefully acknowledge the financial support from the Research Council of Norway and the partners of the SFI NorwAI.

References

1. Boratto, L., Fabbri, F., Fenu, G., Marras, M., Medda, G.: Counterfactual graph augmentation for consumer unfairness mitigation in recommender systems. In: Proceedings of the 32nd ACM International Conference on Information and Knowledge Management, pp. 3753–3757. CIKM '23, Association for Computing Machinery, New York, NY, USA (2023). https://doi.org/10.1145/3583780.3615165
2. Boratto, L., Fabbri, F., Fenu, G., Marras, M., Medda, G.: Fair augmentation for graph collaborative filtering. In: Proceedings of the 18th ACM Conference on Recommender Systems, pp. 158–168. RecSys '24, Association for Computing Machinery, New York, NY, USA (2024). https://doi.org/10.1145/3640457.3688064
3. Bose, A., Hamilton, W.: Compositional fairness constraints for graph embeddings. In: Chaudhuri, K., Salakhutdinov, R. (eds.) Proceedings of the 36th International Conference on Machine Learning. Proceedings of Machine Learning Research, vol. 97, pp. 715–724. PMLR (2019). https://proceedings.mlr.press/v97/bose19a.html
4. Creager, E., et al.: Flexibly fair representation learning by disentanglement. In: Chaudhuri, K., Salakhutdinov, R. (eds.) Proceedings of the 36th International Conference on Machine Learning. Proceedings of Machine Learning Research, vol. 97, pp. 1436–1445. PMLR (2019). https://proceedings.mlr.press/v97/creager19a.html
5. Gajane, P.: On formalizing fairness in prediction with machine learning (2017). http://arxiv.org/abs/1710.03184
6. Ge, Y., et al.: Explainable fairness in recommendation. In: Proceedings of the 45th International ACM SIGIR Conference on Research and Development in Information Retrieval, pp. 681–691. SIGIR '22, Association for Computing Machinery, New York, NY, USA (2022). https://doi.org/10.1145/3477495.3531973
7. Harper, F.M., Konstan, J.A.: The MovieLens datasets: history and context. ACM Trans. Interact. Intell. Syst. **5**(4) (2015). https://doi.org/10.1145/2827872

8. Kingma, D.P., Welling, M.: Auto-Encoding Variational Bayes (2022). https://arxiv.org/abs/1312.6114
9. Li, Y., Chen, H., Xu, S., Ge, Y., Zhang, Y.: Towards personalized fairness based on causal notion. pp. 1054–1063. SIGIR '21, Association for Computing Machinery, New York, NY, USA (2021). https://doi.org/10.1145/3404835.3462966
10. Liang, D., Krishnan, R.G., Hoffman, M.D., Jebara, T.: Variational autoencoders for collaborative filtering. In: Proceedings of the 2018 World Wide Web Conference, pp. 689–698. WWW '18, International World Wide Web Conferences Steering Committee, Republic and Canton of Geneva, CHE (2018). https://doi.org/10.1145/3178876.3186150
11. Melchiorre, A.B., Rekabsaz, N., Parada-Cabaleiro, E., Brandl, S., Lesota, O., Schedl, M.: Investigating gender fairness of recommendation algorithms in the music domain. Inf. Process. Manag. **58**(5), 102666 (2021). https://doi.org/10.1016/j.ipm.2021.102666
12. Slokom, M., Hanjalic, A., Larson, M.: Towards user-oriented privacy for recommender system data: a personalization-based approach to gender obfuscation for user profiles. Inf. Process. Manag. **58**(6), 102722 (2021). https://doi.org/10.1016/j.ipm.2021.102722
13. Vassøy, B., Langseth, H.: Consumer-side fairness in recommender systems: a systematic survey of methods and evaluation. Artif. Intell. Rev. **57**(4), 101 (2024). https://doi.org/10.1007/s10462-023-10663-5
14. Vassøy, B., Langseth, H., Kille, B.: Providing previously unseen users fair recommendations using variational autoencoders. In: Proceedings of the 17th ACM Conference on Recommender Systems, pp. 871–876. RecSys '23, Association for Computing Machinery, New York, NY, USA (2023). https://doi.org/10.1145/3604915.3608842
15. Weinsberg, U., Bhagat, S., Ioannidis, S., Taft, N.: BlurMe: inferring and obfuscating user gender based on ratings. In: Proceedings of the Sixth ACM Conference on Recommender Systems, pp. 195–202. RecSys '12, Association for Computing Machinery, New York, NY, USA (2012). https://doi.org/10.1145/2365952.2365989
16. Zhu, X., Zhang, L., Yang, N.: Adaptive fair representation learning for personalized fairness in recommendations via information alignment. In: Proceedings of the 47th International ACM SIGIR Conference on Research and Development in Information Retrieval, pp. 427–436. SIGIR '24, Association for Computing Machinery, New York, NY, USA (2024). https://doi.org/10.1145/3626772.3657709
17. Zhu, X., Zhang, L., Yang, N.: Adaptive fair representation learning for personalized fairness in recommendations via information alignment. In: Proceedings of the 47th International ACM SIGIR Conference on Research and Development in Information Retrieval, pp. 427–436. SIGIR '24, Association for Computing Machinery, New York, NY, USA (2024). https://doi.org/10.1145/3626772.3657709

Malevolence Attacks Against Pretrained Dialogue Models

Pengjie Ren[1], Ruiqi Li[1], Zhaochun Ren[2], Zhumin Chen[1], Maarten de Rijke[3], and Yangjun Zhang[4](✉)

[1] Shandong University, Qingdao, China
{renpengjie,chenzhumin}@sdu.edu.cn, 202215118@mail.sdu.edu.cn
[2] Leiden University, Leiden, The Netherlands
z.ren@liacs.leidenuniv.nl
[3] University of Amsterdam, Amsterdam, The Netherlands
m.derijke@uva.nl
[4] Queen Mary University of London, London, UK
yangjun.zhang@qmul.ac.uk

Abstract. Pretrained language models (PLMs) are widely used by dialogue systems to generate high-quality responses. Adversarial samples can cause pretrained dialogue models to generate unexpected responses that influence the deployment of the system since malevolent responses can cause the breakdown of the dialogue. Exploring adversarial attacks is important to understand hidden risks and improve victim models. We propose the dialogue malevolence attack task and formulate it as a constrained Markov decision process. We also propose a two-stage reinforcement learning framework to find vulnerabilities of dialogue systems based on pretrained language models (PLMs). Experiments show that the proposed attack framework can effectively attack pretrained dialogue models with high-quality adversarial samples. **Warning**: this paper contains examples that may be offensive or upsetting.

Keywords: Adversarial attack · Dialogue system · Large language model

1 Introduction

Pretrained language models (PLMs) are increasingly employed by dialogue systems, and in particular, by conversational information retrieval systems [11] as they can generate high quality responses [2,50]. These models are pretrained on large-scale datasets that may contain negative views, bias, and other unsafe content [4,31]. The generation process is hard to control. Hence, current pretrained dialogue models are vulnerable to adversarial samples [18,32]. As current models may not truly understand human language [42], it is important to explore adversarial samples and adversarial attacks to reveal the safety weaknesses [34,35] of dialogue responses generated by victim pretrained dialogue models.

Malevolent dialogue responses that may contain offensive or objectionable content [49], have drawn broad attention. Malevolent responses may result in serious consequences for users, including dialogue breakdown, discomfort, and abandonment of

C. Hauff et al. (Eds.): ECIR 2025, LNCS 15572, pp. 372–388, 2025.
https://doi.org/10.1007/978-3-031-88708-6_24

the system [49]. E.g., for DialoGPT [50], a state-of-the-art dialogue model, around 10% of its responses are malevolent [48].

Table 1. DMA example.

Context	Turn 1: And before any funny prick decides to defend their own actions in my mentions i don't want to hear it. Turn 2: Lmao, you're just an absolute twat.	
Original	Turn 3: Why are we twats for having an opinion?	
Adversarial	Turn 3: Why do I always have stupid opinions?	
	Non-malevolent	Malevolent
Response	Because you're not funny.	Because you're stupid.

We present the *dialogue malevolence attack* (DMA), an adversarial attack task that we formulate as a constrained Markov decision process (CMDP) problem, as shown in Table 1. To address the dialogue malevolence attack task, we propose the *iteRative malevOlence atTAcker wiTh constraIned markOv decisioN* (ROTATION) model, illustrated in Fig. 1, which modifies original utterances into adversarial ones that provoke victim models to generate malevolent responses. The *iterative attacker* (IA) for editing original utterances uses four operations: "K," "R," "D," and "I," for keeping, replacing, deleting, and inserting a token, respectively. Unlike existing methods, IA considers correlations among edits and can iteratively edit utterances until convergence. IA training employs reinforcement learning to interact with victim models. This raises two challenges. First, the editing reward for each position in the utterance is unknown; we learn the malevolence reward from the malevolence feedback of the whole utterance with a *reward evaluator* (RE). Second, the attack efficiency is low due to the computationally intensive nature of pretrained dialogue models; we devise a two-stage reinforcement learning strategy that includes an offline warm-up stage and an online fine-tuning stage to train iteRative malevOlence atTAcker wiTh constraIned markOv decisioN (ROTATION). During the offline stage, we warm up the parameters of the iterative attacker and reward evaluator networks with a pseudo trajectory generation process. During online learning, we use the iterative attacker to directly attack the victim models and update the iterative attacker through the reward evaluator.

We conduct experiments on two datasets: (i) the malevolent dialogue response detection and classification (MDRDC) dataset [49] with malevolence labels, and (ii) an unlabelled dialogue dataset that has been collected from Twitter and covers a diverse set of topics. Our experiments show that ROTATION can generate high-quality and more diverse adversarial samples that outperform the best baseline by 126.90% and 56.50% in terms of distinct attack success rate (DASR). We also conduct a human evaluation for ROTATION on the MDRDC dataset, the DASR outperforms the best baseline by 56.14%.

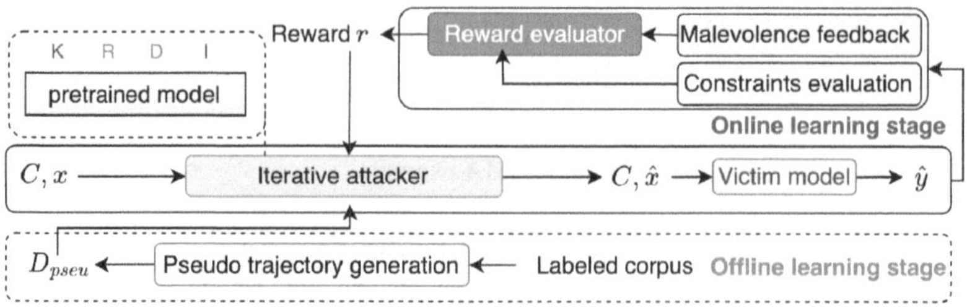

Fig. 1. Model architecture and learning process of the ROTATION.

2 Related Work

Adversarial Attacks. Adversarial attacks intentionally modify the input of a model with perturbations and use the modified inputs to change the outputs of victim models; they can be used to find vulnerabilities of victim models [25,28,45,47]. They come in two types: non-targeted and targeted. Non-targeted attacks attempt to trick victim models into producing wrong outputs without specific constraints on the wrong output [3,17,18,27,51]; e.g., for a classification model, the attacker tries to misguide it to predict any one of the incorrect classes [24,30]; for a generation model, the attacker tries to misguide it to generate any type unpredicted content [23,43]. Prior work on non-targeted attacks either does not consider the semantic consistency and coherence of the adversarial samples or does not target dialogue models. Targeted attacks aim to trick victim models into producing specific wrong outputs, which is more challenging than non-targeted attacks [6,22]. For existing methods on targeted attacks, the number of target words or responses is too small for the dialogue malevolence attack (DMA) task since the space of malevolent content is large. In addition, they edit the tokens of the original input only once without considering multiple edits.

Safety of Pretrained Language Models. Pretrained language models may generate content that is not consistent with humanistic values [1,37]. [35] build a dataset to assess the safety of pretrained dialogue models and conclude that they all have risks of generating unsafe responses. [14] build a dialogue dataset; models pretrained on it may generate socially prejudiced responses. To understand what kind of inputs could increase unsafe generation, previous work focuses on input-agnostic attacks and input-specific attacks [20].

Input-agnostic attacks aim to fool victim models by adding learned universal noise, e.g., a sequence of tokens [12,41,44], to any input or by directly generating adversarial samples without considering the input [29,34]. ToxicBuddy [34] first generates an auxiliary dataset that has a high toxicity score on pretrained dialogue models, and then augments this to attack the pretrained dialogue models to generate more toxic responses. The input generated by ToxicBuddy for attacking a victim model is limited as it focuses on politics, which makes an attack relatively easy to detect as it is often neither coherent with the context nor consistent with the current utterance. ToxicBuddy does not interact with a target dialogue system; it is built for general attacks.

Input-specific attacks generate input-specific adversarial samples by making small changes to the original input [19,22,33]. Limited studies have been done for input-specific attacks because the sparse nature of input-specific attacks makes it hard to learn a model [5] and targeted input-specific attacks for safety-related aspects are easy to overfit without suitable feedback to evaluate the attack output of the victim model [8].

Our work differs from prior work in three ways: (i) We make the adversarial samples imperceptible by formulating dialogue malevolence attack as a constrained Markov decision process and modeling semantic consistency, fluency, and non-malevolency of adversarial samples as constraints. (ii) Most previous attack methods only demonstrate effectiveness on specific domains, where domain knowledge dominates the toxic behavior of dialogue systems. (iii) Our method generates adversarial samples by iteratively editing the input, and considers correlations between edits and the coherence with the dialogue context.

3 Methodology

3.1 Dialogue Malevolence Attack Definition

Given a dialogue context C, current user utterance x, and victim model g, which is a pretrained dialogue model, *dialogue malevolence attack* (DMA) aims to learn an attack model f to perform a perturbation on x to obtain an adversarial version $\hat{x} = f(C, x)$ such that the g generates a malevolent response $\hat{y} = g(C, \hat{x})$ and satisfies these constraints: (i) $\hat{x}$ is consistent with x; (ii) $\hat{x}$ is fluent; and (iii) the attack model f does not introduce extra malevolence for $\hat{x}$ compared to x.

3.2 Constrained Markov Decision Process

We formulate dialogue malevolence attack as a constrained Markov decision process in which the iterative attacker tries to attack the victim model to generate targeted responses that maximize the malevolence reward and do not violate the constraints. The dialogue malevolence attack framework is shown in Fig. 1. The constrained Markov decision process is defined as a tuple $\langle S, A, P, r_\psi, c_{1:3}, \alpha_{1:3}, \gamma \rangle$ where: (i) S is the state space, $s_t \in S$ is the current state that consists of the dialogue context and current user utterance, i.e., $s_t = (C, \hat{x}_t)$ where t is the number of times the original utterance x has been edited and $\hat{x}_t$ is an adversarial sample in t-th step; $\hat{x}_0 = x$. (ii) A is the action space, $a_t \in A$ is a sequence of editing operations that are independent of each other, i.e., $a_t = [a_{t,1}, a_{t,2}, \ldots, a_{t,n}]$, where n is the length of $\hat{x}_t$, $a_{t,j}$ is an editing operation for the j-th token of $\hat{x}_t$. (iii) $P : S \times A \times S \to [0, 1]$ is the transition function mapping a triple (s_t, a_t, s_{t+1}) to a probability by $P(s_t, a_t, s_{t+1}) = p(s_{t+1} \mid s_t, a_t)$, where $p(s_{t+1} \mid s_t, a_t)$ is the probability of transitioning from s_t to s_{t+1} under action a_t at time step t. The transitions are deterministic, once action a_t is selected, the next state s_{t+1} is known. (iv) r_ψ is the reward function and c_i is the i-th constraint function. And (v) γ is the discount factor $\in [0, 1]$.

The goal is to maximize the total expected malevolence reward while ensuring the constraints are under thresholds:

$$\max_{\theta} J_{r_\psi}(\pi_\theta) = \mathbb{E}\left[\sum_{t=0}^{\infty} \gamma^t \left(r_\psi(C, \hat{x}_t, a_t)\right)\right], \tag{1}$$

$$\text{subject to} \begin{cases} J_{c_1}(\pi_\theta) = \mathbb{E}\big[\sum_{t=0}^{\infty} \gamma^t \left(c_1(x, \hat{x}_t)\right)\big] \geq \alpha_1, \\ J_{c_2}(\pi_\theta) = \mathbb{E}\big[\sum_{t=0}^{\infty} \gamma^t \left(c_2(x, \hat{x}_t)\right)\big] \leq \alpha_2, \\ J_{c_3}(\pi_\theta) = \mathbb{E}\big[\sum_{t=0}^{\infty} \gamma^t \left(c_3(x, \hat{x}_t)\right)\big] \leq \alpha_3, \end{cases} \tag{2}$$

where π_θ is the policy network, which is implemented by the iterative attacker; θ is the parameter of iterative attacker; r_ψ is the reward function, which is implemented by the reward evaluator; ψ are the parameters of reward evaluator; c_1, c_2, c_3 are the consistency, fluency, and non-malevolency cost functions, respectively; $\alpha_1, \alpha_2, \alpha_3$ are the pre-defined thresholds for the cost functions. The sum is over an infinite number of time steps, but we use finite horizons for t in practice.

The constraints are explained as follows: (i) $J_{c_1}(\pi_\theta)$ is the consistency constraint. To ensure that adversarial sample $\hat{x}_t$ is consistent with x, we embed x and $\hat{x}_t$ with a BERT model and use the cosine similarity between the BERT embeddings as the consistency score. (ii) $J_{c_2}(\pi_\theta)$ is the fluency constraint. We penalize actions leading to non-fluent outputs using the perplexity score [50] to measure the fluency of $\hat{x}_t$ by $-\frac{1}{n}\sum_{j=1}^{n}\log(P(w_j|w_{j-1}, w_{j-2}, \ldots, w_1))$, where n is the number of tokens in $\hat{x}_t$ and w_j is the j-th token. (iii) $J_{c_3}(\pi_\theta)$ is the non-malevolency constraint. To ensure that adversarial sample $\hat{x}_t$ does not introduce extra malevolence, we use the change of malevolence confidence score between x and $\hat{x}_t$ as the non-malevolency score, i.e., $S(\hat{x}_t) - S(x)$, where S is malevolence confidence score from a binary BERT-based malevolence classifier [49].

We apply Lagrange relaxation to approximate the constrained optimization problem [46] and aim to solve the following problem:

$$\min_{\lambda \geq 0} \max_{\theta} J_{r_\psi}(\pi_\theta) - \lambda_1(\alpha_1 - J_{c_1}(\pi_\theta)) - \lambda_2(J_{c_2}(\pi_\theta) - \alpha_2) - \lambda_3(J_{c_3}(\pi_\theta) - \alpha_3), \tag{3}$$

where $\lambda_1, \lambda_2, \lambda_3$ are the Lagarangian multipliers.

3.3 Iterative Attacker

Unlike previous work, which divides sentence editing into token position prediction and operation prediction [26], our approach merges them into one with four editing operations:"K" for keeping a token, "R" for replacing, "D" for deleting, and "I" for inserting. We build an editing policy network, i.e., iterative attacker (IA), π_θ, based on BERT to generate a sequence of operations. The iterative attacker takes $(C, \hat{x}_t)$ as input and uses separator "[SEP]" after each utterance except the last one. We feed tokens into

BERT to output representations H_t in the original utterance and apply a linear layer with parameter W_θ and b_θ to predict action a_t:

$$\pi_\theta(a_t|s_t = (C, \hat{x}_t)) = \mathrm{softmax}(W_\theta H_t + b_\theta), \tag{4}$$

where W_θ and b_θ are the parameters to be learned during the training process.

3.4 Reward Evaluator

To solve the constrained Markov decision process problem and train iterative attacker with reinforcement learning, we need to get the reward for each editing operation, which is unknown. We only know the malevolence feedback of the final edited utterance $\hat{x}_t$, where the iterative attacker predicts "K" (keep($\hat{x}_t[j]$)) for all tokens of $\hat{x}_t$. Therefore, we learn a reward evaluator r_ψ parameterized by ψ to estimate the reward for each token. We take the hidden state H_t of BERT as input and feed them into the MLP to output a reward for each editing operation of the corresponding token position:

$$r_\psi(r|s = (C, \hat{x}_t), a = a_t) = \mathrm{sigmoid}(W_\psi H_t + b_\psi), \tag{5}$$

where W_ψ and b_ψ are the parameters to be learned during the training process.

3.5 Two-Stage Learning

To speed up training, we warm up the iterative attacker and reward evaluator with an offline learning stage by making use of the existing MDRDC corpus [49]. First, we train iterative attacker, reward evaluator, and λ with trajectories generated by a pseudo trajectory generation process. The parameters of iterative attacker and reward evaluator are passed to the online learning stage. Then, we use the iterative attacker to attack victim models and sample trajectories generated by the iterative attacker to update iterative attacker, reward evaluator, and λ with an online learning stage. To train the reward evaluator, its gradient is given by:

$$\nabla_\psi L(\lambda, \theta, \psi) = -\mathbb{E}_{\tau \in D^+}\left[\nabla_\psi r_\psi(\tau)\right] + \mathbb{E}_{\tau \in D^-}\left[\nabla_\psi r_\psi(\tau)\right], \tag{6}$$

where τ is a trajectory of state-action pairs from $\hat{x}_0$ to $\hat{x}_t$, i.e., $[(C, \hat{x}_0), a_0, (C, \hat{x}_1), a_1, \ldots, (C, \hat{x}_t), a_t, \ldots]$. We use pseudo trajectories as D^+ and D^- in the offline learning stage and use actual trajectories as D^+ and D^- in the online learning stage [10,52]. In practice, we treat malevolent trajectories as D^+ and non-malevolency trajectories as D^-. Reward training by maximizing the distance between positive and negative trajectories may not provide enough guidance for the learning process. Therefore, we introduce a margin as an additional reward to better guide the editor in attacking dialogue models and subtract a minor offset δ from the rewards of all negative instances [13,15,39]. The gradient of the iterative attacker and λ parameters are:

$$\begin{aligned}\nabla_\theta L(\lambda, \theta, \psi) &= \mathbb{E}_{\tau \in D}(r_\psi(C, \hat{x}_t, a_t) + \lambda_1 c_1(x, \hat{x}_t) - \lambda_2 c_2(x, \hat{x}_t) \\ &\qquad - \lambda_3 c_3(x, \hat{x}_t)) \nabla_\theta \log \pi_\theta(C, \hat{x}_t, a_t), \\ \nabla_\lambda L(\lambda, \theta, \psi) &= -\,\mathbb{E}_{\tau \in D}([-c_1(x, \hat{x}_t), c_2(x, \hat{x}_t), c_3(x, \hat{x}_t)]) + [-\alpha_1, \alpha_2, \alpha_3],\end{aligned} \tag{7}$$

Alg. 1: Pseudo trajectory generation

Input: Corpus $\{(C, \hat{x})\}$; M: maximum iteration editing steps; $\tilde{A}$: generation rule containing operations; $f_{mlm}(x)$: masked language model;
Output: Pseudo trajectories D_{pseu}.
for $(C, \hat{x}) \in Corpus$ **do**
    $m \; in \; [1, M]$;
    $\hat{x}_0 \leftarrow \hat{x}$, $\tau' \leftarrow \{\hat{x}_0\}$, $\tau \leftarrow \{\}$;
    for *step* $t' = \{1, \ldots, m\}$ **do**
        Randomly sample $a_{t'} \sim \tilde{A}$;
        $\hat{x}_{t'+1} = f_{mlm}(\hat{x}_{t'}, a_{t'})$;
        $\tau' \leftarrow \tau' \cup \{(C, a_{t'}), \hat{x}_{t'+1}\}$;
    for *step* $t' = \{m-1, \ldots, 0\}$ **do**
        $\tau \leftarrow \tau \cup \{(C, \hat{x}_{t'+1}), \Gamma(a_{t'})\}$
    $\tau \leftarrow \tau \cup \{(C, \hat{x}_0)\}$;
    D_{pseu}.append(τ);

where $D = D^+ \cup D^-$. In both stages, the parameters of iterative attacker, reward evaluator and λ are updated as:

$$\begin{aligned} \theta_{k+1} &= \theta_k + \eta_1 \nabla_\theta L(\lambda_k, \theta_k, \psi_k), \\ \psi_{k+1} &= \psi_k + \eta_2 \nabla_\psi L(\lambda_k, \theta_k, \psi_k), \\ \lambda_{k+1} &= \lambda_k - \eta_3 \nabla_\lambda L(\lambda_k, \theta_k, \psi_k), \end{aligned} \tag{8}$$

where η_1, η_2, η_3 are learning rates.

In the offline learning stage, we use a pseudo trajectory generation (PTG) process to generate pseudo trajectories D_{pseu}. D_{pseu} and a malevolence feedback signal, are used to train the parameters of reward evaluator, iterative attacker, and λ. The pseudo trajectory generation process is shown in Algorithm 1. Γ denotes replacing each editing operation $a_{t'}$ with its opposite editing operation, e.g., if $a_{t'}$ is "D", $\Gamma(a_{t'})$ becomes "I". To improve the policy network training efficiency, we select a subset of available trajectories [21]. We vary the ratio of the number of malevolent and non-malevolent trajectories to achieve optimal performance; the effect of this ratio is discussed in Sect. 5.5. To train reward evaluator, the gradient of the reward evaluator is estimated by Eq. 6 except that we use pseudo trajectories D_{pseu} to replace D^+ and D^-.

In the online learning stage, we train iterative attacker and reward evaluator based on the parameters learned in the offline stage and re-initialize λ. Similar to offline training, we use reward evaluator to assign the reward for each editing operation and calculate malevolence feedback. The difference is that we sample actual trajectories D_{actu} during the online attacking process from the iterative attacker.

4 Experimental Setup

4.1 Datasets and Baselines

We use two datasets: (i) the **MDRDC Dataset** has binary labels of malevolence and comprises 6,000 dialogues, encompassing a total of 28,510 utterances [49], which is an open-sourced dataset; and (ii) the **Unlabeled Dialogue Dataset** includes 6,000 unlabeled dialogues from Twitter for a total of 27,143 utterances, which is collected following Twitter policies.

We choose DialoGPT as the victim model. DialoGPT is a state-of-the-art large-scale pretrained dialogue response generation model for multi-turn conversations. To assess the potential impact of dialogue malevolence attack on larger pretrained models, we also test the performance of the DMA method that is trained with DialoGPT, on LLaMA-based [38] chatbots, i.e., Alpaca [36] and Vicuna [7]. We choose Vicuna-7B and Alpaca-7B as victim models.

We have three baseline models: (i) A naive word replacement baseline that randomly replaces words in the text with a dictionary of malevolent words[1] including "hate" and "offensive" which belong to the second-level malevolence categories of the MDRDC dataset [9,49]. We replace [1, 2, 3] positions in the text for all the datasets and 3 yields the best results. (ii) TDGPN [22], a reinforcement learning (RL)-based method to generate utterances that contain target curse words. To make it comparable with our model, we use all words in [9] as target words. (iii) ToxicBuddy [34], a benchmark for attacking dialogue systems to generate toxic responses. For a fair comparison, the model is fine-tuned with MDRDC.

4.2 Implementation Details

We implement Iterative attacker based on BERT with a vocabulary of 30,522 tokens. We set the max sequence length to 256, the hidden size to 768, and the edit vocabulary to 2,000. We use a two-layer MLP with hidden sizes [768, 768] for the modeling of the Reward evaluator module. α is set to [1.0, 5.834, 0.0] to ensure consistency with the original samples. We use the Adam optimizer to train all models. We use greedy decoding strategy for both ROTATION and ToxicBuddy, except in cases involving multiple attacks, and we use sampling decoding strategy for TDGPN. The models are trained with a maximum of 40 epochs on GeForce RTX 3090 GPUs.

4.3 Evaluation

We conduct both automatic and human evaluations to evaluate ROTATION. First, automatic evaluation concerns attack success rate, attack sample quality, and attack efficiency. Second, human evaluation concerns the former two dimensions.

For attack success rate, we report *redundant attack success rate* (RASR) and *distinct attack success rate* (DASR). RASR follows previous research [16,40], and DASR is used since the models tend to modify the original utterances into the same text and generate the same responses, which is effective but perceptible. redundant attack success rate (RASR) can not reflect the effectiveness considering duplication. RASR is defined as RASR $= \frac{N^{suc}}{N}$, where N^{suc} is the number of adversarial samples that provoke malevolent responses and N is the number of adversarial samples. We use a binary classifier trained on the MDRDC dataset to identify the malevolent response. We also

[1] https://github.com/t-davidson/hate-speech-and-offensive-language.

propose *distinct attack success rate* (DASR), which is the attack success rate with duplicate adversarial samples corrected: DASR $= \frac{N^{cor}}{N}$.

For attack sample quality, we report the results of six metrics. The consistency, fluency, and non-malevolency follow Eq. 2. Distinct-1 and distinct-2 are used to define the diversity of the response. Coherency is used to evaluate whether the generated adversarial sample is coherent with the original utterance.

For attack efficiency, we report *Attack Times* (AT) and attack times (AT)-metrics curves. For each time of attack, different methods input queries to the model. With more attack times, the attack success rate increases. Attack times at a targeted attack success rate threshold (RASR reaches 95%) and the AT-metrics can evaluate the efficiency of the multiple attacks. For a single attack, we iteratively edit utterances at most M steps and feed the adversarial samples in the last step into a victim model. For multiple attacks, we edit the failed adversarial sample to get a new adversarial sample and leave the successful one unchanged. We continue this process till the RASR reaches 95%. Then, we calculate the number of times we feed the adversarial samples to a victim model.

We designed a questionnaire for the human evaluation for the MDRDC test dataset. First, we ask workers to label if the current user utterances are malevolent. Second, we ask workers to label if the attacking samples are malevolent, fluent, and consistent with the current user utterance. Third, we ask workers to label whether the generated response from the chatbot system is malevolent. We randomly choose 500 samples from the ToxicBuddy and ROTATION for labeling. All questionnaire answers are binary. In terms of quality control standards, we use a minimum of 500 HITs and a 97% approval rate for workers in Amazon MTurk. The calculation of human evaluation metrics is similar to automatic metrics.

5 Results and Analyses

5.1 Overall Performance

Performance for a Single Attack. As described in Sect. 4.3, a single attack denotes that we generate an adversarial sample once with at most M editing steps. The results are shown in Table 2.

First, across the two datasets, the DASR of ROTATION, which is trained with both offline and online stages, is the highest. The RASR score of ToxicBuddy is higher than that of ROTATION because ToxicBuddy tends to attack the pretrained model with duplicate adversarial samples. The higher DASR score indicates that ROTATION successfully attacks the pretrained model and outperforms baselines on attack success rate with lower duplication.

Second, across the two datasets, the quality metrics of ROTATION are comparable to the three baseline methods. On the MDRDC dataset and the unlabeled dialogue dataset, ROTATION outperforms all other methods in terms of consistency, non-malevolency, and distinct-1. ROTATION achieves the second-best results in terms of fluency and distinct-2. The reason why the fluency of ToxicBuddy is higher than that of ROTATION is that the fluency of the original dialogue utterance is high and ROTATION tends to satisfy constraints rather than dramatically improving the fluency scores

Table 2. Two-stage single attack results. **Bold** denotes the best results except for the original utterances as input. Original denotes attacking with original utterances. Underline denotes the best results among all methods. * denotes statistically significant improvement compared to all the baselines based on the t-test ($p < 0.05$).

Dataset	Method	DASR	RASR	Con	Flu	Non-mal	Dist-1	Dist-2	Coh
MDRDC	Original	9.18	9.24	1.0000	5.834	0.0000	0.0832	0.5091	0.7136
	Random	10.40	10.44	0.7199	7.560	0.3836	0.0738	**0.5513**	**0.7052**
	TDGPN	9.96	12.25	0.4840	6.576	0.3067	0.0762	0.3814	0.3930
	ToxicBuddy	12.53	**42.52**	0.5018	**4.701**	0.3711	0.0300	0.1356	0.6779
	ROTATION	**28.43***	29.38	**0.7873***	5.787	**0.2852***	**0.0770**	0.4211	0.6503
Unlabeled	Original	8.58	8.65	1.0000	5.281	0.0000	0.0996	0.5589	0.7225
	Random	9.90	9.93	0.7145	7.247	0.4084	0.0874	**0.5890**	**0.7169**
	TDGPN	10.82	14.47	0.4675	5.441	0.2852	0.0904	0.3997	0.4004
	ToxicBuddy	11.31	**43.59**	0.4925	**3.714**	0.3614	0.0392	0.1682	0.7031
	ROTATION	**17.70***	18.82	**0.7687***	5.331	**0.1377***	**0.0942**	0.4939	0.6663

Table 3. Two-stage multiple attack results on the MDRDC dataset when the attack time is 18. Bold face: the best performance.

Method	DASR	RASR	Con	Flu	Non-mal	Dist-1	Dist-2	Coh
Random	48.79	49.96	0.6016	8.122	0.6325	0.0545	0.4163	**0.6441**
TDGPN	60.46	78.48	0.4935	6.441	**0.4398**	**0.0629**	0.3575	0.4068
ToxicBuddy	78.25	81.07	0.5112	**5.768**	0.4623	0.0401	0.3260	0.3835
ROTATION	**93.87**	**95.22**	**0.6197**	6.884	0.5501	0.0457	**0.4465**	0.5243

of generated adversarial samples since we set the fluency threshold α_2 to 5.834, which is the fluency score of attacking with original utterances. The reason that the coherency of Random is higher than that of the ROTATION is that the GRADE score is a keyword-based metric and Random would add new keywords into the original utterance. The results indicate that the quality of adversarial samples from ROTATION drops a bit with regard to the original utterances, but the fluency is better than the original utterances. The reason why the quality metrics of original utterances are slightly superior to ROTATION is that ROTATION is set to satisfy all the constraints with the threshold settings with the score of original utterances, and during training the model reaches a balance of attack success rate and adversarial sample quality.

Performance for Multiple Attacks. To evaluate the attacking efficiency, we attack the victim model multiple times on the MDRDC dataset and end the attack when the RASR of ROTATION achieves 95% where AT $=$ 18 to investigate how the attack performance changes with increasing attack times. For all models, we use a sampling decoding strategy. The results are shown in Fig. 2 and Table 3.

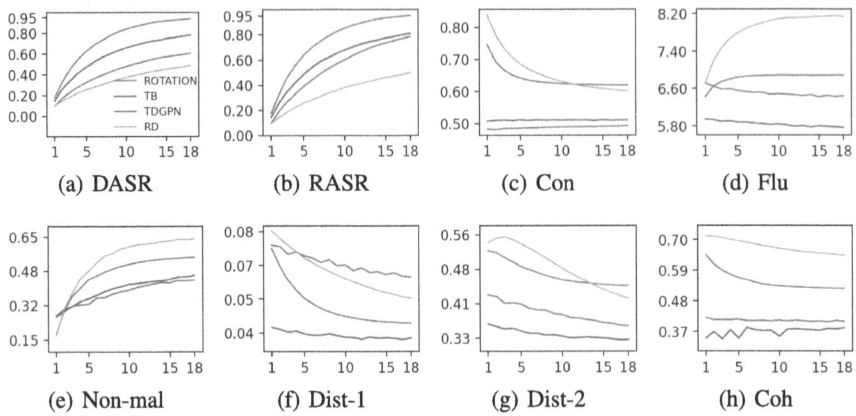

Fig. 2. Performance of TDGPN, ToxicBuddy (TB), Random (RD), and ROTATION (with two-stage learning) over different attack times (x-axis) on the MDRDC dataset. For Flu, lower values are better.

ROTATION has better attack efficiency than ToxicBuddy and TDGPN with acceptable quality. First, ROTATION consistently achieves higher attack success rates at all levels of attack times. The improvement of ROTATION in terms of attack quality and efficiency can be attributed to the consideration of the correlation among edits and the use of constraints to enhance the overall quality.

5.2 Ablation Studies

Ablation Analyses of Two-Stage Learning. To evaluate the impact of two-stage learning, we report the results of ROTATION with only offline learning or only online learning and compare them with ROTATION in Table 4.

First, the results indicate that incorporating online learning benefits both DASR and RASR, while the results of other metrics indicate that incorporating online learning slightly decreases the quality of adversarial samples, but the decrease is acceptable. Compared with ROTATION with two-stage learning, removing online learning decreases the DASR and RASR by 57.09% and 58.07%, respectively; the consistency, fluency, non-malevolency, dictinct-2 and coherency increase by 5.35%, 5.50%, 8.41% and 4.20%, respectively; non-malevolency and distinct-1 decrease by 56.31% and 1.17%, respectively. Second, the results show that incorporating offline learning benefits both DASR and RASR, as well as the quality of adversarial samples. Removing offline learning decreases the DASR and RASR by 80.48% and 81.11%; the fluency increases by 52.86%; consistency, non-malevolency, distinct-1, dictinct-2, coherency decrease by 32.82%, 32.64%, 71.95%, 21.11%, and 29.51%, respectively.

5.3 Attacking Larger Victim Models

We also conduct experiments to see if ROTATION trained by attacking DialoGPT could help to improve the success rate when attacking larger victim models. To this end,

Table 4. Ablation studies of two-stage learning on the MDRDC dataset. Boldface and underlining indicate the best results in each group. For Flu, lower values are better.

Ablation setting	DASR	RASR	Con	Flu	Non-mal	Dist-1	Dist-2	Coh
ROTATION	**28.43**	**29.38**	0.7873	**5.787**	0.2852	**0.0770**	0.4211	0.6503
- online	12.20	12.32	**0.8294**	6.105	**0.1246**	0.0761	**0.4565**	**0.6776**
- offline	5.55	5.55	0.5289	8.846	0.1921	0.0216	0.3322	0.4584

we compare ROTATION trained by attacking DialoGPT with the original utterances when attacking two LLaMA-based victim models, i.e., Alpaca and Vicuna. The results are listed in Table 5. Compared with attacking DialoGPT, the inferior performance of ROTATION on Alpaca and Vicuna can be attributed to two factors: the significant disparity in parameter sizes and the absence of training specifically on Alpaca and Vicuna. The parameter size of Alpaca and Vicuna is ten times larger than that of DialoGPT, resulting in increased complexity and potential difficulties in executing attacks.

Table 5. Attacking Alpaca and Vicuna for MDRDC dataset. ROTATION uses two-stage learning and the quality metrics are the same as in Table 2

Metrics	DASR		RASR	
Model	Alpaca	Vicuna	Alpaca	Vicuna
Original	10.35	12.96	10.35	13.03
ROTATION	14.58	14.40	14.77	14.91

Table 6. Human evaluation results on the MDRDC dataset. "TB" is short for ToxicBuddy; "ROT" is short for "ROTATION."

Method	TB	ROT
DASR-h	0.228	**0.356**
RASR-h	**0.464**	0.358
Consistency	0.396	**0.634**
Fluency	0.572	**0.624**
Non-malevolency	0.278	**0.260**

5.4 Human Evaluation

We conduct a human evaluation to evaluate the performance of ROTATION, and we report the results in Table 6. The results indicate that the ROTATION method has better performance in attacking the pretrained dialogue model to generate malevolent responses, and ensures the quality of adversarial samples at the same time. Compared with ToxicBuddy, the DASR, consistency, and fluency of ROTATION increase by 56.14%, 60.10%, and 9.09% respectively and RASR and non-malevolency decreases by 22.84% and 6.47% respectively. The trend is similar to the automatic results in Table 2.

5.5 Hyperparameter Analyses

Effect of Maximum Iteration Editing Steps. To study the influence of maximum number of iteration editing steps M, we report the results of different values of M for

ROTATION. First, when M increases, the attack success rate first increases and then decreases. When $M = 2$, the model gets the best attack success rate. With overly large M, the pseudo trajectory generation module excessively edits the original dialogue utterance and decreases the attack success rate; with smaller M, editing is not enough to achieve successful attacks. Second, when M increases, the quality first decreases and then increases. The reason is that the model maintains a trade-off between the attack success rate of adversarial attacks and the quality of the generated adversarial samples. As the attack success rate increases, the quality of adversarial samples tends to decrease.

Table 7. Results of ROTATION on the MDRDC dataset (with margin).

	DASR	RASR	Con	Flu	Non-mal	Dist-1	Dist-2	Coh
1:0	28.43	29.38	0.7873	5.787	0.2852	**0.0770**	0.4211	0.6503
1:0.01	**31.09**	**33.84**	0.7737	5.674	0.3228	**0.0770**	0.4279	0.6411
1:0.1	28.40	30.46	0.7686	**5.562**	0.3106	0.0769	0.4143	0.6410
1:1	9.98	10.03	**0.8124**	6.720	**0.0620**	0.0765	**0.4600**	**0.6764**

Effect of the Ratio of Malevolent and Non-malevolent Trajectories. We modify the ratio of malevolent and non-malevolent trajectories ($D^+ : D^-$) during training to explore how it affects performance. First, as the ratio decreases, the attack success rate decreases. Non-malevolent trajectories are not very helpful in demonstrating how to successfully attack the victim models. Second, with a decrease in the ratio, the consistency of adversarial samples increases, and the non-malevolency of adversarial samples decreases. Compared with a ratio of "1:0", training with a ratio of "1:0.01", "1:0.01", and "1:1" has a better quality of adversarial samples. Non-malevolent trajectories may exhibit lower rewards for most editing operations, thereby pushing the optimization process towards satisfying constraints first.

Effect of Adding Margin Score to the Reward. Since the ratio of malevolent and non-malevolent trajectories with "1:0" has the best attack success rate, we assume that non-malevolent trajectories are not very helpful in training; the reward maximizing the distance between positive and negative trajectories does not provide perfect guidance for the training process. To improve the reward, we add reward margins to the reward function of positive and negative trajectories. We consider [0.15, 0.2, 0.25] for the offset δ and choose the best result. Table 7 shows the results of adding a margin to the reward. It improves the attack success rate of ROTATION by providing better guidance for the non-malevolent trajectories. First, with a margin added, compared with ratio "1:0", the ratios of "1:0.01" and "1:0.1" achieve better DASR, RASR, indicating enhanced dialogue malevolence attack ability. Compared with the best result of no margin (ratio "1:0"), adding the margin score improves the results. The improvement in attack success rate comes with a decline in consistency, non-malevolency, and coherency; fluency, distinct-1, and distinct-2 do not decline.

6 Conclusion and Future Work

We have introduced the dialogue malevolence attack task, formulated it as a constrained Markov decision process problem, and proposed ROTATION, a pseudo trajectory generation-based two-stage reinforcement framework to identify pretrained dialogue model vulnerabilities and generate high-quality adversarial samples. ROTATION uses an iterative attacker considering the correlation between edits. The effectiveness and imperceptible adversarial sample generation ability of our method are confirmed through extensive experiments and analyses.

Our work confirms that a proportion of utterances with small perturbations would trigger malevolent responses that may cause great harm to users. The attack model helps understand the vulnerability of victim models and provides a basis for developing defense mechanisms. The goal of our work is not to attack existing systems but to discover the possible malevolent content generation risks of pretrained dialogue models and provide improvement hints for the deployment of safe dialogue system applications. This work can be used to discover potential malevolence risk instances; these instances can be used to enhance the training effect of the malevolent dialogue detection model and to detect whether a model reply is safe in the actual system. Nevertheless, the outcomes of this work could be misused to produce malevolent content in online dialogue systems and harm users. We believe the risks are outweighed by the benefits as there have been similar studies in the past [22,34].

In the future, we plan to attack victim models with evaluation based on more fine-grained classifiers. We will also work on the creation of defense mechanisms to fix the vulnerability of the victim models.

Acknowledgments. This research was (partially) supported by the Key R&D Program of Shandong Province with grant 2024CXGC010108, the Natural Science Foundation of China (62472261, 62102234, 62372275, 62272274, 62202271, T2293773, 62072279), the National Key R&D Program of China with grant No.2022YFC3303004, the Dutch Research Council (NWO), under project numbers 024.004.022, NWA.1389.20.183, and KICH3.LTP.20.006, and the European Union's Horizon Europe program under grant agreement No 101070212, and by the EPSRC grant EP/W032473/1 (AP4L: Adaptive PETs to Protect & emPower People during Life Transitions). All content represents the opinion of the authors, which is not necessarily shared or endorsed by their respective employers and/or sponsors.

Reproducibility. The code and the Unlabeled Dialogue Dataset used to obtain the results are available at https://github.com/ruiqili001/DMA. The binary classifier trained on the MDRDC dataset is available at https://huggingface.co/mjmm/mal_classifier. The MDRDC dataset is available at https://github.com/repozhang/malevolent_dialogue.

References

1. Abid, A., Farooqi, M., Zou, J.: Large language models associate muslims with violence. Nat. Mach. Intell. **3**(6), 461–463 (2021)
2. Azzopardi, L., et al.: Report on the Search Futures workshop at ECIR 2024. SIGIR Forum **58**(1) (2024)

3. Belinkov, Y., Bisk, Y.: Synthetic and natural noise both break neural machine translation. In: International Conference on Learning Representations (2018)
4. Bertsch, A., Oh, A., Natu, S., Gangu, S., Black, A.W., Strubell, E.: Evaluating gender bias transfer from film data. In: Proceedings of the 4th Workshop on Gender Bias in Natural Language Processing (GeBNLP), pp. 235–243 (2022)
5. Biju, E., Sriram, A., Kumar, P., Khapra, M.M.: Input-specific attention subnetworks for adversarial detection. In: Findings of the Association for Computational Linguistics: ACL 2022, pp. 31–44 (2022)
6. Cheng, M., Yi, J., Chen, P.Y., Zhang, H., Hsieh, C.J.: Seq2sick: evaluating the robustness of sequence-to-sequence models with adversarial examples. In: Proceedings of the AAAI Conference on Artificial Intelligence, vol. 34, pp. 3601–3608 (2020)
7. Chiang, W.L., et al.: Vicuna: an open-source chatbot impressing GPT-4 with 90%* ChatGPT quality (2023). https://lmsys.org/blog/2023-03-30-vicuna/
8. Co, K.T., Muñoz-González, L., de Maupeou, S., Lupu, E.C.: Procedural noise adversarial examples for black-box attacks on deep convolutional networks. In: Proceedings of the 2019 ACM SIGSAC Conference on Computer and Communications Security, pp. 275–289 (2019)
9. Davidson, T., Warmsley, D., Macy, M., Weber, I.: Automated hate speech detection and the problem of offensive language. In: Proceedings of the International AAAI Conference on Web and Social Media, vol. 11, pp. 512–515 (2017)
10. Finn, C., Levine, S., Abbeel, P.: Guided cost learning: deep inverse optimal control via policy optimization. In: Proceedings of the 33nd International Conference on Machine Learning, ICML 2016, New York City, NY, USA, 19–24 June 2016, pp. 49–58 (2016)
11. Gao, J., Xiong, C., Bennett, P., Craswell, N.: Neural Approaches to Conversational Information Retrieval. Springer (2023)
12. Gehman, S., Gururangan, S., Sap, M., Choi, Y., Smith, N.A.: Realtoxicityprompts: evaluating neural toxic degeneration in language models. In: Findings of the Association for Computational Linguistics: EMNLP 2020, pp. 3356–3369 (2020)
13. Grzes, M., Kudenko, D.: Theoretical and empirical analysis of reward shaping in reinforcement learning. In: 2009 International Conference on Machine Learning and Applications, pp. 337–344 (2009)
14. Gu, Y., et al.: EVA2.0: investigating open-domain Chinese dialogue systems with large-scale pre-training. Mach. Intell. Res. 1–13 (2023)
15. Henderson, P., Islam, R., Bachman, P., Pineau, J., Precup, D., Meger, D.: Deep reinforcement learning that matters. In: McIlraith, S.A., Weinberger, K.Q. (eds.) Proceedings of the Thirty-Second AAAI Conference on Artificial Intelligence, (AAAI-18), the 30th Innovative Applications of Artificial Intelligence (IAAI-18), and the 8th AAAI Symposium on Educational Advances in Artificial Intelligence (EAAI-18), New Orleans, Louisiana, USA, 2–7 February 2018. AAAI Press (2018)
16. Huang, Z., Zhang, T.: Black-box adversarial attack with transferable model-based embedding. In: International Conference on Learning Representations (2019)
17. Iyyer, M., Wieting, J., Gimpel, K., Zettlemoyer, L.: Adversarial example generation with syntactically controlled paraphrase networks. In: Proceedings of the 2018 Conference of the North American Chapter of the Association for Computational Linguistics: Human Language Technologies, Volume 1 (Long Papers), pp. 1875–1885 (2018)
18. Jia, R., Liang, P.: Adversarial examples for evaluating reading comprehension systems. In: Proceedings of the 2017 Conference on Empirical Methods in Natural Language Processing, pp. 2021–2031 (2017)
19. Li, L., Ma, R., Guo, Q., Xue, X., Qiu, X.: Bert-attack: adversarial attack against bert using bert. In: Proceedings of the 2020 Conference on Empirical Methods in Natural Language Processing (EMNLP), pp. 6193–6202 (2020)

20. Li, M., Yang, Y., Wei, K., Yang, X., Huang, H.: Learning universal adversarial perturbation by adversarial example. In: Proceedings of the AAAI Conference on Artificial Intelligence, vol. 36, pp. 1350–1358 (2022)
21. Li, Z., Kiseleva, J., de Rijke, M.: Dialogue generation: from imitation learning to inverse reinforcement learning. In: The Thirty-Third AAAI Conference on Artificial Intelligence, AAAI 2019, pp. 6722–6729 (2019)
22. Liu, H., Wang, Z., Derr, T., Tang, J.: Chat as expected: learning to manipulate black-box neural dialogue models. arXiv preprint arXiv:2005.13170 (2020)
23. Liu, J., Nogueira, M., Fernandes, J., Kantarci, B.: Adversarial machine learning: a multi-layer review of the state-of-the-art and challenges for wireless and mobile systems. IEEE Commun. Surv. Tutor. **24**(1), 123–159 (2021)
24. Liu, S., Lu, N., Chen, C., Tang, K.: Efficient combinatorial optimization for word-level adversarial textual attack. IEEE/ACM Trans. Audio Speech Lang. Process. **30**, 98–111 (2021)
25. Liu, Y.A., Zhang, R., Guo, J., de Rijke, M., Fan, Y., Cheng, X.: Multi-granular adversarial attacks against black-box neural ranking models. In: Proceedings of the 47th International ACM SIGIR Conference on Research and Development in Information Retrieval, pp. 1391–1400. Association for Computing Machinery (2024)
26. Liu, Z., Ren, P., Chen, Z., Ren, Z., de Rijke, M., Zhou, M.: Learning to ask conversational questions by optimizing levenshtein distance. In: Proceedings of the 59th Annual Meeting of the Association for Computational Linguistics and the 11th International Joint Conference on Natural Language Processing (Volume 1: Long Papers), pp. 5638–5650 (2021)
27. Niu, T., Bansal, M.: Adversarial over-sensitivity and over-stability strategies for dialogue models. In: CoNLL 2018, p. 486 (2018)
28. Parry, A., Fröbe, M., MacAvaney, S., Potthast, M., Hagen, M.: Analyzing adversarial attacks on sequence-to-sequence relevance models. In: European Conference on Information Retrieval, pp. 286–302. Springer (2024)
29. Perez, E., et al.: Red teaming language models with language models. arXiv preprint arXiv:2202.03286 (2022)
30. Raj, C., Mukherjee, A., Purohit, H., Anastasopoulos, A., Zhu, Z.: Salsa: salience-based switching attack for adversarial perturbations in fake news detection models. In: European Conference on Information Retrieval, pp. 35–49. Springer (2024)
31. Schick, T., Udupa, S., Schütze, H.: Self-diagnosis and self-debiasing: a proposal for reducing corpus-based bias in NLP. Trans. Assoc. Comput. Linguist. **9**, 1408–1424 (2021)
32. Shah, D.S., Schwartz, H.A., Hovy, D.: Predictive biases in natural language processing models: a conceptual framework and overview. In: Proceedings of the 58th Annual Meeting of the Association for Computational Linguistics, pp. 5248–5264 (2020)
33. Shen, L., et al.: Textdefense: adversarial text detection based on word importance entropy. arXiv preprint arXiv:2302.05892 (2023)
34. Si, W.M., et al.: Why so toxic? Measuring and triggering toxic behavior in open-domain chatbots. In: Proceedings of the 2022 ACM SIGSAC Conference on Computer and Communications Security, CCS 2022, pp. 2659–2673. Association for Computing Machinery (2022)
35. Sun, H., et al.: On the safety of conversational models: taxonomy, dataset, and benchmark. arXiv preprint arXiv:2110.08466 (2021)
36. Taori, R., et al.: Stanford Alpaca: An instruction-following LLaMA model (2023). https://github.com/tatsu-lab/stanford_alpaca
37. Thoppilan, R., et al.: LaMDA: language models for dialog applications. arXiv preprint arXiv:2201.08239 (2022)
38. Touvron, H., et al.: LLaMA: open and efficient foundation language models. arXiv preprint arXiv:2302.13971 (2023)
39. Touvron, H., et al.: Llama 2: open foundation and fine-tuned chat models. arXiv preprint arXiv:2307.09288 (2023)

40. Tsai, Y.T., Yang, M.C., Chen, H.Y.: Adversarial attack on sentiment classification. In: Proceedings of the 2019 ACL Workshop BlackboxNLP: Analyzing and Interpreting Neural Networks for NLP, pp. 233–240 (2019)
41. Wallace, E., Feng, S., Kandpal, N., Gardner, M., Singh, S.: Universal adversarial triggers for attacking and analyzing NLP. In: Proceedings of the 2019 Conference on Empirical Methods in Natural Language Processing and the 9th International Joint Conference on Natural Language Processing (EMNLP-IJCNLP), pp. 2153–2162 (2019)
42. Wallace, E., Rodriguez, P., Feng, S., Yamada, I., Boyd-Graber, J.: Trick me if you can: human-in-the-loop generation of adversarial examples for question answering. Trans. Assoc. Comput. Linguist. **7**, 387–401 (2019)
43. Wang, J., Bao, R., Zhang, Z., Zhao, H.: Rethinking textual adversarial defense for pre-trained language models. IEEE/ACM Trans. Audio Speech Lang. Process. **30**, 2526–2540 (2022)
44. Yu, D., Sagae, K.: Automatically exposing problems with neural dialog models. In: Proceedings of the 2021 Conference on Empirical Methods in Natural Language Processing, pp. 456–470 (2021)
45. Zhang, P.F., Huang, Z., Bai, G.: Universal adversarial perturbations for vision-language pre-trained models. In: Proceedings of the 47th International ACM SIGIR Conference on Research and Development in Information Retrieval, pp. 862–871. Association for Computing Machinery (2024)
46. Zhang, R., Yu, T., Shen, Y., Jin, H., Chen, C., Carin, L.: Reward constrained interactive recommendation with natural language feedback. In: NeurIPS (2020)
47. Zhang, W.E., Sheng, Q.Z., Alhazmi, A., Li, C.: Adversarial attacks on deep-learning models in natural language processing: a survey. ACM Trans. Intell. Syst. Technol. **11**(3), 1–41 (2020)
48. Zhang, Y.: Malevolent dialogue response detection and evaluation. Ph.D. thesis, University of Amsterdam (2022)
49. Zhang, Y., Ren, P., de Rijke, M.: A taxonomy, data set, and benchmark for detecting and classifying malevolent dialogue responses. J. Am. Soc. Inf. Sci. **72**(12), 1477–1497 (2021)
50. Zhang, Y., et al.: DIALOGPT: large-scale generative pre-training for conversational response generation. In: Proceedings of the 58th Annual Meeting of the Association for Computational Linguistics: System Demonstrations, pp. 270–278 (2020)
51. Zhao, Z., Dua, D., Singh, S.: Generating natural adversarial examples. In: International Conference on Learning Representations (2018)
52. Ziebart, B.D., Maas, A.L., Bagnell, J.A., Dey, A.K.: Maximum entropy inverse reinforcement learning. In: Proceedings of the Twenty-Third AAAI Conference on Artificial Intelligence, AAAI 2008, Chicago, Illinois, USA, 13–17 July 2008, pp. 1433–1438 (2008)

Zero-Shot and Efficient Clarification Need Prediction in Conversational Search

Lili Lu[1(✉)], Chuan Meng[2], Federico Ravenda[1], Mohammad Aliannejadi[2], and Fabio Crestani[1]

[1] Università della Svizzera italiana, Lugano, Switzerland
{lili.lu,federico.ravenda,fabio.crestani}@usi.ch
[2] University of Amsterdam, Amsterdam, The Netherlands
{c.meng,m.aliannejadi}@uva.nl

Abstract. Clarification need prediction (CNP) is a key task in conversational search, aiming to predict whether to ask a clarifying question or give an answer to the current user query. However, current research on CNP suffers from the issues of limited CNP training data and low efficiency. In this paper, we propose a zero-shot and efficient CNP framework (OUR), in which we first prompt LLMs in a zero-shot manner to generate two sets of synthetic queries: ambiguous and specific (unambiguous) queries. We then use the generated queries to train efficient CNP models. OUR eliminates the need for human-annotated clarification-need labels during training and avoids the use of LLMs with high query latency at query time. To further improve the generation quality of synthetic queries, we devise a topic-, information-need-, and query-aware CoT prompting strategy (PROMPT). Moreover, we enhance PROMPT with counterfactual query generation (SEQ), which guides LLMs first to generate a specific/ambiguous query and then sequentially generate its corresponding ambiguous/specific query. Experimental results show that OUR achieves superior CNP effectiveness and efficiency compared with zero- and few-shot LLM-based CNP predictors.

Keywords: Clarification need prediction · Conversational search · Mixed-initiative

1 Introduction

Recent years have witnessed considerable progress in mixed-initiative conversational search [49], where the user and system can both take the initiative at different times in a conversation [33]. Since users often fail to express their information needs, asking clarifying questions is a crucial system-initiative action to address ambiguity in user queries [2,4,14]. Most existing work has focused on the selection or generation of clarifying questions [4,36,42], with relatively less attention given to determining when to ask them. In this paper, we focus on the clarification need prediction (CNP) task, which is to predict whether to ask a clarifying question to the current user query [3,5,10,13,34,40,41,43]. CNP is a

C. Hauff et al. (Eds.): ECIR 2025, LNCS 15572, pp. 389–404, 2025.
https://doi.org/10.1007/978-3-031-88708-6_25

critical task because asking clarifying questions at an inappropriate time would hurt user experience [25].

Motivation. Existing studies on CNP have two main limitations:

(i) **Limited or no training data for CNP**. Manually annotating clarification-need labels is labor-intensive, which restricts the scale of CNP training data. For example, ClariQ, a widely-used CNP dataset [3], contains only 187 human-annotated clarification-need labels for training. In some domain-specific conversational search scenarios, there is no training data for CNP.
(ii) **Low efficiency**. Although existing work [5] proposes an unsupervised method for CNP, which analyzes retrieved documents for the current user query, this type of method is inefficient because it performs CNP after retrieval. Large language models (LLMs) have achieved promising results on various information retrieval (IR) tasks [6,26]. One might argue that zero-shot prompting of LLMs has the potential to achieve promising results on CNP [14,23,48,49]. However, LLMs with billions of parameters result in a substantial increase in computational overhead, making it hard to apply them in practice [27].

A New Framework for CNP. In this paper, we aim to address these limitations by proposing a **z**ero-shot and **ef**ficient clarification need prediction (CNP) framework (Zef-CNP), in which we first use zero-shot prompting with large language models (LLMs) to generate synthetic specific queries (which do not require clarification) and ambiguous queries. These pseudo queries are then utilized to train efficient CNP models, which do not rely on retrieval or LLMs. We only employ the trained efficient CNP models during inference. Zef-CNP allows us to avoid using human-annotated clarification-need labels during training, and avoid using retrieval/LLMs introducing high latency at query time.

Challenges. However, Zef-CNP presents new challenges. Prompting LLMs to generate synthetic queries in a single attempt results in limited generation quality. Especially without the context of topics or information needs, it is challenging for humans to identify specific and ambiguous queries.

Solution. To address the challenge, we devise a **t**opic-, **i**nformation-need-, and **q**uery-aware chain-of-thought (CoT) prompting strategy (TIQ-CoT). In TIQ-CoT, we instruct LLMs to generate a topic, an information need, step by step, before generating synthetic queries. Furthermore, instead of generating specific and ambiguous queries separately, we enhance TIQ-CoT with **co**unterfactual **qu**ery generation (CoQu), which guides LLMs to first generate a specific/ambiguous query and then sequentially generate its corresponding ambiguous/specific query, based on the same topic and information need.

Experiments. In Zef-CNP, we prompt 2 LLMs (i.e., Llama-3.1-8B-Instruct, and GPT-4o-mini) for synthetic query generation. With each LLM, we generate 2,500 specific queries and 2,500 ambiguous ones. We then fine-tune BERT [15] on synthetic specific and ambiguous queries for CNP. Finally, we evaluate the fine-tuned BERT on two CNP datasets, ClariQ [3], and AmbigNQ [32]. Experimental results show that Zef-CNP (fully zero-shot) achieves superior CNP effectiveness and efficiency compared with baselines directly using LLMs as CNP predictors. Our analysis indicates that our devised TIQ-CoT and CoQu markedly improve data generation quality.

Our main contributions are as follows:

1. We propose Zef-CNP, a zero-shot and efficient CNP framework, which does not rely on any human-annotated clarification-need labels.
2. We devise TIQ-CoT, a topic-, information-need- and query-aware CoT prompting strategy, which instructs LLMs to first generate a topic and information need before generating queries. Moreover, we improve TIQ-CoT with counterfactual query generation (CoQu), which guides LLMs first to generate a specific/ambiguous query and then sequentially generate its corresponding ambiguous/specific query.
3. Experimental results on two CNP benchmark datasets show that Zef-CNP achieves superior CNP quality compared to baselines directly using LLMs as CNP predictors.
4. We release our generated synthetic data, called Counterfactual Queries for Clarification in Conversational Search (CoQu-CS), to the public. Codes and data are available at https://github.com/lulili0963/Zef-CNP.

2 Related Work

Clarification Need Prediction. Conversational search has attracted increasing attention [24,28–30,47]. Mixed initiative is a key part of conversational search, which means that both the system and the user can take the initiative at any time during the conversation [33]. The system can take the initiative of performing various actions, such as asking for clarification [4] or eliciting user preferences [22]. In this paper, we focus on clarification. The literature focuses on two main tasks in clarification: clarification need prediction (CNP) and clarifying question selection/generation [4,17,36,37]. Our work focuses on CNP in mixed-initiative conversational search, which is to predict whether to ask a clarifying question to the current user query in an information-seeking conversation.

Most studies [3,10,21,25,43] rely on training data to train an effective CNP model. For example, Aliannejadi et al. [3] and Chen et al. [10] fine-tuned pre-trained language models (PLMs) (e.g., BERT) and LLMs on human-annotated clarification-need labels, respectively. However, manually annotating clarification-need labels is labor-intensive, restricting the scale of CNP training data. For example, Aliannejadi et al. [3] curated the ClariQ dataset, involving human annotators to have clarification-need labels, however, ClariQ only contains 187 annotated queries for training CNP. Methods trained on these limited datasets could become biased towards the datasets and lack generalizability [5].

Some studies do not rely on human-annotated clarification-need labels. For example, the two works of Wang and Ai [40,41] focused on training CNP purely via reinforcement learning, where rewards are based on these human-designed simulators, however, the use of human-designed simulators limits the generalizability of CNP methods. Arabzadeh et al. [5] proposed an unsupervised method, which predicts CNP by analyzing the retrieved documents for the current user query; this type of method is inefficient because it performs CNP after retrieval.

Recently, some work [14,23,48,49] focused on prompting LLMs in a zero/few-shot manner for CNP. However, directly using LLMs as clarification-need predictors is highly inefficient and impractical due to their substantial computational overhead, as they have billions of parameters. In particular, Deng et al. [14] and Zhang et al. [48] used a CoT prompting strategy for CNP, i.e., they first prompted LLMs to generate descriptive reasoning before deciding whether clarification is necessary. The additional tokens produced by CoT further degrade the CNP efficiency.

Our work differs as (i) we aim at fully zero-shot CNP, where we do not rely on any human-annotated clarification-need labels, (ii) we focus on efficient CNP, where we do not rely on retrieval or LLMs at query time.

Note that unlike the work from Deng et al. [12], Guo et al. [19] and Xu et al. [43], focusing on CNP scenarios where the ambiguity of a query depends on a given document or entities, we address a more general CNP setting in conversational search, where CNP models only have access to the current user query without additional contextual information.

Counterfactual Generation. Leveraging LLMs for counterfactual generation has gained popularity and proven to be impactful in various IR and natural language processing (NLP) applications [1,7,8,11,16]. It refers to generating a query/utterance/question with a specific label that is the opposite to a given label. For example, Abolghasemi et al. [1] proposed using LLMs to generate counterfactual dialogues in which the simulated user is not satisfied with the system's response. They showed that this approach leads to a more balanced distribution of labels, resulting in higher satisfaction prediction performance in task-oriented dialogue systems. Calderon et al. [9] generated counterfactual utterances for multi-label intent prediction. They produced domain-counterfactual samples similar to the original ones in every aspect but different in their domain. Bhattacharjee et al. [8] introduced a zero-shot pipeline for generating high-quality counterfactuals, without any additional training or fine-tuning, indicating the effectiveness of the pipeline in a zero-shot manner.

Our work is aligned with the idea of counterfactual generation using LLMs, however, we take a different angle and focus on generating both specific queries and their counterfactual ambiguous ones in conversational search.

Zero-Shot Synthetic Data Generation. Given the high cost of obtaining human-annotated data, zero-shot dataset generation has become a popular approach. It often contains 3 steps, which are the generation of synthetic data by PLMs or LLMs, the fine-tuning of PLMs or smaller models on the generated data, and the evaluation of fine-tuned models on various downstream tasks [18,31,44,45,50]. Meng et al. [31] presented a method, called SuperGen, leveraging different PLMs for data generation and fine-tuning, and evaluating various classification tasks. However, the generative capabilities of PLMs are still limited, compared to LLMs. Ye et al. [44] performed an extremely similar pipeline, called ZEROGEN, and it relies on smaller models (e.g., DistilBERT [35]) for fine-tuning.

However, it differs as (i) we focus on CNP in conversational search, a domain where zero-shot dataset generation has not been explored, (ii) to improve synthetic query generation quality, we propose TIQ-CoT to ask LLMs to generate topics and information needs before generating synthetic queries, and further introduce CoQu to generate specific and ambiguous queries in a counterfactual way, and (iii) beyond generating synthetic specific queries, we explore generating *synthetic ambiguous queries*, which has not been well investigated yet.

3 Task Definition

Previous studies [3,14] generally define CNP as a binary classification task. Given the current user query q, a clarification-need predictor f determines whether the query is ambiguous or not. Formally, $l^* = f(q)$, where $l^* \in \{0, 1\}$, l^* is a predicted clarification-need label. A label of 1 indicates that the query q is ambiguous and requires clarification, while a label of 0 implies that the query is specific and does not need clarification.

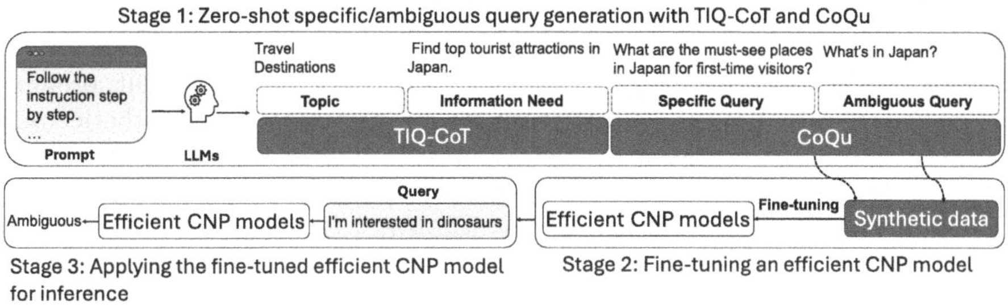

Fig. 1. Our proposed zero-shot and efficient CNP framework (Zef-CNP). Please be aware that we implement TIQ-CoT and CoQu in the same prompt (see in Fig. 2).

4 Methodology

We propose Zef-CNP, a zero-shot and efficient CNP framework. As illustrated in Fig. 1, it consists of three main stages: (i) zero-shot specific/ambiguous query generation with TIQ-CoT and CoQu, (ii) fine-tuning efficient CNP models, and (iii) applying fine-tuned efficient CNP models for inference.

Instruction: Follow the instruction step by step. 1. Come up with 10 topics. 2. Based on the topics, generate 10 corresponding user information needs. 3. Based on the user information needs, generate 10 corresponding specific queries. 4. Based on the user information needs, generate 10 corresponding ambiguous queries.
The output format should be Topic: {topic}, User information need: {user information need}, Specific query: {specific query}, Ambiguous query: {ambiguous query}.

Fig. 2. A topic-, information-need-, and query-aware CoT prompting strategy (TIQ-CoT) enhanced with counterfactual query generation (CoQu).

4.1 Zero-Shot Specific/Ambiguous Query Generation with TIQ-CoT and CoQu

This stage aims to generate M synthetic training examples for CNP, i.e., $\mathcal{D} = \{(\hat{q}_i, \hat{l}_i)\}_{i=1}^{M}$, where if $\hat{q}_i$ is a specific query (denoted as sq), then $\hat{l}_i = 0$; if $\hat{q}_i$ is an ambiguous query (denoted as aq), $\hat{l}_i = 1$.

We completely leverage LLMs' self-knowledge to generate synthetic specific/ambiguous queries in a purely zero-shot manner. To accomplish effective specific/ambiguous query generation, we propose a *topic-, information-need-, and query-aware CoT prompting strategy* (TIQ-CoT), which is accompanied by our devised *counterfactual query generation* (CoQu).

TIQ-CoT. Specifically, TIQ-CoT aims to first prompt (the input prompt denoted as p) an LLM to generate a synthetic topic t, and a synthetic user information need in, before generating a specific (sq) or an ambiguous query (aq). Formally:

$$t, in, sq/aq = \text{LLM}(p). \tag{1}$$

CoQu. CoQu further improves TIQ-CoT by instructing the LLM to generate a specific query (sq), and then generating its corresponding ambiguous query (aq) in a counterfactual manner. CoQu can potentially help the LLM more clearly differentiate between the two types of queries. We formulate TIQ-CoT with CoQu in the following:

$$t, in, sq, aq = \text{LLM}(p). \tag{2}$$

In this way, we can collect a pair of specific and ambiguous queries (sq, aq) based on the same topic and information need. We hypothesize that generating specific and ambiguous queries in a counterfactual manner helps LLMs better distinguish between the two types, to improve generation quality for both query types. We also examine the generation order of specific and ambiguous queries in Table 3.

Specifically, we use the input prompt p shown in Fig. 2 to instruct an LLM to generate ten pairs of specific and ambiguous queries per call. After data generation, we conduct data filtering: we remove duplicates, empty values, and special characters.

4.2 Fine-Tuning an Efficient CNP Model

We use synthetic training data $\mathcal{D} = \{(\hat{q}_i, \hat{l}_i)\}_{i=1}^{M}$ to train an efficient CNP model f, which does not rely on LLMs or take retrieved information as inputs, ensuring high efficiency. The learning objective is defined as:

$$\mathcal{L}(\theta) = -\frac{1}{M}\sum_{i=1}^{M} \log P(\hat{l}_i \mid \hat{q}_i), \tag{3}$$

where θ are all the parameters of f.

4.3 Applying the Fine-Tuned Efficient CNP Model for Inference

After training the efficient CNP model, we apply it to a test user query q for inference. The model predicts whether the query requires clarification, denoted as $l^* = f(q)$, where l^* is the predicted clarification-need label.

5 Experiments

Research Questions. Our work is steered by the following research questions:

RQ1 To what extent does Zef-CNP improve the *quality* and *efficiency* of CNP, compared to baselines that use LLMs as zero- or few-shot predictors for clarification needs?

RQ2 To what extent does the choice of LLMs for synthetic data generation in Zef-CNP affect CNP quality?

RQ3 To what extent do our devised TIQ-CoT and CoQu improve CNP quality? Does the generation order of specific and ambiguous queries in CoQu affect CNP quality?

RQ4 To what extent does the scale of synthetic specific/ambiguous queries generated by Zef-CNP affect CNP quality?

Table 1. Statistics of ClariQ, and AmbigNQ for evaluation.

	ClariQ	AmbigNQ
#of ambiguous queries	262	1,172
#of unambiguous queries	37	830
avg tokens of queries	9.36	12.75

Datasets. Following [3,5], we use two datasets: ClariQ [3], AmbigNQ [32]. During experiments, we only consider test sets for inference. Due to the small ClariQ test set (62 annotated queries) and the imbalanced feature, we use the entire ClariQ dataset, including train, validation, and test sets, for inference (299 annotated queries in total). As for AmbigNQ, we use its validation set for inference since its test set is not publicly available. Table 1 illustrates the statistics of the two datasets.

Note that we focus on a more general CNP setting in conversational search, using only user queries without relying on any additional context for CNP. Hence, we do not consider the PACIFIC [12] and Abg-CoQA [19] datasets, in which the ambiguity of a user query depends on a given document. We do not use CLAQUA [43], which requires entities and conversational history for CNP. Additionally, we do not consider the datasets of MIMICS [46] and MIMICS-Duo [39], which are not in the domain of conversational search and lacking clarification-need labels.

ClariQ[1] is a mixed-initiative conversational search dataset with clarification-need labels annotated via crowdsourcing [3]. The labels are ranged from 1 (no need for clarification) to 4 (clarification is necessary). We follow the work from Meng et al. [25] to binarize the labels, with a label of 1 as "not asking" and labels of 2, 3, and 4 as "asking" a clarifying question.

AmbigNQ[2] is an open-domain question answering dataset [32]. The original one does not contain clarification-need labels, however, each question is accompanied by possible pairs of disambiguated questions and answers. Arabzadeh et al. [5] took advantage of the number of possible question-answer pairs to classify the level of ambiguity, and hence, we follow their work to collect pseudo-clarification-need labels. Specifically, if a question is linked to more than one question-answer pair, it is classified as ambiguous; otherwise, it is considered as unambiguous.

Instruction: Given the user query:{query}, predict if the query is vague or not. Please only return yes or no.

Fig. 3. Prompt for LLMs as clarification-need predictors.

Baselines. We focus on a zero-shot setting, and hence, we compare Zef-CNP with baselines using LLMs as zero-shot clarification-need predictors. We use two open-source LLMs: (i) Phi-3-mini-128k-instruct, and (ii) Llama-3.1-8B-Instruct. We use three commercial LLMs: (i) GPT-3.5-turbo,(ii) GPT-4, and (iii) GPT-4omini. Figure 3 shows the prompt we use. Note that we do not consider using

[1] https://github.com/aliannejadi/ClariQ.
[2] https://huggingface.co/datasets/sewon/ambig_qa.

CoT to prompt LLMs directly for CNP [14] because we focus on efficient CNP in this work and using CoT for CNP significantly reduces inference efficiency.

We also consider an unsupervised CNP model, MiniLm-ANC [5]. It first retrieves documents for a user query and then utilizes the coherency of documents retrieved for the user query: the more coherent the retrieved items are, the less ambiguous the query is, and the need for clarification decreases.

Moreover, we consider baselines accessing limited human-annotated clarification-need labels. We consider Llama-3.1-8B-Instruct, GPT-4, and GPT-4o-mini using the few-shot prompting, with 2, 4, or 6 demonstration examples, since they perform better in our preliminary experiments. As mentioned before, we use the entire ClariQ for evaluation. Therefore, we randomly sample demonstration examples from the AmbigNQ training set for few-shot prompting, inferring on ClariQ and AmbigNQ.

Note that we do not use CNP baselines from [12,14,19] that focus on the CNP scenarios where the ambiguity of a query depends on a given document.

Evaluation Metrics. As illustrated in Table 1, our datasets show significant label imbalance. As suggested by [3][3], we adopt weighted Precision, Recall and F1.[4] In the following sections, we will refer to them as Precision, Recall and F1 for simplification. To evaluate CNP efficiency, we measure CNP inference latency averaged per query.

Implementation Details. In Zef-CNP, we use Llama-3.1-8B-Instruct or GPT-4o-mini for generating synthetic queries, considering the budget limitation. We set the temperature of the LLMs to 1 to enhance the output diversity. Overall, we generate 5,000 unique synthetic data, with 2,500 specific and ambiguous queries, respectively. We use a BERT model (bert-base-uncased[5]) as our efficient CNP predictor and train it on the whole generated queries. We use 5e-5 as our learning rate, and a batch size of 64, with 3 training epochs.[6] For baselines using LLMs as zero- and few-shot clarification-need predictors, we set the temperature of the LLMs to 0, making predicted results deterministic. For MiniLM-ANC, we adhere to the setup shown in its original paper [5].

6 Results and Analysis

To answer RQ1, Table 2 shows CNP quality and efficiency of Zef-CNP, and baselines using LLMs as zero/few-shot clarification-need predictors, on ClariQ and AmbigNQ. We have three main observations.

---

[3] https://github.com/aliannejadi/ClariQ.

[4] Note that the score of weighted F1 may not fall between precision and recall. Please refer to https://scikit-learn.org/1.5/modules/generated/sklearn.metrics.f1_score.html.

[5] https://huggingface.co/google-bert/bert-base-uncased.

[6] We use Hugging Face Trainer to implement our model fine-tuning. Other unmentioned hyperparameters are in the default settings.

Table 2. Results of Zef-CNP and baselines using LLMs as zero/few-shot clarification-need predictors, on ClariQ and AmbigNQ, in terms of weighted Precision (P), Recall (R) and F1. CNP inference latency is measured in seconds on a single GPU (A40), averaged per query. The best value in each column is marked in **bold**, and the second best is underlined.

Methods	ClariQ (%)				AmbigNQ (%)			
	P	R	F1	Latency	P	R	F1	Latency
MiniLm-ANC [5]	81.25	69.13	70.32	0.30	53.15	47.06	45.46	0.30
Phi-3-mini-128k-instruct (0-shot)	**89.17**	13.38	4.73	0.21	43.18	41.41	24.63	0.22
Llama-3.1-8B-Instruct (0-shot)	79.00	66.56	71.53	0.08	54.01	55.19	54.28	0.08
GPT-3.5-turbo (0-shot)	79.49	56.19	63.66	0.42	54.25	49.00	47.52	0.45
GPT-4 (0-shot)	86.28	29.77	32.85	0.58	57.14	42.31	27.34	0.65
GPT-4o-mini (0-shot)	85.90	56.86	63.92	0.51	52.05	42.96	32.03	0.47
Llama-3.1-8B-Instruct (2-shot)	78.72	75.25	76.88	0.08	55.42	**57.84**	54.19	0.07
Llama-3.1-8B-Instruct (4-shot)	79.20	67.56	72.26	0.08	52.79	53.25	52.98	0.08
Llama-3.1-8B-Instruct (6-shot)	79.52	62.88	68.98	0.09	51.78	51.65	51.71	0.08
GPT-4 (2-shot)	87.02	50.17	57.33	2.37	**59.28**	44.06	32.20	1.44
GPT-4 (4-shot)	86.95	49.50	56.67	0.52	52.96	44.56	37.36	0.55
GPT-4 (6-shot)	86.57	57.53	64.48	1.45	49.49	43.96	39.74	1.50
GPT-4o-mini (2-shot)	86.78	59.20	65.97	0.60	53.68	43.16	31.70	0.48
GPT-4o-mini (4-shot)	86.62	72.58	76.99	0.48	55.36	44.26	34.52	0.48
GPT-4o-mini (6-shot)	85.85	73.24	77.41	0.48	55.25	43.86	33.28	0.51
Zef-CNP (Llama-3.1-8B-Instruct) specific → ambiguous	86.65	**88.63**	**86.99**	0.01	54.98	48.85	46.68	0.01
Zef-CNP (GPT-4o-mini) specific → ambiguous	82.40	76.59	79.02	0.01	56.41	55.94	**56.13**	0.01

First, Zef-CNP using GPT-4o-mini for the generation of synthetic queries outperforms all baselines in terms of F1 in both datasets. Specifically, it exceeds the best baseline in ClariQ, GPT-4o-mini (6 shots), by 1.61 F1 points, and surpasses the best baseline in AmbigNQ, Llama-3.1-8B-Instruct (0 shots), by 1.85 F1 points.

Second, baselines demonstrate a large variation in CNP quality across datasets. For example, GPT-4o-mini (6-shot) achieves an F1 score of 77.41 on ClariQ, but its performance drops significantly to 33.28 on AmbigNQ. However, Zef-CNP with GPT-4o-mini performs consistently well across both datasets, indicating it has better generalizability.

Third, Zef-CNP is much more efficient than all baselines in terms of CNP latency. For example, Zef-CNP using GPT-4o-mini for synthetic query generation is approximately 80 times more efficient than the zero-shot baseline using the same LLM, on ClariQ. Moreover, it is around 90 times faster than the 2-shot baseline using GPT-4o-mini. However, we notice that there is inconsistency regarding the latency of OpenAI's API, which is also indicated by other research [20].

6.1 Impact of the Choice of LLMs on Synthetic Data Generation

To answer RQ2, Table 2 also presents CNP results of Zef-CNP using open- and closed-source LLMs for synthetic data generation. We find that Zef-CNP with

Llama-3.1-8B-Instruct performs much better than Zef-CNP with GPT-4o-mini on ClariQ, but performs worse than many baselines on AmbigNQ. In contrast, Zef-CNP with GPT-4o-mini consistently shows advantages over baselines. This finding indicates that different LLMs can excel in specific datasets, emphasizing the need to choose a versatile model like GPT-4o-mini for Zef-CNP. Moreover, combining synthetic query data generated from various LLMs might be able to further improve Zef-CNP's performance, but we leave it to the future research.

Table 3. Results of Zef-CNP in different settings, on ClariQ and AmbigNQ, in terms of weighted Precision (P), Recall (R), and F1. The best value in each column for each category is marked in **bold**.

Methods	ClariQ (%)			AmbigNQ (%)		
	P	R	F1	P	R	F1
Zef-CNP (Llama-3.1-8B-Instruct) specific → ambiguous	**86.65**	**88.63**	**86.99**	54.98	48.85	46.68
Zef-CNP (Llama-3.1-8B-Instruct) ambiguous → specific	79.61	84.62	81.67	**56.64**	**58.99**	52.88
w/o CoQu	83.42	87.63	83.01	54.69	57.19	**53.81**
w/o CoQu, w/o TIQ-CoT	80.28	84.62	82.05	54.04	53.55	53.75
Zef-CNP (GPT-4o-mini) specific → ambiguous	**82.40**	76.59	79.02	**56.41**	55.94	**56.13**
Zef-CNP (GPT-4o-mini) ambiguous → specific	81.32	**85.28**	**82.83**	55.89	57.94	55.22
w/o CoQu	79.93	79.93	79.93	55.86	**58.34**	53.85
w/o CoQu, w/o TIQ-CoT	81.06	79.26	80.11	55.83	49.50	47.37

6.2 Analysis of TIQ-CoT and CoQu

To answer RQ3, Table 3 shows the CNP quality of Zef-CNP in four settings of synthetic query generation: (i) generating a specific query first, followed by an ambiguous one in CoQu with TIQ-CoT, (ii) generating an ambiguous query first, followed by a specific one in CoQu with TIQ-CoT, (iii) generating both queries separately without CoQu but with TIQ-CoT (Eq. 1), and (iv) further removing TIQ-CoT. We still use Llama-3.1-8B-Instruct and GPT-4o-mini for data generation. We have two key findings.

First, the generation order of specific/ambiguous queries impacts Zef-CNP's performance. Different LLMs on different datasets prefer different generation orders. For example, Llama-3.1-8B-Instruct performs better when generating a specific query first on ClariQ but prefers generating an ambiguous query first on AmbigNQ. In contrast, GPT-4o-mini shows the opposite trend.

Second, removing CoQu, or further removing TIQ-CoT, results in poorer performance in most cases. For example, for ClariQ, when using Llama-3.1-8B-Instruct, removing CoQu results in a significant drop of 3.98 F1 points, compared to generating specific queries first with both TIQ-CoT and CoQu; further removing TIQ-CoT leads to an additional drop of 0.96 F1 points. It indicates the effectiveness of TIQ-CoT and CoQu in improving the synthetic specific/ambiguous query generation in a zero-shot manner.

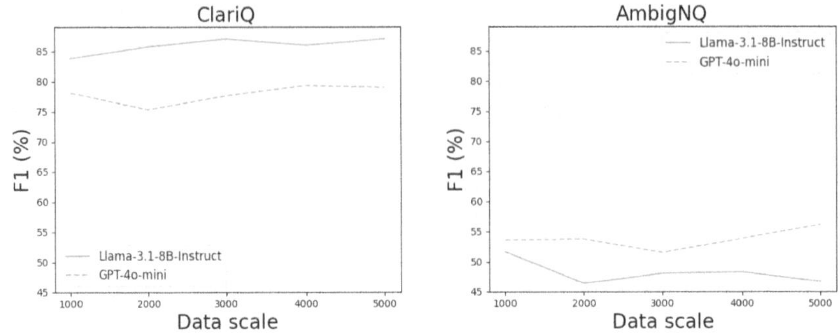

Fig. 4. CNP results of Zef-CNP in terms of weighted F1 w.r.t. the impact of different scales of generated data by Llama-3.1-8B-Instruct and GPT-4o-mini, on ClariQ and AmbigNQ.

6.3 Impact of Synthetic Data Scale on CNP Quality

To answer RQ4, Fig. 4 illustrates CNP quality on ClariQ and AmbigNQ, in terms of F1 w.r.t. the impact of different scales of synthetic data generated by Llama-3.1-8B-Instruct and GPT-4o-mini, and each data scale consists of an equal split between specific and ambiguous queries. We observe that CNP quality improves when our efficient CNP model (BERT) is trained with larger amounts of synthetic data, despite some fluctuations, except when using BERT fine-tuned on synthetic queries generated by Llama-3.1-8B-Instruct, and inferring on AmbigNQ. Surprisingly, only using 1000 synthetic queries already results in a decent performance in some cases. For example, training in 1,000 queries generated by Llama-3.1-8B-Instruct achieves an F1 score of 83.80 in ClariQ, which already surpasses all baselines listed in Table 2.

6.4 Room for Improvement

To show the room for improvement of Zef-CNP, Table 4 presents the CNP results of PLMs fine-tuned on the training set of AmbigNQ (8,028 queries) and our 5,000 synthetic queries with clarification-need labels (referred to as CoQu-CS), inferring on AmbigNQ. We observe that training on large-scale in-domain data still results in better CNP quality than using our synthetic data. It suggests that there is a space for the improvement of our work.

7 Conclusions

In this paper, we have proposed Zef-CNP, a zero-shot and efficient CNP framework for mixed-initiative conversational search, which uses LLMs to generate synthetic CNP training data in a zero-shot way and then trains efficient CNP predictors on the synthetic data. Zef-CNP eliminates the need for human-annotated

Table 4. CNP results of PLMs fine-tuned on human-annotated clarification-need labels (the training set of AmbigNQ with 8,028 examples) and our synthetic clarification-need labels (referred to as data) on AmbigNQ. The best value in each column is marked in **bold**.

Methods	Training source	AmbigNQ (%)		
		P	R	F1
roberta-base	AmbigNQ training set	**67.24**	**66.68**	**66.86**
bert-base-uncased	AmbigNQ training set	66.21	65.33	65.56
Zef-CNP (Llama-3.1-8B-Instruct) specific → ambiguous	CoQu-CS	54.98	48.85	46.68
Zef-CNP (GPT-4o-mini) specific → ambiguous	CoQu-CS	56.41	55.94	56.13

clarification-need labels during training and avoids relying on LLMs or retrieved documents at query time. To improve the quality of zero-shot generation of specific and ambiguous queries, we devise TIQ-CoT and CoQu. TIQ-CoT guides LLMs to generate topics and information needs, step by step, before generating synthetic queries. CoQu enhances TIQ-CoT by enabling counterfactual generation of specific and ambiguous queries. Experimental results show that Zef-CNP (fully zero-shot) outperforms baselines using LLMs as zero/few-shot clarification-need predictors, in terms of both CNP effectiveness and efficiency, on two benchmark datasets, ClariQ and AmbigNQ.

Broader Impact. We believe that our method has two broader impacts. First, Zef-CNP offers a solution for domain-specific scenarios where no human-annotated CNP training data are available. Second, Zef-CNP operates without using LLMs or retrieval at query time, significantly reducing latency and improving CNP's real-world applicability in practice.

Limitations and Future Work. Our work has three limitations. First, as shown in Table 4, there is room to further improve the quality of synthetic queries. One factor that may limit the quality of synthetic data is that we fine-tune efficient CNP models using synthetic data generated by a single LLM. Only relying on one LLM might introduce biases in the data. To mitigate it, we plan to fine-tune CNP models on data generated by multiple LLMs. Second, we only use BERT as our efficient CNP model. Although BERT is widely used, it is also worthwhile to investigate the results of simpler models fine-tuned on our synthetic data. Third, we only focus on English datasets and it is interesting to explore the suitability and robustness of our proposed method on multi-lingual aspect [38].

Acknowledgement. We want to thank our reviewers for their feedback. Our work is supported by the Swiss National Science Foundation (SNSF), under the project PACINO (Personality And Conversational INformatiOn Access).

References

1. Abolghasemi, A., Ren, Z., Askari, A., Aliannejadi, M., de Rijke, M., Verberne, S.: CAUSE: counterfactual assessment of user satisfaction estimation in task-oriented dialogue systems. In: ACL, pp. 14623–14635 (2024)
2. Aliannejadi, M., Gwizdka, J., Zamani, H.: Interactions with generative information retrieval systems. arXiv preprint arXiv:2407.11605 (2024)
3. Aliannejadi, M., Kiseleva, J., Chuklin, A., Dalton, J., Burtsev, M.: Building and evaluating open-domain dialogue corpora with clarifying questions. In: EMNLP, pp. 4473–4484 (2021)
4. Aliannejadi, M., Zamani, H., Crestani, F., Croft, W.B.: Asking clarifying questions in open-domain information-seeking conversations. In: SIGIR, pp. 475–484 (2019)
5. Arabzadeh, N., Seifikar, M., Clarke, C.L.: Unsupervised question clarity prediction through retrieved item coherency. In: CIKM, pp. 3811–3816 (2022)
6. Askari, A., Meng, C., Aliannejadi, M., Ren, Z., Kanoulas, E., Verberne, S.: Generative retrieval with few-shot indexing. arXiv preprint arXiv:2408.02152 (2024)
7. Bhattacharjee, A., Moraffah, R., Garland, J., Liu, H.: Towards LLM-guided causal explainability for black-box text classifiers. In: AAAI 2024 Workshop on Responsible Language Models, Vancouver, BC, Canada (2024)
8. Bhattacharjee, A., Moraffah, R., Garland, J., Liu, H.: Zero-shot LLM-guided counterfactual generation for text. arXiv preprint arXiv:2405.04793 (2024)
9. Calderon, N., Ben-David, E., Feder, A., Reichart, R.: Docogen: domain counterfactual generation for low resource domain adaptation. In: ACL, pp. 7727–7746 (2022)
10. Chen, M., Sun, R., Arık, S.Ö., Pfister, T.: Learning to clarify: multi-turn conversations with action-based contrastive self-training. arXiv preprint arXiv:2406.00222 (2024)
11. Chen, Z., Gao, Q., Bosselut, A., Sabharwal, A., Richardson, K.: Disco: distilling counterfactuals with large language models. arXiv preprint arXiv:2212.10534 (2022)
12. Deng, Y., Lei, W., Zhang, W., Lam, W., Chua, T.S.: Pacific: towards proactive conversational question answering over tabular and textual data in finance. In: EMNLP, pp. 6970–6984 (2022)
13. Deng, Y., Li, S., Lam, W.: Learning to ask clarification questions with spatial reasoning. In: SIGIR, pp. 2113–2117 (2023)
14. Deng, Y., Liao, L., Chen, L., Wang, H., Lei, W., Chua, T.S.: Prompting and evaluating large language models for proactive dialogues: clarification, target-guided, and non-collaboration. arXiv preprint arXiv:2305.13626 (2023)
15. Devlin, J., Chang, M.W., Lee, K., Toutanova, K.: Bert: pre-training of deep bidirectional transformers for language understanding. In: NAACL, pp. 4171–4186 (2019)
16. Dixit, T., Paranjape, B., Hajishirzi, H., Zettlemoyer, L.: Core: a retrieve-then-edit framework for counterfactual data generation. In: EMNLP, pp. 2964–2984 (2022)
17. Erbacher, P., Nie, J.Y., Preux, P., Soulier, L.: Paqa: toward proactive open-retrieval question answering. arXiv preprint arXiv:2402.16608 (2024)
18. Gao, J., et al.: Zerogen: self-guided high-quality data generation in efficient zero-shot learning. CoRR (2022)
19. Guo, M., Zhang, M., Reddy, S., Alikhani, M.: Abg-coqa: clarifying ambiguity in conversational question answering. In: AKBC (2021)
20. Irugalbandara, C., et al.: Scaling down to scale up: a cost-benefit analysis of replacing openai's GPT-4 with self-hosted open source SLMs in production. arXiv preprint arXiv:2312.14972 (2023)

21. Kim, J.K., Wang, G., Lee, S., Kim, Y.B.: Deciding whether to ask clarifying questions in large-scale spoken language understanding. In: ASRU, pp. 869–876 (2021)
22. Kostric, I., Balog, K., Radlinski, F.: Generating usage-related questions for preference elicitation in conversational recommender systems. TORS **2**(2), 1–24 (2024)
23. Kuhn, L., Gal, Y., Farquhar, S.: Clam: selective clarification for ambiguous questions with generative language models. arXiv preprint arXiv:2212.07769 (2022)
24. Meng, C., Aliannejadi, M., de Rijke, M.: Performance prediction for conversational search using perplexities of query rewrites. In: QPP++2023, pp. 25–28 (2023)
25. Meng, C., Aliannejadi, M., de Rijke, M.: System initiative prediction for multi-turn conversational information seeking. In: CIKM, pp. 1807–1817 (2023)
26. Meng, C., Arabzadeh, N., Askari, A., Aliannejadi, M., de Rijke, M.: Query performance prediction using relevance judgments generated by large language models. arXiv preprint arXiv:2404.01012 (2024)
27. Meng, C., Arabzadeh, N., Askari, A., Aliannejadi, M., de Rijke, M.: Ranked list truncation for large language model-based re-ranking. In: SIGIR, pp. 141–151 (2024)
28. Meng, C., Ren, P., Chen, Z., Monz, C., Ma, J., de Rijke, M.: Refnet: a reference-aware network for background based conversation. In: AAAI (2020)
29. Meng, C., Ren, P., Chen, Z., Ren, Z., Xi, T., Rijke, M.D.: Initiative-aware self-supervised learning for knowledge-grounded conversations. In: SIGIR, pp. 522–532 (2021)
30. Meng, C., et al.: Dukenet: a dual knowledge interaction network for knowledge-grounded conversation. In: SIGIR, pp. 1151–1160 (2020)
31. Meng, Y., Huang, J., Zhang, Y., Han, J.: Generating training data with language models: towards zero-shot language understanding. In: NeurIPS (2024)
32. Min, S., Michael, J., Hajishirzi, H., Zettlemoyer, L.: AmbigQA: answering ambiguous open-domain questions. In: EMNLP, pp. 5783–5797 (2020)
33. Radlinski, F., Craswell, N.: A theoretical framework for conversational search. In: CHIIR, pp. 117–126 (2017)
34. Roitman, H., Erera, S., Feigenblat, G.: A study of query performance prediction for answer quality determination. In: ICTIR, pp. 43–46 (2019)
35. Sanh, V.: Distilbert, a distilled version of bert: smaller, faster, cheaper and lighter. arXiv preprint arXiv:1910.01108 (2019)
36. Sekulić, I., Aliannejadi, M., Crestani, F.: Towards facet-driven generation of clarifying questions for conversational search. In: SIGIR, pp. 167–175 (2021)
37. Sekulić, I., Lajewska, W., Balog, K., Crestani, F.: Estimating the usefulness of clarifying questions and answers for conversational search. In: ECIR, pp. 384–392 (2024)
38. Sun, W., et al.: Conversations powered by cross-lingual knowledge. In: SIGIR, pp. 1442–1451 (2021)
39. Tavakoli, L., Trippas, J.R., Zamani, H., Scholer, F., Sanderson, M.: Mimics-duo: offline & online evaluation of search clarification. In: SIGIR, pp. 3198–3208 (2022)
40. Wang, Z., Ai, Q.: Controlling the risk of conversational search via reinforcement learning. In: WWW, pp. 1968–1977 (2021)
41. Wang, Z., Ai, Q.: Simulating and modeling the risk of conversational search. TOIS **40**(4), 1–33 (2022)
42. Wang, Z., Tu, Y., Rosset, C., Craswell, N., Wu, M., Ai, Q.: Zero-shot clarifying question generation for conversational search. In: WWW, pp. 3288–3298 (2023)
43. Xu, J., et al.: Asking clarification questions in knowledge-based question answering. In: EMNLP-IJCNLP, pp. 1618–1629 (2019)

44. Ye, J., et al.: Zerogen: efficient zero-shot learning via dataset generation. arXiv preprint arXiv:2202.07922 (2022)
45. Ye, J., Gao, J., Wu, Z., Feng, J., Yu, T., Kong, L.: ProGen: progressive zero-shot dataset generation via in-context feedback. In: EMNLP, pp. 3671–3683 (2022)
46. Zamani, H., Lueck, G., Chen, E., Quispe, R., Luu, F., Craswell, N.: Mimics: a large-scale data collection for search clarification. In: CIKM, pp. 3189–3196 (2020)
47. Zamani, H., Trippas, J.R., Dalton, J., Radlinski, F., et al.: Conversational information seeking. Found. Trends® Inf. Retrieval **17**(3-4), 244–456 (2023)
48. Zhang, T., et al.: Clamber: a benchmark of identifying and clarifying ambiguous information needs in large language models. arXiv preprint arXiv:2405.12063 (2024)
49. Zhang, X., Deng, Y., Ren, Z., Ng, S.K., Chua, T.S.: Ask-before-plan: proactive language agents for real-world planning. arXiv preprint arXiv:2406.12639 (2024)
50. Zou, T., Liu, Y., Li, P., Zhang, J., Liu, J., Zhang, Y.Q.: Fusegen: PLM fusion for data-generation based zero-shot learning. arXiv preprint arXiv:2406.12527 (2024)

Decoding the Hierarchy: A Hybrid Approach to Hierarchical Multi-label Text Classification

Fatos Torba[1,2(✉)], Christophe Gravier[2], Charlotte Laclau[3], Abderrhammen Kammoun[1], and Julien Subercaze[1]

[1] AItenders, Saint Etienne, France
{fatos.torba,abderrhammen.kammoun,julien.subercaze}@aitenders.com
[2] Laboratoire Hubert Curien, UMRCNRS 5516, Saint-Etienne, France
christophe.gravier@univ-st-etienne.fr
[3] Télécom Paris, Institut Polytechnique de Paris, Paris, France
charlotte.laclau@telecom-paris.fr

Abstract. Hierarchical multi-label text classification (HMTC) aims to predict multiple labels from a tree-like hierarchy for a given input text. Recent approaches frame HMTC as a seq2seq problem, where the objective is to predict the sequence of associated labels, regardless of their order or position in the hierarchy. Despite promising results, these approaches rely solely on attention mechanisms from previously generated tokens. This limit prevents them from acquiring information about the global hierarchy and may lead to the accumulation of errors as the model learns hierarchical cues among labels. We propose a novel HMTC model based on a hybrid version of the encoder-decoder architecture where the decoder is pre-populated with the entire label embeddings. By leveraging the decoder's Cross-Attention and Hierarchical Self-Attention mechanisms, we achieve a label representation that benefits from instance and global label-wise information. Empirical experiments on four HMTC benchmark datasets demonstrated the effectiveness of our approach by settling new state-of-the-art results. Code (https://github.com/FatosTorba/HLPD) and datasets are made available to facilitate the reproducibility and future work.

Keywords: Hierarchical Multi-label Text Classification · Hierarchical Self-Attention Mechanism · Seq2Seq · Reproducibility

1 Introduction

Hierarchical multi-label text classification (HMTC) aims to classify text into multiple labels organized as a tree-like hierarchy, with an arbitrary depth and width. Each label in the hierarchy corresponds to a node and is usually referred to as a surface form composed of multiple tokens (see Taxonomy in Fig. 1). HMTC is considered compulsory in diverse industrial domains, including e-commerce applications [5,8], medical recording [3], and finance [10], among others. Despite advancements in natural language processing, especially with pre-trained large language models (PLMs), HMTC remains challenging for two reasons. First, HMTC assigns a variable number of categories to a

C. Hauff et al. (Eds.): ECIR 2025, LNCS 15572, pp. 405–420, 2025.
https://doi.org/10.1007/978-3-031-88708-6_26

document, and these categories can be taken from one or multiple, potentially incomplete, pathways drawn from the label taxonomy. Consequently, the classifier must learn to disentangle hierarchical cues from co-occurrences between labels for multiple paths. Second, the training data are imbalanced as annotations for the leaves in the class tree are inevitably sparse [9]. Consequently, classifiers trained for this task perform poorly on lower levels of the hierarchy and tend to overfit on higher ones [26].

To tackle these challenges, two main strategies were explored in the literature. The first one consists in directly learning the hierarchical information and the correlation between labels. To achieve these goals we can differentiate between Graph Neural Network (GNN) based models [4,28,35], contrastive approaches [33,34], and NLP inspired approaches [7,11,29]. The second strategy involves capturing the semantic information between label tokens and the text to classify, which can also be paired with label metadata [19,23]. In addition, several recent approaches propose to solve the problem HMTC task as a seq2seq prediction task [9,32]. The aim is then, given the tokens predicted in the current level in the label hierarchy, to predict the tokens of the node further down the hierarchy. In that case, the decoder aims to capture both the semantic information from the input label tokens and the hierarchical information by applying attention to previously generated tokens in an auto-regressive manner. With promising results, these approaches use an attention mechanism only on the previously generated tokens, possibly an incomplete path that does not necessarily reaches a leaf, and thus lack the global information of the hierarchy. Starting from these observations, our work is at the crossroads between previous approaches proposed for solving HMTC: we also cast HMTC as a seq2seq problem on the one hand, and we model to capture both the semantic information from the input label tokens and the hierarchical information on the other hand. For this purpose, we propose to expand the decoder input with label representations and we propose a Hierarchical Self-Attention mechanism to allow the model to capture global information regarding the hierarchy of labels.

To summarize, our contributions are threefold:

(1) We develop a label encoder, based on the T5 [18] model and exploit it in a non-regressive manner for HTMC by pre-populating the decoder with label embeddings. Since the label embeddings derive from the same encoder as the text encoder, the model captures potential semantic cues between the input text and the labels.
(2) We propose a Hierarchical Self-Attention mechanism that benefits from the label taxonomy hierarchical positional information.
(3) We conduct thorough experiments showing that our framework learns both hierarchical information and label co-occurrence, enabling it to outperform the other 16 baselines.

The rest of this paper is organized as follows. Section 2 presents recent advances in HMTC with a separation between discriminative and generative methods. Section 3 introduces the problem definition and Sect. 4 our proposal to solve this task. Section 5 provides a thorough and comprehensive set of experiments to support our claims and Sect. 6 concludes.

2 Related Work

Existing HMTC works can be divided between discriminative methods and generative methods.

Discriminative Methods. Discriminative methods frame HMTC as a classification task. Depending on the way hierarchical information is used, discriminative methods are traditionally categorized into local vs. global approaches. The former considers isolated classifiers learned for each node, for each parent node, or for each class tree level [2,24,30]. Since the final prediction is a combination of these classifiers the major shortcoming is that the number of classifiers would grow as the class tree grows and so the error accumulation would. Contrary to local approaches, the global ones learn one model to capture the hierarchical information between classes. Since the labels are naturally organized as a tree-like hierarchy, [17,28,35] adopt graph-based models for label representations which are fused into the text encoder. [11,29] use only a text encoder by directly incorporating the label hierarchy, modeled as a sequence, where the Masked Language task is performed. Leveraging the hierarchical information between classes is one advantage, but the label representations also contain semantic information that guides the model toward more consistent and easier decisions [4,6,23]. To exploit the possible semantic cues between label tokens and the text, [4] cast the problem into a matching problem while [6] utilizes a text-label mutual information maximization, which is even sometimes expanded with label metadata [17,19]. To jointly learn hierarchical and semantic information, [17,28,33] suggested contrastive learning methods.

Generative Methods. Generative methods cast the sequence-to-sequence (seq2seq) task into a HMTC task by predicting the label names as a target sequence [31] that can be built in different ways depending on the choice of visiting the class tree [26]. [9] builds the target sequence with a Breadth-First Search algorithm and proposes a Path-Adaptive Attention Loss to focus the decoder self-attention only on the previously predicted tokens of the path where the currently generated label is located, i.e. ancestors tokens, inhibiting the noise from other paths. [32] uses the Depth-First Search algorithm to build the target sequence and forces the generation to be coherent with respect to the hierarchy by distributing the next token probability only on the possible next token set.

Recent research took a new course by framing HMTC as a seq2seq task and exploited the breakthrough brought by generative models to propose competitive solutions. Most generative methods perform well at preserving the class hierarchy when making predictions [26] which makes this issue irrelevant for generative methods. These solutions have however, by design, set aside what made the strength of discriminative methods, that is to make predictions based on the entire hierarchical information and to capture correlation from input text and labels for better classifications. In this paper, we reconcile both approaches by investigating how to condition the decoder output by (i) class labels representations, and (ii) global hierarchy cues through a new hierarchical attention mechanism.

3 Problem Definition and Background

3.1 Problem Definition

The goal of HMTC is to assign multiple labels from a taxonomic hierarchy to a text document. The number of labels to predict per level is arbitrary. The classification can stop before reaching the hierarchy leaves. Formally, our input is a set of texts $\mathcal{X} = \{x_1, x_2, \ldots, x_{|\mathcal{X}|}\}$ and a graph representing the taxonomic hierarchy $\mathcal{T} = \{\mathcal{V}, \mathcal{E}\}$. Each $v \in \mathcal{V}$ is a label in the taxonomy, and each edge $(v_i, v_j) \in \mathcal{E}$ corresponds to the relationship between a parent v_i and a child v_j. In addition, we assume that each label v_i can be described by a short text description l_i. HMTC aims to predict $Y = \{y_1, y_2, \ldots, y_{|\mathcal{V}|}\}$ where $y_i = 1$ if x_i belongs to the class v_i, $y_i = 0$ otherwise. Note that the prediction is considered coherent within the taxonomic hierarchy if, for each predicted node, all of its ancestors have also been predicted. For example, in Fig. 1, the input text is classified as *Economics, Monetary, Government, European Community* etc. For consistency with respect to the class hierarchy, a document not classified as *Government* could not be classified as *European Community* for instance.

3.2 Background

We now briefly introduce the T5 model, an encoder-decoder architecture widely used for seq2seq tasks [20,36]. We recall that the encoder of T5 mainly consists of Multi-Head Attention layers [27], denoted by *MHAtt(.)* in the following, and a Feed-Forward Network. In what follows, we note n: the number of tokens in a text, d: the hidden states dimension, h: the number of attention heads, d_q: the query weights dimension, and d_v: the value weights dimension. Consequently, the *MHAtt(.)* layer is defined by:

$$\mathit{MHAtt}(Q, K, V) = Concat(A_1, A_2, \ldots, A_h)W^O, \tag{1}$$

where $\forall i \in \{0, h\}$, A_i stands for the ith attention head (*Attn(.)*). The later is defined as:

$$A_i(Q, K, V) = \mathit{Attn}(QW_i^Q, KW_i^K, VW_i^V), \tag{2}$$

where $W_i^Q \in R^{d \times d_q}$, $W_i^K \in R^{d \times d_q}$, $W_i^V \in R^{d \times d_v}$ and $W^O \in R^{hd_v \times d_q}$ are learnable parameters in the attention head. In practice, we use $d_k = d_v = d/h$.
The attention mechanism is calculated as follows:

$$\mathit{Attn}(Q, K, V) = \mathit{Score}(Q, K)V = \mathit{Softmax}\left(\frac{QK^T}{\sqrt{d_k}} + R + M\right)V, \tag{3}$$

where $Q, K, V \in \mathbb{R}^{n \times d}$ and $R, M \in \mathbb{R}^{n \times n}$. R stands for the relative position embeddings [21] and M for the attention mask which prevents attending to future tokens or to pad tokens. The output of the Multi-head Attention is then passed into a Feed-Forward Network and Batch Normalisation layers. The decoder, in addition to the encoder, contains Multi-Head Cross-Attention layers consisting of applying attention mechanism where the query Q is derived from the decoder input while the key matrix K and value matrix V are derived from encoder output.

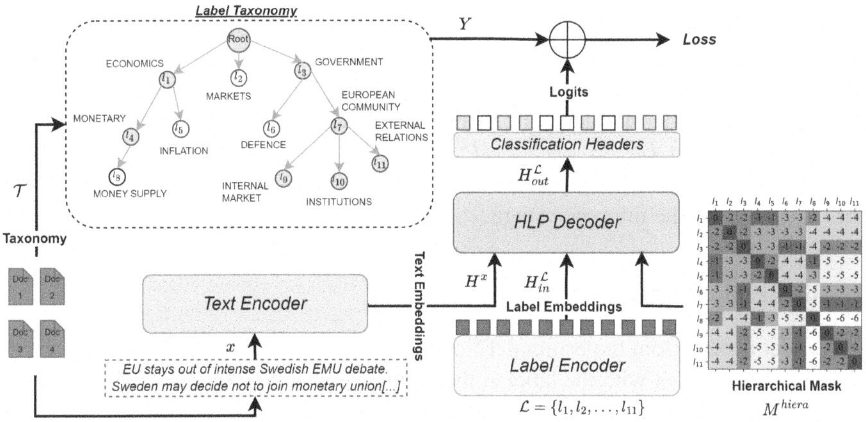

Fig. 1. HLPD architecture. The Text Encoder produces text embeddings. The Label Encoder embeds description of the nodes. HLP Decoder applies Hierarchical Self-Attention to the label embeddings using the Hierarchical Mask and performs Cross-Attention between the labels and the text This process generates a text-wise and label-wise representation, on top of which classification headers are applied. Green nodes denote correct labels. (Color figure online)

4 Methodology

Generative models like T5 [18] can also be used in a non-autoregressive manner for classification task [12, 16]. We propose here a hybrid version of T5 where the decoder is pre-populated with label embeddings (Sect. 4.1) as well as a new Hierarchical Self-Attention mechanism (Sect. 4.2) to capture the correlation between labels, in order to guide the decoding task (Sect. 4.3). In the rest of the paper, we refer to this model as Hierarchical Label Pre-Populated Decoder (HLPD). The overall architecture of HLPD is depicted in Fig. 1 and detailed hereafter.

4.1 Encoders of HLPD

The label description consists of multiple tokens. Thus, apart from the text encoder, HLPD requires also a label encoder to achieve label representation.

Text Encoder. To obtain the text representation we use the default T5 encoder. Given an input text $x \in \mathcal{X}$ of length n, the *text encoder* output the text embedding as: $H^x = TextEncoder(x)$ where $H^x \in R^{n \times d}$, with d being the model hidden states dimension.

Label Encoder. Our label encoder comprises one Multi-Head Self-Attention layer initialized with the T5 encoder's last layer followed by a projection linear layer. We denote by $\mathcal{L} = \{l_1, l_2, \ldots, l_{|\mathcal{V}|}\}$ the set of text descriptions, where each l_i is the text description of the ith label. For the first level, the text description corresponds to the name of the label, e.g., $l_1 =$ *Economics* and $l_2 =$ *Government* in Fig. 1. For subsequent levels, we use the template[1] "Child is a sub-topic of parent" to enhance the propagation of the parent-child relationship, e.g., l_4 = *Monetary is a sub-topic of Economics* in Fig. 1. Then, we obtain the label embedding for l_i with: $H^{\mathcal{L}}_{in,i} = LabelEncoder(l_i)$, where $H^{\mathcal{L}}_{in,i} \in R^d$ represents the outputted hidden state of the first label token. Thus, one can obtain the label embeddings set $H^{\mathcal{L}}_{in} = \{H^{\mathcal{L}}_{in,1}, H^{\mathcal{L}}_{in,2}, \ldots, H^{\mathcal{L}}_{in,|\mathcal{V}|}\}$ which will be further used as input for the decoder. The label encoder is jointly trained with the other modules but is discarded during the inference stage.

4.2 Hierarchical Self-attention

Our decoder differs from the original T5 decoder in several respects: (1) we replace the token embedding layer with the label embeddings described in Sect. 4.1; (2) we replace the relative positions encodings [21] with a hierarchical positions encodings. Our motivation for proposing a novel Hierarchical Self-Attention mechanism stems from the limitations of the vanilla T5 decoder, which uses relative position embeddings [21] to adjust attention based on token order. While this is crucial for text generation, it poses two problems for HMTC: (1) T5's embeddings, with dimensions of 32 and logarithmically increasing ranges up to an offset of 128, are inadequate for classification tasks with over 128 labels, common in HMTC (see Appendix, Table 3). (2) These embeddings prioritize closer tokens, which is beneficial for natural language tasks but problematic for our context, where label order should not affect results. Instead, we aim to capture label positioning within the tree hierarchy $\mathcal{T}$, which we address with our novel Hierarchical Mask.

Hierarchical Self-attention Layer. As explained above, in the Self-Attention layer, we replace the traditional attention mask and relative positions encodings with a learnable hierarchical positional mask. The underlying idea of this layer is to help the model learn both the hierarchical dependencies and the co-occurrence between labels. We define the attention layer as follows:

$$A_{self} = Attn_{hiera}(H^{\mathcal{L}}_{in}W^Q, H^{\mathcal{L}}_{in}W^K, H^{\mathcal{L}}_{in}W^V), \tag{4}$$

where $Attn_{hiera}(.)$ is calculated as follows:

$$Attn_{hiera}(Q, K, V) = \mathit{Softmax}\left(\frac{QK^T}{\sqrt{d_k}} + W \circ M^{hiera}\right)V, \tag{5}$$

where $W \in R^{|\mathcal{V}|\times|\mathcal{V}|}$ are learnable parameters, $\circ$ stands for the Hadamard product. One should note that W is shared across all heads and layers to speed up training and inference. $M^{hiera} \in R^{|\mathcal{V}|\times|\mathcal{V}|}$ corresponds to a hierarchical positional mask that guides the model to assign more attention between labels that are hierarchically closer. Formally, $M^{hiera}_{i,j} = -dist(v_i, v_j)$, where $dist$ can be any distance. In this paper, we consider the length of the shortest path connecting a node v_i with a node v_j For instance, in Fig. 1,

[1] This template showed a slight improvement for BGC and RCV1-V2, but it's not significant enough for an ablation study. We believe augmenting label descriptions with external knowledge could further enhance performance and recommend this for future work.

the shortest path between l_1 and l_8 is $l_1 \rightarrow l_4 \rightarrow l_8$, hence $M^{hiera}_{1,8} = -2$. By making this mask learnable (with W), the attention mechanism can also adapt to learn the label co-occurrence.

4.3 Decoder of HLPD

To summarize, the decoder of HLPD, henceforth *HLPDecoder*, takes as an input the text embedding and the label embedding, and is composed of both Multi-Head Hierarchical Self-Attention layers and Multi-Head Cross-Attention layers Using Multi-Head Cross-Attention layers allows us to capture fine-grained details from different parts of the input text and their associated labels. We recall that the cross-attention layer is defined as:

$$A_{cross} = Attn(H^{\mathcal{L}}_{in}W^Q, H^xW^K, H^xW^V), \tag{6}$$

where $Attn(.)$ is the usual attention mechanism defined in Eq. 3. Thus, combining the advantages of both the Hierarchical Self-Attention mechanism and the Cross-Attention mechanism we obtain the decoder's final output: $H^{\mathcal{L}}_{out} = HLPDecoder(H^x, H^{\mathcal{L}}_{in})$, with $H^{\mathcal{L}}_{out} \in R^{|\mathcal{V}| \times d}$, that incorporate both instance-wise and label-wise information. Finally, we apply classification headers to $H^{\mathcal{L}}_{out}$ to obtain the logits for the loss function. HLPD uses the *Zero-bounded Logsum-exp & Pairwise Rank-based* (ZLPR) [25] proposed for multi-label classification (see Appendix B for details). We also provide a computational complexity analysis in Appendix A.

5 Experiments

In this section, we present our evaluation protocol and compare the performance of HLPD with a broad range of SOTA methods for HMTC.

5.1 Evaluation Protocol

Datasets. We conduct experiments on four public datasets, including Reuters corpus (RCV1V2) [15], Blurb Genre Collection (BGC) [1], Arxiv Academic Papers Dataset (AAPD) [31] and Web-Of-Science (WOS) [13]. The labels of all these datasets are organized as a tree-like hierarchy. RCV1V2, AAPD and BGC are multi-paths while WOS is single-path. The appendix provides Table 3 with dataset statistics.

Metrics. Following [26,32], we evaluate our model using F1 scores: Micro-F1 and Macro-F1, as well as constrained F1 scores: C-Micro-F1 and C-Macro-F1, which correspond to the Micro-F1 and Macro-F1 of constrained predictions obtained through post-processing. A node is considered `true` only if all its ancestor nodes are predicted as `true`.

Baselines. We compare our model against several robust and recent HMTC baselines to demonstrate its effectiveness. The baselines include three Hierarchy-Aware models, namely `HiAGM` [35], `HTCInfoMax` [6] and `HiMatch` [4], where the two former ones originally used `TextRCNN` [14] as backbone model. We also compare with three Generative Pretrained Language Models (PLMs), namely `SGM` [31], `Seq2Tree` [32] and

Table 1. Results on the four datasets. The best results are in bold and the second-best is underlined. We report the mean results (± standard deviation, reported when available in the original paper) across 4 runs with random seeds. The ↑ stands for the improvement to the second best model; the * indicates our implementation and ♣, † denote results reported by [28] and [32], respectively.

Model	BGC		AAPD		RCV1-V2		WOS	
	Micro-F1	Macro-F1	Micro-F1	Macro-F1	Micro-F1	Macro-F1	Micro-F1	Macro-F1
Hierarchy-Aware Models								
TextRCNN	-	-	-	-	81.57	59.25	83.55	76.99
HiAGM	77.22	57.91	-	-	83.96	63.35	85.82	80.28
HTCInfoMax	76.84	58.01	79.64	54.48	83.51	62.71	85.58	80.05
HiMatch	76.57	58.34	80.74	56.16	84.73	64.11	86.20	80.53
Discriminative Pretrained Language Models (BERT)								
BERT	78.84	61.19	80.88	57.17	85.65	67.02	85.63	79.07
HiAGM ♣	79.48	62.84	80.68	59.47	85.58	67.93	86.04	80.19
HTCInfoMax ♣	79.16	62.94	80.76	59.46	85.83	67.09	86.30	79.97
HiMatch	78.89	63.19	80.42	59.23	86.33	68.66	86.70	81.06
HGCLR	79.22	64.04	80.95	59.34	86.49	68.31	87.11	81.20
HPT	-	-	-	-	87.26	69.53	87.16	81.93
K-HTC	80.52	65.99	-	-	-	-	87.29	81.69
HBGL	-	-	-	-	87.23	71.07	87.36	82.00
HJCL	$81.30_{\pm0.29}$	$66.77_{\pm0.37}$	$81.91_{\pm0.18}$	$\mathbf{61.59}_{\pm0.23}$	$87.04_{\pm0.24}$	$70.49_{\pm0.32}$	$87.03^{*}_{\pm0.21}$	$81.51^{*}_{\pm0.32}$
Generative Pretrained Language Models (T5)								
SGM	77.86*	60.88*	80.51*	58.46*	84.52*	65.07*	85.94*	80.81*
Seq2Tree	79.72	63.96	80.55	59.58	86.88	70.01	87.20	82.50
PAAM-HiA	-	-	-	-	87.22	70.02	90.36	81.64
HLPD	$\mathbf{82.00}^{\uparrow0.70}_{\pm0.11}$	$\mathbf{67.78}^{\uparrow1.01}_{\pm0.39}$	$\mathbf{82.02}^{\uparrow0.11}_{\pm0.24}$	$60.91^{\downarrow0.68}_{\pm0.67}$	$\mathbf{88.22}^{\uparrow0.96}_{\pm0.12}$	$\mathbf{71.19}^{\uparrow0.12}_{\pm0.55}$	$\mathbf{90.88}^{\uparrow0.52}_{\pm0.25}$	$\mathbf{86.78}^{\uparrow4.28}_{\pm0.08}$

PAAM-HiA-T5 [9]. These models use T5 as a backbone. Finally, we include five Discriminative PLMs, namely HGCLR [28], HPT [29], HBGL [11], K-HTC [17] and HJCL [33]. All these models use BERT as a backbone model.

Implementation Details. Following [16] we chose a pre-trained version of T5 [18] as a backbone model. For the optimizer, we use Adafactor [22] with a learning rate of $5e^{-4}$ and a batch size of 16. To accelerate model training, we cache the second-to-last layer output of the T5 encoder, which is used as an input for the *LabelEncoder*'s Multi-Head Self-Attention layer. The latter is trained during 2 epochs for BGC which has a more complex taxonomy. For the other datasets, we only trained the linear projection layer. We suspend training after Macro-F1 in the validation dataset does not increase for 6 epochs. To reproduce the results for other baselines, we use the hyperparameters from the original paper except for HJCL which uses large batches. We implemented the latter for WOS with a batch size of 32 instead of 80 due to GPU power limitations. For a fair comparison, we replaced the LSTM encoder-decoder of SGM with the T5 ones, as suggested by [32].

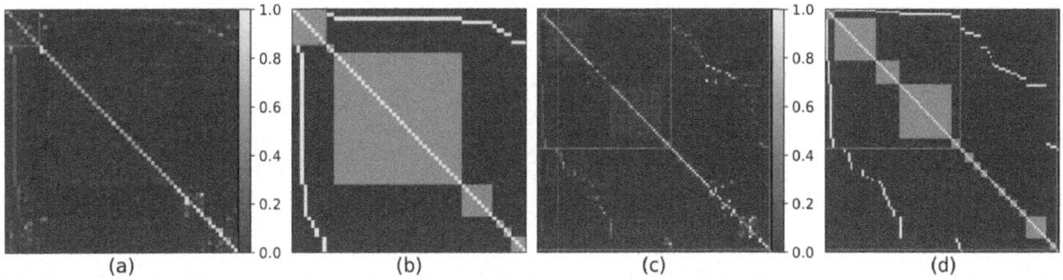

Fig. 2. Hierarchical Self-Attention scores matrix. (a) and (c) display scores for AAPD and RCV1-V2 datasets. They represent the attention score a label on the vertical axis assigns to a label on the horizontal one. (b) and (d) show the Parent-Child-Siblings relationship matrix for AAPD and RCV1-V2, respectively, with values of 1 and 0.5 indicating parent-child and sibling relationships, and 0 for others. The red lines separate the levels of labels: two levels for AAPD and four levels for RCV1-V2, with the latter's last level having only one label. Labels are sorted from the first to the last level. Values below the diagonal represent attention assigned to left labels (ancestors and some from the same level), while values above the diagonal correspond to attention assigned to right labels (descendants and some from the same level). Attention a label has to itself is located on the diagonal. (Color figure online)

5.2 Main Results

Table 1 presents the global F1-micro and F1-macro scores on the four HMTC datasets. Results for the constrained metrics are presented in Table 2. As one can see, our model outperforms the other baselines in three out of four datasets. We also note that HJCL model achieves a superior Macro-F1 score on the AAPD dataset. This comes as no surprise as this latter was initially pre-processed into uncased text to align with BERT's uncased models. Let us take a closer look at the results.

HLPD vs. Generative PLMs. Compared to the Generative PLMs baselines, SGM [31], Seq2Tree [32] and PAAM-HiA-T5 [9], our model yields significant improvements across all the datasets, more than 3.5% in Macro-F1 score for BGC and WOS as well as more than 1.5% in Micro-F1 score for BGC and AAPD, despite using the same text encoder which demonstrates the effectiveness of our proposed design.

HLPD vs. Discriminative PLMs. Compared to the Discriminative PLMs baselines HPT, HBGL, K-HTC, HJCL are the most competitive ones with our model. Both our model and PAAM-HiA-T5, which use the decoder attention mechanism, show significant improvement in Micro-F1 on the WOS dataset. Unlike Seq2Tree, PAAM-HiA-T5 uses path adaptive loss to focus attention on ancestors, aiding in hierarchy learning. WOS, a single-path dataset, requires predicting one label per level down to the leaves. We verified single-path consistency in predictions, achieving 98.23% for HLPD and 96.27% for HJCL. Removing the Self-Attention mechanism from our model drops performance by 4.55% in Micro-F1 and 5.91% (see Table 1). These results highlight the decoder attention mechanism's effectiveness in learning hierarchy. Our performance on AAPD is comparable to the state-of-the-art HJCL. On RCV1-V2, our model outperforms all baselines.

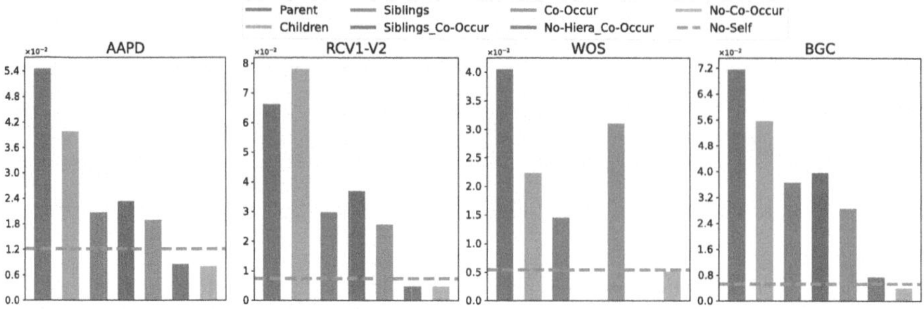

Fig. 3. Hierarchical Self-Attention computes the mean score based on label relationships and co-occurrences, excluding self-scores. «Parent», «Children», and «Siblings» represent the mean attention score a label gives to its parent, children, and siblings, respectively. «Siblings_Co-Occur», «Co-Occur», «No-Hiera_Co-Occur», and «No-Co-Occur» denote the mean attention score a label assigns to siblings with co-occurrence, labels with co-occurrence, non-hierarchical labels with co-occurrence, and labels without co-occurrence, respectively. «No-Self» represents the global mean score excluding self-scores. For the WOS dataset, «Siblings_Co-Occur» and «No-Hiera_Co-Occur» cannot be defined due to its single-path nature.

5.3 Ablation Study

Now, we investigate the independent effect of each component of `HLPD`.

Impact of the Self-attention Mechanism. Firstly, we show that the attention mechanism (hierarchical of traditional) between the labels is a major asset for HMTC. Removing this component leads to an approximate performance drop of 1.5% and 2% in Micro-F1 and Macro-F1 scores across AAPD, BGC, and RCV1-V2 which is even more pronounced for WOS (see Table 2). Without the Self-Attention mechanism, the model fails to learn both hierarchical information and label correlations.

Impact of the Hierarchical Self-attention Mechanism. Secondly, to study the impact of our proposed Hierarchical Self-Attention mechanism we present the results without this component while retaining the decoder's vanilla Self-Attention. Its major impact is particularly noticeable on the Macro-F1 score, especially for BGC and RCV1-V2 which have deeper levels than WOS and AAPD, which shows that the Hierarchical Self-Attention mechanism contributes to learning labels that are rare in the training dataset which are usually located in the deeper level. By assigning the label attention score primarily to its parent and children, as shown in Fig. 3, the Hierarchical Self-Attention mechanism guides the model toward accurate predictions on fine-grained labels.

Impact of Coherence Constraints. Finally, `Seq2Tree` [32] guarantees coherent prediction with respect to the label hierarchy. They noticed that the performance of some baselines including `SGM` [31], `HiAGM` [35], `HTCInfoMax` [6] and `HiMatch` [4], were negatively impacted by such constraints. However, in our case, we notice a very slight improvement or comparable results (see `HLPD` Cons. in Table 2) when applying such post-processing. This indicates that our model naturally captures the hierarchical coherence.

Table 2. Ablation Models. `no Self-Attn` means the decoder is free of Self-Attention Mechanism, `Self->H-Attn` stands for replacing the original Self-Attention with Hierarchical Self-Attention. `HLPD Cons.` represents results after post-processing to ensure coherent prediction. In this case Micro-F1 and Macro-F1 are equal to C-Micro-F1 and C-Macro-F1.

Ablation Models	BGC		AAPD		RCV1-V2		WOS	
	Micro-F1	Macro-F1	Micro-F1	Macro-F1	Micro-F1	Macro-F1	Micro-F1	Macro-F1
`HLPD`	82.00	67.78	82.02	60.91	88.22	71.19	90.88	**86.78**
`no Self-Attn`	80.59	65.38	80.79	59.02	86.72	68.26	86.33	80.87
`Self->H-Attn`	81.52	66.01	81.81	60.15	88.24	70.50	90.88	86.64
`HLPD Cons.`	**82.03**	**67.83**	**82.04**	**60.92**	**88.27**	**71.20**	**90.90**	86.76

5.4 Hierarchical Self-attention Score Analysis

We analyze the values of the Hierarchical Self-Attention coefficients in the last layer. The attention scores align with the ground relationships (Fig. 2). To investigate further, we differentiate the scores based on label relationship (see Fig. 3). Overall, the attention scores given to the parents are higher than to others. This is beneficial from a hierarchical coherence perspective, as the model must first predict the parent before it can accurately predict the child. The only exception is the RCV1-V2 dataset, where a label gives more attention to its children than its parent. To better understand this behaviour, we investigated performance at different levels. Contrary to the other datasets, for the RCV1-V2 dataset our model shows higher Micro-F1 scores for the third level (86.51%) compared to the second level (83.66%), which might explain this exception. Furthermore, while a label may not always co-occur with all its siblings, the attention it assigns to its siblings is higher when they do co-occur. This illustrates that our model learns both hierarchical information and co-occurrence patterns.

6 Conclusion and Future Work

In this paper, we design a novel encoder-decoder architecture to address the HMTC task. Our model leverages the power of the decoder enhanced by the Hierarchical Self-Attention mechanism. We use our architecture in a non-autoregressive fashion, leading to state-of-the-art results compared to previous encoder-decoder baselines. When compared to other encoder-only baselines, our model still outperforms them in three out of four datasets. In future work, we plan to explore methods for enriching label descriptions, considering our *LabelEncoder*'s capacity to handle up to 512 tokens. This could benefit industrial applications, therefore allowing experts to enhance label descriptions.

A Computational Complexity Analysis

Number of Parameters. We compare the parameter counts of `HLPD` with the top-performing Discriminative PLMs, `HJCL`, and T5-based PLMs that share architecture but differ in training. The number of parameters in HLPD depends on the taxonomy's label count $|\mathcal{V}|$, with classification headers proportional to $|\mathcal{V}|$. In our benchmark, HLPD contains roughly 232M parameters, similar to the T5-based PLMs with 220M parameters. HJCL's number of parameters varies from 176M for the AAPD dataset up to 239M for the BGC dataset. HJCL uses a classification header whose number of parameters is proportional to $|\mathcal{V}|^2$. Overall, HLPD and HJCL have comparable number of parameters.

Inference Speed. Unlike generative baseline models that predict one token at a time, HLPD decodes all labels in one step. We compared HLPD with a standard T5 model and BERT using the BGC test set on an NVIDIA L40 GPU with a batch size of 16. For the seq2seq model we generated the target labels with the greedy strategy. The label names remain pristine. BERT predicted the BGC test dataset in 126,06 s while HLPD took 195.86 s. In other words BERT is 1.55 times faster. Compared to the seq2seq model, HLPD shows a significant inference speed improvement, being in average, 3.51 times faster. Under this condition, our design is more appropriate for the HMTC task than the seq2seq design.

Table 3. Statistics of datasets: $|V|$ is the total number of labels, $Depth$ is the maximum level, $m|V|$, mD, mW_{max} are the respective averages of the number, the depth and the maximum width of labels in each sample.

Dataset	$\|V\|$	$Depth$	$m\|V\|$	mD	mW_{max}	Train	Val	Test
RCV1-V2	103	4	3.24	2.56	1.45	20833	2316	781265
WOS	141	2	2	2	1	30070	7518	9397
AAPD	61	2	4.12	1.62	2.41	53840	1000	1000
BGC	146	4	3.01	2.34	1.51	58715	14785	18394

Table 4. ZLPR vs. BCE. For each loss, best results are in bold and the second-best is underlined.

Model	BGC		AAPD		RCV1-V2		WOS	
	Micro-F1	Macro-F1	Micro-F1	Macro-F1	Micro-F1	Macro-F1	Micro-F1	Macro-F1
ZLPR Loss								
HJCL	81.30	66.77	81.91	**61.59**	87.04	70.49	87.03*	81.51*
HPT	-	-	-	-	87.26	69.53	87.16	81.93
HLPD	**82.00**	**67.78**	**80.02**	60.91	**88.22**	**71.19**	**90.88**	**86.78**
BCE Loss								
HJCL	80.48	65.87	80.95	59.73	86.42	69.24	86.82	81.43
HBGL	-	-	-	-	87.23	71.07	87.36	82.00
HLPD	**81.74**	**67.73**	**81.59**	**60.03**	**87.84**	**71.47**	**90.61**	**86.49**

B Impact of the Loss

Traditionally, HMTC baselines use Binary Cross-Entropy (BCE) as the objective function. Recently, [25] proposed a novel loss function for multi-label classification called *Zero-bounded Logsum-exp & Pairwise Rank-based* (ZLPR), which writes as follows:

$$\mathcal{L}_{\mathrm{Z}} = \log\left(1 + \sum_{i \in \Omega_{pos}} e^{-s_i}\right) + \log\left(1 + \sum_{j \in \Omega_{neg}} e^{s_j}\right) \tag{7}$$

where $s_i,\ s_j \in \mathbb{R}$ are the logits for label $l_i,\ l_j$ and $\Omega_{pos},\ \Omega_{neg}$ stand for the set of positive labels and negative labels, respectively. As HPT [29] and HJCL [33] baselines, we decided to adopt ZLPR as our default objective function. For a fair comparison, we also conducted empirical experiments with the BCE loss. In this case, as shown in Appendix, Table 4, we observe that HJCL exhibits a more significant improvement with the ZLPR loss than HLPD. This suggests that our good performances are attributable not only to the ZLPR loss but also to our Hierarchical Self-Attention mechanism.

References

1. Aly, R., Remus, S., Biemann, C.: Hierarchical multi-label classification of text with capsule networks. In: Alva-Manchego, F., Choi, E., Khashabi, D. (eds.) Proceedings of the 57th Annual Meeting of the Association for Computational Linguistics: Student Research Workshop, Florence, Italy, pp. 323–330. Association for Computational Linguistics (2019). https://doi.org/10.18653/v1/P19-2045
2. Banerjee, S., Akkaya, C., Perez-Sorrosal, F., Tsioutsiouliklis, K.: Hierarchical transfer learning for multi-label text classification. In: Korhonen, A., Traum, D., Màrquez, L. (eds.) Proceedings of the 57th Annual Meeting of the Association for Computational Linguistics, Florence, Italy, pp. 6295–6300. Association for Computational Linguistics (2019). https://doi.org/10.18653/v1/P19-1633
3. Cao, P., Chen, Y., Liu, K., Zhao, J., Liu, S., Chong, W.: HyperCore: hyperbolic and co-graph representation for automatic ICD coding. In: Jurafsky, D., Chai, J., Schluter, N., Tetreault, J. (eds.) Proceedings of the 58th Annual Meeting of the Association for Computational Linguistics, pp. 3105–3114. Association for Computational Linguistics (2020). https://doi.org/10.18653/v1/2020.acl-main.282
4. Chen, H., Ma, Q., Lin, Z., Yan, J.: Hierarchy-aware label semantics matching network for hierarchical text classification. In: Zong, C., Xia, F., Li, W., Navigli, R. (eds.) Proceedings of the 59th Annual Meeting of the Association for Computational Linguistics and the 11th International Joint Conference on Natural Language Processing (Volume 1: Long Papers), pp. 4370–4379. Association for Computational Linguistics (2021). https://doi.org/10.18653/v1/2021.acl-long.337
5. Chen, L., Chou, H., Zhu, X.: Developing prefix-tuning models for hierarchical text classification. In: Li, Y., Lazaridou, A. (eds.) Proceedings of the 2022 Conference on Empirical Methods in Natural Language Processing: Industry Track, Abu Dhabi, UAE, pp. 390–397. Association for Computational Linguistics (2022). https://doi.org/10.18653/v1/2022.emnlp-industry.39
6. Deng, Z., Peng, H., He, D., Li, J., Yu, P.: HTCInfoMax: a global model for hierarchical text classification via information maximization. In: Toutanova, K., et al. (eds.) Proceedings of

the 2021 Conference of the North American Chapter of the Association for Computational Linguistics: Human Language Technologies, pp. 3259–3265. Association for Computational Linguistics (2021). https://doi.org/10.18653/v1/2021.naacl-main.260

7. Devlin, J., Chang, M.W., Lee, K., Toutanova, K.: BERT: pre-training of deep bidirectional transformers for language understanding. In: Burstein, J., Doran, C., Solorio, T. (eds.) Proceedings of the 2019 Conference of the North American Chapter of the Association for Computational Linguistics: Human Language Technologies, Volume 1 (Long and Short Papers), Minneapolis, Minnesota, pp. 4171–4186. Association for Computational Linguistics (2019). https://doi.org/10.18653/v1/N19-1423
8. Goumy, S., Mejri, M.: Ecommerce product title classification. In: eCOM@SIGIR (2018). https://api.semanticscholar.org/CorpusID:67866207
9. Huang, W., et al.: Exploring label hierarchy in a generative way for hierarchical text classification. In: Calzolari, N., et al. (eds.) Proceedings of the 29th International Conference on Computational Linguistics, Gyeongju, Republic of Korea, pp. 1116–1127. International Committee on Computational Linguistics (2022). https://aclanthology.org/2022.coling-1.95
10. Jiang, H., et al.: Financial news annotation by weakly-supervised hierarchical multi-label learning. In: Chen, C.C., Huang, H.H., Takamura, H., Chen, H.H. (eds.) Proceedings of the Second Workshop on Financial Technology and Natural Language Processing, Kyoto, Japan, pp. 1–7 (2020). https://aclanthology.org/2020.finnlp-1.1
11. Jiang, T., Wang, D., Sun, L., Chen, Z., Zhuang, F., Yang, Q.: Exploiting global and local hierarchies for hierarchical text classification. In: Goldberg, Y., Kozareva, Z., Zhang, Y. (eds.) Proceedings of the 2022 Conference on Empirical Methods in Natural Language Processing, Abu Dhabi, United Arab Emirates, pp. 4030–4039. Association for Computational Linguistics (2022). https://doi.org/10.18653/v1/2022.emnlp-main.268
12. Kementchedjhieva, Y., Chalkidis, I.: An exploration of encoder-decoder approaches to multi-label classification for legal and biomedical text. In: Rogers, A., Boyd-Graber, J., Okazaki, N. (eds.) Findings of the Association for Computational Linguistics: ACL 2023, Toronto, Canada, pp. 5828–5843. Association for Computational Linguistics (2023). https://doi.org/10.18653/v1/2023.findings-acl.360
13. Kowsari, K., Brown, D.E., Heidarysafa, M., Meimandi, K.J., Gerber, M.S., Barnes, L.E.: Hdltex: hierarchical deep learning for text classification. In: 2017 16th IEEE International Conference on Machine Learning and Applications (ICMLA), pp. 364–371 (2017). https://doi.org/10.1109/ICMLA.2017.0-134
14. Lai, S., Xu, L., Liu, K., Zhao, J.: Recurrent convolutional neural networks for text classification. In: AAAI Conference on Artificial Intelligence (2015). https://api.semanticscholar.org/CorpusID:16756501
15. Lewis, D.D., Yang, Y., Rose, T.G., Li, F.: RCV1: a new benchmark collection for text categorization research. J. Mach. Learn. Res. **5**, 361–397 (2004). https://api.semanticscholar.org/CorpusID:11027141
16. Liu, F., Shakeri, S., Yu, H., Li, J.: Enct5: fine-tuning T5 encoder for non-autoregressive tasks. CoRR abs/2110.08426 (2021). https://arxiv.org/abs/2110.08426
17. Liu, Y., et al.: Enhancing hierarchical text classification through knowledge graph integration. In: Rogers, A., Boyd-Graber, J., Okazaki, N. (eds.) Findings of the Association for Computational Linguistics: ACL 2023, Toronto, Canada, pp. 5797–5810. Association for Computational Linguistics (2023). https://doi.org/10.18653/v1/2023.findings-acl.358
18. Raffel, C., et al.: Exploring the limits of transfer learning with a unified text-to-text transformer. J. Mach. Learn. Res. **21**(140), 1–67 (2020). http://jmlr.org/papers/v21/20-074.html
19. Rivas Rojas, K., Bustamante, G., Oncevay, A., Sobrevilla Cabezudo, M.A.: Efficient strategies for hierarchical text classification: external knowledge and auxiliary tasks. In: Jurafsky, D., Chai, J., Schluter, N., Tetreault, J. (eds.) Proceedings of the 58th Annual Meeting of the

Association for Computational Linguistics, pp. 2252–2257. Association for Computational Linguistics (2020). https://doi.org/10.18653/v1/2020.acl-main.205
20. Rothe, S., Maynez, J., Narayan, S.: A thorough evaluation of task-specific pretraining for summarization. In: Moens, M.F., Huang, X., Specia, L., Yih, S.W.T. (eds.) Proceedings of the 2021 Conference on Empirical Methods in Natural Language Processing, Punta Cana, Dominican Republic, pp. 140–145. Association for Computational Linguistics (2021). https://doi.org/10.18653/v1/2021.emnlp-main.12
21. Shaw, P., Uszkoreit, J., Vaswani, A.: Self-attention with relative position representations. In: Walker, M., Ji, H., Stent, A. (eds.) Proceedings of the 2018 Conference of the North American Chapter of the Association for Computational Linguistics: Human Language Technologies, Volume 2 (Short Papers), New Orleans, Louisiana, pp. 464–468. Association for Computational Linguistics (2018). https://doi.org/10.18653/v1/N18-2074
22. Shazeer, N., Stern, M.: Adafactor: adaptive learning rates with sublinear memory cost. In: Dy, J., Krause, A. (eds.) Proceedings of the 35th International Conference on Machine Learning. Proceedings of Machine Learning Research, vol. 80, pp. 4596–4604. PMLR (2018). https://proceedings.mlr.press/v80/shazeer18a.html
23. Shen, J., Qiu, W., Meng, Y., Shang, J., Ren, X., Han, J.: TaxoClass: hierarchical multi-label text classification using only class names. In: Toutanova, K., et al. (eds.) Proceedings of the 2021 conference of the North American Chapter of the Association for Computational Linguistics: Human Language Technologies, pp. 4239–4249. Association for Computational Linguistics (2021). https://doi.org/10.18653/v1/2021.naacl-main.335
24. Shimura, K., Li, J., Fukumoto, F.: HFT-CNN: learning hierarchical category structure for multi-label short text categorization. In: Riloff, E., Chiang, D., Hockenmaier, J., Tsujii, J. (eds.) Proceedings of the 2018 Conference on Empirical Methods in Natural Language Processing, Brussels, Belgium, pp. 811–816. Association for Computational Linguistics (2018). https://doi.org/10.18653/v1/D18-1093
25. Su, J., Zhu, M., Murtadha, A., Pan, S., Wen, B., Liu, Y.: ZLPR: a novel loss for multi-label classification. arXiv abs/2208.02955 (2022). https://api.semanticscholar.org/CorpusID:251371673
26. Torba, F., Gravier, C., Laclau, C., Kammoun, A., Subercaze, J.: A study on hierarchical text classification as a seq2seq task. In: European Conference on Information Retrieval (2024). https://api.semanticscholar.org/CorpusID:268752649
27. Vaswani, A., et al.: Attention is all you need. In: Neural Information Processing Systems (2017). https://api.semanticscholar.org/CorpusID:13756489
28. Wang, Z., Wang, P., Huang, L., Sun, X., Wang, H.: Incorporating hierarchy into text encoder: a contrastive learning approach for hierarchical text classification. In: Proceedings of the 60th Annual Meeting of the Association for Computational Linguistics (Volume 1: Long Papers), Dublin, Ireland, pp. 7109–7119. Association for Computational Linguistics (2022). https://aclanthology.org/2022.acl-long.491
29. Wang, Z., et al.: HPT: hierarchy-aware prompt tuning for hierarchical text classification. In: Goldberg, Y., Kozareva, Z., Zhang, Y. (eds.) Proceedings of the 2022 Conference on Empirical Methods in Natural Language Processing, Abu Dhabi, United Arab Emirates, pp. 3740–3751. Association for Computational Linguistics (2022). https://doi.org/10.18653/v1/2022.emnlp-main.246
30. Wehrmann, J., Cerri, R., Barros, R.: Hierarchical multi-label classification networks. In: Dy, J., Krause, A. (eds.) Proceedings of the 35th International Conference on Machine Learning. Proceedings of Machine Learning Research, vol. 80, pp. 5075–5084. PMLR (2018). https://proceedings.mlr.press/v80/wehrmann18a.html
31. Yang, P., Sun, X., Li, W., Ma, S., Wu, W., Wang, H.: SGM: sequence generation model for multi-label classification. In: Bender, E.M., Derczynski, L., Isabelle, P. (eds.) Proceedings of the 27th International Conference on Computational Linguistics,Santa Fe, New

Mexico, USA, pp. 3915–3926. Association for Computational Linguistics (2018). https://aclanthology.org/C18-1330
32. Yu, C., Shen, Y., Mao, Y.: Constrained sequence-to-tree generation for hierarchical text classification. In: Proceedings of SIGIR, pp. 1865–1869. ACM (2022)
33. Yu, S.C.L., He, J., Basulto, V., Pan, J.: Instances and labels: hierarchy-aware joint supervised contrastive learning for hierarchical multi-label text classification. In: Bouamor, H., Pino, J., Bali, K. (eds.) Findings of the Association for Computational Linguistics: EMNLP 2023, Singapore, pp. 8858–8875. Association for Computational Linguistics (2023). https://doi.org/10.18653/v1/2023.findings-emnlp.594
34. Zheng, L., Xiong, J., Zhu, Y., He, J.: Contrastive learning with complex heterogeneity. In: Proceedings of the 28th ACM SIGKDD Conference on Knowledge Discovery and Data Mining (2021). https://api.semanticscholar.org/CorpusID:250886991
35. Zhou, J., et al.: Hierarchy-aware global model for hierarchical text classification. In: Jurafsky, D., Chai, J., Schluter, N., Tetreault, J. (eds.) Proceedings of the 58th Annual Meeting of the Association for Computational Linguistics, pp. 1106–1117. Association for Computational Linguistics (2020). https://doi.org/10.18653/v1/2020.acl-main.104
36. Zhou, X., Gupta, A., Upadhyay, S., Bansal, M., Faruqui, M.: Can sequence-to-sequence transformers naturally understand sequential instructions? In: Palmer, A., Camacho-collados, J. (eds.) Proceedings of the 12th Joint Conference on Lexical and Computational Semantics (*SEM 2023), Toronto, Canada, pp. 527–534. Association for Computational Linguistics (2023). https://doi.org/10.18653/v1/2023.starsem-1.45

A Multi-modal Recipe for Improved Multi-domain Recommendation

Zixuan Yi(✉) and Iadh Ounis

University of Glasgow, Glasgow, UK
z.yi.1@research.gla.ac.uk, iadh.ounis@glasgow.ac.uk

Abstract. Multi-domain recommendation (MDR) aims to enhance recommendation accuracy by leveraging knowledge from various source domains to benefit a target domain using a universal model. However, traditional MDR approaches often fail in leveraging the complex users' preferences spanning multiple domains, especially in the case of users that are active in more than one domain. Hence, a key problem in MDR is how to better leverage the diverse but disparate user/item interaction data available across multiple domains. We argue that the concept of "model soups", which involves combining multiple models through averaging, presents a promising method for unifying the users' profiles and item representations from diverse domains, thereby potentially improving the multi-domain recommendation performance. We propose AdapterSoupRec, a novel approach that employs the "model soup" technique to enrich the multi-modal item contents. Our approach involves averaging the weights of Multi-modal Large Language Models (MLLMs) trained across diverse source domains in order to enhance the recommendation performance in the target domain. To enable a universal item multi-modal representation learning for recommendation purposes, we train an MLLM using the items' images and descriptions. Then, we aggregate the weights of several MLLMs, each trained with various hyper-parameter configurations. This process is akin to blending various soup ingredients. Furthermore, we incorporate a graph neural network as an adapter into AdapterSoupRec to effectively combine the universal item representations obtained from the MLLM and the users' preferences across multiple domains. We conduct extensive experiments on three e-commerce datasets and show that AdapterSoupRec consistently improves performance in the target domains over nine existing state-of-the-art recommendation approaches. In particular, AdapterSoupRec outperforms the best baseline, BM3, significantly by up to 18.35%. We also show that the model soup technique further benefits modality fusion for item representation learning and exhibits strong out-of-domain performance in a video recommendation scenario.

1 Introduction

Recommender systems have traditionally focused on single-domain recommendation approaches, optimising recommendations within a single context or product

C. Hauff et al. (Eds.): ECIR 2025, LNCS 15572, pp. 421–436, 2025.
https://doi.org/10.1007/978-3-031-88708-6_27

category [3,22]. However, the growth of e-commerce has led users to increasingly interact with diverse content and product categories [14], prompting the need for recommendations that span multiple interconnected domains. Multi-domain recommender (MDR) systems aim to leverage overlapping user interests and item associations across various domains to enhance the recommendation performance on the target domains [43].

MDR systems typically leverage the shared knowledge obtained from the users' interactions across various domains to construct a more complete estimation of the users' preferences and item characteristics. These approaches allow to produce more detailed and personalised recommendations by capturing the commonalities in the users' interests, for example, when a user likes both science fiction books and the latest technology gadgets. This interplay between domains enriches the data and provides an abundant ground for cross-domain representation learning and knowledge transfers [18]. However, the complexity of the MDR systems stems from their need to balance the domain-specific characteristics of items with the generalisation of the users' preferences across multiple domains. Indeed, simply applying a universal model may overlook the important differences that define or characterise each domain. Conversely, maintaining distinct models for each domain is resource-intensive and less scalable [34,43]. Therefore, the challenge in developing an effective MDR system is to design an approach that retains both the depth and specificity of domain-focused models while ensuring the broad applicability of a universal model.

In addition, existing MDR approaches [19,27,39] primarily rely on using IDs as user/item features. However, relying solely on IDs might lead to an inadequate item representation learning because of the sparsity of interaction data [18,21,35,36]. This sparsity prevents the MDR models from sharing transferable knowledge between multiple similar domains, thereby constraining the effectiveness of IDs-based models. In this paper, we argue that incorporating multi-modal item contents, such as the item images and the item descriptions, offers richer semantic information, enabling a better shared knowledge across multiple domains. With recent advances in computer vision [8] and natural language processing (NLP) [28], large-scale vision and language models now possess the ability to extract semantic content from items with a high degree of generalisation. Inspired by the recent success of the *model soup* technique [5], which enhances the large language models for an improved out-of-domain performance, we use the expressive power of Multi-modal Large Language Models (MLLMs) [24,30] to model the item multi-modal representations in MDR. To better leverage the diverse but disparate user/item data available in MDR, we propose AdapterSoupRec, a novel approach that applies the model soup technique [31] to MLLMs (namely, CLIP [24] and LLaVA [15]) in order to enhance performance in MDR systems. In particular, we train these 'soup-enhanced' MLLMs across various source domains to effectively learn item representations for recommendation purposes. To ensure that the resulting model retains the domain-specific knowledge, we additionally employ an adapter to effectively enrich the obtained generalised item features, enhancing their adaptability to each target domain.

To summarise, our contributions in this paper are fivefold: (1) We propose AdapterSoupRec, which leverages the soup-enhanced MLLMs and a graph adapter to ensure that the resulting MDR model retains both the domain-specific and common shared knowledge across multiple domains. To the best of our knowledge, we are the first to apply the model soup technique on a Multi-modal Large Language Model so as to model universal item representations in the top-K (MDR) task; (2) We use two distinct MLLMs, namely CLIP and LLaVA, to model the items' multi-modal contents instead of only their IDs and provide more generalisable semantic item features across multiple domains; (3) We conduct extensive experiments on three target e-commerce datasets and demonstrate the superiority of AdapterSoupRec on the MDR task. We show that AdapterSoupRec outperforms nine strong baselines, and yields improvements of up to 18.35% over the best baseline. We also examine different recipes such as a uniform soup or a greedy soup for an improved performance; (4) We also experiment with different AdapterSoupRec variants and conclude that the greedy soup variant effectively fuses multi-modal item representations, thereby improving the MDR performance; (5) We extend the use of AdapterSoupRec, originally trained in an e-commerce context, to a video recommendation scenario. Our findings show that AdapterSoupRec exhibits a superior out-of-domain performance compared to the best-performing baseline, BM3.

2 Related Work

2.1 Multi-domain Recommendation

Shifting from the concept of Cross-Domain Recommendation (CDR), which focuses on transferring knowledge from a data-rich source domain to enhance a target domain, Multi-Domain Recommendation (MDR) systems optimise performance across various domains using a unified model [42]. This shift motivates the need for a universal recommendation model that effectively generalises across multiple target domains, thereby simplifying the integration of domain-specific knowledge without the necessity for separate models for each domain [43]. Existing MDR approaches such as MMOE [19], PLE [27], and STAR [39] typically incorporate ID features shared across multiple domains and task-specific components for domain representation. However, these MDR approaches are affected by domain conflicts where overlapping or contradictory domain characteristics can disrupt the model's learning process. They are also prone to generalisation issues, especially in sparse domains where limited data leads to overfitting [23]. Different from the above MDR models, we use multi-modal large language models (MLLMs) to enable generalisation by obtaining universal representations for the multi-modal items' contents. After training the MLLMs from the source domains, we leverage a model soup technique to further enhance the MLLMs' performance on a target domain outside of these sources, thereby improving the recommendation performance in a target domain.

2.2 Model Soups of LLMs

The linear interpolation of neural network weights has recently garnered attention [1,2,20]. Such an interpolation results in an improved accuracy across various domains [13]. The model soup technique, first proposed by Wortsman et al. [31], includes either a uniform or greedy mechanism for averaging weights of multiple large language models with varying hyper-parameter configurations. The greedy soup selectively combines the best-performing model parameters based on the validation set's accuracy, whereas the uniform soup averages parameters across various model configurations. These model soup techniques have been shown to significantly improve tasks such as image classification and text summarisation. Later works [6,13] found that averaging fine-tuned language models improves out-of-domain generalisation in the text summarisation task. The model soup technique involves blending multiple model variants, each variant optimised under different conditions, to generate a more adaptable model ensemble. Such a procedure, which aggregates the best aspects of different configurations, prevents overfitting to specific domains and enhances the used model's ability to generalise to a new domain. Hence, we argue that the model soup technique can be particularly suitable for MDR scenarios, since it helps prevent overfitting within specific source domains and improves the MLLMs' capacity to adapt to a new target domain. In this work, we apply the model soup technique to two MLLMs (CLIP [24] and LLaVA [15]) so as to enable generalisable item representation learning in the MDR task. Additionally, we leverage a graph neural network as an adapter with the MLLM to incorporate personalised information (i.e., user-item interactions), thus retaining the domain-specific knowledge for each target domain. To the best of our knowledge, we propose the first soup-enhanced MLLMs for enabling effective multi-domain recommendations.

3 Methodology

In this section, we present our AdapterSoupRec approach for the top-K multi-domain recommendation task. First, we state the multi-domain recommendation problem. Next, in Sect. 3.2, we present how to have removing the mtem embeddings using two distinct MLLMs. Section 3.3 describes AdapterSoupRec with different model soups and the personalised adapter for each target domain.

3.1 Problem Statement

Given a multi-modal large language model f_θ adapted to n source domains $\{D_1, \ldots, D_n\}$, the Multi-Domain Recommendation (MDR) task aims to return the top-K items for a target user u in a target domain D_{n+1}. For each domain D, we use $m \in \mathcal{M} = \{v, t\}$ as the modality indicator, where v and t represent the visual and textual modalities[1] of each item, respectively. Note that there

[1] We use these two modalities, namely item images and textual descriptions, since most available datasets have only these two types of raw data.

is no constraint on the availability of overlapping users between the source and target domains. Instead, we devise a universal MDR approach to learn multi-modal item representations $\mathcal{I}_m$ and user preferences $\mathcal{U}_m$ from all the domains. This ensures that AdapterSoupRec effectively functions with both overlapping and non-overlapping users, thereby allowing its applicability in diverse domains.

3.2 Universal Multi-modal Item Representation Learning

As discussed in Sect. 1, we extract universal and shared knowledge from the source domains. Hence, we train a Multi-modal Large Language Model (MLLM), such as CLIP or LLaVA, with multi-modal item contents from the source domains. More specifically, we obtain the textual item representations as follows:

$$x_i^{(l)} = \text{LN}\left(x_i^{(l-1)} + \text{MSA}(x_i^{(l-1)}, x_i^{(l-1)}, m_i)\right), \tag{1}$$

where $x_i^{(l)}$ is the hidden state of the i-th item in the l-th layer, while LN and MSA are the layer normalisation operation and the multi-head self-attention mechanism, respectively. m_i corresponds to the textual embedding. Similarly, we obtain the visual item representations. As such, the used MLLM learns the semantic representations of items from multiple source domains, thereby mapping all learned items into a unified space. Following [15,29,32], we select two configurations of our AdapterSoupRec model: CLIP ViT-B/16 and LLaVA-7b. We encode the item images along with the item descriptions using these two configurations and train each configuration using a loss function that linearly combines the MLLM's own loss function with a BPR loss [25]: $\mathcal{L} = \mathcal{L}_{bpr} + \lambda_1 \mathcal{L}_{mllm} + \lambda_2 \|\Theta\|_2$, where λ_1 and λ_2 are hyper-parameters that control the strengths of the MLLM's loss and the L_2 regularisation, respectively, whereas Θ is the set of model's parameters.

3.3 AdapterSoupRec: A Model Soup Approach

Model Soups: We now describe the recipes for the uniform and greedy soups using the obtained MLLMs' parameters. After pre-training with the multi-modal item contents from n source domains, we obtain an MLLM model $f(x, \theta)$ with data x and parameters $\theta \in \mathbb{R}^d$. Let $\theta = \text{PreTrain}\ (\theta_0, p)$ denote the parameters obtained with a pre-trained initialisation θ_0 and a hyper-parameter configuration p. The hyper-parameter configuration includes the learning rate, training iterations, and a random seed, which will determine the data order [31]. To obtain multiple configurations of MLLMs' parameters, we define $\theta_i = \text{PreTrain}\ (\theta_0, p_i)$ where each p_i denotes a different hyper-parameter configuration. In traditional machine learning practices, we select the parameters θ_i that achieve the highest accuracy on a held-out validation set and discard the remaining parameters [31]. Instead, the model soup technique averages the MLLM's configurations with all parameters θ_i – the so-called uniform soup – as follows:

$$\text{AdapterSoupRec}(x) = f\left(x, \frac{1}{n}\sum_{i=1}^{n}\theta_i\right), \tag{2}$$

Algorithm 1. Greedy Interpolation Soup Procedure

Input: Configurations of Model Weight $\Theta = \{\theta_1, \ldots, \theta_n\}$
\# sort the ingredients in decreasing order of validation accuracy ValAcc (θ_i)
$\Theta_{\text{sorted}} \leftarrow \text{SORT}_{\text{ValAcc}}(\Theta)$
soup $\leftarrow \Theta_{\text{sorted}}[0]$
for $i = 1$ to n **do**
    if ValACC (interpolate(soup, Θ_i, α)) $\geq$ ValACC (soup) **then**
        soup $\leftarrow$ interpolate(soup, Θ_i, α)
end for
Return soup

where n is the number of the MLLM's configurations, and x is the resulting user/item embedding. As such, we obtain an MLLM by interpolating different configurations of the MLLM's parameters with the original model size. Another recipe for enhancing the model soup involves creating a greedy soup by sequentially adding the MLLM with different parameters as potential ingredients. Algorithm 1 presents the procedure of the greedy soup. The greedy soup only retains the configurations that show an improved performance on a held-out validation set in the soup, which is disjoint from the training and test sets: $textAdapterSoupRec(x) = f\left(x, \frac{1}{n'} \sum_{i=1}^{n'} \theta_i\right)$, where n' is the number of selected MLLM's parameters to obtain a new set of parameters for an MLLM in the target domain and $n' <= n$. If $n' = n$, this model becomes akin to a uniform soup. As such, the greedy soup prevents overfitting to specific source domains and enhances the used model's ability to generalise to a new target domain by using diverse configurations of MLLM parameters.

Personalised Adapter: Adapters [5,9] and mixture of experts [4,26] have been shown by [7] to be effective in enhancing domain adaptation. In this work, we enhance domain adaptation by adding an adapter into the used MLLM in each target domain. Specifically, we use a lightweight graph neural network [12] as a graph adapter to incorporate the user-item interactions in a target domain. This graph adapter enables the MLLM to learn domain-specific knowledge, thereby enriching the multi-modal item features produced by the MLLM. We define a graph adapter with parameters θ_α and fine-tune these parameters in a target domain. Notably, we only fine-tune the adapters' parameters θ_α without updating the MLLM's parameter θ, ensuring an efficient adaption to the target domain. We leverage the resulting graph adapter to tailor the MLLM for a target domain by aggregating the multi-modal features obtained from the MLLM:

$$x_{u-m}^{(l_\alpha)} = \sum_{i \in \mathcal{N}_u} \frac{x_{i-m}^{(l_\alpha - 1)}}{\sqrt{|\mathcal{N}_i|\,|\mathcal{N}_u|}}, \tag{3}$$

where l_α is the number of layers. $\mathcal{N}_i$ and $\mathcal{N}_u$ denote the set of neighbours for user u and item i, respectively, while $|\mathcal{N}_u|$ and $|\mathcal{N}_i|$ represent the size of $\mathcal{N}_u$ and

$\mathcal{N}_i$, respectively. As a result, we aggregate the multi-modal features according to the interactions between the users and items, thereby incorporating personalised information in the obtained features from the used MLLM. We optimise the graph adapter using the BPR loss to effectively capture domain-specific knowledge.

Once we obtain both visual and textual user embeddings from the graph adapter for the target domain, we then concatenate them for the later top-K multi-domain recommendation task:

$$x_{u-vt} = x_{u-v} \parallel x_{i-t}, \tag{4}$$

where $\parallel$ represents the concatenation operation. Analogously, we obtain the multi-modal embeddings of the items. By applying a graph adapter on top of the MLLM, we incorporate domain-specific knowledge from the user-item interactions of the target domain, thereby enhancing the recommendation accuracy.

Table 1. Statistics of the used datasets.

Datasets	#Users	#Items	#Interactions	Sparsity
Food	115,349	39,670	1,027,413	99.99%
Home	731,913	185,552	6,451,926	99.99%
Clothing	39,387	23,033	237,488	99.97%
Office	87,436	25,986	684,837	99.97%
Pantry	13,101	4,898	126,962	99.80%
Electronics	192,403	63,001	1,689,188	99.99%
Sports	87,436	25,986	684,837	99.97%

4 Experiments

To demonstrate the effectiveness of our AdapterSoupRec approach, and illustrate the reasons for its effectiveness, we conduct experiments to answer the following research questions: **RQ1**: Does AdapterSoupRec outperform the baselines in top-K multi-domain recommendation, and which variant – MLLMs, uniform or greedy soup – is more effective? **RQ2**: What is the performance impact of our AdapterSoupRec model's key components – the model soup method and the graph adapter – along with its pre-training strategy? **RQ3**: Do the uniform and greedy model soup techniques enhance multi-modal fusion in item representations more effectively than non-soup variants of AdapterSoupRec? **RQ4**: Does AdapterSoupRec demonstrate a superior out-of-domain performance compared to the best-performing multi-modal baselines?

4.1 Experimental Setups

Datasets: To evaluate our AdapterSoupRec method, we conduct experiments on 7 domains of a widely used multi-domain benchmark dataset, namely the Amazon Review dataset [10]. Table 1 shows the statistics of the used datasets. Specifically, we pre-train the Food, Home, Clothing, and Office datasets as the source domain datasets and evaluate the effectiveness of multi-domain recommendation on the Pantry, Electronics and Sports datasets as target datasets. For each dataset in the source domains, we pre-train the MLLMs using the URLs of the items' images as well as the detailed items' descriptions including their titles, categories and brands.

Baselines: We compare our proposed AdapterSoupRec[2] approach with three major groups of recommendation baselines: (1) Multi-modal methods: **VBPR** [11] is a recommender, which exclusively integrates the visual features with the user/item IDs into a matrix factorisation to allow recommendation; **MMGCL** [37] introduces a modality edge dropout and modality masking augmentations to the concatenated multi-modal user/item embeddings, enhancing multi-modal representation learning through a self-supervised learning paradigm; **BM3** [41] bootstraps the latent user/item representations in contrastive learning by reconstructing the user-item interaction graph to enhance the recommendation performance; (2) Cross-domain methods: **PPGN** [40]**:** PPGN fuses the interaction information of the two domains into a graph, and shares the features of users learned from the joint interaction graph by stacking multiple graph convolution layers to obtain the user/item embeddings; **BiTGCF** [16]**:** This approach extends LightGCN for cross-domain recommendation tasks by employing dual linear graph encoders to generate user and item representations in each domain. Subsequently, a feature transfer layer is used to fuse user representations across domains, capturing the underlying relationships and facilitating enhanced recommendations in the target domain; (3) Multi-domain methods: **MOME** [19] uses multiple expert networks and a gating network that selects a relevant subset of experts for each target recommendation domain, thereby enhancing the recommendation accuracy in those target domains; **PLE** [27] distinguishes between the task-shared and task-specific experts and uses a progressive routing mechanism to dynamically route the target domain recommendations through the appropriate experts; **MGFN** [39] uses Graph Attention Networks to learn both intra-domain and inter-domain knowledge, enhancing the model's ability to facilitate recommendations across multiple domains.

---

[2] Our source code is available at: https://github.com/zxy-ml84/SoupRec.

Evaluation Protocol and Hyper-Parameter Settings: We randomly split the given datasets into training, validation, and testing sets with a 7:1:2 ratio. We adopt two widely used evaluation metrics, namely Recall@K and NDCG@K, to evaluate the performance of top-K recommendation. Following [38], we set $K = 20$ and report the average performance achieved for all users in the testing set. To ensure a fair comparison between AdapterSoupRec and the baselines, we use the Xavier initialisation and the Adam optimiser with a batch size of 2048 for all tested models.

We apply a grid search to tune the hyper-parameters on both the AdapterSoupRec's CLIP variant (i.e., AdapterSoupRec (CLIP)) and all used baselines on the validation set, with the learning rate ranging within $\{10^{-2}, 10^{-3}..., 10^{-6}\}$ and the batch size varying within $\{32, 64, ..., 512, 1024\}$. For tuning the hyper-parameters specific to the AdapterSoupRec's LLaVA variant (i.e., AdapterSoupRec (LLaVA)), which configures a Large Language Model (e.g., Llama) with a higher number of specific parameters that are not used in the baseline models, we use the following settings: we used a batch size of 8, a token length of 2048, varied the learning rate within $\{0.001, 0.002, ..., 0.005\}$, the number of epochs within $\{1, 2, 3\}$, the warm-up ratio within $\{0.01, 0.02, ..., 0.10\}$ and the weight decay within $\{0, 0.01, ..., 0.05\}$. In addition, we use the Multi-Layer Perceptrons to reduce the dimensionality of the MLLMs' multi-modal item embeddings from 768 to 128, so as to match the multi-modal baselines, thereby ensuring a fair comparison.

4.2 Performance Comparison (RQ1)

We compare our proposed AdapterSoupRec method with all the baselines. Table 2 presents the comparison results. In Table 2, "Uni" and "Grd" denote the uniform and greedy soup, respectively. From the observed results on the three datasets, we observe that all AdapterSoupRec variants (CLIP-Uni, CLIP-Grd, LLaVA-Uni, LLaVA-Grd) significantly outperform all the baselines in most instances (except in 2 out of 54 instances) on both used metrics as confirmed by a paired t-test with the Holm-Bonferroni correction. These results indicate that AdapterSoupRec is a promising approach for enabling an effective transfer of both domain-specific and common shared knowledge in MDR. Comparing the multi-modal baselines (VBPR, MMGCL, LATTICE, BM3), which combine multi-modal features with ID features against the cross-domain baselines (PPGN, BiTGCF) that only use IDs, we observe that the former group of models incorporating multi-modal item contents are overall more effective, suggesting a more successful knowledge transfer. Notably, AdapterSoupRec only use multi-modal features without IDs, simplifying the model structure by avoiding the complexity of aligning multi-modal and ID embeddings. These approaches consistently demonstrate a superior performance against the multi-modal baselines.

Table 2. Comparison of AdapterSoupRec (greedy) with the baselines. * denotes a significant difference with a baseline using the corrected paired t-test with $p < 0.05$.

Dataset	Pantry		Electronics		Sports	
Methods	Recall@20	NDCG@20	Recall@20	NDCG@20	Recall@20	NDCG@20
VBPR	0.0723*	0.0326*	0.0442*	0.0196*	0.0771*	0.0349*
MMGCL	0.0907*	0.0377*	0.0627*	0.0304*	0.0913*	0.0428*
LATTICE	0.0918*	0.0409*	0.0618*	0.0296*	0.0944*	0.0424*
BM3	0.0932*	0.0417*	0.0638	0.0310*	0.0970*	0.0438*
PPGN	0.0774*	0.0341*	0.0425*	0.0173*	0.0836*	0.0366*
BiTGCF	0.0823*	0.0360*	0.0552*	0.0247*	0.0870*	0.0383*
MOME	0.0797*	0.0352*	0.0573*	0.0261*	0.0749*	0.0318*
PLE	0.0862*	0.0384*	0.0595*	0.0278*	0.0866*	0.0367*
MGFN	0.0891*	0.0413*	0.0623*	0.0305*	0.0894*	0.0383*
AdapterSoupRec (CLIP-Uni)	0.1021*	0.0441*	0.0633*	0.0304*	0.0975*	0.0450*
AdapterSoupRec (LLaVA-Uni)	0.1061*	<u>0.0455</u>*	0.0642	0.0317*	<u>0.1023</u>*	<u>0.0464</u>*
AdapterSoupRec (CLIP-Grd)	<u>0.1074</u>*	0.0452*	<u>0.0646</u>	<u>0.0319</u>*	0.0985*	0.0461*
AdapterSoupRec (LLaVA-Grd)	**0.1103**	**0.0477**	**0.0656**	**0.0328**	**0.1068**	**0.0480**

We also evaluate the performance of the multi-domain methods (MOME, PLE, MMGFN, AdapterSoupRec) in comparison to the cross-domain methods. The improved performance observed in Table 2 for the multi-domain methods indicates that training on multiple source domains, as opposed to one source domain in cross-domain setting, indeed enhances the recommendation performance on the target datasets. Comparing all the AdapterSoupRec variants in Table 2, we observe that the AdapterSoupRec (LLaVA) variants outperform the CLIP variants by a large margin across all used datasets. These results suggest that on the used datasets, LLaVA is more effective than CLIP as an MLLM for modelling the universal user/item multi-modal representations in the MDR task. Table 2 also shows that the greedy variants of AdapterSoupRec, i.e., LLaVA-Grd CLIP-Grd, outperform all their uniform (Uni) variants on the used datasets. These results suggest that the greedy soup, which selectively interpolates effective weights based on the recommendation accuracy on the validation set, is a better choice for the MDR task on the used datasets. This is also consistent with the model soup's observed generalisation in the text summarisation task [13].

In answer to RQ1, we conclude that our AdapterSoupRec approach, which includes a soup-enhanced MLLM and a graph adapter for each target domain, is overall promising in enabling an effective knowledge transfer across all target domains. In particular, we found that LLaVA is a more effective MLLM than CLIP when used along with AdapterSoupRec. Moreover, we showed that the greedy soup is the most effective model soup method for MDR on the used datasets.

4.3 Ablation Study (RQ2)

Table 3. Results for ablating the key components of CLIP/LLaVA AdapterSoupRec. * indicates a significant difference using a paired t-test with $p < 0.05$.

Dataset	Pantry		Electronics		Sports	
Variants	Recall@20	NDCG@20	Recall@20	NDCG@20	Recall@20	NDCG@20
AdapterSoupRec (CLIP-Uni) w/o ModelSoup	0.0987*	0.0421*	0.0620	0.0287*	0.0954*	0.0417*
AdapterSoupRec (CLIP-Uni) w/o Adapter	0.0794*	0.0336*	0.0521*	0.0223*	0.0817*	0.0357*
AdapterSoupRec (CLIP-Uni) w/o PreTrain	0.0940*	0.0426*	0.0618*	0.0292	0.0890*	0.0383*
AdapterSoupRec (CLIP-Uni)	**0.1021**	**0.0441**	**0.0633**	**0.0304**	**0.0975**	**0.0450**
AdapterSoupRec (CLIP-Grd) w/o ModelSoup	0.0946*	0.0434*	0.0624*	0.0306	0.0891*	0.0389*
AdapterSoupRec (CLIP-Grd) w/o Adapter	0.0823*	0.0350*	0.0534*	0.0232*	0.0803*	0.0323*
AdapterSoupRec (CLIP-Grd) w/o PreTrain	0.0914*	0.0398*	0.0621*	0.0303*	0.0883*	0.0386*
AdapterSoupRec (CLIP-Grd)	**0.1074**	**0.0452**	**0.0646**	**0.0319**	**0.0924**	**0.0410**
AdapterSoupRec (LLaVA-Uni) w/o ModelSoup	0.1042*	0.0438*	0.0624*	0.0303	0.0975*	0.0442*
AdapterSoupRec (LLaVA-Uni) w/o Adapter	0.0806*	0.0330*	0.0534*	0.0245*	0.0841*	0.0368*
AdapterSoupRec (LLaVA-Uni) w/o PreTrain	0.0917*	0.0396*	0.0626*	0.0293*	0.0968*	0.0440*
AdapterSoupRec (LLaVA-Uni)	**0.1061**	**0.0455**	**0.0642**	**0.0317**	**0.1023**	**0.0464***
AdapterSoupRec (LLaVA-Grd) w/o ModelSoup	0.0971*	0.0443*	0.0634*	0.0311	0.0942*	0.0421*
AdapterSoupRec (LLaVA-Grd) w/o Adapter	0.0814*	0.0336*	0.0552*	0.0265*	0.0824*	0.0354*
AdapterSoupRec (LLaVA-Grd) w/o PreTrain	0.0933*	0.0414*	0.0610*	0.0288*	0.0945*	0.0426*
AdapterSoupRec (LLaVA-Grd)	**0.1103**	**0.0477**	**0.0656**	**0.0328**	**0.1068**	**0.0461**

In this section, we conduct an ablation study to assess the impact of the main components in our AdapterSoupRec method, namely the model soup (ModelSoup) and the adapter (Adaptor). We also evaluate the pre-training strategy (PreTrain), which refers to pre-training AdapterSoupRec with multi-modal item contents from the source datasets. We assess the performance of each component within four AdapterSoupRec variants (CLIP-Uni, CLIP-Grd, LLaVA-Uni and LLaVA-Grd). First, to gauge the effectiveness of the model soup technique using MLLM, we conduct a comparative analysis by removing the model soup procedure from each AdapterSoupRec's CLIP or LLaVA variant, in both their 'Uni' and 'Grd' configurations. From Table 3, we observe that all AdapterSoupRec variants (CLIP-Uni, CLIP-Grd, LLaVA-Uni and LLaVA- Grd) significantly outperform their "w/o ModelSoup" variants, where the model soup technique has been removed, in 20 out of 24 instances. This result confirms that blending the parameters of multiple MLLMs results in an improved multi-domain recommendation performance. Next, we ablate the adapter in our AdapterSoupRec method, so as to examine its usefulness. By comparing the CLIP/LLaVA variants in both their uniform (Uni) and greedy (Grd) configurations with their corresponding variants without the adapter (w/o Adapter) in Table 3, we observe that there is a significant decrease in recommendation performance when removing the graph adapter. This confirms the necessity of including personalised information such as user-item interactions in order to enable domain-specific knowledge in the target domain. Moreover, Table 3 shows that pre-training these MLLMs on recommendation data enables a more effective item modelling than their variants without pre-training (w/o PreTrain). Such an observation indicates that these MLLMs become better adapted to the recommendation domain when trained with the recommendation datasets.

Overall, in response to RQ2, we conclude that our AdapterSoupRec method successfully leverages each of its two key components – the model soup and graph adapter. In addition, pre-training is an effective strategy that significantly enhances the performance of multi-domain recommendation.

4.4 Multi-modal Fusion (RQ3)

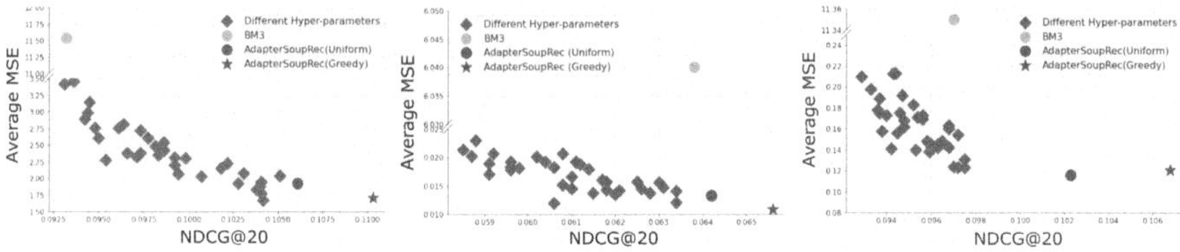

Pantry Electronic Sports

Fig. 1. Effectiveness of the AdapterSoupRec model compared to multi-modal recommendation baselines.

We have already shown in Table 3 that the model soup variants outperform the non-model soup variants (i.e., their single hyper-parameter configurations). In this section, we analyse whether the improved performance is reflected in the quality of features produced by the model soup techniques (i.e., uniform soup and greedy soup) in comparison to those single hyper-parameter configuration variants. To assess the quality of the generated item features, we measure the average Mean Squared Error (MSE) between the visual and textual embeddings produced by the AdapterSoupRec variants. This metric allows us to assess the quality of the item embeddings, with smaller MSE values of visual and textual features [33] indicating that the corresponding model enables an effective fusion of item representations. Figure 1 shows the MSE values and NDCG@20 scores of each AdapterSoupRec variant. From Fig. 1, we observe that a lower average MSE value (y-axis) always leads to a better recommendation performance (x-axis) for all AdapterSoupRec variants across all three used datasets. In particular, AdapterSoupRec (Greedy) achieves the best recommendation performance while maintaining a lower average MSE value (e.g., 1.71 on the Pantry dataset). AdapterSoupRec (Uniform) also exhibits comparable recommendation performance with lower average MSE values across all datasets used. Moreover, most AdapterSoupRec variants outperform the best-performing baseline BM3 with markedly lower MSE values (e.g., 11.54 on the Pantry dataset). These consistent performances of AdapterSoupRec (Greedy) and AdapterSoupRec (Uniform) indicate that our proposed model soup techniques effectively integrate multi-modal features into item representations, thereby improving performance in the top-K recommendation task.

In response to RQ3, we conclude that the model soup techniques, namely uniform and greedy soups, result in a lower average MSE value between the MLLM's generated visual and textual embeddings compared to the best-performing baseline, BM3, thereby improving the top-K recommendation performance.

4.5 Out-of-Domain Performance (RQ4)

In this section, we evaluate our AdapterSoupRec model's out-of-domain performance, particularly its top-K recommendation performance in another recommendation scenario. This analysis allows us to investigate the generalisability of AdapterSoupRec by training only on previously used Amazon e-commerce datasets and performing inference on other domain datasets, specifically video recommendation datasets. In particular, we perform this generalisability study of AdapterSoupRec on three video recommendation datasets, namely Douyin, Bilibili and Kuaishou, where each video is associated with a cover image and its video description. We compare the performance of AdapterSoupRec (LLaVA-Grd), our best-performing variant as shown in Table 2, against multi-modal recommendation baselines, which are selected due to their demonstrated stronger generalisation in the cross-domain recommendation task [17]. Our aim is to determine if our AdapterSoupRec model trained on e-commerce datasets still has a comparable recommendation performance on another domain. Fig. 2 reports the NDCG@20 performance of both AdapterSoupRec and the multi-modal baselines since we observe the same conclusion using Recall@20 across all three video recommendation datasets. From Fig. 2, we observe that AdapterSoupRec outperforms all the multi-modal recommendation baselines on both the Douyin and Kuaishou datasets while maintaining a comparable performance on the Bilibili dataset. This result indicates that our AdapterSoupRec model effectively generalises the multi-modal knowledge from an e-commerce recommendation scenario to a video recommendation scenario. This generalisability is attributed to the MLLM's training on a large web corpus and our model souping strategy in adapting the MLLM to the multi-modal ranking task.

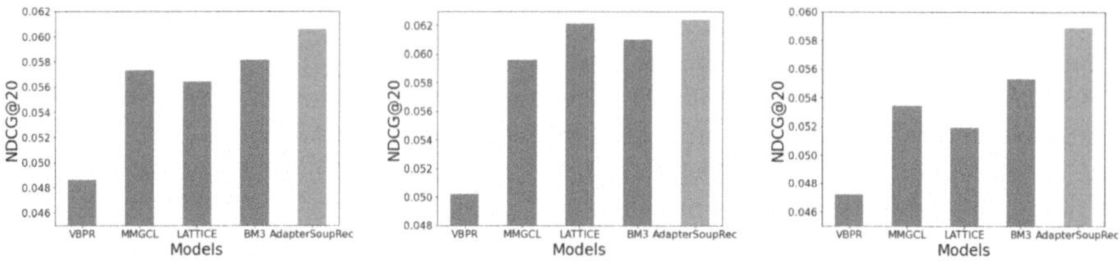

Douyin Bilibili Kuaishou

Fig. 2. Effectiveness of the AdapterSoupRec model compared with multi-modal recommendation baselines.

In response to RQ4, we conclude that AdapterSoupRec effectively generalises from the e-commerce scenario to the video recommendation scenario when compared to the multi-modal recommendation baselines.

5 Conclusions

We introduced AdapterSoupRec, a novel model soup approach for top-K multi-domain recommendation, which enables effective knowledge transfer from diverse source domains. Specifically, AdapterSoupRec leverages a single multi-modal large language model (i.e., CLIP or LLaVA) to enable effective multi-modal representation learning for recommendation purposes. It also leverages a graph adapter to incorporate domain-specific knowledge in a target domain. Our extensive experiments on 4 source domains and 3 target datasets showed that AdapterSoupRec significantly outperformed 9 strong existing baselines, including the state-of-the-art multi-modal recommendation models. In addition, we showed that the greedy soup effectively enhanced multi-domain recommendation performance compared to the uniform soup. Furthermore, LLaVA proved to be a more effective MLLM within AdapterSoupRec compared to CLIP. Our ablation study showed that the model soup, the graph adapter and the pre-training strategy significantly contribute to a more effective multi-domain recommendation. In addition, we measured the average Mean Squared Error (MSE) distances between the visual and textual embeddings of items from AdapterSoupRec and its non-soup variants. We found that the uniform and greedy soups produce lower average MSE values, further confirming the effectiveness of the model soup techniques in recommendations. Finally, we found that our AdapterSoupRec model, initially trained for e-commerce, still performs more effectively than the multi-modal recommendation baselines in an out-of-domain video recommendation scenario.

Conflict of Interest. The authors have no competing interests to declare that are relevant to the content of this article.

References

1. Alayrac, J.B., et al.: Flamingo: a visual language model for few-shot learning. In: Proceedings of NeurIPS (2022)
2. Bolya, D., Fu, C.Y., Dai, X., Zhang, P., Feichtenhofer, C., Hoffman, J.: Token merging: Your ViT but faster. In: Proceedings of ICML (2022)
3. Cantador, I., Fernández-Tobías, I., Berkovsky, S., Cremonesi, P.: Cross-domain recommender systems. Recommender systems handbook (2015)
4. Chronopoulou, A., Peters, M.E., Dodge, J.: Efficient hierarchical domain adaptation for pretrained language models. In: Proceedings of ACL (2022)
5. Chronopoulou, A., Peters, M.E., Fraser, A., Dodge, J.: AdapterSoup: weight averaging to improve generalisation of pretrained language models. In: Proceedings of ACL (Findings) (2023)
6. Croce, F., Rebuffi, S.A., Shelhamer, E., Gowal, S.: Seasoning model soups for robustness to adversarial and natural distribution shifts. In: Proceedings of CVPR (2023)

7. Diao, S., Xu, T., Xu, R., Wang, J., Zhang, T.: Mixture-of-domain-adapters: decoupling and injecting domain knowledge to pre-trained language models' memories. In: Proceedings of ACL (2023)
8. Dosovitskiy, A., et al.: An image is worth 16×16 words: transformers for image recognition at scale. In: Proceedings of ICLR (2020)
9. Gururangan, S., et al.: Don't stop pretraining: Adapt language models to domains and tasks. In: Proceedings of ACL (2020)
10. He, R., McAuley, J.: Ups and downs: Modeling the visual evolution of fashion trends with one-class collaborative filtering. In: Proceedings of WWW (2016)
11. He, R., McAuley, J.: VBPR: visual bayesian personalized ranking from implicit feedback. In: Proceedings of AAAI (2016)
12. He, X., Deng, K., Wang, X., Li, Y., Zhang, Y., Wang, M.: LightGCN: simplifying and powering graph convolution network for recommendation. In: Proceedings of SIGIR (2020)
13. Jaiswal, A.K., Liu, S., Chen, T., Ding, Y., Wang, Z.: Instant soup: Cheap pruning ensembles in a single pass can draw lottery tickets from large models. In: Proceedings of ICML (2023)
14. Jing, M., Zhu, Y., Zang, T., Wang, K.: Contrastive self-supervised learning in recommender systems: a survey. Transactions on Information Systems (2023)
15. Liu, H., Li, C., Wu, Q., Lee, Y.J.: Visual instruction tuning. In: Proceedings of NeurIPS (2024)
16. Liu, M., Li, J., Li, G., Pan, P.: Cross domain recommendation via Bi-directional transfer graph collaborative filtering networks. In: Proceedings of CIKM (2020)
17. Liu, Q., et al.: Multimodal recommender systems: a survey. Computing Surveys (2024)
18. Luo, L., et al.: MAMDR: a model agnostic learning method for multi-domain recommendation. In: Proceedings of ICDE (2023)
19. Ma, J., Zhao, Z., Yi, X., Chen, J., Hong, L., Chi, E.H.: Modeling task relationships in multi-task learning with multi-gate mixture-of-experts. In: Proceedings of SIGKDD (2018)
20. Matena, M.S., Raffel, C.A.: Merging models with fisher-weighted averaging. In: Proceedings of NeurIPS (2022)
21. Meng, Z., Ounis, I., Macdonald, C., Yi, Z.: Knowledge graph cross-view contrastive learning for recommendation. In: Proceedings of ECIR (2024)
22. Meng, Z., Yi, Z., Ounis, I.: Knowledge-enhanced multi-behaviour contrastive learning for effective recommendation. In: Proceedings of RecSys (2024)
23. Ning, W., Yan, X., Liu, W., Cheng, R., Zhang, R., Tang, B.: Multi-domain recommendation with embedding disentangling and domain alignment. In: Proceedings of CIKM (2023)
24. Radford, A., et al.: Learning transferable visual models from natural language supervision. In: Proceedings of ICML (2021)
25. Rendle, S., Freudenthaler, C., Gantner, Z., Schmidt-Thieme, L.: BPR: bayesian personalized ranking from implicit feedback. In: Proceedings of UAI (2009)
26. Stickland, A.C., Bérard, A., Nikoulina, V.: Multilingual domain adaptation for NMT: decoupling language and domain information with adapters. In: Proceedings of ACL (2021)
27. Tang, H., Liu, J., Zhao, M., Gong, X.: Progressive layered extraction (PLE): a novel multi-task learning model for personalized recommendations. In: Proceedings of RecSys (2020)
28. Vaswani, A., et al.: Attention is all you need. In: Proceedings of NeurIPS (2017)

29. Wang, J., et al.: MISSRec: pre-training and transferring multi-modal interest-aware sequence representation for recommendation. In: Proceedings of MM (2023)
30. Wang, W., et al.: Image as a foreign language: Beit pretraining for all vision and vision-language tasks. In: Proceedings of CVPR (2022)
31. Wortsman, M., et al.: Model soups: averaging weights of multiple fine-tuned models improves accuracy without increasing inference time. In: Proceedings of ICML (2022)
32. Yi, Z., Long, Z., Ounis, I., Macdonald, C., Mccreadie, R.: Large multi-modal encoders for recommendation. Transactions on Recommender Systems (2025)
33. Yi, Z., Ounis, I.: A unified graph transformer for overcoming isolations in multi-modal recommendation. In: Proceedings of RecSys (2024)
34. Yi, Z., Ounis, I., Macdonald, C.: Contrastive graph prompt-tuning for cross-domain recommendation. Transactions on Information Systems (2023)
35. Yi, Z., Ounis, I., Macdonald, C.: Graph contrastive learning with positional representation for recommendation. In: Proceedings of ECIR (2023)
36. Yi, Z., Wang, X., Ounis, I.: A directional diffusion graph transformer for recommendation. arXiv preprint arXiv:2404.03326 (2024)
37. Yi, Z., Wang, X., Ounis, I., Macdonald, C.: Multi-modal graph contrastive learning for micro-video recommendation. In: Proceedings of SIGIR (2022)
38. Yu, J., Yin, H., Xia, X., Chen, T., Cui, L., Nguyen, Q.V.H.: Are graph augmentations necessary? Simple graph contrastive learning for recommendation. In: Proceedings of SIGIR (2022)
39. Zhang, F., et al.: Multi-graph based multi-scenario recommendation in large-scale online video services. In: Proceedings of WWW (2022)
40. Zhao, C., Li, C., Fu, C.: Cross-domain recommendation via preference propagation GraphNet. In: Proceedings of CIKM (2019)
41. Zhou, X., et al.: Bootstrap latent representations for multi-modal recommendation. In: Proceedings of WWW (2023)
42. Zhu, F., Chen, C., Wang, Y., Liu, G., Zheng, X.: DTCTR: a framework for dual-target cross-domain recommendation. In: Proceedings of CIKM (2019)
43. Zhu, F., Wang, Y., Chen, C., Zhou, J., Li, L., Liu, G.: Cross-domain recommendation: challenges, progress, and prospects. In: Proceedings of IJCAI (2021)

Towards Identity-Aware Cross-Modal Retrieval: A Dataset and a Baseline

Nicola Messina[1(✉)], Lucia Vadicamo[1], Leo Maltese[2], and Claudio Gennaro[1]

[1] Institute of Information Science and Technologies, CNR, Pisa, Italy
{nicola.messina,lucia.vadicamo,claudio.gennaro}@isti.cnr.it
[2] University of Pisa, Pisa, Italy
l.maltese1@studenti.unipi.it

Abstract. Recent advancements in deep learning have significantly enhanced content-based retrieval methods, notably through models like CLIP that map images and texts into a shared embedding space. However, these methods often struggle with domain-specific entities and long-tail concepts absent from their training data, particularly in identifying specific individuals. In this paper, we explore the task of *identity-aware cross-modal retrieval*, which aims to retrieve images of persons in specific contexts based on natural language queries. This task is critical in various scenarios, such as for searching and browsing personalized video collections or large audio-visual archives maintained by national broadcasters. We introduce a novel dataset, COCO Person FaceSwap (COCO-PFS), derived from the widely used COCO dataset and enriched with deepfake-generated faces from VGGFace2. This dataset addresses the lack of large-scale datasets needed for training and evaluating models for this task. Our experiments assess the performance of different CLIP variations repurposed for this task, including our architecture, Identity-aware CLIP (Id-CLIP), which achieves competitive retrieval performance through targeted fine-tuning. Our contributions lay the groundwork for more robust cross-modal retrieval systems capable of recognizing long-tail identities and contextual nuances. Data and code are available at https://github.com/mesnico/IdCLIP.

Keywords: Personalized Retrieval · Vision and Language · Cross-modal Person Retrieval · Long-tail Concepts Understanding

1 Introduction

In modern content-based retrieval literature, deep learning techniques that map images and texts into the same common space – like CLIP [22] – are gaining increasing interest and have been used in many critical downstream multi-modal scenarios. These models are particularly effective for cross-modal retrieval, enabling efficient searches for relevant images based on natural language queries using straightforward k-nearest neighbor techniques in the forged common space.

C. Hauff et al. (Eds.): ECIR 2025, LNCS 15572, pp. 437–452, 2025.
https://doi.org/10.1007/978-3-031-88708-6_28

Consequently, they are widely used in large-scale image or video retrieval, allowing users to easily explore large visual collections using natural language descriptions [31].

However, despite their widespread adoption, these methods exhibit significant limitations. A major challenge is their inability to handle queries containing domain-specific entities and long-tail concepts that were not present within their training set. For instance, CLIP struggles to identify specific individuals, objects, or buildings not included during training, which can hinder performance in scenarios like the query, "*Gianni Morandi sings and dances at the Sanremo Festival in Italy, black and white footage*". This limitation becomes particularly important in personalization scenarios, where users seek to locate specific individuals or places within their video collections, as well as in cultural heritage and audiovisual content preservation applications that require retrieving footage of distinct monuments or historical figures in large archives maintained, for example, by national broadcasters.

Among all the possible long-tail concepts, the ability to recognize people's identities is one of the most critical, with great applicability in many of the above-cited scenarios. Although some studies [4,12] have already taken some steps in this direction, the main obstacle remains the lack of sufficient high-quality data to fine-tune a multimodal model and make it more discriminative with respect to identity features. Additionally, evaluating these models poses its own challenges, as it is not enough to merely retrieve images of a particular person; the context specified in the query must also be taken into account. In the above-mentioned example, we would like not just to retrieve all the possible images depicting *Gianni Morandi*, but only the ones in which he is *singing and dancing* in black and white footage.

To tackle these challenges, we propose a novel dataset, COCO-PFS, derived from COCO [16] and containing persons whose faces have been replaced with entities from VGGFace2 [2] employing deepfake generation techniques. Together with the dataset, we introduce suitable metrics able to monitor both the retrieval of the scene context and the person's identity. We experiment with different CLIP variations to show the challenges this benchmark poses. We demonstrate that the original CLIP struggles to differentiate between identities. We also examine CLIP-PAD [12], a recent CLIP variation that represents a significant step toward creating a domain-independent identity-aware model, capable of taking both visual features of a person's face and text describing the context as input to generate similarity scores with each dataset image. On top of this solution, we devise our Identity-aware CLIP (Id-CLIP), which features an effective, clever fine-tuning strategy for the visual backbone to further boost the results, obtaining a strong baseline for this challenging task.

To summarize, the contributions of this paper are as follows:

- We introduce a novel dataset, COCO-PFS, featuring images with faces replaced by a controlled set of public figures.
- We identify a suitable set of metrics able to assess the system's ability to retrieve both the person's identity and the context.

- We run multiple experiments on various CLIP models, also introducing Id-CLIP which utilizes fine-tuning of the visual backbone in order to create a strong baseline and capture insights on this novel challenging task.

2 Related Work

In recent years, vision-language models, and especially image-text matching methods, have gained significant attention for their application in cross-modal retrieval scenarios. Many of this methods adopt hinge-based triplet ranking loss with hard-negative mining using late-fusion approaches [6,10,14,18,21,22,25,27, 34] or effective yet less efficient early-fusion methods [13,15,17,28,32,40]. Among late-fusion methods, CLIP [22] has become particularly influential for image-text matching and a strong backbone to solve many downstream vision-language tasks. However, these methods struggle with long-tail or out-of-distribution concepts unless extensively fine-tuned.

Most of the personalized vision-language research has focused on personalized image generation. For instance, [7,20] employed textual inversion to improve personalized feature, while [8,33] proposed encoder-based methods for efficient and accurate customization. Perfusion [30] integrated multiple concepts while preventing overfitting, improving inference speed. Other approaches employ multi-concept extraction from a single image [1], user history-based prompt rewriting [3], and fine-tuning-free method [38].

Some works also started exploring personalized text-image retrieval. Cohen et al. [4] introduced a PerVL setup that leverages few-shot learning to reason about personalized concepts. Differently from our work, they update an already existing generic model, fine-tuning it on a small set of specific concepts. Yeh et al. [35] developed instance-specific embeddings without requiring annotated data, enabling instance retrieval in personal videos through vision-language similarity. The work most similar to ours is by [12], which proposed a method (*CLIP-PAD*) and a dataset (*Celebrities in Action*) for in-context person retrieval. Nevertheless, the proposed dataset collects only very famous actors that may be already present in the CLIP training dataset. Furthermore, our approach builds upon their method by adding a targeted fine-tuning of the CLIP visual backbone to enhance the fine-grained facial understanding abilities.

Entity-Aware Datasets. Most existing computer vision datasets lack named entities in the images or captions, limiting their relevance for identity-aware cross-modal retrieval tasks. While some efforts have leveraged knowledge bases to connect textual names with corresponding entities in multimedia content, they usually involve generic entities, such as *woman*, rather than specific, named individuals [5]. The VELD dataset [41] provides scene graphs linking entities from text to the depicted visual content using DBpedia, but it is not publicly available and covers broader categories such as locations and organizations, which are beyond our scope. WikiPerson [29] offers a dataset of over 48k annotated images with bounding boxes and corresponding Wikidata IDs, but it lacks descriptive captions, limiting its applicability for identity-aware retrieval. COFAR [9] explores *visual* named entities like business brands and world landmarks, linking them through external knowledge sources, yet it does not include named entities in its captions, hindering its applicability in our context. The recently introduced *This-Is-My* dataset [35] includes annotated video segments with captions for contextualized retrieval, but the available captions are limited to a small query-time dataset split, which only includes 30 instances related to 15 entities. Rosasco et al. [24] proposed *ConCon-Chi*, a benchmark designed to assess the compositional capabilities of models on novel chimeric objects within diverse contexts, revealing limitations in handling new concept-context combinations. However, this is a dataset created in controlled scenarios and has limited variability.

Several other datasets have attempted to link textual names to entities in visual content. For example, [23] presented a dataset with manually annotated names and captions in TV episode videos, which would have been relevant to our task but not publicly available. Similarly, [12] used *Celebrities in Places* and *Celebrities in Action* datasets, containing images and videos of celebrities in various contexts. While highly relevant to our task, these datasets are currently unavailable for download due to privacy concerns.

3 COCO Person FaceSwap (COCO-PFS) Dataset

Current widely-used datasets for text-to-image retrieval [16,26,37] generally contain images with generic descriptions, which are not suitable for identity-aware tasks. These datasets often feature random web-scraped images, making it difficult to find multiple images of the same person, and rarely include names in captions due to privacy concerns. These elements are of critical importance for our purpose, which is to understand to which extent state-of-the-art text-to-image retrieval models are able to find the same identity in different contexts.

To address this gap, we developed a novel dataset, COCO Person FaceSwap, which transform an existing text-image dataset into an identity-aware benchmark. Our approach includes an automatic pipeline that: (1) selects from the provided dataset only the images containing at least one person, (2) swaps faces with identities from a public face database using deepfake techniques, and (3) enriches captions to include the substituted identity's name. This section describes in detail the stages of the pipeline we utilized to create our benchmark that better suits the evaluation of identity-aware cross-modal retrieval.

Image Pre-selection. The first step involves selecting images containing individuals, as our focus is solely on retrieving scenes with identifiable persons. We selected images from COCO where only a single person appears, given that it is difficult to automatically modify captions to refer to different identities and to ensure a balanced representation of identities across various contexts. To this scope, we first detected faces in the images using MTCNN [39] pretrained on VGGFace2 [2] and CASIA-Webface [36]. We then filtered out images with multiple detected faces and cross-verified the results using COCO's annotations to

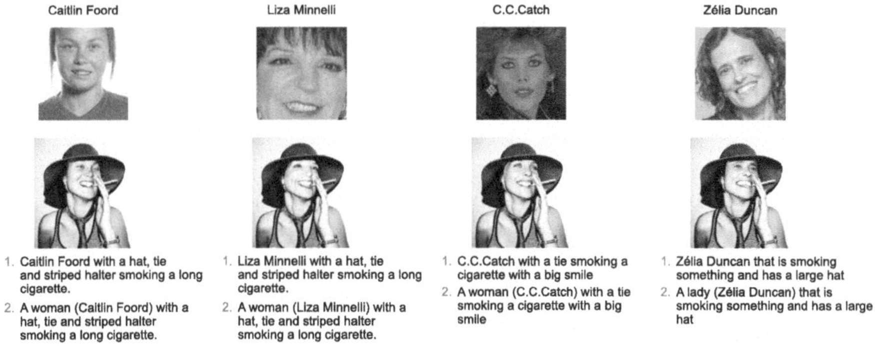

(a) One context, multiple identities.

(b) One identity, multiple contexts.

Fig. 1. Examples from our COCO-PFS dataset showing images obtained by substituting (a) different identities in the same context or (b) the same identity in different contexts. The captions depicted are obtained through two different templates, starting from one of the five original COCO captions.

minimize false positives. For each selected image, we extracted the facial features of the depicted individual using an Inception Resnet (V1), also trained on VGGFace2 and CASIA-Webface.

Gender-Ethnicity Detection and Face-Swapping. In this stage, we substitute the faces in COCO images with those from a VIP face database, using the VGGFace2 test set, which includes 500 identities. To ensure a balanced dataset, we ensure that each identity is present in at least two different contexts, and each image is swapped with multiple faces to depict various people. We used *Roop* tool[1], an advanced, easy-to-use face-swapping software, to reliably swap the faces. To make these swaps realistic, we don't select faces randomly. Instead, we use *DeepFace tool*[2] – which implements advanced face-analysis models – to detect the gender and ethnicity of the original faces in COCO, selecting a suitable matching face from VGGFace2 for the swap. Before the actual face-swapping phase, we keep only the images in which the individuals' faces are large in proportion to the image size. This filtering step is crucial to ensure the opti-

[1] https://github.com/s0md3v/roop.
[2] https://github.com/serengil/deepface.

mal performance of tools such as Deepface and Roop, particularly when dealing with face swapping. In fact, these tools tend to produce superior results when the facial features are prominently visible, and the face occupies a significant portion of the image.

Caption Enrichment. To create more personalized and realistic captions, we transform COCO's generic descriptions using a template-based approach. For each image I_i, where N is the swapped person's name, we modify the five original COCO captions $(C_{i1}, \ldots, C_{i5})$ by replacing generic noun referring to the person in the image (e.g., *person, man, woman, kid*, etc.) with the specific name N using k templates, $\{\mathbf{T}^t\}_{t=1}^k$. As a results, we generated k alternative versions of each caption, $C_{ij}^t = \mathbf{T}^t(C_{ij}, N)$, for $t = 1, \ldots, k$.

For example, assuming a set of three different templates, the original caption *"A person is running at the park"* and the name $N =$ *"Gianni Morandi"* can be transformed into: (i) "*Gianni Morandi is running at the park*", (ii) "*An image with Gianni Morandi. The person is running at the park*"; (iii) "*The famous person Gianni Morandi is running at the park*". Our dataset has been constructed using $k = 11$ different templates for each caption. For ease of notation, we flatten the indexes j and t to obtain again C_{ij}, where j this time spans from 1 to $5 \cdot k$.

We also release the captions with a [ENTITY] placeholder on behalf of the actual name so that methods can directly employ this information without having to rely on external NER modules for re-extracting the name.

Dataset Overview. We applied the proposed pipeline to the training, validation, and test sets of the COCO dataset. The resulting training set contains 49,957 images with 500 distinct entities, resulting in 2,747,635 captions. The validation and test sets each contain 1,760 images, with 466 and 467 entities, respectively, and 96,800 captions per set. We report an example in Fig. 1.

4 Identity-Aware Cross-Modal Retrieval

Identity-aware cross-modal retrieval extends the classic text-to-image retrieval task, where the goal is to find the most relevant images in a potentially large database given a natural language query. In our scenario, textual queries are enriched with person names, and the task is to retrieve all images featuring the specified person in the described context.

To train or evaluate an identity-aware model, we assume the availability of a *dataset* $\mathcal{D} = \{(I_i, \{C_{ij}\}_j), i = 1 \ldots, n\}$, which consists of n images I_i, each associated with one or more descriptive natural language captions $\{C_{ij}\}$. Additionally, we consider a *face gallery* $\mathcal{G} = \{(F_l, N_l)\}_{l=1}^m$ containing m identities. Each person in the gallery is represented by an image of their face F_l and their name N_l (e.g., *Mario Rossi*). Each image I_i should contain individuals from $\mathcal{G}$ in various contexts, with the corresponding captions C_{ij} explicitly mentioning the person's name (e.g., *Mario Rossi is riding a bike down the city streets*). This is

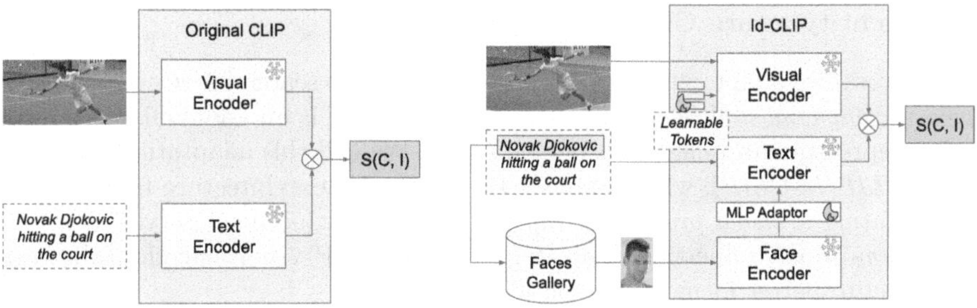

(a) Original CLIP. (b) Identity-Aware CLIP (Id-CLIP).

Fig. 2. (a). The original CLIP model. In order to digest entity names, should be properly finetuned and specialized to a given gallery $\mathcal{G}$. (b) Id-CLIP solves the problem by factoring out the entity from the caption. Through $\mathcal{G}$, the name is used to lookup the face, which becomes an additional input to the system. The learnable tokens in input to the Visual Encoder implement *visual prompt tuning.*

exactly how our COCO-PFS dataset is designed. The face gallery – VGGFace2 in our case – serves as an external knowledge base, allowing us to retrieve a face F_l given a person's name N_l. In this setting, given any of the descriptions C_{ij} as a query, the identity-aware retrieval objective is to rank all n images of the dataset by decreasing relevance, ensuring that I_i – the *correct* image containing the person mentioned in C_{ij} – is ranked as high as possible.

A straightforward solution to this task involves training a text-to-image retrieval model on a dataset containing all identities from $\mathcal{G}$. Once trained, the model would contain all the entity-specific knowledge in its weights (Fig. 2a). However, this would require re-training the model for each gallery to transfer knowledge of specific person identities into the model weights. While feasible, this is impractical for real-world applications where users may want to fine-tune the system with their own galleries, such as smartphone photo collections.

A much more application-friendly formulation of this problem would use a single general-purpose trained model that adapts, at inference time, to any user-provided gallery of identities $\mathcal{G}$. The idea is to leverage the general knowledge that cross-modal models like CLIP [22] have about the world – which helps in understanding generic image-text correspondences – together with the domain-specific knowledge present in the user-provided gallery $\mathcal{G}$ – necessary to distinguish between different identities. To achieve this, during training, we remove the person's name from the query C_{ij}, substituting it with its face instead. This can be done by extracting the person's name N_i from C_{ij} using off-the-shelf named entities recognition methods and using it to look up the corresponding face F_i in $\mathcal{G}$. This is exactly the idea behind Id-CLIP, presented in the next section and summarized in Fig. 2b.

4.1 Identity-aware CLIP (Id-CLIP)

In this work, we adapt existing CLIP-based models to search for relevant images given a *compound multimodal query* $(\tilde{C}, F)$, where $\tilde{C}$ is an anonymized caption and F represents an image of a person's face. We call this adaptation *Identity-aware CLIP(Id-CLIP)*, which extends the CLIP-PAD architecture from [12] by incorporating a *visual prompt tuning* stage. This enhancement serves as a targeted, non-invasive fine-tuning strategy for the CLIP vision encoder, allowing the model to better focus on person identities.

Overview. Id-CLIP is inspired by the work in [12], in which the authors produce an anonymized caption $\tilde{C}$ by substituting the person name in C with a special token called [TOK]. This special token is not associated with a learnable embedding, as is the case with normal word tokens. Instead, its embedding $\mathbf{w}^{[\text{TOK}]}$ is obtained as a multi-layer perceptron (MLP) projection of the visual embedding extracted from the person's face F through a state-of-the-art face recognition network (FRN): $\mathbf{w}^{[\text{TOK}]} = \text{MLP}(\text{FRN}(F))$. The whole network is then subject to the same CLIP contrastive learning objective used to pull relevant image-text pairs and repel irrelevant ones. The network is trained using a suitable dataset $\mathcal{D}^{\text{train}}$ containing images of persons in specific contexts. During training, the CLIP-PAD network only updates the MLP projection network, while the rest of the CLIP weights are kept frozen. The MLP network learns to map the facial features into an actual word token that the frozen pre-trained text encoder can process alongside the other caption tokens. The training phase's sole purpose is to learn meaningful semantic mapping between the face feature space and the CLIP token embedding space. Therefore, we obtain the caption embedding $\mathbf{c}$ by applying the CLIP text encoder to the sequence of tokens containing the special token $\mathbf{w}^{[\text{TOK}]}$ together with the original sentence tokens $\mathbf{w}_i$: $\mathbf{c} = \text{CLIP}^{\text{txt}}([\mathbf{w}_1, \mathbf{w}_2, \ldots, \mathbf{w}^{[\text{TOK}]}, \ldots, \mathbf{w}_n])$. Notice that $\mathbf{w}^{[\text{TOK}]}$ is inserted in specific positions inside the caption $\tilde{C}$, as better explained in Sect. 5.

Visual Prompt Tuning. The rationale behind Id-CLIP is that the person's identities are discriminated solely through their visual features, both at the query and image encoding stages. At query encoding time, the model can just leverage learning a suitable MLP projection network to discriminate between two different queries that carry the same context but different identities – relying on the discriminative ability of the off-the-shelf face recognition network. From the image dataset encoding perspective, instead, CLIP-PAD assumes that the visual encoder is already able to discriminate between different identities, assuming equal context. Considering that CLIP has not been directly trained to recognize fine-grained facial differences and that this information is not present in the generic image-text training datasets, it is likely that the CLIP visual encoder fails in such a task. In order to compensate for this potential drawback, we enrich the visual encoder of Id-CLIP with some additional learnable visual tokens $\tilde{\mathbf{v}}_i$, employing the approach presented in [11]. These additional visual tokens are purposed to learn the discriminative characteristics of the person's faces. Specifically, the final image feature $\mathbf{v}$ is obtained through the application of the CLIP visual encoder to the original image tokens $\{\mathbf{v}_i\}_{i=1}^{wh}$ concatenated to the series of the special learnable prompt tuning tokens $\{\tilde{\mathbf{v}}_i\}_{i=1}^{p}$: $\mathbf{v} = \text{CLIP}^{\text{vis}}([\tilde{\mathbf{v}}_1, \ldots, \tilde{\mathbf{v}}_p, \mathbf{v}_1, \ldots, \mathbf{v}_{wh}])$, where p is the number of prompt tuning tokens (we empirically found that $p = 5$ works best in our scenario), and wh is the total number of visual patches of the image. Notice that, in this setup, the MLP and the added visual tokens are learned together during the training procedure, while the core CLIP weights remain frozen.

Objective. As our image-text contrastive objective, we employ an Info-NCE loss [19], formulated as follows:

$$\mathcal{L}_{\text{InfoNCE}} = -\frac{1}{B}\sum_{i=1}^{B}\log\frac{\exp(\text{sim}(\mathbf{c}_i, \mathbf{v}_i))}{\sum_{j=1}^{B}\exp(\text{sim}(\mathbf{c}_i, \mathbf{v}_j))}, \tag{1}$$

where $\mathbf{c}_i$ and $\mathbf{v}_j$ are the textual and visual features in output from CLIP, as explained in the previous paragraphs, $\text{sim}(\cdot, \cdot)$ is the cosine similarity, and B is the batch size employed during training.

5 Experiments

We evaluate the performance of Id-CLIP mainly on the targeted identity-aware cross-modal retrieval task, which we refer to as the *entity-in-context retrieval* – that focuses on retrieving images that contain both the requested person and the specified context. However, we can probe the same model also on a second related task, which we call *entity-only retrieval.* This secondary task aims at finding all images depicting a given entity, regardless of the specific context in which they appear. This second task is mainly crafted to understand the abilities of the trained model to actually understand the person's features without the potential noise introduced by the context.

Inference Strategies. We apply two different inference strategies depending on the task we are tackling, i.e., *entity-in-context* retrieval or *entity-only* retrieval.

Concerning the entity-in-context retrieval subtask, our proposed COCO-PFS dataset already comes with the anonymized captions $\tilde{C}_{ij}$, which contain the [ENTITY] placeholder instead of the actual name[3]. In the most simple case, [ENTITY] could be substituted with the [TOK] token expected by Id-CLIP. However, other meaningful strategies may be applied, such as replacing [ENTITY] solely with the original text name (e.g., *Gianni Morandi*), or a combination of textual and visual features – like *[TOK] Gianni Morandi*, or *Gianni*

[3] In a real scenario, we should use NER techniques to extract the entity from the query. However, in this setup, we already have this information available considering how we built the dataset.

Morandi [TOK]. Other strategies are possible, like placing the visual feature at the beginning of the string and then substituting [ENTITY] to the sole textual name – e.g., *The famous kid [ENTITY] is running at the park* may become *[TOK]. The famous kid Gianni Morandi is running at the park.*. For entity-in-context retrieval, we consider each caption $\tilde{C}_{ij}$ as a separate query with which the whole image database is searched, and we tested different templates to substitute [ENTITY] with [TOK] and the original person's name.

Differently, for entity-only retrieval, the sole query is the name of the person to be searched. In this context, we treat each person as a different categorical class. We can solve this task using Id-CLIP by employing [ENTITY] as the sole input to the model. Since we are interacting with a language model that understands natural language sentences, we surround the [ENTITY] tag with some pre-defined templates, like *An image with [ENTITY]* or *The famous [ENTITY]*, performing *prompt ensembling* like in CLIP [22]. [ENTITY] is then expanded in the same way as for the entity-in-context retrieval, using a combination of [TOK] and the original person's name.

5.1 Metrics

In the experiments, we report the standard metric commonly employed to evaluate the cross-modal retrieval models, the average recall@k over all the tested queries, varying $k \in \{1, 5, 10, 50\}$.

$$\text{Recall@k} = \frac{|\text{Relevant instances} \cap \text{Retrieved top-k instances}|}{\min(k, |\text{Relevant instances}|)} \tag{2}$$

In the case of *entity-in-context retrieval*, we have only one relevant image in the dataset for each query text. Therefore, the average recall@k corresponds to the percentage of text queries that successfully retrieve the correct image containing the appropriate entity in the right context within the top k results.

We also assess model performance on the *entity-only retrieval* task, where the query text includes only the entity's name without any context. In this case, the recall@k measures the percentage of items in the top-k results that match the specified identity, regardless of the image's context.

We also report Rsum $= \sum_{k \in \{1,5,10,50\}}$ Recall@k as an aggregated measure.

5.2 Implementation Details

We optimized Id-CLIP using the Adam optimizer. We employed a learning rate of 5e-5. The MLP used to convert face features in a textual token is composed of one hidden layer without bias weights and ReLU non-linearities. The model is trained for a maximum of 10 epochs and validated on the main entities-in-context retrieval task. During training, as a robust augmentation strategy, we employ the face crops of the face-swapped persons instead of the original ones as F. During testing, instead, F is a real face taken from $\mathcal{G}$. Concerning captions, at training time, a single template is chosen randomly among all the available ones. This

Table 1. Entity-in-context Retrieval Results

Model	VPT	Recall@k				Rsum
		k=1	k=5	k=10	k=50	
CLIP (Original)	–	12.10	56.20	69.10	92.20	229.60
CLIP N		26.20	58.20	70.10	91.80	246.30
CLIP-PAD T (All $\mathbf{T}_i$)		36.30	66.60	77.80	95.20	275.90
CLIP-PAD T+N (All $\mathbf{T}_i$)		36.40	65.00	75.80	94.40	271.50
CLIP-PAD T ($\mathbf{T}_1$)		39.30	69.70	80.80	96.50	286.20
CLIP-PAD T+N ($\mathbf{T}_1$)		38.00	67.10	78.30	95.50	278.90
CLIP (Original)	✓	12.40	57.90	70.60	93.20	234.10
CLIP N		27.90	61.70	72.90	93.00	255.50
Id-CLIP T (All $\mathbf{T}_i$)		39.20	68.20	78.50	95.70	281.60
Id-CLIP T+N (All $\mathbf{T}_i$)		37.50	66.10	76.30	94.30	274.10
Id-CLIP T ($\mathbf{T}_1$)		**43.20**	**71.20**	**80.80**	**96.70**	**291.90**
Id-CLIP T+N ($\mathbf{T}_1$)		39.90	68.60	78.10	95.00	281.60

can be considered as an additional text augmentation procedure, which enables the network to generalize to different phrasings of the same (context, identity) pair, in turn stabilizing the learning procedure.

5.3 Results

For the entity-in-context retrieval, we report the results in Table 1, which is split into two equal sets of experiments: one without Visual Prompt Tuning (VPT) and one with VPT engaged. The first and the second lines report the two trivial baselines for this task, which are represented by the original CLIP prompted with the original caption and by the original CLIP prompted with COCO-PFS captions containing the person names (CLIP N).

By focusing on the recall@1 metric, we can notice how CLIP is already able to exploit some biases when prompted with the person's names – most likely, certain names imply specific ethnical groups, which imply certain facial features. We then first re-evaluate CLIP-PAD (which does not have visual prompt tuning active) on our benchmark, considering the most successful inference strategies. The performance improves significantly with CLIP-PAD (All T_i), which averages recall across 11 different templates evaluated independently, resulting in an improvement of more than 38% compared to CLIP N. The two different reported versions, T and T+N, relate to how [ENTITY] is substituted – with only [TOK] in the first case and with [NAME] [TOK] in the second one, where the name is the person's name. Interestingly, CLIP-PAD seems not to improve as much as the CLIP baselines when presented with original names. In fact, the difference between CLIP-PAD T and CLIP-PAD T+N is quite marginal with respect to the huge improvement observed in the baselines. This may be due

Table 2. Entity-only Retrieval Results. In configuration (1), [ENTITY] is expanded as "[NAME] [TOK]", while in (2), the sentence always starts with "An image with [TOK]." and is then followed by the original caption where [ENTITY] is substituted with [NAME].

Model	VPT	Recall@k				Rsum
		k = 1	k = 5	k = 10	k = 50	
CLIP N	–	12.80	6.00	8.40	17.50	44.80
CLIP-PAD T		15.40	11.10	13.50	28.00	68.00
CLIP-PAD T+N (1)		20.40	11.80	13.90	28.00	74.10
CLIP-PAD T+N (2)		21.80	11.80	14.20	28.20	75.90
CLIP N	✓	15.40	7.30	9.20	19.10	51.00
Id-CLIP T		21.20	11.50	14.20	**31.90**	78.90
Id-CLIP T+N (1)		22.60	11.60	14.60	29.70	78.60
Id-CLIP T+N (2)		**24.30**	**13.70**	**16.20**	31.40	**85.60**

to the ethical biases present in the pre-trained CLIP transforming into unuseful noise when the facial feature is injected through the [TOK] token. We also report CLIP-PAD (T_1) – coming as well in the T and in T+N configurations, which, differently from CLIP-PAD (All T_i), only tests the best-performing template, which is T_1 = "$[ENTITY] in the image. [CAPTION]$". We can notice that the best template behaves particularly better than the "All T_i" configuration, meaning that, on average, some reasonable query behaves particularly worse than the average. Shifting the attention to the results where visual prompt tuning is active, we can appreciate how much our proposed Id-CLIP improves over CLIP-PAD for all the probed inference schemas. In particular, Id-CLIP obtains an improvement on recall@1 of 55% with respect to CLIP N, with an average increase over CLIP-PAD of 6%, proving the effectiveness of this targeted fine-tuning strategy for the visual encoder. Notice also that the improvement on the recall@k for higher values for k is limited given that, at a certain point, the model will have retrieved all the pictures of the different identities in a given context[4], thus in the limit behaving similarly to the original CLIP model. We obtain the best overall results over all the metrics with the Id-CLIP T (T_1) model.

Concerning the entity-only retrieval, we report the results in Table 2. Again, we can immediately notice how the original CLIP model prompted with names (CLIP N) behaves poorly on this subtask. Nevertheless, differently from the results in entity-in-context retrieval, we can notice how the original name plays a more important role, as far as CLIP-PAD is concerned. In fact, the configurations denoted as CLIP-PAD T+N obtain a significant improvement over CLIP-PAD T, meaning that the name biases present in CLIP help in bringing reasonable identities to the top of the ranked list. This effect is actually damped in Id-

[4] We remind that, in our dataset, each image has been swapped with multiple identities, and this applies to the test set as well.

CLIP, where visual prompt tuning effectively increases the importance of the visual features over the textual biases. For completeness, we report two different [ENTITY] expansion strategies: T+N (1) represents the case in which [ENTITY] is expanded as "[NAME] [TOK]"; (2) represents the case where the sentence is always started with "An image with [TOK]." and then followed by the original caption where [ENTITY] is substituted with only [NAME]. The difference between the two versions is not so pronounced, meaning that the model seems robust to different entity expansion formulations. Even in this case, we reach the best result through Id-CLIP, again proving the importance of VPT. Specifically, the best-performing model results to be Id-CLIP T+N (2), which increases the recall@1 by 57% with respect to CLIP N and by 11% with respect to the corresponding configuration of CLIP-PAD.

6 Conclusions

In this work, we tackled the *identity-aware cross-modal retrieval* task, introducing a dataset and some strong baselines that are designed to avoid a deep fine-tuning of the model on specific identity galleries. More in detail, we first introduced the COCO-PFS dataset, obtained by replacing the faces in the COCO dataset with controlled, identity-specific entities from VGGFace2. We have shown that this task is particularly challenging for existing multimodal models like CLIP, which struggle to handle domain-specific entities and long-tail concepts such as unique identities not present in the training data. As a second step, we also proposed Id-CLIP, a strong baseline that demonstrated significant improvements over standard CLIP models. In particular, we showed that while original CLIP can leverage biases related to names and ethnic groups, it is still limited in its ability to accurately retrieve images that both match the identity and the context described by the text query. Our experiments revealed that modifying CLIP augmenting the query with the face features of the person taken from an external gallery and incorporating a visual prompt tuning leads to notable enhancements in both identity and context retrieval. As a future work, we plan to explore different pre-trained CLIP models and to augment Id-CLIP to actively weigh the importance of context and identity directly at inference time.

Acknowledgments. This work was partially funded by MUCES - a MUltimedia platform for Content Enrichment and Search in audiovisual archive project (CUP B53D23026090001), and by PNRR-M4C2 (PE00000013) "FAIR-Future Artificial Intelligence Research" - Spoke 1 "Human-centered AI", funded under the NextGeneration EU program.

Disclosure of Interests. The authors have no competing interests to declare that are relevant to the content of this article.

References

1. Avrahami, O., Aberman, K., Fried, O., Cohen-Or, D., Lischinski, D.: Break-a-scene: extracting multiple concepts from a single image. In: SIGGRAPH Asia 2023 Conference Papers, pp. 1–12 (2023)
2. Cao, Q., Shen, L., Xie, W., Parkhi, O.M., Zisserman, A.: VGGFace2: a dataset for recognising faces across pose and age. In: 2018 13th IEEE International Conference on Automatic Face & Gesture Recognition (FG 2018), pp. 67–74. IEEE (2018)
3. Chen, Z., Zhang, L., Weng, F., Pan, L., Lan, Z.: Tailored visions: enhancing text-to-image generation with personalized prompt rewriting. In: Proceedings of the IEEE/CVF Conference on Computer Vision and Pattern Recognition, pp. 7727–7736 (2024)
4. Cohen, N., Gal, R., Meirom, E.A., Chechik, G., Atzmon, Y.: "this is my unicorn, fluffy": Personalizing frozen vision-language representations. In: European Conference on Computer Vision, pp. 558–577. Springer (2022). https://doi.org/10.1007/978-3-031-20044-1_32
5. Dost, S., Serafini, L., Rospocher, M., Ballan, L., Sperduti, A.: Jointly linking visual and textual entity mentions with background knowledge. In: Métais, E., Meziane, F., Horacek, H., Cimiano, P. (eds.) NLDB 2020. LNCS, vol. 12089, pp. 264–276. Springer, Cham (2020). https://doi.org/10.1007/978-3-030-51310-8_24
6. Faghri, F., Fleet, D.J., Kiros, J.R., Fidler, S.: VSE++: improving visual-semantic embeddings with hard negatives (2018). https://github.com/fartashf/vsepp
7. Gal, R., et al.: An image is worth one word: Personalizing text-to-image generation using textual inversion. In: The Eleventh International Conference on Learning Representations, ICLR 2023, Kigali, Rwanda, 1-5 May 2023 (2023). https://openreview.net/pdf?id=NAQvF08TcyG
8. Gal, R., Arar, M., Atzmon, Y., Bermano, A.H., Chechik, G., Cohen-Or, D.: Encoder-based domain tuning for fast personalization of text-to-image models. ACM Trans. Graph. (TOG) **42**(4), 1–13 (2023)
9. Gatti, P., Penamakuri, A.S., Teotia, R., Mishra, A., Sengupta, S., Ramnani, R.: COFAR: commonsense and factual reasoning in image search. arXiv preprint arXiv:2210.08554 (2022)
10. Jia, C., et al.: Scaling up visual and vision-language representation learning with noisy text supervision. In: International Conference on Machine Learning, pp. 4904–4916. PMLR (2021)
11. Jia, M., Tang, L., Chen, B.C., Cardie, C., Belongie, S., Hariharan, B., Lim, S.N.: Visual prompt tuning. In: European Conference on Computer Vision, pp. 709–727. Springer (2022). https://doi.org/10.1007/978-3-031-19827-4_41
12. Korbar, B., Zisserman, A.: Personalised clip or: how to find your vacation videos (2022)
13. Li, J., Selvaraju, R., Gotmare, A., Joty, S., Xiong, C., Hoi, S.: Align before fuse: vision and language representation learning with momentum distillation. Adv. Neural. Inf. Process. Syst. **34**, 9694–9705 (2021)
14. Li, K., Zhang, Y., Li, K., Li, Y., Fu, Y.: Visual semantic reasoning for image-text matching. In: Proceedings of the IEEE/CVF International Conference on Computer Vision, pp. 4654–4662 (2019)
15. Li, X., et al.: OSCAR: Object-semantics aligned pre-training for vision-language tasks. In: Vedaldi, A., Bischof, H., Brox, T., Frahm, J.-M. (eds.) ECCV 2020. LNCS, vol. 12375, pp. 121–137. Springer, Cham (2020). https://doi.org/10.1007/978-3-030-58577-8_8

16. Lin, T.-Y., et al.: Microsoft COCO: common objects in context. In: Fleet, D., Pajdla, T., Schiele, B., Tuytelaars, T. (eds.) ECCV 2014. LNCS, vol. 8693, pp. 740–755. Springer, Cham (2014). https://doi.org/10.1007/978-3-319-10602-1_48
17. Lu, J., Batra, D., Parikh, D., Lee, S.: VilBERT: pretraining task-agnostic visiolinguistic representations for vision-and-language tasks. In: Advances in Neural Information Processing Systems, vol. 32 (2019)
18. Messina, N., et al.: Aladin: Distilling fine-grained alignment scores for efficient image-text matching and retrieval. In: International Conference on Content-based Multimedia Indexing, pp. 64–70 (2022)
19. Oord, A.v.d., Li, Y., Vinyals, O.: Representation learning with contrastive predictive coding. arXiv preprint arXiv:1807.03748 (2018)
20. Pang, L., Yin, J., Xie, H., Wang, Q., Li, Q., Mao, X.: Cross initialization for face personalization of text-to-image models. In: Proceedings of the IEEE/CVF Conference on Computer Vision and Pattern Recognition, pp. 8393–8403 (2024)
21. Qu, L., Liu, M., Cao, D., Nie, L., Tian, Q.: Context-aware multi-view summarization network for image-text matching. In: Proceedings of the 28th ACM International Conference on Multimedia, pp. 1047–1055 (2020)
22. Radford, A., et al.: Learning transferable visual models from natural language supervision. In: Proceedings of the 38th International Conference on Machine Learning. Proceedings of Machine Learning Research, vol. 139, pp. 8748–8763. PMLR (2021). https://proceedings.mlr.press/v139/radford21a.html
23. Ramanathan, V., Joulin, A., Liang, P., Fei-Fei, L.: Linking people in videos with "their" names using coreference resolution. In: Computer Vision–ECCV 2014: 13th European Conference, Zurich, Switzerland, 6-12 September 2014, Proceedings, Part I 13, pp. 95–110. Springer (2014). https://doi.org/10.1007/978-3-319-10590-1_7
24. Rosasco, A., et al.: ConCon-Chi: concept-context chimera benchmark for personalized vision-language tasks. In: Proceedings of the IEEE/CVF Conference on Computer Vision and Pattern Recognition, pp. 22239–22248 (2024)
25. Sarafianos, N., Xu, X., Kakadiaris, I.A.: Adversarial representation learning for text-to-image matching. In: Proceedings of the IEEE/CVF International Conference on Computer Vision, pp. 5814–5824 (2019)
26. Sharma, P., Ding, N., Goodman, S., Soricut, R.: Conceptual captions: a cleaned, hypernymed, image alt-text dataset for automatic image captioning. In: Proceedings of the 56th Annual Meeting of the Association for Computational Linguistics (Volume 1: Long Papers), pp. 2556–2565 (2018)
27. Stefanini, M., Cornia, M., Baraldi, L., Cucchiara, R.: A novel attention-based aggregation function to combine vision and language. In: 2020 25th International Conference on Pattern Recognition (ICPR), pp. 1212–1219. IEEE (2021)
28. Su, W., et al.: Vl-BERT: pre-training of generic visual-linguistic representations. In: International Conference on Learning Representations (2020). https://openreview.net/forum?id=SygXPaEYvH
29. Sun, W., Fan, Y., Guo, J., Zhang, R., Cheng, X.: Visual named entity linking: a new dataset and a baseline. arXiv preprint arXiv:2211.04872 (2022)
30. Tewel, Y., Gal, R., Chechik, G., Atzmon, Y.: Key-locked rank one editing for text-to-image personalization. In: ACM SIGGRAPH 2023 Conference Proceedings, pp. 1–11 (2023)
31. Vadicamo, L., et al.: Evaluating performance and trends in interactive video retrieval: insights from the 12th VBS competition. IEEE Access **12**, 79342–79366 (2024)

32. Wang, W., et al.: Image as a foreign language: BEiT pretraining for vision and vision-language tasks. In: Proceedings of the IEEE/CVF Conference on Computer Vision and Pattern Recognition (2023)
33. Wei, Y., Zhang, Y., Ji, Z., Bai, J., Zhang, L., Zuo, W.: Elite: Encoding visual concepts into textual embeddings for customized text-to-image generation. In: Proceedings of the IEEE/CVF International Conference on Computer Vision, pp. 15943–15953 (2023)
34. Wen, K., Gu, X., Cheng, Q.: Learning dual semantic relations with graph attention for image-text matching. IEEE Trans. Circ. Syst. Video Technol. **31**(7), 2866–2879 (2020)
35. Yeh, C.H., Russell, B., Sivic, J., Heilbron, F.C., Jenni, S.: Meta-personalizing vision-language models to find named instances in video. In: Proceedings of the IEEE/CVF Conference on Computer Vision and Pattern Recognition, pp. 19123–19132 (2023)
36. Yi, D., Lei, Z., Liao, S., Li, S.Z.: Learning face representation from scratch. arXiv preprint arXiv:1411.7923 (2014)
37. Young, P., Lai, A., Hodosh, M., Hockenmaier, J.: From image descriptions to visual denotations: new similarity metrics for semantic inference over event descriptions. Trans. Assoc. Comput. Linguist. **2**, 67–78 (2014)
38. Zeng, Y., et al.: Jedi: Joint-image diffusion models for finetuning-free personalized text-to-image generation. In: Proceedings of the IEEE/CVF Conference on Computer Vision and Pattern Recognition, pp. 6786–6795 (2024)
39. Zhang, K., Zhang, Z., Li, Z., Qiao, Y.: Joint face detection and alignment using multitask cascaded convolutional networks. IEEE Sig. Process. Lett. **23**(10), 1499–1503 (2016)
40. Zhang, P., et al.: VinVL: revisiting visual representations in vision-language models. In: Proceedings of the IEEE/CVF Conference on Computer Vision and Pattern Recognition, pp. 5579–5588 (2021)
41. Zheng, Q., Wen, H., Wang, M., Qi, G.: Visual entity linking via multi-modal learning. Data Intell. **4**(1), 1–19 (2022)

Corpus Subsampling: Estimating the Effectiveness of Neural Retrieval Models on Large Corpora

Maik Fröbe[1(✉)], Andrew Parry[2], Harrisen Scells[3], Shuai Wang[4], Shengyao Zhuang[4,5], Guido Zuccon[4], Martin Potthast[6], and Matthias Hagen[1]

[1] Friedrich-Schiller-Universität Jena, Jena, Germany
maik.froebe@uni-jena.de
[2] University of Glasgow, Glasgow, Scotland
[3] University of Kassel and hessian.AI, Kassel, Germany
[4] University of Queensland, Brisbane, Australia
[5] CSIRO, Pullenvale, Australia
[6] University of Kassel, hessian.AI, and ScaDS.AI, Kassel, Germany

Abstract. Due to their low efficiency, neural retrieval models are usually evaluated on small corpora (e.g. MS MARCO or BEIR subsets) or in re-ranking scenarios using a more efficient first-stage retriever. To estimate their effectiveness on larger corpora independently of a first-stage retriever, we propose a new corpus subsampling strategy based on the top-k results of the pooled systems that contributed to the relevance judgments of a corpus. Our experiments on nine TREC tasks covering different corpus sizes show that using the top-1,000 or even only the top-100 pools provides a reliable effectiveness estimate for neural models. This reduces the required experimental resources for large corpora by a factor of up to 1,000 and enables a "green" IR evaluation.

Keywords: Evaluation · Neural Retrieval · Green IR Evaluation

1 Introduction

Over time, the document collections used in information retrieval research have become larger to make laboratory experiments more realistic. For example, the ClueWeb09 and the ClueWeb12[1] each comprise around one billion documents, while the ClueWeb22 [44] comprises ten billion documents. To evaluate a given set of systems on large collections, a considerable number of judgments is required (e.g., 20 per topic for MTC [1]), whereas the validity of an evaluation is typically low with only a few judgments. Few (binary) judgments per topic impair the discriminatory power of evaluation measures like nDCG [30], which discriminate systems better the more (graded) judgments are available [41]. The

[1] https://lemurproject.org/clueweb09/ and https://lemurproject.org/clueweb12/.

C. Hauff et al. (Eds.): ECIR 2025, LNCS 15572, pp. 453–471, 2025.
https://doi.org/10.1007/978-3-031-88708-6_29

Table 1. Our recommended subsamples for reliable evaluation at substantially reduced computational effort by reducing documents not on topic ($\not\in_J$).

Corpus	Tracks	Judgm.	Complete			Subsampled		
			Docs.	$\not\in_J$	Size	Docs.	$\not\in_J$	Size
ClueWeb09	Web 09–12 [13–16]	84 366	1.0 b	99 %	4.0 TB	0.3 m	73 %	0.9 GB
ClueWeb12	Web 13/14 [17,18]	28 906	0.7 b	99 %	4.5 TB	0.1 m	72 %	0.5 GB
Disks 4/5	Robust04 [64]	311 410	0.5 m	41 %	0.6 GB	0.4 m	31 %	0.5 GB
MS MARCO	DL 19/20 [21,22]	20 646	8.8 m	99 %	2.9 GB	0.3 m	97 %	42.1 MB

ClueWeb09/12 were frequently used in TREC-like evaluations, so a large number of graded relevance judgments are available for many topics. Evaluations on collections with many judgments are more likely valid when they build on pools of different systems [66], but also failed validity tests have been reported [67].

Still, evaluating neural retrievers (e.g., LLMs [36]) on large collections is not really practical: indexing 50% of the ClueWeb09 with an efficient neural retrieval model already takes 71 days on an Nvidia V100 GPU [33]. More feasible are smaller collections like the ones from the BEIR meta-dataset [60] (19 collections with median size of 530,000 documents) but they have only a limited number of relevant documents (median of 3 per topic). At the same time, when evaluating a new retriever in post-hoc experiments (i.e., after a collection has been judged), many of the retrieved results may be unjudged and the common practice of then assuming them to be non-relevant or measuring effectiveness on condensed lists [49] often considerably under- or overestimates effectiveness [25,50]. As a way out, subsampling (subsets of) the judged documents and combining them with random documents has been proposed [34,57,58]. But for large collections, random documents are "easy negatives" and unlikely to be retrieved. This, too, leads to an overestimation of effectiveness as in the case of using condensed lists.

In this paper, we compare different subsampling approaches to systematically study which strategies allow for reliable post-hoc experiments at a reasonable compute budget. But in contrast to previous sampling approaches [34,57,58], we use the runs that contributed to the judgment pool. The runs' top results below a collection's pooling depth include a diversity of hard negative unjudged documents, minimizing the overestimation of effectiveness. We evaluate eight sampling strategies (Sect. 3) using leave-one-group-out retrieval (Sect. 4) on diverse collections (between 500,000 and 1 billion documents) and find that evaluation becomes reliable when subsamples are pooled to a depth of 100 up to 1,000. Table 1 provides an overview of our recommended subsamples that we found in our experiments for the nine TREC tasks on four corpora, yielding dataset sizes of only a few hundred Megabytes. As these subsamples are way easier to index and to process at retrieval time than whole collections, the subsampling-based evaluation fulfills one of the goals of Green IR [53]: minimizing the resources of IR evaluationf.[2]

2 Related Work

Reliability of IR Experiments. IR experiments are reliable if observations (e.g., System A>System B) transfer to similar scenarios with a high probability [66]. Two main aspects impact reliability [66]: subjectiveness and incompleteness of relevance judgments. TREC-style test collections are constructed as a community effort. Teams submit runs that are subsequently pooled and query–document pairs from the judgment pool are independently labeled by assessors that are unaware of the originating retrieval system [23,51]. Relevance judgments are subjective and, therefore, might vary among different assessors [40]. Low inter-assessor agreement can indicate low reliability [51]. To obtain high agreement among assessors, topics usually have a narrative describing what separates relevance from non-relevance and assessors undergo extensive training [28].

Once relevance judgments have been collected, TREC-style test collections are re-used with the assumption that the judgment pool is "essentially complete" [66], i.e., unjudged documents retrieved by a new system are considered not relevant. In cases where this assumption is true (e.g., recent findings on TREC-8 [68]), the system rankings do not change substantially depending on whether the systems contributed to the judgments or not [66]. However, especially if corpora are large or diverse or manual submissions cannot be pooled (e.g., during the time–constrained construction of TREC-COVID [48,60]), this assumption might not hold. These limitations motivate dedicated evaluation procedures using either specialized measures [6,42,65,74,75] or predicted relevance labels [2,3,7,8,37]. Importantly, corpus subsampling strategies might limit which of those procedures can be applied, e.g., for subsamples that resemble re-ranking only (parts of) the judgment pool [34,57,58], none of these specialized methods can be applied as no retrievable unjudged documents remain in the corpus.

Overall, the high effort to construct TREC-style test collections and make them reliable, usually involving work by multiple expert assessors for several weeks [54], motivates us to study how the underlying corpora can be subsampled to allow reliable evaluation of expensive modern neural retrieval models, while still ensuring that the subsampled corpora allow evaluation setups that account for the uncertainties of post-hoc experimentation due to unjudged documents.

Corpus Subsampling for Validation. Validation queries are used to stop neural model training when the model converges before over–fitting [77]. As the validation step compares model checkpoints, we include the subsampling strategies employed there in our experiments. During the test phase, dense retrieval models create document embeddings for the corpus for efficient retrieval [76]. Encoding the corpus would be an unrealistic expense for validation [77], especially as often only one relevant document is available per query [76]. Different types of

[2] Our code and data are available at: https://github.com/webis-de/ECIR-25.

subsampling make validation more efficient [77], either by re-instantiating the training objective on the validation data [31,72] or by re-ranking candidate documents [29,47]. While the training objective permits efficient validation [31,72], it can not be applied during test, as this would require knowledge of document relevance. Re-ranking documents retrieved by a first-stage retriever [47], like BM25, makes the validation more realistic, as the documents in the comparison are harder to distinguish from the relevant document. Still, this re-ranking can hide problems in neural retrieval models (like missing length normalization [59]), but as this strategy can be realistically used during test, we include it in our experiments. In sum, training and validation often differ greatly from the test phase. Consequently, subsampling strategies that work well during validation may not be reliable when used to evaluate using TREC-style test collections.

Corpus Subsampling for Evaluation. The general rule of thumb that "software is getting slower more rapidly than hardware becomes faster" [70] can also be observed for research-oriented IR systems. Approaches that either store the corpus within the weights of transformers [58], embed document corpora with large language models [36], or expand documents with transformers [46], can rarely be evaluated on large corpora. Enriching MS MARCO v2 with DocT5Query expansions took 2 500 h and cost 6 000 USD [35], highlighting the importance of sharing such resources, but also showing that it is not feasible to apply it to substantially larger corpora like the ClueWebs. Consequently, there have been experiments on subsampled corpora [34,57,58] to evaluate computationally expensive neural models using relevance judgments from large TREC-style collections. LOFT's subsampling strategy [34] includes all documents that are judged as relevant for a given query and subsequently adds random documents. The subsampling strategy by Tay et al. [58], recently re-used by Lee et al. [34], includes the complete judgment pool and subsequently adds random documents. While both sampling strategies allow to evaluate computationally expensive retrieval approaches, it remains unclear whether the resulting subsamples provide reliable evaluation. Given that both scenarios greatly increase the proportion of relevant documents, they might overestimate retrieval effectiveness or favor some techniques over others, similar to observations on condensed lists [25,50]. Therefore, we include this family of sampling strategies into our experiments; they also form the basis of our own sampling approach. Rather than sampling random documents, we sample from runs that contributed to the judgment pool.

Green and Efficiency-Oriented IR. Both efficiency and utilization play equal yet separate roles during development and operation of IR systems. Concerns of excessive utilization by large language models were first raised by Strubell et al. [55] by measuring energy and CO_2e usage. From there, Scells et al. [53] and Zuccon et al. [78] investigated energy, water, and CO_2e usage in IR systems. Despite several studies having investigated energy usage of search systems in CPU-bound environments [5,9–12], few studies have focused on search systems in GPU-based environments. The popularity of leaderboard-driven effectiveness-focused benchmarks may be one reason for this [52]. Despite several studies noting the importance of utilization [24,27,32,45,56], there is still no generally accepted measurement [26]. However, we are the first to focus on tackling the problem of utilization from a data perspective. Despite the rich history of efficiency IR research [20,61,71], we found no studies investigating the evaluation efficiency.

3 Post-Hoc Subsampling From Pooled Retrieval Runs

In TREC-style evaluation campaigns, organizers publish a document corpus D and a set of topics T. Research groups then submit retrieval runs R that retrieve documents $d \in D$ for each topic $t \in T$. A subset of query—document pairs $J \subset T \times D$ form the judgment pool (typically, the top-10 results of each system) that is assessed by experts to form relevance labels $rel(t, d)$ with $(t, d) \in J$.

In this section, we describe prior subsampling strategies and introduce pooling as a new strategy to create reliable subcorpora. As most corpora are already a sample (e.g., a web crawl), we refer to this "second" round of deriving a subcorpus as subsampling that can be applied either before or after relevance assessment.

3.1 Subsampling Before the Relevance Assessments

Cascading re-ranking pipelines can subsample large corpora without needing the relevance assessments of topics. A cascading re-ranking pipeline starts with an initial efficient retrieval model to reduce the corpus to a few hundred or thousands of documents. Progressively more expensive retrieval models, e.g., neural retrieval models, are applied to re-rank the documents [38]. Formally, a retrieval model retrieves the top results for each topic $t \in T$ from the document corpus D. The subsampled corpus D' is the union of all retrieval results.

In practice, BM25 is predominantly used as the first stage retrieval system in such scenarios (e.g., the TREC Deep Learning tracks provided an official BM25 subsample [21,22]). Subsampling with BM25 has the advantage that it is implemented in a wide range of research retrieval systems [39,73], can be used efficiently, and has a natural order to adjust the size of the subsample. However, as the judgments and submitted runs are not used, the resulting subcorpus might miss difficult relevant documents. All documents in the subcorpus must have a lexical overlap with the query. The subcorpus thereby imposes a view that semantic overlap without lexical overlap is not existent (especially manual runs in TREC-campaigns aim to identify those documents) favoring a certain type of systems. Both factors might negatively impact reliability of BM25 subcorpora.

3.2 Subsampling After the Relevance Assessments

The judgment pool J and the submitted runs R provide valuable resources for reliable subsampling. We first describe existing subsampling strategies that only incorporate the judgment pool J and then describe our pooling approach that additionally uses the submitted runs R to include more realistic hard negatives.

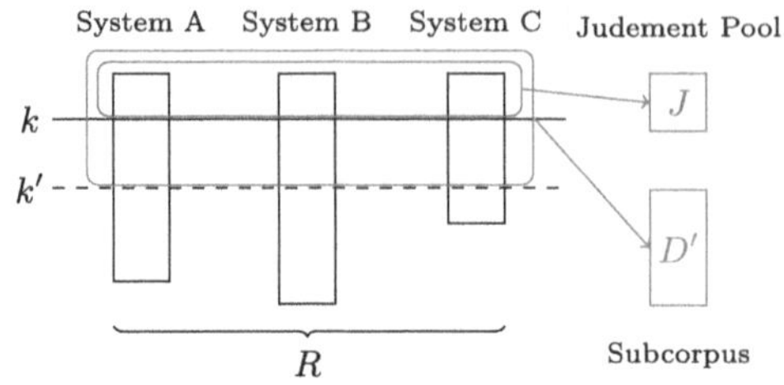

Fig. 1. Corpus subsampling via run pooling. k is the depth of the judgement pool J on pooled runs R. The subcorpus D' has depth k'.

Algorithm 1 LOFT Sampling

Input: T set of topics
J relevance judgments
k target corpus size
Output: set of documents C

$C \leftarrow \emptyset$, $i \leftarrow 0$
$m \leftarrow \min(\{\forall j \in D^T_{rel} : |j|\})$
repeat $i \leftarrow i + 1$
  if $i < m$ **then**
    for all topics $t \in T$ **do**
      $C \leftarrow d_i \in D^T_{rel}$
  else
    $C \leftarrow d \in \{D \setminus D_{rel}\}$
until $C = k$

LOFT Subsampling. The LOFT subsampling approach [34] was used to create small subcorpora that can fit into the input of long context large language models with a context size of up to one million tokens. LOFT subsampling takes all documents that are relevant for some topic and adds random documents from the corpus until a desired size is reached. Formally, if k is the desired size of the subcorpus, $D_{rel} = \{d \mid (t, d) \in J \wedge rel(t, d) > 0\}$ are all relevant documents, and $D_{rand} \subset D \setminus D_{rel}$ are random documents with $|D_{rand}| = k - |D_{rel}|$, LOFT sampling yields $D' = D_{rel} \cup D_{rand}$ as a subcorpus. Given that the document length can differ among corpora, the LOFT benchmark has an average k of 885 documents for a subcorpus (minimum: 328; maximum: 3306) to fit documents of different sizes into the fixed size context. We instantiate LOFT sampling twice, with $k = 1\,000$ and $k = 10\,000$. As some of our test collections have more than 1 000 relevant documents, we include the relevant documents round-robin per query, ensuring that each query has at most a difference of one relevant document. Algorithm 1 details how we use LOFT for corpus subsampling.

A potential problem of LOFT subsampling is that it may be too easy to distinguish the relevant documents from random documents. The subsampling strategy by Tay et al. [58] reduces this effect by taking the complete judgment pool $D_J = \{d \mid (t, d) \in J\}$ as corpus that is then expanded with random documents until the target size. However, as the judgment pools are usually rather small, random documents make the majority of the subcorpus, thereby the tasks still remains substantially more easy than more realistic subsampling strategies.

Corpus Subsampling via Pooling. In our corpus sampling strategy, we perform a second pooling round after the initial pooling for relevance judgments but at a higher depth. This allows us to incorporate a diverse set of hard-negative documents. Previous subsampling approaches have not used the submitted retrieval runs R. We argue that they form an excellent source to construct reliable subcorpora as the submitted runs were produced as a community effort from independent retrieval pipelines and should, therefore, cover diverse notions of non-relevant documents, making the subcorpus more realistic and also more accurately biased towards the topics than random sampling. As we pool to a higher depth than the judgment pool, all judged documents are included, and the subcorpus size can be varied naturally via the depth of the second pooling.

Figure 1 shows corpus subsampling via pooling. The input is a set of topics T and the set of pooled runs R. We assume the judgment pool J is an top-k pool (e.g., k=10) of $r \in R$ with the subcorpus pooling size k' being substantially larger than k. Let $r_{k'} \subset D$ denote the set of documents within the top-k' results of $r \in R$. Then, pooling subsampling yields $D' = \bigcup_{r \in R} r_{k'}$ as subsample.

4 Evaluation

We compare all subsampling strategies in theoretical and empirical experiments on four corpora with 9 TREC tracks using nDCG@10 as evaluation measure. For theoretical validation, we use simulated experiments on all runs submitted to the 9 tracks. For empirical validation, we run 20 diverse retrieval systems on all subsampled corpora while contrasting energy usage with reliability.

Subsampling Strategies in our Experiments. We experiment with 9 subsampling strategies: (1) the top-1000 BM25 results, (2) two LOFT subsamples at k=1 000 and k=10 000, and five pooling variants ($k \in \{10 = |J|, 25, 50, 100, 1000\}$). Where applicable, we use the full-corpus (i.e., not removing any documents) as baseline.

Corpora and Tracks. We use ClueWeb09, ClueWeb12, MS MARCO passage v1, and Disks 4/5 (between 0.5 m and 1 b documents; cf. Table 1). We use all runs submitted to the 2009–2014 Web tracks (30 to 71 runs), the Robust04 track (110 runs), and the MS MARCO v1 Deep Learning tracks (37 to 59 runs). Following previous setups that use TREC runs [4], we sort all runs by their nDCG@10 score and remove the least effective 25 % to mitigate the effect of potentially erroneous runs. MS MARCO comes with passage-level relevance judgments. Still, document level relevance judgments are frequently used for neural evaluations (e.g., BEIR has pre-neural datasets with document-level relevance judgments like Robust04 and Touché, truncating documents by default [60]).

Subsampled Corpora. Table 2 overviews the subcorpora sizes (averaged across the tracks). Even the largest subcorpus of pooling to 1 000 reduces the corpus size by more than 1 500 for large corpora like the ClueWeb09, and still by 8.9 for smaller corpora like MS MARCO (61 times for a top-100 pool), making it feasible to share and re-use subcorpora. Furthermore, Table 2 already hints how reliable subsampling strategies might be: the size of the judgment pool (Pool_J) shows how many documents are retrieved at top positions, allowing to calculate the probability that a random document makes it into the top results, which is negligible (e.g., $\frac{9303}{1e+09} = 9.3e - 06$ for the ClueWeb09). Therefore, random sampling might be unlikely to include documents that are retrieved into the top positions that are important for evaluation, contrary to our corpus pooling.

Table 2. Overview of the subsamples (documents and bytes) for all datasets.

Sampling	ClueWeb09		ClueWeb12		MS MARCO		Robust04	
	Docs.	Size	Docs.	Size	Docs.	Size	Docs.	Size
BM25	48091	134.3 MB	48875	241.2 MB	46392	5.9 MB	165165	209.2 MB
LOFT_{1k}	1000	3.9 MB	1000	5.9 MB	1000	0.1 MB	1000	3.3 MB
LOFT_{10k}	10000	28.9 MB	10000	45.8 MB	10000	1.4 MB	10000	22.2 MB
Pool_J	9303	26.5 MB	6971	33.1 MB	15864	2.2 MB	27239	59.4 MB
Pool_{25}	21469	60.2 MB	15416	71.0 MB	38244	5.3 MB	57579	110.2 MB
Pool_{50}	40520	111.9 MB	28032	126.5 MB	74388	10.2 MB	98653	170.9 MB
Pool_{100}	77013	211.2 MB	52477	232.0 MB	143083	19.6 MB	160825	253.8 MB
Pool_{1000}	629140	1.7 GB	435859	1.9 GB	983765	135.9 MB	449371	529.5 MB

4.1 Theoretical Validation of Corpus Subsampling

We validate the subsampling strategies using simulated incompleteness in three experiments. First, we analyze how nDCG@10 scores differ. Second, we analyze the impact on system rankings when a team that did not participate is compared with systems that participated in judgment and subsampling. Third, we analyze the impact on system rankings when multiple teams did not participate.

Simulated Incompleteness. We group all runs by their submitting team for leave-one-group-out experiments to simulate that a team did not participate in a track [63]. If the corpus is not changed, leave-one-group-out experiments impact only the judgment pool. However, as we subsample the corpus, leave-one-group-out experiments additionally impact the subsampled corpora, as subsampling strategies can not include documents solely retrieved by a left-out group. We simulate judgment incompleteness for top-10 judgment pools, i.e., removing documents from the judgment pool solely contributed by the left-out group (i.e., does not occur in the top-10 results of any system of other groups). All subsampling strategies only have access to the modified judgment pool and the remaining contributing runs. This yields one incomplete subsampled corpus and judgment pool per group, where runs of other groups fully contributed to the subsampled corpus and remain fully judged. In all cases, we use the nDCG@10 on the complete corpus on all judgments as ground-truth. For systems that retrieve on the subsampled corpora, we assume that subsampling does not affect the order of documents (we validate this in Sect. 4.2), and thereby remove documents that are not in a subsample from a run to simulate retrieval from the subcorpus.

The Impact of Corpus Subsampling on nDCG@10. We compare the nDCG@10 scores of left-out systems with their ground-truth to show the behavior of the subsampling strategies. Two factors contribute to wrong nDCG@10 scores: (1) unjudged documents and (2) missing documents. A document is unjudged if retrieved only by the left-out group, potentially underestimating nDCG@10 [25,

Table 3. Leave-one-group-out differences in nDCG@10 on ClueWeb09 (C09), ClueWeb12 (C12), MS MARCO (MSM), and Robust04 (R04). We report the RMSE without and with post-judgments (RMSE vs. RMSE_{Judged}), and the difference of subsampled vs. correct nDCG@10 ($\Delta_{\text{nDCG@10}} \pm$ std dev.). * indicates Bonferroni corrected significant differences to the full corpus.

Sampl.	RMSE				RMSE_{Judged}				$\Delta_{\text{nDCG@10}}$			
	C09	C12	MSM	R04	C09	C12	MSM	R04	C09	C12	MSM	R04
BM25	.057	.084	.096*	.007	.083*	.052*	.110*	.010	-.013 $\pm$.06	-.053 $\pm$.07	.049* $\pm$.08	-.005 $\pm$.00
LOFT_{1k}	.372*	.333*	.529*	.042*	.372*	.333*	.529*	.047*	.365* $\pm$.08	.311* $\pm$.12	.528* $\pm$.04	.015* $\pm$.04
LOFT_{10k}	.381*	.342*	.110*	.263*	.386*	.349*	.110*	.271*	.375* $\pm$.07	.325* $\pm$.11	.062* $\pm$.09	.259* $\pm$.05
Pool_J	.041	**.039***	.009	.014*	.042*	.040*	.009*	.018*	.030 $\pm$.03	.031* $\pm$.02	.005 $\pm$.01	.011* $\pm$.01
Pool_{25}	**.025***	.040*	**.006**	**.005***	.037*	.029*	.007	.011*	**-.011*** $\pm$.02	**-.028*** $\pm$.03	**-.002** $\pm$.01	**-.002*** $\pm$.00
Pool_{50}	.032*	.060	.007	.006*	.034*	.019*	.006	.008	-.023* $\pm$.02	-.050 $\pm$.03	-.004 $\pm$.01	-.005* $\pm$.00
Pool_{100}	.039	.071	**.006**	.008	.030*	.015*	.006	**.007**	-.030 $\pm$.02	-.060 $\pm$.04	-.004 $\pm$.00	-.007 $\pm$.00
Pool_{1000}	.047	.082	**.006**	.009	**.023**	**.008***	**.005**	**.007**	-.039 $\pm$.03	-.070 $\pm$.04	-.005 $\pm$.00	-.007 $\pm$.00
Full	.050	.085	.006	.009	.000	.000	.000	.000	-.041 $\pm$.03	-.072 $\pm$.04	-.005 $\pm$.00	-.007 $\pm$.00

50]. A document is missing if a left-out system would retrieve this document from the full corpus that is not in the subsample. Without post-judgments under the default assumption that unjudged documents are not relevant [66], missing documents would be unjudged and assumed not relevant, thereby overestimating nDCG@10 (a frequent scenario as post-hoc experiments rarely add judgments).

To show the impact of unjudged and missing documents independently, in Table 3 we report the root mean squared error without post-judgments (RMSE), with post-judgments (RMSE_{Judged}), and the average difference of the subsampled versus the actual nDCG@10 score to show which subsampling strategies overestimate (positive $\Delta_{nDCG@10}$) or underestimate (negative $\Delta_{nDCG@10}$). Evaluations on the full corpus underestimate nDCG@10 scores the most, by 0.005 for MS MARCO up to 0.072 for the ClueWeb12. With post-judgments, nDCG@10 scores are correct ($\text{RMSE}_{Judged} = 0.0$). LOFT sampling substantially overestimates nDCG@10 scores, in the worst case by 0.528 for MS MARCO. Using only the judgment pool (Pool_J) also overestimates the effectiveness, but much less than LOFT as more realistic non-relevant documents are included. BM25 may over- or underestimate the effectiveness, having a much higher standard deviation than pooling. Pooling to a higher depth than the original judgment pool produces the most reliable subcorpora, as nDCG@10 scores are underestimated, but to a lesser extent as processing the complete corpus as fewer unjudged documents are included but still enough to not overestimate. Without post-judgments, pooling to 25 produces reliable scores, whereas the depth-100 and depth-1000 scores are statistically indistinguishable from the full corpus. With post-judgments, pooling beyond 100 only yields negligible improvements, making a top-100 pool a good choice for all scenarios.

Corpus Subsampling for Comparisons Against Participating Systems. We study the scenario where systems of the left-out group are compared with systems that contributed to the judgments and the subsampling. All contributing systems are correctly evaluated, whereas only the left-out systems could introduce changes in the system rankings. To compare system

Table 4. Overview of the reliability of system rankings measured as τ_{AP} without post-judgments (τ), with condensed evaluation (τ_C), and with post judgments (τ_{PJ}) for comparing systems of a non-participating group with participating systems. *: Bonferroni corrected significant differences to the full corpus.

Sampl.	ClueWeb09			ClueWeb12			MS MARCO			Robust04		
	τ	τ_C	τ_{PJ}	τ	τ_C	τ_{PJ}	τ	τ_C	τ_{PJ}	τ	τ_C	τ_{PJ}
BM25	.919	.807	.936*	.894	.922	.938	.847*	.827*	.836*	.980	.945	.994
LOFT_{1k}	.793	.690*	.793*	.775	.775*	.775*	.776*	.774*	.776*	.940	.904	.940*
LOFT_{10k}	.799	.693*	.799*	.763	.762*	.765*	.790*	.786*	.789*	.794*	.764*	.794*
Pool_J	.944	**.832**	.944*	**.941**	**.939**	.941*	**.983**	**.980**	.983	.978	**.948**	.978*
Pool_{25}	**.956**	**.832**	.967*	.917	**.939**	.959	.978	**.980**	.992	**.987**	**.948**	.993*
Pool_{50}	.940	**.832**	.974	.902	**.939**	.976	.976	**.980**	.993	.981	**.948**	.997*
Pool_{100}	.934	**.832**	.980	.898	**.939**	.987	.974	**.980**	.995	.979	**.948**	.999
Pool_{1000}	.931	**.832**	**.994**	.896	**.939**	**.997**	.974	**.980**	**.996**	.978	**.948**	**1.00**
Full	.930	.832	1.00	.895	.939	1.00	.972	.980	1.00	.978	.948	1.00

rankings we use τ_{AP} variant of Kendall's correlation with handling of ties [62] (1: perfect, 0: random, -1: perfect inverse correlation), as τ_{AP} gives more importance to top-ranked systems. We report changes in the system rankings in three scenarios, without post-judgments (τ), with condensed lists that remove all unjudged documents (τ_C), and with post-judgments (τ_{PJ}) in Table 4. Across all scenarios, LOFT yields the most changes in system rankings, often significantly different to the full corpus. BM25 subsamples are more reliable, but pooling yields even better results in all cases, statistically not distinguishable from evaluating on the complete corpus for top-100 and top-1000 pools. Still, comparisons against participating systems are rarely used, as research often compares systems across multiple corpora, and the same system rarely is pooled on different corpora. Hence, we next study scenarios where multiple non-participating systems are compared.

Corpus Subsampling for Comparisons of Non-participating Systems. To simulate a setting where systems of different groups that did not contribute to the judgments and subsampling are compared, we contrast the system ranking obtained from only leave-one-group-out nDCG@10 scores with the ground-truth system ranking in Table 5 (no significance tests are possible in this scenario, as no averages are reported). Every system can have unjudged and missing documents. Consequently, system ranking correlations for the full corpus without post judgments (τ) are rather low for large corpora with few pooled runs (e.g., $\tau = 0.302$ for the ClueWeb12 with 30 runs) whereas reliable for corpora with many pooled runs (e.g., $\tau = 0.884$ for Robust04 with 110 runs). However, with post-judgments (τ_{PJ}), the full corpora produce perfect correlations, whereas the top-100 and top-1000 pooled subcorpora also achieve high ranking correlations; in case of condensed lists (τ_C) even indistinguishable from the full corpora. Consistent with the previous experiments, BM25 subsampling is better than LOFT but less reliable than pooling (especially for the condensed and the post-judgment setting). Altogether, our three experiments show that pooling to a depth of 100 or 1000 is often indistinguishable from processing the full corpora, at only a tiny fraction of the compute.

Table 5. The reliability of system rankings as τ_{AP} without post-judgments (τ), with condensed evaluation (τ_C), and with post judgments (τ_{PJ}) when comparing systems that all did not participate in the judgment and subsampling.

Sampling	ClueWeb09			ClueWeb12			MS MARCO			Robust04		
	τ	τ_C	τ_{PJ}	τ	τ_C	τ_{PJ}	τ	τ_C	τ_{PJ}	τ	τ_C	τ_{PJ}
BM25	.265	.298	.435	.211	.541	.542	.601	.593	.627	.888	.859	.949
LOFT_{1k}	.278	.281	.278	.329	.329	.329	.294	.294	.294	.416	.392	.416
LOFT_{10k}	.373	.360	.372	.366	.337	.373	.496	.497	.497	.312	.259	.305
Pool_J	.611	.579	.611	**.699**	**.699**	.699	**.872**	**.872**	.872	.909	**.899**	.909
Pool_{25}	**.706**	.583	.734	.560	**.699**	.732	.854	**.872**	.926	**.940**	**.899**	.951
Pool_{50}	.658	.584	.781	.470	**.699**	.816	.841	**.872**	.940	.912	**.899**	.971
Pool_{100}	.618	.585	.827	.376	**.699**	.889	.835	**.872**	.952	.888	**.899**	.991
Pool_{1000}	.574	**.586**	**.946**	.337	**.699**	**.965**	.830	**.872**	**.957**	.884	**.899**	**1.00**
Full	.565	.586	1.00	.302	.699	1.00	.869	.869	1.00	.884	.899	1.00

4.2 Empirical Validation of Corpus Subsampling for Retrieval

While our previous experiments assumed that corpus subsampling does not change the relative order of retrieval results to validate theoretically the effect of incompleteness and subsampling, in practice, removing documents from the collection changes the corpus statistics and thereby potentially the ranking. Therefore, we evaluate how rankings of 20 diverse first-stage retrieval systems change over all subcorpora monitoring their energy consumption to identify which subsampling yield reliable retrieval results at acceptable energy consumption.

Retrieval Systems. We include retrieval systems of three popular paradigms: 10 lexical retrieval models in PyTerrier [39], 7 bi-encoder models in BEIR [60], and 3 late interaction models in PLAID-X [43].[3] All systems operate in their default configuration. We do not include re-rankers like monoT5 or duoT5 [46] as they solely depend on the first-stage, not on global aspects like corpus statistics.

Experimental Setup. We run all retrieval systems on all subcorpora on the same machine (four A100 GPUs with 96 CPU cores) sequentially measuring their energy consumption with codecarbon [19]. To make the experiments comparable across the diverse corpora, we use the retrieval results of a system for the top-1000 pool as the ground-truth ranking and report the similarities for all smaller subsamples measured as RBO [69] as this allows to compare rankings of different documents. The energy consumption includes indexing and retrieval, i.e., the footprint of the complete experiment.

Results. Figure 2 shows the energy consumption in kWh for all approaches across the different corpus subsamples. We extrapolate the energy consumption beyond the top-1000 pools to the full corpora and indicate the consumption for our recommended top-100 and top-1000 pooled

[3] Bi-Encoder: ANCE 🔗, DistilBERT 🔗, MiniLM-L6 🔗, MiniLM-L12 🔗, TAS-B 🔗, MultiQA (DistilBERT) 🔗, MultiQA (MpNet) 🔗; Late Interaction: ColBERT v1.9 🔗, ColBERT v2.0 🔗, PLAID-X English Large 🔗; Lexical: BM25 🔗, DirichletLM 🔗, DFIZ 🔗, DLH 🔗, DPH 🔗, Hiemstra LM 🔗, IFB2 🔗, InB2 🔗, PL2 🔗, TF IDF 🔗.

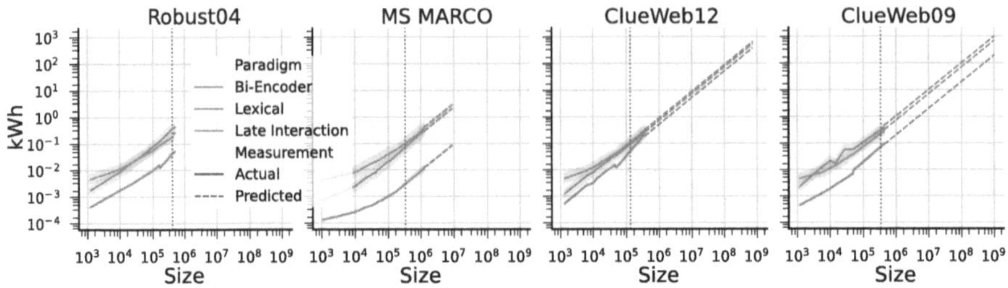

Fig. 2. Energy consumption in kWh for all paradigms and subsamples interpolated to the full corpora. The vertical line indicates the recommended subsample.

Table 6. RBO similarity of Bi-Encoder, Late-Interaction, and Lexical models to themself when retrieving from a subsample versus the largest Pool_{1000} subcorpus.

Sampling	ClueWeb09			ClueWeb12			MS MARCO			Robust04		
	Bi-E.	Late	Lex.	Bi-E.	Late	Lex.	Bi-E.	Late	Lex.	Bi-E.	Late	Lex.
BM25	.139	.192	.037	.146	.159	.0837	.89	.749	.858	.921	.758	**.963**
LOFT_{1k}	.070	.090	.040	.115	.103	.176	.68	.634	.513	.113	.13	.138
LOFT_{10k}	.096	.111	.056	.185	.167	.308	.777	.693	.601	.431	.459	.495
Pool_J	.297	.263	.295	.289	.256	.606	.908	.756	.880	.712	.682	.885
Pool_{25}	.405	.333	.449	.407	.358	.686	.954	.761	.901	.824	.726	.910
Pool_{50}	.495	.394	.550	.506	.44	.752	.973	.781	.914	.888	.755	.927
Pool_{100}	**.600**	**.481**	**.660**	**.613**	**.524**	**.792**	**.981**	**.787**	**.933**	**.936**	**.759**	.946

subsamples. For large corpora, our subsamples reduce the energy consumption by more than 1 000, and even for small corpora, energy savings of around a factor of 10 are reached.

Table 6 shows how similar the rankings on the subsampled corpora are towards the top-1000 pool for the three retrieval paradigms. Consistently, the LOFT and BM25 subcorpora do not yield similar rankings, whereas a top-25 pool already shows substantial improvement compared to just re-ranking the judgment pool. Still, using the judgment pool shows reasonable ranking similarities for small corpora, as there the judgment pools cover a substantial part of the complete collection, whereas huge corpora such as the ClueWeb09 and ClueWeb12 show a very low ranking similarity for judged documents only (e.g., RBO of 0.295 for lexical models on ClueWeb09 vs. 0.880 on MS MARCO). The top-100 pool yields the most similar rankings. Overall, our experiments were consistent across all tested scenarios, providing evidence that pooling allows the construction of reliable subcorpora, whereas top-100 pools can be recommended as they already perform as reliably as top-1000 pools.

5 Conclusion and Future Work

We introduced pooling as an approach to produce subcorpora that allows reliable evaluation of expensive retrieval approaches on large corpora. We reduced the computational requirements by up to 1000 times while providing statistically indistinguishable evaluations from processing the complete corpus. The resulting subcorpora allow exploration of many expensive re-ranking cascades more systematically. Even corpora that are already small could be substantially reduced, yielding much faster and more Green experiments. An interesting direction for future work could be to study how corpus subsampling can be incorporated up-front into the design of evaluation campaigns. Retrieval from huge, noisy corpora is difficult; thereby, running a campaign on such huge, ClueWeb-style corpora might be able to construct a realistic and diverse set of hard negatives, while subsequent corpus subsampling can ensure that the post-hoc experiments with the fixed set of queries only need a small computational budget.

Acknowledgements. This publication has received funding from the European Union's Horizon Europe research and innovation programme under grant agreement No 101070014 (OpenWebSearch.EU, https://doi.org/10.3030/101070014).

References

1. Allan, J., Aslam, J.A., Pavlu, V., Kanoulas, E., Carterette, B.: Million Query track 2008 overview. In: Voorhees, E.M., Buckland, L.P. (eds.) Proceedings of The Seventeenth Text REtrieval Conference, TREC 2008, Gaithersburg, Maryland, USA, 18–21 November 2008. NIST Special Publication, vol. 500–277. National Institute of Standards and Technology (NIST) (2008)
2. Aslam, J.A., Pavlu, V., Yilmaz, E.: A statistical method for system evaluation using incomplete judgments. In: Efthimiadis, E.N., Dumais, S.T., Hawking, D., Järvelin, K. (eds.) SIGIR 2006: Proceedings of the 29th Annual International ACM SIGIR Conference on Research and Development in Information Retrieval, Seattle, Washington, USA, 6–11 August 2006, pp. 541–548. ACM (2006)
3. Aslam, J.A., Yilmaz, E.: Inferring document relevance from incomplete information. In: Silva, M.J., Laender, A.H.F., Baeza-Yates, R.A., McGuinness, D.L., Olstad, B., Olsen, Ø.H., Falcão, A.O. (eds.) Proceedings of the Sixteenth ACM Conference on Information and Knowledge Management, CIKM 2007, Lisbon, Portugal, 6–10 November 2007, pp. 633–642. ACM (2007)
4. Bernstein, Y., Zobel, J.: Redundant documents and search effectiveness. In: Herzog, O., Schek, H., Fuhr, N., Chowdhury, A., Teiken, W. (eds.) Proceedings of the 2005 ACM CIKM International Conference on Information and Knowledge Management, Bremen, Germany, October 31 – November 5, 2005, pp. 736–743. ACM (2005)
5. Blanco, R., Catena, M., Tonellotto, N.: Exploiting green energy to reduce the operational costs of multi-center web search engines. In: Bourdeau, J., Hendler, J., Nkambou, R., Horrocks, I., Zhao, B.Y. (eds.) Proceedings of the 25th International Conference on World Wide Web, WWW 2016, Montreal, Canada, 11–15 April 2016, pp. 1237–1247. ACM (2016)
6. Buckley, C., Voorhees, E.M.: Retrieval evaluation with incomplete information. In: Sanderson, M., Järvelin, K., Allan, J., Bruza, P. (eds.) SIGIR 2004: Proceedings of the 27th Annual International ACM SIGIR Conference on Research and Development in Information Retrieval, Sheffield, UK, July 25–29, 2004, pp. 25–32. ACM (2004)
7. Büttcher, S., Clarke, C.L.A., Yeung, P.C.K., Soboroff, I.: Reliable information retrieval evaluation with incomplete and biased judgements. In: Kraaij, W., de Vries, A.P., Clarke, C.L.A., Fuhr, N., Kando, N. (eds.) SIGIR 2007: Proceedings of the 30th Annual International ACM SIGIR Conference on Research and Development in Information Retrieval, Amsterdam, The Netherlands, 23–27 July 2007, pp. 63–70. ACM (2007)

8. Carterette, B., Jones, R.: Evaluating search engines by modeling the relationship between relevance and clicks. In: Platt, J.C., Koller, D., Singer, Y., Roweis, S.T. (eds.) Advances in Neural Information Processing Systems 20, Proceedings of the Twenty-First Annual Conference on Neural Information Processing Systems, Vancouver, British Columbia, Canada, 3–6 December 2007, pp. 217–224. Curran Associates, Inc. (2007)
9. Catena, M.: Energy efficiency in web search engines. In: Azzopardi, L., Wilson, M.L. (eds.) Sixth BCS-IRSG Symposium on Future Directions in Information Access, FDIA 2015, 31 August - 4 September 2015, Thessaloniki. Greece. Workshops in Computing, BCS (2015)
10. Catena, M., Frieder, O., Tonellotto, N.: Efficient energy management in distributed web search. In: Cuzzocrea, A., et al. (eds.) Proceedings of the 27th ACM International Conference on Information and Knowledge Management, CIKM 2018, Torino, Italy, October 22–26, 2018, pp. 1555–1558. ACM (2018)
11. Catena, M., Macdonald, C., Tonellotto, N.: Load-sensitive CPU power management for web search engines. In: Baeza-Yates, R., Lalmas, M., Moffat, A., Ribeiro-Neto, B.A. (eds.) Proceedings of the 38th International ACM SIGIR Conference on Research and Development in Information Retrieval, Santiago, Chile, August 9–13, 2015, pp. 751–754. ACM (2015)
12. Catena, M., Tonellotto, N.: A study on query energy consumption in web search engines. In: Boldi, P., Perego, R., Sebastiani, F. (eds.) Proceedings of the 6th Italian Information Retrieval Workshop, Cagliari, Italy, 25–26 May 2015. CEUR Workshop Proceedings, vol. 1404. CEUR-WS.org (2015)
13. Clarke, C.L.A., Craswell, N., Soboroff, I.: Overview of the TREC 2009 Web track. In: Voorhees, E.M., Buckland, L.P. (eds.) Proceedings of The Eighteenth Text REtrieval Conference, TREC 2009, Gaithersburg, Maryland, USA, 17–20 November 2009. NIST Special Publication, vol. 500–278. National Institute of Standards and Technology (NIST) (2009)
14. Clarke, C.L.A., Craswell, N., Soboroff, I., Cormack, G.V.: Overview of the TREC 2010 Web track. In: Voorhees, E.M., Buckland, L.P. (eds.) Proceedings of The Nineteenth Text REtrieval Conference, TREC 2010, Gaithersburg, Maryland, USA, 16–19 November 2010. NIST Special Publication, vol. 500–294. National Institute of Standards and Technology (NIST) (2010)
15. Clarke, C.L.A., Craswell, N., Soboroff, I., Voorhees, E.M.: Overview of the TREC 2011 Web track. In: Voorhees, E.M., Buckland, L.P. (eds.) Proceedings of The Twentieth Text REtrieval Conference, TREC 2011, Gaithersburg, Maryland, USA, 15–18 November 2011. NIST Special Publication, vol. 500–296. National Institute of Standards and Technology (NIST) (2011)
16. Clarke, C.L.A., Craswell, N., Voorhees, E.M.: Overview of the TREC 2012 Web track. In: Voorhees, E.M., Buckland, L.P. (eds.) Proceedings of The Twenty-First Text REtrieval Conference, TREC 2012, Gaithersburg, Maryland, USA, 6–9 November 2012. NIST Special Publication, vol. 500–298. National Institute of Standards and Technology (NIST) (2012)
17. Collins-Thompson, K., Bennett, P.N., Diaz, F., Clarke, C., Voorhees, E.M.: TREC 2013 Web track overview. In: Voorhees, E.M. (ed.) Proceedings of The Twenty-Second Text REtrieval Conference, TREC 2013, Gaithersburg, Maryland, USA, 19–22 November 2013. NIST Special Publication, vol. 500–302. National Institute of Standards and Technology (NIST) (2013)
18. Collins-Thompson, K., Macdonald, C., Bennett, P.N., Diaz, F., Voorhees, E.M.: TREC 2014 Web track overview. In: Voorhees, E.M., Ellis, A. (eds.) Proceed-

ings of The Twenty-Third Text REtrieval Conference, TREC 2014, Gaithersburg, Maryland, USA, 19–21 November 2014. NIST Special Publication, vol. 500–308. National Institute of Standards and Technology (NIST) (2014)
19. Courty, B.: mlco2/codecarbon: v2.4.1 (2024). https://doi.org/10.5281/zenodo.11171501
20. Crane, M., Culpepper, J.S., Lin, J., Mackenzie, J.M., Trotman, A.: A comparison of document-at-a-time and score-at-a-time query evaluation. In: de Rijke, M., Shokouhi, M., Tomkins, A., Zhang, M. (eds.) Proceedings of the Tenth ACM International Conference on Web Search and Data Mining, WSDM 2017, Cambridge, United Kingdom, 6–10 February 2017, pp. 201–210. ACM (2017)
21. Craswell, N., Mitra, B., Yilmaz, E., Campos, D.: Overview of the TREC 2020 Deep Learning Track. In: Voorhees, E.M., Ellis, A. (eds.) Proceedings of the 29th Text REtrieval Conference, TREC 2020, Virtual Event, Gaithersburg, MD, USA, November 16–20, 2020. NIST Special Publication, vol. 1266. National Institute of Standards and Technology (NIST) (2020)
22. Craswell, N., Mitra, B., Yilmaz, E., Campos, D., Voorhees, E.M.: Overview of the TREC 2019 deep learning track. In: Voorhees, E., Ellis, A. (eds.) 28th International Text Retrieval Conference, TREC 2019, Gaithersburg, Maryland. USA. NIST Special Publication, National Institute of Standards and Technology (NIST) (Nov (2019)
23. Ferro, N., Maistro, M.: Evaluation of IR systems. In: Alonso, O., Baeza-Yates, R. (eds.) Information Retrieval: Advanced Topics and Techniques, ACM Books, vol. 60, pp. 111–191. ACM (2024)
24. Frayling, E., MacAvaney, S., Macdonald, C., Ounis, I.: Effective adhoc retrieval through traversal of a query-document graph. In: Goharian, N., Tonellotto, N., He, Y., Lipani, A., McDonald, G., Macdonald, C., Ounis, I. (eds.) Advances in Information Retrieval - 46th European Conference on Information Retrieval, ECIR 2024, Glasgow, UK, March 24–28, 2024, Proceedings, Part III. Lecture Notes in Computer Science, vol. 14610, pp. 89–104. Springer (2024). https://doi.org/10.1007/978-3-031-56063-7_6
25. Fröbe, M., Gienapp, L., Potthast, M., Hagen, M.: Bootstrapped nDCG estimation in the presence of unjudged documents. In: Kamps, J., Goeuriot, L., Crestani, F., Maistro, M., Joho, H., Davis, B., Gurrin, C., Kruschwitz, U., Caputo, A. (eds.) Advances in Information Retrieval - 45th European Conference on Information Retrieval, ECIR 2023, Dublin, Ireland, April 2–6, 2023, Proceedings, Part I. Lecture Notes in Computer Science, vol. 13980, pp. 313–329. Springer (2023). https://doi.org/10.1007/978-3-031-28244-7_20
26. Fröbe, M., Mackenzie, J., Mitra, B., Nardini, F.M., Potthast, M.: ReNeuIR at SIGIR 2024: The third workshop on reaching efficiency in neural information retrieval. In: Yang, G.H., Wang, H., Han, S., Hauff, C., Zuccon, G., Zhang, Y. (eds.) Proceedings of the 47th International ACM SIGIR Conference on Research and Development in Information Retrieval, SIGIR 2024, Washington DC, USA, 14–18 July 2024, pp. 3051–3054. ACM (2024)
27. Gienapp, L., Deckers, N., Potthast, M., Scells, H.: Learning effective representations for retrieval using self-distillation with adaptive relevance margins. arXiv:2407.21515 (2024)
28. Harman, D.: TREC-style evaluations. In: Agosti, M., Ferro, N., Forner, P., Müller, H., Santucci, G. (eds.) PROMISE 2012. LNCS, vol. 7757, pp. 97–115. Springer, Heidelberg (2013). https://doi.org/10.1007/978-3-642-36415-0_7

29. Hofstätter, S., Lin, S., Yang, J., Lin, J., Hanbury, A.: Efficiently teaching an effective dense retriever with balanced topic aware sampling. In: Diaz, F., Shah, C., Suel, T., Castells, P., Jones, R., Sakai, T. (eds.) SIGIR 2021: The 44th International ACM SIGIR Conference on Research and Development in Information Retrieval, Virtual Event, Canada, 11–15 July 2021, pp. 113–122. ACM (2021)
30. Järvelin, K., Kekäläinen, J.: Cumulated gain-based evaluation of IR techniques. ACM Trans. Inf. Syst. **20**(4), 422–446 (2002)
31. Karpukhin, V., et al.: Dense passage retrieval for open-domain question answering. In: Webber, B., Cohn, T., He, Y., Liu, Y. (eds.) Proceedings of the 2020 Conference on Empirical Methods in Natural Language Processing, EMNLP 2020, Online, 16–20 November 2020, pp. 6769–6781. ACL (2020)
32. Khandel, P., Yates, A., Varbanescu, A.L., de Rijke, M., Pimentel, A.D.: Distillation vs. sampling for efficient training of learning to rank models. In: Oosterhuis, H., Bast, H., Xiong, C. (eds.) Proceedings of the 2024 ACM SIGIR International Conference on Theory of Information Retrieval, ICTIR 2024, Washington, DC, USA, 13 July 2024, pp. 51–60. ACM (2024)
33. Lawrie, D.J., Kayi, E.S., Yang, E., Mayfield, J., Oard, D.W.: PLAID SHIRTTT for large-scale streaming dense retrieval. In: Yang, G.H., Wang, H., Han, S., Hauff, C., Zuccon, G., Zhang, Y. (eds.) Proceedings of the 47th International ACM SIGIR Conference on Research and Development in Information Retrieval, SIGIR 2024, Washington DC, USA, 14–18 July 2024, pp. 2574–2578. ACM (2024)
34. Lee, J., et al.: Can long-context language models subsume retrieval, RAG, SQL, and more? arXiv:2406.13121 (2024)
35. Ma, X., Pradeep, R., Nogueira, R.F., Lin, J.: Document expansion baselines and learned sparse lexical representations for MS MARCO V1 and V2. In: Amigó, E., Castells, P., Gonzalo, J., Carterette, B., Culpepper, J.S., Kazai, G. (eds.) SIGIR '22: The 45th International ACM SIGIR Conference on Research and Development in Information Retrieval, Madrid, Spain, 11-15 July 2022, pp. 3187–3197. ACM (2022)
36. Ma, X., Wang, L., Yang, N., Wei, F., Lin, J.: Fine-tuning LLaMA for multi-stage text retrieval. In: Yang, G.H., Wang, H., Han, S., Hauff, C., Zuccon, G., Zhang, Y. (eds.) Proceedings of the 47th International ACM SIGIR Conference on Research and Development in Information Retrieval, SIGIR 2024, Washington DC, USA, 14–18 July 2024, pp. 2421–2425. ACM (2024)
37. MacAvaney, S., Soldaini, L.: One-shot labeling for automatic relevance estimation. In: Chen, H., Duh, W.E., Huang, H., Kato, M.P., Mothe, J., Poblete, B. (eds.) Proceedings of the 46th International ACM SIGIR Conference on Research and Development in Information Retrieval, SIGIR 2023, Taipei, Taiwan, 23–27 July 2023, pp. 2230–2235. ACM (2023)
38. MacAvaney, S., Tonellotto, N., Macdonald, C.: Adaptive re-ranking with a corpus graph. In: Hasan, M.A., Xiong, L. (eds.) Proceedings of the 31st ACM International Conference on Information & Knowledge Management, Atlanta, GA, USA, 17-21 October 2022, pp. 1491–1500. ACM (2022)
39. Macdonald, C., Tonellotto, N., MacAvaney, S., Ounis, I.: PyTerrier: declarative experimentation in Python from BM25 to dense retrieval. In: Demartini, G., Zuccon, G., Culpepper, J.S., Huang, Z., Tong, H. (eds.) CIKM 2021: The 30th ACM International Conference on Information and Knowledge Management, Virtual Event, Queensland, Australia, 1–5 November 2021, pp. 4526–4533. ACM (2021)
40. Mizzaro, S.: Relevance: the whole history. J. Am. Soc. Inf. Sci. **48**(9), 810–832 (1997)

41. Moffat, A., Mackenzie, J.: How much freedom does an effectiveness metric really have? J. Am. Soc. Inf. Sci. **75**(6), 686–703 (2024)
42. Moffat, A., Zobel, J.: Rank-biased precision for measurement of retrieval effectiveness. ACM Trans. Inf. Syst. **27**(1), 2:1–2:27 (2008)
43. Nair, S., et al.: Transfer learning approaches for building cross-language dense retrieval models. In: Hagen, M., et al. (eds.) ECIR 2022. LNCS, vol. 13185, pp. 382–396. Springer, Cham (2022). https://doi.org/10.1007/978-3-030-99736-6_26
44. Overwijk, A., Xiong, C., Callan, J.: ClueWeb22: 10 billion web documents with rich information. In: Amigó, E., Castells, P., Gonzalo, J., Carterette, B., Culpepper, J.S., Kazai, G. (eds.) SIGIR '22: The 45th International ACM SIGIR Conference on Research and Development in Information Retrieval, Madrid, Spain, 11-15 July 2022, pp. 3360–3362. ACM (2022)
45. Pasin, A., Cunha, W., Gonçalves, M.A., Ferro, N.: A quantum annealing instance selection approach for efficient and effective transformer fine-tuning. In: Oosterhuis, H., Bast, H., Xiong, C. (eds.) Proceedings of the 2024 ACM SIGIR International Conference on Theory of Information Retrieval, ICTIR 2024, Washington, DC, USA, 13 July 2024, pp. 205–214. ACM (2024)
46. Pradeep, R., Nogueira, R.F., Lin, J.: The expando-mono-duo design pattern for text ranking with pretrained sequence-to-sequence models. arXiv 2101.05667 (2021)
47. Qu, Y., et al.: RocketQA: An optimized training approach to dense passage retrieval for open-domain question answering. In: Toutanova, K., Rumshisky, A., Zettlemoyer, L., Hakkani-Tür, D., Beltagy, I., Bethard, S., Cotterell, R., Chakraborty, T., Zhou, Y. (eds.) Proceedings of the 2021 Conference of the North American Chapter of the Association for Computational Linguistics: Human Language Technologies, NAACL-HLT 2021, Online, 6–11 June 2021, pp. 5835–5847. ACL (2021)
48. Roberts, K., et al.: TREC-COVID: rationale and structure of an information retrieval shared task for COVID-19. J. Am. Med. Inform. Assoc. **27**(9), 1431–1436 (2020)
49. Sakai, T.: Alternatives to BPREF. In: Kraaij, W., de Vries, A.P., Clarke, C.L.A., Fuhr, N., Kando, N. (eds.) SIGIR 2007: Proceedings of the 30th Annual International ACM SIGIR Conference on Research and Development in Information Retrieval, Amsterdam, The Netherlands, July 23–27, 2007, pp. 71–78. ACM (2007)
50. Sakai, T.: Comparing metrics across TREC and NTCIR: the robustness to system bias. In: Shanahan, J.G., Amer-Yahia, S., Manolescu, I., Zhang, Y., Evans, D.A., Kolcz, A., Choi, K., Chowdhury, A. (eds.) Proceedings of the 17th ACM Conference on Information and Knowledge Management, CIKM 2008, Napa Valley, California, USA, October 26–30, 2008, pp. 581–590. ACM (2008)
51. Sakai, T.: How to run an evaluation task. In: Information Retrieval Evaluation in a Changing World. TIRS, vol. 41, pp. 71–102. Springer, Cham (2019). https://doi.org/10.1007/978-3-030-22948-1_3
52. Santhanam, K., et al.: Moving beyond downstream task accuracy for information retrieval benchmarking. In: Rogers, A., Boyd-Graber, J.L., Okazaki, N. (eds.) Findings of the Association for Computational Linguistics: ACL 2023, Toronto, Canada, July 9–14, 2023, pp. 11613–11628. ACL (2023)
53. Scells, H., Zhuang, S., Zuccon, G.: Reduce, reuse, recycle: Green information retrieval research. In: Amigó, E., Castells, P., Gonzalo, J., Carterette, B., Culpepper, J.S., Kazai, G. (eds.) SIGIR 2022: The 45th International ACM SIGIR Conference on Research and Development in Information Retrieval, Madrid, Spain, July 11 – 15, 2022, pp. 2825–2837. ACM (2022)

54. Soboroff, I.: Don't use LLMs to make relevance judgments. arXiv:2409.15133 (2024)
55. Strubell, E., Ganesh, A., McCallum, A.: Energy and policy considerations for deep learning in NLP. In: Korhonen, A., Traum, D.R., Màrquez, L. (eds.) Proceedings of the 57th Conference of the Association for Computational Linguistics, ACL 2019, Florence, Italy, July 28 – August 2, 2019, Volume 1: Long Papers, pp. 3645–3650. ACL (2019)
56. Sun, S., Zhuang, S., Wang, S., Zuccon, G.: An investigation of prompt variations for zero-shot LLM-based rankers. arXiv 2406.14117 (2024)
57. Tang, Y., Zhang, R., Guo, J., de Rijke, M., Chen, W., Cheng, X.: Listwise generative retrieval models via a sequential learning process. ACM Trans. Inf. Syst. **42**(5), 133:1–133:31 (2024)
58. Tay, Y., et al.: Transformer memory as a differentiable search index. In: Koyejo, S., Mohamed, S., Agarwal, A., Belgrave, D., Cho, K., Oh, A. (eds.) Advances in Neural Information Processing Systems 35: Annual Conference on Neural Information Processing Systems 2022, NeurIPS 2022, New Orleans, LA, USA, November 28 – December 9, 2022 (2022)
59. Thakur, N., et al.: Systematic evaluation of neural retrieval models on the Touché 2020 argument retrieval subset of BEIR. In: Yang, G.H., Wang, H., Han, S., Hauff, C., Zuccon, G., Zhang, Y. (eds.) Proceedings of the 47th International ACM SIGIR Conference on Research and Development in Information Retrieval, SIGIR 2024, Washington DC, USA, 14–18 July 2024, pp. 1420–1430. ACM (2024)
60. Thakur, N., Reimers, N., Rücklé, A., Srivastava, A., Gurevych, I.: BEIR: a heterogeneous benchmark for zero-shot evaluation of information retrieval models. In: Vanschoren, J., Yeung, S. (eds.) Proceedings of the Neural Information Processing Systems Track on Datasets and Benchmarks 1, NeurIPS Datasets and Benchmarks 2021, December 2021, virtual (2021)
61. Trotman, A.: Compressing inverted files. Inf. Retrieval **6**(1), 5–19 (2003)
62. Urbano, J., Marrero, M.: The treatment of ties in AP correlation. In: Kamps, J., Kanoulas, E., de Rijke, M., Fang, H., Yilmaz, E. (eds.) Proceedings of the ACM SIGIR International Conference on Theory of Information Retrieval, ICTIR 2017, Amsterdam, The Netherlands, 1–4 October 2017, pp. 321–324. ACM (2017)
63. Voorhees, E.M.: The philosophy of information retrieval evaluation. In: Peters, C., Braschler, M., Gonzalo, J., Kluck, M. (eds.) CLEF 2001. LNCS, vol. 2406, pp. 355–370. Springer, Heidelberg (2002). https://doi.org/10.1007/3-540-45691-0_34
64. Voorhees, E.M.: Overview of the TREC 2004 Robust track. In: Voorhees, E.M., Buckland, L.P. (eds.) Proceedings of the Thirteenth Text REtrieval Conference, TREC 2004, Gaithersburg, Maryland, USA, 16–19 November 2004. NIST Special Publication, vol. 500–261. National Institute of Standards and Technology (NIST) (2004)
65. Voorhees, E.M.: The effect of sampling strategy on inferred measures. In: Geva, S., Trotman, A., Bruza, P., Clarke, C.L.A., Järvelin, K. (eds.) The 37th International ACM SIGIR Conference on Research and Development in Information Retrieval, SIGIR '14, Gold Coast , QLD, Australia - 06–11 July 2014, pp. 1119–1122. ACM (2014)
66. Voorhees, E.M.: The evolution of Cranfield. In: Ferro, N., Peters, C. (eds.) Information Retrieval Evaluation in a Changing World – Lessons Learned from 20 Years of CLEF, The Information Retrieval Series, vol. 41, pp. 45–69. Springer (2019). https://doi.org/10.1007/978-3-030-22948-1_2
67. Voorhees, E.M., Craswell, N., Lin, J.: Too many relevants: whither Cranfield test collections? In: Amigó, E., Castells, P., Gonzalo, J., Carterette, B., Culpepper,

J.S., Kazai, G. (eds.) SIGIR '22: The 45th International ACM SIGIR Conference on Research and Development in Information Retrieval, Madrid, Spain, 11–15 July 2022, pp. 2970–2980. ACM (2022)
68. Voorhees, E.M., Soboroff, I., Lin, J.: Can old TREC collections reliably evaluate modern neural retrieval models? arXiv 2201.11086 (2022)
69. Webber, W., Moffat, A., Zobel, J.: A similarity measure for indefinite rankings. ACM Trans. Inf. Syst. **28**(4), 20:1–20:38 (2010)
70. Wirth, N.: A plea for lean software. IEEE Comput. **28**(2), 64–68 (1995)
71. Witten, I.H., Moffat, A., Bell, T.C.: Managing Gigabytes: Compressing and Indexing Documents and Images, 2nd edn. Morgan Kaufmann (1999)
72. Xiong, L., et al.: Approximate nearest neighbor negative contrastive learning for dense text retrieval. In: 9th International Conference on Learning Representations, ICLR 2021, Virtual Event, Austria, 3–7 May 2021. OpenReview.net (2021)
73. Yang, P., Fang, H., Lin, J.: Anserini: enabling the use of Lucene for information retrieval research. In: Kando, N., Sakai, T., Joho, H., Li, H., de Vries, A.P., White, R.W. (eds.) Proceedings of the 40th International ACM SIGIR Conference on Research and Development in Information Retrieval, Shinjuku, Tokyo, Japan, 7–11 August 2017, pp. 1253–1256. ACM (2017)
74. Yilmaz, E., Aslam, J.A.: Estimating average precision with incomplete and imperfect judgments. In: Yu, P.S., Tsotras, V.J., Fox, E.A., Liu, B. (eds.) Proceedings of the 2006 ACM CIKM International Conference on Information and Knowledge Management, Arlington, Virginia, USA, 6–11 November 2006, pp. 102–111. ACM (2006)
75. Yilmaz, E., Kanoulas, E., Aslam, J.A.: A simple and efficient sampling method for estimating AP and NDCG. In: Myaeng, S., Oard, D.W., Sebastiani, F., Chua, T., Leong, M. (eds.) Proceedings of the 31st Annual International ACM SIGIR Conference on Research and Development in Information Retrieval, SIGIR 2008, Singapore, 20–24 July 2008, pp. 603–610. ACM (2008)
76. Zhao, W.X., Liu, J., Ren, R., Wen, J.: Dense text retrieval based on pretrained language models: a survey. ACM Trans. Inf. Syst. **42**(4), 89:1–89:60 (2024)
77. Zhuang, S., Zuccon, G.: Asyncval: a toolkit for asynchronously validating dense retriever checkpoints during training. In: Amigó, E., Castells, P., Gonzalo, J., Carterette, B., Culpepper, J.S., Kazai, G. (eds.) SIGIR 2022: The 45th International ACM SIGIR Conference on Research and Development in Information Retrieval, Madrid, Spain, 11-15 July 2022, pp. 3235–3239. ACM (2022)
78. Zuccon, G., Scells, H., Zhuang, S.: Beyond CO2 emissions: the overlooked impact of water consumption of information retrieval models. In: Yoshioka, M., Kiseleva, J., Aliannejadi, M. (eds.) Proceedings of the 2023 ACM SIGIR International Conference on Theory of Information Retrieval, ICTIR 2023, Taipei, Taiwan, 23 July 2023, pp. 283–289. ACM (2023)

Author Index

C. Hauff et al. (Eds.): ECIR 2025, LNCS 15572, pp. 473–474, 2025.
https://doi.org/10.1007/978-3-031-88708-6

The manufacturer's authorised representative in the EU is Springer Nature Customer Service Centre GmbH, Europaplatz 3, 69115 Heidelberg, Germany. If you have any concerns regarding our products, please contact ProductSafety@springernature.com

Printed and bound by CPI Group (UK) Ltd, Croydon, CR0 4YY
01/10/2026
02225834-0017